Ribosomes

CELLULAR ORGANELLES

Series Editor: PHILIP SIEKEVITZ
Rockefeller University, New York, New York

CHLOROPLASTS
J. Kenneth Hoober

CYTOSKELETON
Alexander D. Bershadsky and Juri M. Vasiliev

LYSOSOMES
Eric Holtzman

MITOCHONDRIA
Alexander Tzagoloff

RIBOSOMES
Alexander S. Spirin

Ribosomes

Alexander S. Spirin
Russian Academy of Sciences
Moscow, Russia

Kluwer Academic / Plenum Publishers
New York, Boston, Dordrecht, London, Moscow

Library of Congress Cataloging-in-Publication Data

Spirin, A. S. (Aleksandr Sergeevich)
Ribosomes / Alexander S. Spirin.
p. cm. -- (Cellular organelles)
Includes bibliographical references and index.
ISBN 0-306-46145-5 (hardbound). -- ISBN 0-306-46146-3 (paperback)
1. Ribosomes. I. Title. II. Series.
QH603.R5S6833 1999
571.6'58--dc21
99-42890
CIP

ISBN 0-306-46145-5

233 Spring Street, New York, N.Y. 10013

10 9 8 7 6 5 4 3 2 1

A C.I.P. record for this book is available from the Library of Congress

Printed in the United States of America

Preface

This book is based on an advanced course of lectures on ribosome structure and protein biosynthesis that I offer at the Moscow State University. These lectures have been part of a general course on molecular biology for almost three decades, and they have undergone considerable evolution as knowledge has been progressing in this field. The progress continues, and readers should be prepared that some facts, statements, and ideas included in the book may be incomplete or out-of-date. In any case, this is primarily a textbook, but not a comprehensive review. It provides a *background* of knowledge and current ideas in the field and gives *examples* of observations and their interpretations. I understand that some interpretations and generalizations may be tentative or disputable, but I hope that this will stimulate thinking and discussing better than if I left white spots.

The book has a prototype: it is my monograph *"Ribosome Structure and Protein Biosynthesis"* published by the Benjamin/Cummings Publishing Company, Menlo Park, California, in 1986. Here I have basically kept the former order of presentation of the topics and the subdivision into chapters. The contents of the chapters, however, have been significantly revised and supplemented. The newly written chapters on translational control in prokaryotes (Chapter 16) and eukaryotes (Chapter 17) are added. The chapters on morphology of the ribosome (Chapter 5), ribosomal RNA (Chapter 6), and cotranslational folding and transmembrane transport of proteins (Chapter 18) are completely rewritten in the co-authorship with Dr. V. D. Vasiliev, Prof. A. A. Bogdanov, and Prof. V. N. Luzikov, respectively. The concluding chapter on general principles of ribosome structure and function is appended.

The literature references in this book, as in the previous one, are given mainly for teaching purposes, so the reference lists at the end of each chapter are far from complete. To give an insight into the histories of discoveries I cited preferentially pioneer studies in the fields discussed. To provide information on the present state of knowledge, I have referred the reader to some of the recent publications. In addition, many illustrations, specifically those which are borrowed from other authors, are supplied with corresponding references. The book contains also many original illustrations made due to invaluable help of my colleagues at the Institute of Protein Research, Pushchino, especially P. G. Kuzin, A. Kommer, and V. A. Kolb. The assistance of L. N. Rozhanskaya, the secretary, M. G. Dashkevitch, Computers and Communication Department, and T. B. Kuvshinkina and M. S. Shelestova, Scientific Information Department, in preparing the manuscript is also greatly appreciated.

I am grateful to all my colleagues, as well as other scientists, who have read parts of the manuscripts and made their comments.

Alexander S. Spirin

Pushchino and Moscow

Contents

Part I
Historical and Fundamental Introduction

Chapter 4
Ribosomes and Translation

Part II
Structure of the Ribosome

Chapter 5
Morphology of the Ribosome

Chapter 6
Ribosomal RNA

Chapter 7
Ribosomal Proteins

Chapter 8
Mutual Arrangement of Ribosomal RNA and Proteins (Quaternary Structure)

Part III
Function of the Ribosome

Chapter 9
Functional Activities and Functional Sites of the Ribosome

Chapter 10

Elongation Cycle, Step I: Aminoacyl-tRNA Binding

Chapter 11

Elongation Cycle, Step II: Transpeptidation (Peptide Bond Formation)

Chapter 16
Translational Control in Prokaryotes

Chapter 17
Translational Control in Eukaryotes

Chapter 18

Cotranslational Folding and Transmembrane Transport of Proteins

Chapter 19

Conclusion: General Principles of Ribosome Structure and Function

I

Historical and Fundamental Introduction

1

Protein Biosynthesis
Summary and Definitions

The proteins of all living cells are synthesized by *ribosomes*. The ribosome is a large macromolecule consisting of ribonucleic acids *(ribosomal RNAs)* and proteins; it has a complex asymmetric quaternary structure. In order to synthesize protein, the ribosome must be supplied with (1) a program determining the sequence of amino acid residues in the polypeptide chain of a protein, (2) the amino acid substrate from which the protein is to be made, and (3) chemical energy. The ribosome itself plays a catalytic role and is responsible for forming peptide bonds, that is, for the polymerization of amino acid residues into the polypeptide chain.

The program that sets the sequence of amino acid residues in a polypeptide chain comes from *deoxyribonucleic acid* (DNA), that is, from the cell genome. Sections of the double-stranded DNA, which are called genes, serve as templates for synthesizing single-stranded RNA molecules. The synthesized RNA species are complementary replicas of just one of the DNA chains and therefore are faithful copies of the nucleotide sequence of the other DNA chain. This process of gene copying, accomplished by the enzyme RNA polymerase, is called *transcription*. In eukaryotic cells, and to a lesser extent in prokaryotic cells, nascent RNA may undergo a number of additional changes—called *processing;* as a result, certain parts of the nucleotide sequence may be excised from RNA, and in some cases altered, or edited. The mature RNA becomes associated with the ribosomes and serves as a program, or template, which determines the amino acid sequence in the synthesized protein. This template RNA is usually called *messenger RNA* (mRNA). In other words, the flow of information from DNA to ribosomes is mediated by gene transcription and RNA processing, resulting in the formation of mRNA.

In the eukaryotic cell the production of mRNA, that is, transcription and most events of processing, is compartmentalized in the nucleus. At the same time all functioning ribosomes are localized in the cytoplasm. Hence, the *transport* of mRNA from the nucleus to the cytoplasm is a necessary step in the flow of information from DNA to ribosomes. In prokaryotes, as well as in eukaryotic cytoplasmic organelles (mitochondria and chloroplasts), DNA and ribosomes are present in the same compartment, so that the ribosomes can reach mRNA and start to synthesize proteins during transcription; this is the so-called *coupled transcription–translation*.

Proteins consist of amino acids. Free amino acids, however, are not used in the synthetic machinery of the ribosome. To become a substrate for protein synthesis, an amino acid must be *activated* by coupling with the adenylic moiety of ATP and then *accepted* by (covalently linked to) a special RNA molecule called *transfer RNA* (tRNA); this process is performed by the enzyme aminoacyl-tRNA synthetase. The resulting aminoacyl-tRNA is used by the ribosome as a substrate for protein synthesis, and the energy of the chemical bond between the amino acid residue and tRNA is used to form a peptide bond. Thus, the activation of amino acids and formation of aminoacyl-tRNAs provide both material and energy for protein synthesis.

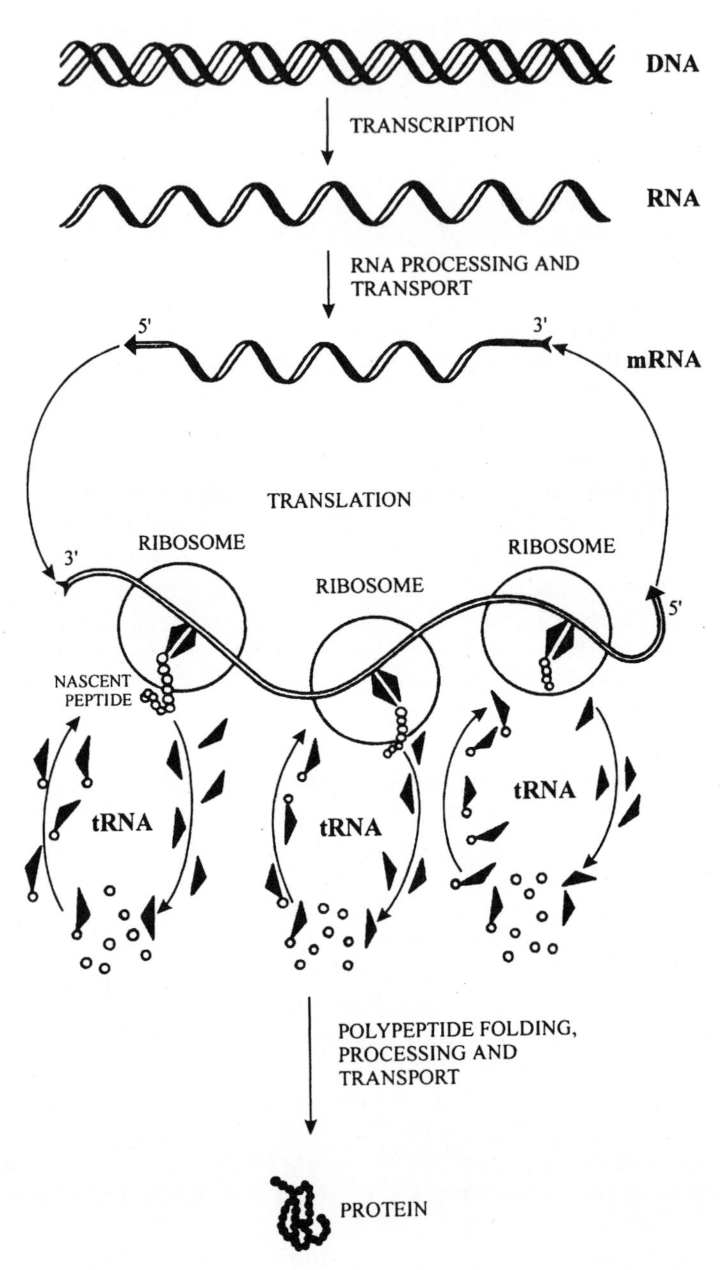

FIG. 1.1. General schematic model of protein biosynthesis (DNA → RNA → protein).

Using mRNA as a program and aminoacyl-tRNAs as energy-rich substrates, the ribosome *translates* genetic information from the nucleotide language of mRNA into the amino acid language of polypeptide chains. In molecular terms this implies that while moving along the mRNA, the ribosome consecutively selects appropriate aminoacyl-tRNA species from the medium. The specificity of the aminoacyl residue of a corresponding aminoacyl-tRNA selected by the ribosome is defined by the combination of nucleotides in a corresponding stretch of mRNA associated with the ribosome. This brings us to the problem of *genetic coding*—the nucleotide combinations that determine, or code, each of the 20 natural amino acids. These combinations are known to be nucleotide triplets, which are called codons.

Hence, the movement of the ribosome along the mRNA chain (or the passing of mRNA through the ribosome) establishes a temporal order of entering the various aminoacyl-tRNA species into the ribosome. This order depends on the sequence of codons along the mRNA. The aminoacyl residue of each selected aminoacyl-tRNA is attached covalently to a growing polypeptide chain by the ribosomal machinery. Deacylated tRNA is released by the ribosome into solution. In each act of aminoacyl-tRNA selection and deacylated tRNA release, an additional energy in the form of GTP hydrolysis is consumed by the ribosome. All this results in the step-by-step formation of the polypeptide chain, according to the program of mRNA.

The general model of protein biosynthesis outlined above is schematically presented in Fig. 1.1.

Historical review articles, on protein biosynthesis and ribosome research were presented by P. Siekevitz and P. C. Zamecnik in 1981, and by Nomura in 1990.

References

Nomura, M. (1990) History of ribosome research: a personal account, in *The Ribosome: Structure, Function, and Evolution* (W. E. Hill, A. Dahlbery, R. A. Garrett, P. B. Moore, D. Schlesinger, and J. R. Warner, eds.) pp. 3–55, ASM Press, Washington, DC.

Siekevitz, P., and Zamecnik, P. C. (1981) Ribosomes and protein synthesis, *J. Cell Biol.* **91**:665–765.

2

Messenger RNA and the Genetic Code

2.1. Discovery of mRNA

After the discovery and final recognition of the genetic function of DNA (Avery, MacLeod, and McCarty, 1944; Hershey and Chase, 1952; Watson and Crick, 1953a,b), it rapidly became clear that DNA itself does not serve as a direct template for protein synthesis. In addition, a number of early observations suggested that ribonucleic acid was closely connected to cellular protein synthesis (Caspersson, Landström-Hydea, and Aquilonius, 1941; Brachet, 1941–1942). These ideas were developed and resulted in the concept that RNA is the intermediate responsible for the transfer of genetic information from DNA to proteins; in particular, it has been suggested that RNA serves as a template upon which amino acid residues are polymerized (DNA → RNA → protein) (Crick, 1959).

This conceptual advance coincided with the discovery of protein-synthesizing ribonucleoprotein particles of the cell which were later called ribosomes (see Chapter 4). It had also been established that the RNA of these particles accounted for the main bulk of cellular RNA. Hence, it was naturally assumed that genes are transcribed into ribosomal RNA species, which in turn serve as templates for protein synthesis. This led to a "one gene–one ribosome–one protein" hypothesis.

During 1956 to 1958, in order to test this hypothesis, a comparative analysis of DNA and RNA base composition in a large number of microorganisms was conducted (Belozersky and Spirin, 1958). DNA base compositions can be rather different in different groups of microorganisms, and it was hypothesized that if the "DNA→RNA→protein" model was correct, the base composition of total RNA would strongly correlate with the DNA base composition in bacteria. The experimental results, however, were unexpected. Despite great differences of DNA base composition in various bacterial species, the composition of total RNA was found to be similar in all of the studied bacteria, and did not mimic DNA base composition. These results implied that the bulk of cellular RNA, most likely ribosomal RNA, could not serve as a direct informational intermediate between DNA and proteins.

At the same time, RNA base composition was shown to vary slightly for different bacterial species, and to be positively correlated with the base composition of DNA. The conclusion based on this correlation was that cells may contain a special *minor RNA fraction* which imitates DNA base composition and could possibly serve as an intermediate between genes and protein-synthesizing particles (Belozersky and Spirin, 1958).

Earlier Volkin and Astrachan (1956) studied RNA synthesis in bacteria infected with DNA-containing T2 bacteriophage. Bacterial protein synthesis ceases soon after infection, and the entire cellular protein-synthesizing machinery is switched over to producing phage proteins. Most of the cellular RNA does not undergo any change during this process, but the cell begins to synthesize a small fraction of metabolically unstable short-lived RNA, the nucleotide composition of which is similar to the base composition of phage DNA.

Several years later, in 1961, the minor RNA fraction, termed *DNA-like RNA,* was separated from the total cellular RNA. Its function as messenger, carrying information from DNA to the ribosomes, was demonstrated in the direct experiments of Brenner, Jacob, and Meselson (1961), and those of Gros, Watson, and co-workers (1961); similar observations have been made by Spiegelman (1961). It has been demonstrated that DNA-like RNA formed after the T4 phage infection binds to the preexisting host ribosomes (no new ribosomes are synthesized after phage infection), and the ribosomes associated with the phage-specific RNA synthesize the phage proteins. This RNA could be detached easily from the ribosomes *in vitro* without destroying the particles. It has been shown that this RNA is indeed complementary to one of the phage DNA chains.

On the basis of their results on genetic regulation in bacteria, Jacob and Monod (1961) advanced the idea that a special short-lived RNA transfers information from genes to ribosomes and serves as a direct template for protein synthesis. The term *messenger RNA* was accepted in all subsequent studies.

2.2. Deciphering the Code

The first step after the discovery of mRNA (1956–1961) was to elucidate the code by which amino acid sequences of proteins are written in the nucleotide sequences of mRNA and correspondingly in the nucleotide sequence of one of the two DNA chains (see Gamov, Rich, and Ycas, 1956). Even before the discovery of mRNA, theoretical considerations led to the assumption that each amino acid had to be coded by a combination of at least three nucleotides. Indeed, proteins are composed of 20 types of natural amino acids (Fig. 2.1), whereas nucleic acids contain only four types of nucleotide residues; the nitrogenous bases of nucleic acids are adenine (A), guanine (G), cytosine (C), and either uracil (U) for RNA or thymine (T) for DNA. It was obvious that one nucleotide could not code for one amino acid (4 *vs.* 20). There could be 16 dinucleotide combinations, or doublets, a number again insufficient to code for 20 amino acids. Thus, the minimum number of nucleotide residues in a combination coding for one amino acid had to be three; in other words, amino acids most probably had to be coded by *nucleotide triplets.* The number of possible triplets is 64, more than enough for the coding of 20 amino acids.

There were two possible explanations for excessive triplets: either only 20 triplets are "meaningful," that is, may code for one or another amino acid, while the other 44 are nonsense ones, or amino acids may be coded by more than one triplet, in which case the code would be *degenerate.*

Furthermore, the triplet code could be overlapping when a given nucleotide is part of three strongly overlapping or two less overlapping coding triplets; alternatively, it could be nonoverlapping when independent coding triplets are adjacent to each other in the template nucleic acid or even separated by noncoding nucleotides. The observation that point mutations (i.e., changes of a single nucleotide in the nucleic acid molecule) usually lead to a change of only one amino acid in the corresponding protein provided evidence against the idea of an overlapping code. Moreover, the overlapping code would inevitably result in the possible neighbors of a given amino acid residue being restricted, a situation that has never been observed in actual protein sequences. Therefore, a *nonoverlapping* code appeared more likely.

Finally, it had to be demonstrated whether the coding triplets were separated by noncoding residues, or commas, or whether they were read along the chain without any punctuation; in other words, whether the code was *comma-free* or not. The comma-free case leads to the problem of the reading frame of the template nucleic acid: only a strict triplet-by-triplet readout from a fixed point on the polynucleotide chain could result in an unambiguous amino acid sequence.

The classic experiments of Crick and associates published at the end of 1961 established that the code was triplet, degenerate, nonoverlapping, and comma-free. In these experiments, numerous mutants were obtained in the rII region of

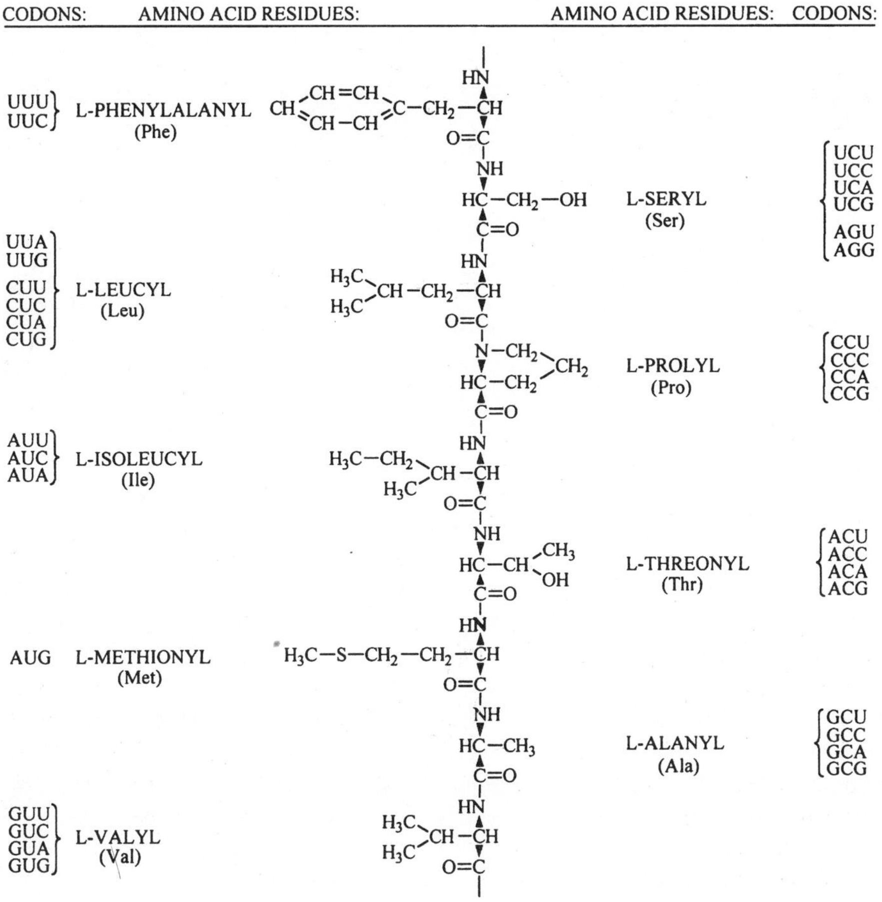

CODONS:	AMINO ACID RESIDUES:	AMINO ACID RESIDUES:	CODONS:
UUU, UUC	L-PHENYLALANYL (Phe)		
		L-SERYL (Ser)	UCU, UCC, UCA, UCG, AGU, AGG
UUA, UUG, CUU, CUC, CUA, CUG	L-LEUCYL (Leu)		
		L-PROLYL (Pro)	CCU, CCC, CCA, CCG
AUU, AUC, AUA	L-ISOLEUCYL (Ile)		
		L-THREONYL (Thr)	ACU, ACC, ACA, ACG
AUG	L-METHIONYL (Met)		
		L-ALANYL (Ala)	GCU, GCC, GCA, GCG
GUU, GUC, GUA, GUG	L-VALYL (Val)		

FIG. 2.1. Amino acid residues from which proteins are synthesized, and the corresponding codons.

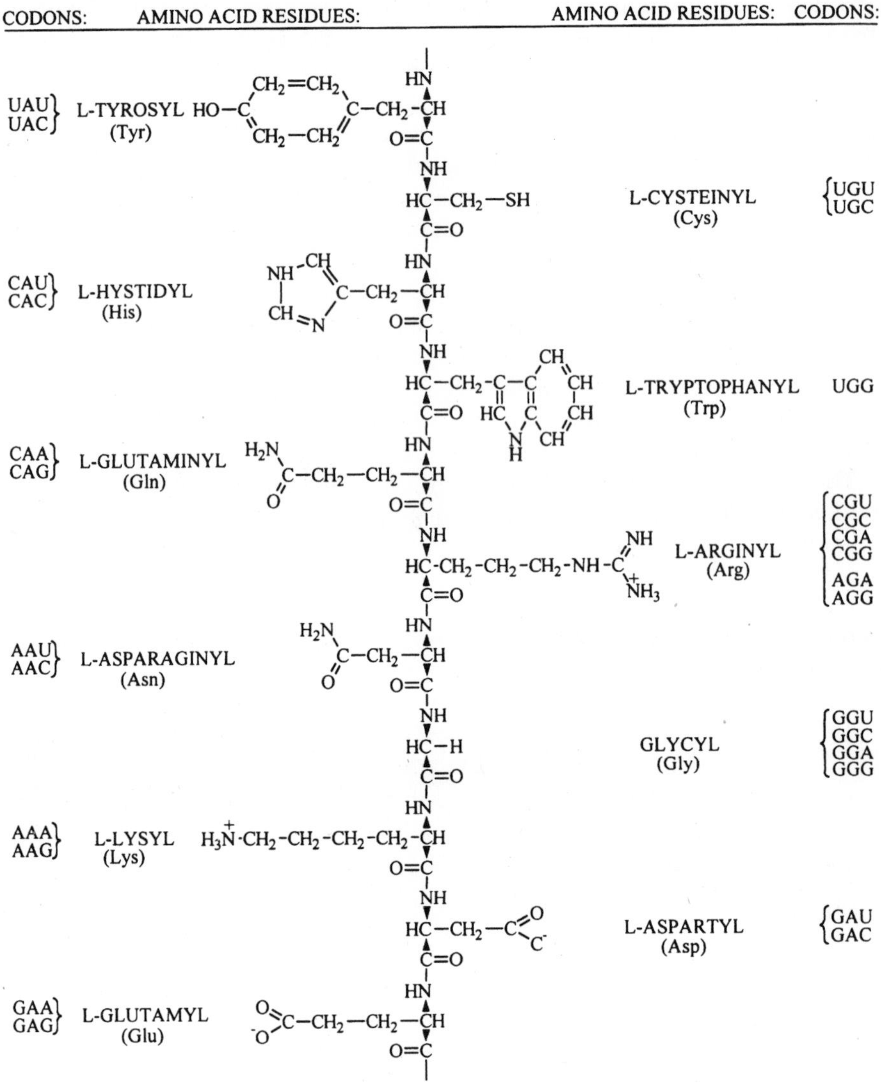

CODONS:	AMINO ACID RESIDUES:	AMINO ACID RESIDUES:	CODONS:
UAU, UAC	L-TYROSYL (Tyr)		
		L-CYSTEINYL (Cys)	UGU, UGC
CAU, CAC	L-HYSTIDYL (His)		
		L-TRYPTOPHANYL (Trp)	UGG
CAA, CAG	L-GLUTAMINYL (Gln)		
		L-ARGINYL (Arg)	CGU, CGC, CGA, CGG, AGA, AGG
AAU, AAC	L-ASPARAGINYL (Asn)		
		GLYCYL (Gly)	GGU, GGC, GGA, GGG
AAA, AAG	L-LYSYL (Lys)		
		L-ASPARTYL (Asp)	GAU, GAC
GAA, GAG	L-GLUTAMYL (Glu)		

FIG. 2.1. (*continued*)

the T4 bacteriophage gene B using chemical agents which produced either insertions or deletions of one nucleotide residue during DNA replication. Proflavine and other acridine dyes were used for this purpose. Nucleotide insertions or deletions close to the gene origin resulted in a loss of gene expression. By recombining different mutant phages in *Escherichia coli* cells, phenotypic revertants showing normal gene expression were obtained. An analysis of the revertants demonstrated that gene expression was restored if the region with the deletion was located near the region with the insertion, or *vice versa.* Gene expression could also be restored if two additional insertions (or deletions) were introduced near the region with the initial insertion (or, respectively, deletion). The following conclusions were drawn: (1) Insertion or deletion of a single nucleotide at the beginning of the coding region appeared to result in a loss of all the coding potential of the corresponding gene instead of simply a point mutation; the inactivation could be the result of a shift of the reading frame. (2) Deletion or insertion located close to the initial insertion or deletion, respectively, restored the coding potential of the sequence because the original reading frame was restored. (3) Three, but no fewer, closely located insertions or deletions also restored the initial coding potential of the nucleotide sequence. From the results of these experiments, it follows that the code is triplet, and that triplets are read sequentially without commas from a strictly fixed point in the same frame. These experiments also provided additional evidence that the code is degenerate: if many of the 64 possible triplets were nonsense ones, it was highly probable that at least one nonsense triplet appeared in the region between the insertion and deletion or between the three insertions where the readout occurs with a shift of frame; this would lead to an interruption of the polypeptide chain synthesis.

Deciphering the nucleotide triplets also began in 1961 when Nirenberg and Matthaei discovered the coding properties of synthetic polyribonucleotides in cell-free translation systems. The possibility of preparing synthetic polyribonucleotides of various compositions using a special enzyme, polynucleotide phosphorylase, was first demonstrated by Grunberg-Manago and Ochoa several years earlier (1955). The composition of polynucleotides synthesized in the system that they described depended only on the selection of ribonucleoside diphosphates supplied as substrates; homopolynucleotides such as polyuridylic acid, polyadenylic acid, and polycytidylic acid prepared from UDP, ADP, and CDP, respectively, were the simplest polyribonucleotides synthesized. Using poly(U) as a template polynucleotide for *E. coli* ribosomes, Nirenberg and Matthaei (1961) demonstrated that this template directs synthesis of polyphenylalanine. It has been concluded that the triplet UUU codes for phenylalanine. Similarly, experiments with polyadenylic and polycytidylic acid have shown that AAA codes for lysine, and CCC for proline.

Further elucidation of the genetic code was based on the use of synthetic statistical heteropolynucleotides of a different composition, which was set by the number and ratio of substrate nucleoside diphosphates in the polynucleotide phosphorylase reaction (Nirenberg *et al.*, 1963; Speyer *et al.*, 1963). Thus, it was demonstrated that the statistical poly(U,C) copolymer directed the incorporation of four amino acids into the polypeptide chain; these were phenylalanine, leucine, serine, and proline. If the U-to-C ratio in the polynucleotide was 1:1, then all four amino acids were incorporated into the polypeptide with equal probabilities. If the U-to-C ratio was 5:1, the probabilities of amino acid incorporation were as follows: Phe $>$ Leu $=$ Ser $>$ Pro. Thus phenylalanine should be coded by triplets consisting of three Us or of two Us and one C. Leucine and serine are coded by triplets consisting of two Us and one C or of two Cs and one U. Proline is coded by triplets consisting of three Cs or of two Cs and one U. Unfortunately, this approach could provide only the composition of the coding triplets, not their nucleotide sequence, since the nucleotide sequence of the template polynucleotide used was statistical.

Due to the invention of a new technique by Nirenberg and Leder (1964), the nucleotide sequences of the coding triplets were soon determined. They found that individual trinucleotides possessed coding properties: after association with the ribosome they supported the selective binding of aminoacyl-tRNA species with the ribosome. For example, UUU and UUC triplets stimulated the binding of phenylalanyl-tRNA, UCU and UCC the binding of seryl-tRNA, CUU and CUC the binding of leucyl-tRNA, and CCU and CCC the binding of prolyl-tRNA. By 1964, methods for synthesizing trinucleotides with the desired sequence were available. In the subsequent two years a wide variety of trinucleotides were tested and, as a result, virtually the whole code was deciphered (Fig. 2.2).

The end of the story was marked by the use of synthetic polynucleotides with a regular nucleotide sequence as templates in the cell-free ribosomal systems of polypeptide synthesis. Methods allowing regular polynucleotides to be synthesized have been developed by Khorana, who has also verified the genetic code by directly using these polynucleotides as templates (Khorana *et al.*, 1966). In complete agreement with the previously established code dictionary, the use of poly$(UC)_n$ as a template resulted in the synthesis of a polypeptide consisting of alternating serine and leucine residues, while poly$(UG)_n$ directed synthesis of the regular copolymer with alternating valine and cysteine residues. Poly$(AAG)_n$ directed the synthesis of three homopolymers: polylysine, polyarginine, and polyglutamic acid.

2.3. Some Features of the Code Dictionary

The complete code dictionary is given in Fig. 2.2. Of the 64 triplets termed *codons,* 61 are meaningful or sense ones: they code for 20 amino acids of natural polypeptides and proteins. Three codons—UAG ("amber"), UAA ("ochre"), and UGA ("opal")—normally do not code for amino acids and therefore are sometimes called *nonsense codons.* The nonsense triplets play an important part in translation, since in mRNA these codons serve as signals for the termination of polypeptide chain synthesis; at present they are usually referred to as *termination* or *stop codons.*

At the same time UGA triplet may also code for the 21st amino acid of a number of proteins, selenocysteine (Chambers *et al.*, 1986; Zinoni *et al.*, 1987). This, however, requires the presence in mRNA of an additional structural element, either immediately adjacent to UGA from its 3′-side (in the case of prokaryotes), or located beyond the coding sequence, in the 3′-proximal untranslated region of mRNA (in eukaryotes) (see Chapter 10, Section 10.2.2).

Second letter

First letter	U	C	A	G	Third letter
U	UUU Phe	UCU Ser	UAU Tyr	UGU Cys	U
	UUC Phe	UCC Ser	UAC Tyr	UGC Cys	C
	UUA Leu	UCA Ser	UAA Ochre	UGA Opal	A
	UUG Leu	UCG Ser	UAG Amber	UGG Trp	G
C	CUU Leu	CCU Pro	CAU His	CGU Arg	U
	CUC Leu	CCC Pro	CAC His	CGC Arg	C
	CUA Leu	CCA Pro	CAA Gin	CGA Arg	A
	CUG Leu	CCG Pro	CAG Gin	CGG Arg	G
A	AUU Ile	ACU Thr	AAU Asn	AGU Ser	U
	AUC Ile	ACC Thr	AAC Asn	AGC Ser	C
	AUA Ile	ACA Thr	AAA Lys	AGA Arg	A
	AUG Met	ACG Thr	AAG Lys	AGG Arg	G
G	GUU Val	GCU Ala	GAU Asp	GGU Gly	U
	GUC Val	GCC Ala	GAC Asp	GGC Gly	C
	GUA Val	GCA Ala	GAA Glu	GGA Gly	A
	GUG Val	GCG Ala	GAG Glu	GGG Gly	G

FIG. 2.2. Codon dictionary. See Crick, F. H. C. (1996) *Cold Spring Harbor Symp. Quant. Biol.* **31**:1–9.

As seen from Fig. 2.2, the degeneracy of the code does not extend to all 20 main amino acids. Two amino acids, methionine and tryptophan, are coded by one codon each, by AUG and UGG, respectively. On the contrary, three amino acids, specifically leucine, serine, and arginine, have six codons each. The remaining amino acids, with the exception of isoleucine, are coded either by two or by four codons; only isoleucine is coded by three codons.

It should be emphasized that the triplets coding for a given amino acid differ in most cases only in the third base. Only when the amino acid is coded by more than four codons do differences occur in the first and second positions of the triplet as well. A group of four codons differing only in the third nucleotide and coding for one and the same amino acid is often called the *codon family*. The code dictionary contains eight such codon families for leucine, valine, serine, proline, threonine, alanine, arginine, and glycine.

The code presented in Fig. 2.2 is universal for the protein-synthesizing systems of most bacteria and for the cytoplasmic extraorganellar protein-synthesizing systems of eukaryotes, that is, animals, fungi, plants, and Protozoa.

2.4. Deviations from the Universal Code

By the end of the 1970s and during the 1980s it was discovered that the universality of the genetic code is not absolute, and some exceptions are possible (Barrell, Bankier, and Drouin, 1979; Yamao *et al.*, 1985). Among living organisms, two genera of eubacteria, *Mycoplasma* and *Spiroplasma*, are now known to have two codons for tryptophan, the universal UGG and the "neighboring" UGA, which is a stop codon in other organisms. In one genus of ciliates (Protozoa), *Euplotes*, UGA codes for cysteine. Two other universal stop codons, UAA and UAG, were reported to code for glutamine in other genera of Ciliates (*Tetrahymena, Paramecium, Stylonychia, Oxytricha*) and in at least one genus of unicellular green algae (*Acetabularia*). Also, in some yeast (*Candida*) the universal leucine codon CUG codes for serine. The known cases of variations in the genetic code are summarized in Table 2.1 (Watanabe and Osawa, 1995). Further exceptions to the universal genetic code may be discovered in the future, especially in unicellular eukaryotes (Protozoa, algae, and fungi).

Organelles of eukaryotic cells, including mitochondria, possess their own protein-synthesizing systems. The protein-synthesizing systems of animal and fungal (but not plant) mitochondria typically show a number of significant deviations from the universal code (Table 2.2). Tryptophan in these mitochondria is coded by both UGG and UGA; UGA is therefore not used as a termination codon. In mitochondria of all vertebrates, most (but not all) invertebrates, and some fungi the universal isoleucine codon AUA codes for methionine, so that methionine is determined by two triplets, the universal AUG and the neighboring AUA. The triplets AGA and AGG do not code for arginine in mitochondria of most animals; they are stop codons in vertebrate mitochondria and codons for serine in mitochondria of many invertebrates (echinoderms, insects, mollusks, nematodes, Platyhelminthes). In yeast mitochondria (*Saccharomyces, Torulopsis*) the whole codon family CUU, CUC, CUA, and CUG codes for threonine but not for leucine, although in other fungi, such as *Neurospora* and *Aspergillus*, these codons correspond to leucine as given by the universal code.

TABLE 2.1. Variations in Eubacterial and in Nuclear Genetic Code from "Universal" Genetic Code[a]

Organism	UGA (Stop)	UAA UAG (Stop)	CUG (Leu)
Eubacteria:			
Mycoplasma	Trp	—	—
Spiroplasma	Trp	—	—
Yeasts:			
Candida	—	—	Ser
Ciliates			
Tetrahymena	—	Gln	—
Paramecium	—	Gln	—
Stylonychia	—	Gln	—
Oxytricha	—	Gln	—
Euplotes	Cys	—	—
Unicellular green algae			
Acetabularia	—	Gln	—

[a]Modified from Watanabe, K., and Osawa, S. (1995) in *tRNA: Structure, Biosynthesis, and Function* (D. Söll and U. RajBhandary, eds.), ASM Press, Washington DC, 1995.

2.5. Structure of mRNA

2.5.1. Primary Structure

In contrast to DNA, mRNA, as well as other cellular RNA species, is a *single-stranded* polynucleotide. It consists of four kinds of linearly arranged ribonucleoside residues—adenosine (A), guanosine (G), cytidine (C), and uridine (U)—sequentially connected by phosphodiester bonds between the 3′-position of the ribose of one nucleoside and the 5′-position of the adjacent one (Fig. 2.3). The terminal nucleoside, the 5′-position of which does not participate in forming the internucleotide bond, is referred to as the 5′-end of RNA. The terminal nucleoside with free 3′-hydroxyl is referred to as the 3′-end. It is accepted practice to read and write RNA nucleotide sequences from the 5′- to the 3′-end, that is, in the direction

TABLE 2.2. Variations in Mitochondrial Genetic Code[a]

Organism	UGA Stop	AUA Ile	AAA Lys	AAA AGG Arg	CYN Leu	UAA Stop
Vertebrates	Trp	Met	—	Stop	—	—
Tunicates	Trp	Met	—	Gly	—	—
Echinoderms	Trp	—	Asn	Ser	—	—
Arthropods	Trp	Met	—	Ser	—	—
Molluscs	Trp	Met	—	Ser	—	—
Nematodes	Trp	Met	—	Ser	—	—
Platyhelminths	Trp	—	Asn	Ser	—	Thr?
Coelenterates	Trp	ND	ND	—	ND	ND
Yeasts	Trp	Met	—	—	Thr	—
Euascomycetes	Trp	—	—	—	—	—
Protozoa	Trp	—	—	—	—	—

ND, not determined; —, same as universal code.
[a]Modified from Watanabe, K., and Osawa, S. In *tRNA: Structure, Biosynthesis, and Function* (D. Söll and U. RajBhandary, eds.), ASM Press, Washington DC.

of the internucleotide phosphodiester bond from the 3′-position to the 5′-position of the neighbor (3′-P-5’ bond direction). This direction corresponds to the polarity of mRNA readout by the ribosome.

The terminal 5′-position in natural mRNAs is always substituted. In prokaryotic organisms this end is either simply phosphorylated (Fig. 2.3) or carries the triphosphate group. Eukaryotic mRNAs generally have a special group, the so-called *cap,* at the terminal 5′-position (Furuichi and Miura, 1975; Furuichi *et al.,* 1975). The cap is the N′-methylated residue of guanosine 5′-triphosphate linked with the 5′-terminal nucleoside by the 5′–5′ pyrophosphate bond (Fig. 2.4). Eukaryotic cells possess a special system including guanylyl transferase and methyl transferase, enzymes that are responsible for mRNA capping. In addition, the capping is usually accompanied by methylation of the 2′-hydroxyl group of ribose and the base in the 5′-terminal nucleoside adjacent to the cap. Often the 5′-terminal residue in mRNA is a purine nucleoside, either G or A.

The 3′-terminal hydroxyl of natural mRNA remains unsubstituted. Thus, this end possesses two hydroxyl groups in *cis*-position *(cis*-glycol group) (see Fig. 2.3).

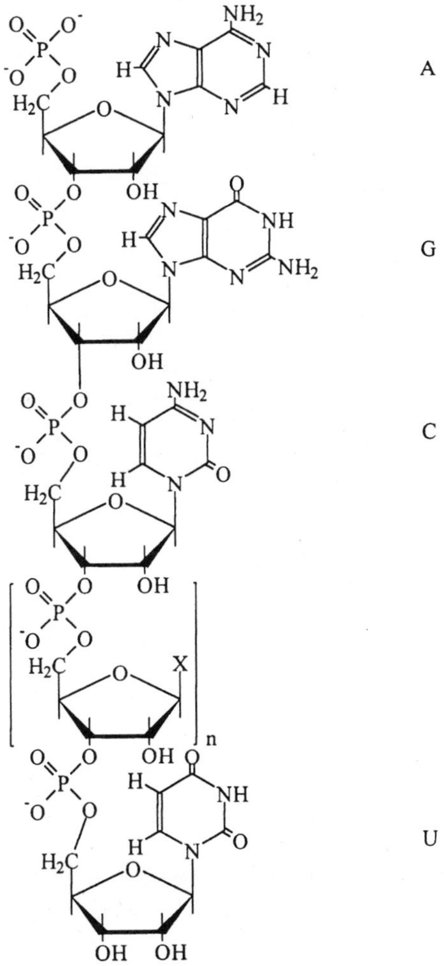

FIG. 2.3. Nucleotide residues in RNA.

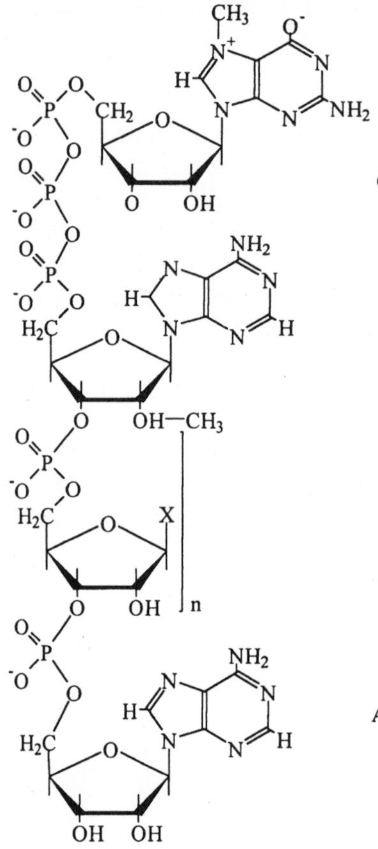

FIG. 2.4. Cap structure at the 5′-end of eukaryotic mRNA.

2.5.2. Functional Regions

The physical length of the mRNA chain is always greater than the length of its coding sequence. The coding sequence includes only part of the total mRNA length. The first codon is preceded by a noncoding (untranslated) 5′-terminal sequence (5′-UTR) the length of which varies for different mRNAs. Furthermore, the terminal codon is never located at the 3′-end of an mRNA chain, but is always followed by a noncoding 3′-terminal sequence (3′-UTR). In addition, most eukaryotic mRNAs contain a long noncoding sequence of adenylic acid residues at their 3′-end. This poly(A) tract (tail) is added to mRNA after the end of transcription by a special enzyme, polyadenylate polymerase.

Identifying the factors that determine the starting point of the coding nucleotide sequence within an mRNA chain is an important problem. Each polypeptide is known to begin with an N-terminal methionine residue, and therefore the first codon in the coding sequence should be that of methionine. In most cases AUG, and less frequently GUG or UUG (in prokaryotes), play the role of the *initiation codon* (see Chapter 15). The codon AUG codes for methionine both when it is the first codon of the mRNA coding sequence and when it occurs in internal positions. The codon GUG, however, codes for valine in internal positions and for the initiator methionine only if it occupies the first position in the coding sequence. The same is true for codon UUG coding for leucine in internal positions. In some exceptional cases, AUU or AUA in prokaryotes and ACG or UUG in eukaryotes may also serve as initiation codons for the first methionine in the chain. The identification of the initiation codons, however, does not solve the starting point problem of the coding sequence. The difficulty is that by no means does every AUG or GUG or UUG triplet become an initiation codon. Generally, translation cannot be initiated from internal AUG, GUG, or UUG triplets. If an mRNA chain is scanned from its 5′-end, AUG as well as GUG and UUG triplets may be found repeatedly both in frame with the subsequent coding sequence and out of frame, but they cannot initiate translation. Finally, many AUG, GUG, and UUG triplets located within the coding sequence but out of the reading frame fortunately do not initiate synthesis of erroneous polypeptides. Thus, in contrast to all other codons, both sense and nonsense ones, the choice of a given codon as an initiation point depends not only on the codon structure, that is, its nucleotide composition and sequence, but also on the position of the codon in the mRNA. Certain structural elements in mRNA confer the capacity to serve as an initiation codon to a given AUG (or GUG or UUG). Specifically, the nucleotide sequence preceding the initiation codon, as well as the particular secondary and tertiary structures of this mRNA region, are vital for the corresponding triplet to be exposed as an initiation codon (Chapter 15).

A given mRNA polynucleotide chain does not necessarily contain just one coding sequence. In prokaryotic mRNAs it is common for one polynucleotide chain to contain coding sequences for several proteins. Such mRNAs are usually called *polycistronic mRNAs.* (This term comes from the word *cistron,* which S. Benzer introduced as an equivalent of a gene). Different coding sequences or cistrons, within a given mRNA chain are usually separated by internal noncoding sequences. Such an internal noncoding sequence begins from the termination codon of the preceding cistron. The next cistron begins from an initiation codon such as AUG or GUG.

In contrast to prokaryotes, in eukaryotic organisms mRNAs are as a rule *monocistronic,* that is, they code for just one polypeptide chain. The eukaryotic mRNA coding sequence is flanked both at the 5′-end and at the 3′-end by noncoding (untranslated) sequences (5′- and 3′-UTRs), the 3′-UTR being typically very long (comparable to the length of the coding sequence). It has already been mentioned that the vast majority of eukaryotic mRNAs have also poly(A) tracts of various lengths at the 3′-end. The 5′-end is usually modified by the cap (Fig. 2.4), which appears to be essential for the association between the mRNA and the ribosome prior to initiation.

It is appropriate to emphasize here that the mechanisms responsible for searching for the initiation codon in prokaryotic and eukaryotic translation systems are different. Prokaryotic ribosomes form a complex with mRNA and recognize the initiation codon independently of the 5′-end; it is for this reason that they can initiate from internal sites in the polycistronic mRNA. In contrast, eukaryotic ribosomes usually need the mRNA 5′-end to form the association complex; the cap contributes to such an association (see Chapter 15). With eukaryotic mRNA it is the first AUG from the 5′-end that in most cases serves as an initiation codon, although there are exceptions to this rule. At the same time some special eukaryotic mRNAs use the alternative mechanism of internal initiation which is also intrinsic to the eukaryotic protein-synthesizing system; in such a case the initiation codon (AUG) situated far away from the 5′-end is preceded by a massive structural element, the so-called *internal ribosome entry site,* or IRES (see Section 15.3.3).

2.5.3. Folding

The three-dimensional structures of mRNAs have yet to be determined. Measurements of various physical parameters of several mRNAs have demonstrated that these molecules may possess extensively folded structures with a large number of intrachain interactions due to the Watson–Crick complementary base-pair formation, as well as noncanonical hydrogen bonding between nucleotides. Although mRNAs are not double helices of the DNA type, they do have a well-developed secondary structure because of the complementary pairing of different regions of the same chain with each other; this results in a large number of relatively short double-helical regions being formed. About 70% of all the nucleotide residues in the chain may typically participate in the complementary pairing and, correspondingly, in the formation of intramolecular helices. Most of the double-helical regions appear to be formed by the complementary pairing of adjacent sections in the polynucleotide chain; the model of the formation of such short helices is given schematically in Fig. 2.5. The complementary pairing of distant chain sections may result in the additional folding of the structure. These interactions are based mainly on A:U and G:C pairing (Watson–Crick pairs), as well as on G:U pairing (see Section 3.2.2).

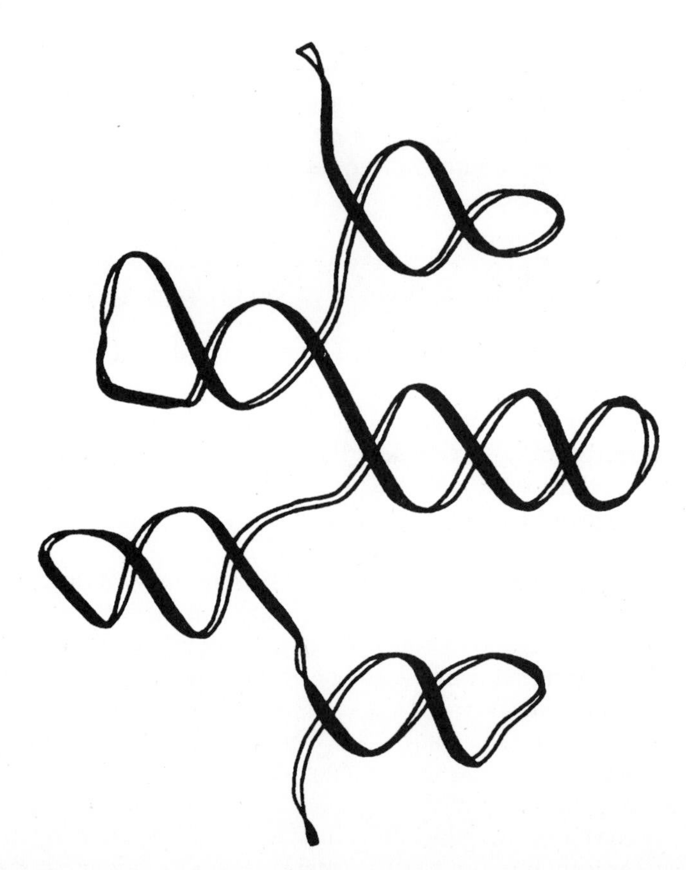

FIG. 2.5. Schematic model illustrating the formation of a secondary structure (double-helical hairpins) by pairing of the adjacent sections of the RNA polynucleotide chain. See Doty, P., Boedtker, H., Fresco, J. R., Haselkorn, R., and Litt, M. (1959) *Proc. Natl. Acad. Sci. USA* **45**:482–499; Spirin, A. S. (1960) *J. Mol. Biol.* **2**:436–446.

There is evidence suggesting that the secondary and tertiary structures of mRNA may play an important role in translation and its regulation. This problem will be considered in Chapters 15 through 17.

It should be emphasized that after initiation of translation the ribosomes may perform a readout more or less independently on the secondary and tertiary structure of mRNA. It is likely that they sequentially unfold the mRNA chain while moving along (of course, the chain sections refold after the ribosomes have moved away). At present, very little is known about the effect of the secondary and tertiary mRNA structures on the rate at which ribosomes move along the RNA chain—the rate of polypeptide elongation. It is known that this rate is non-uniform and it may well be that it depends on the presence and stability of the secondary and tertiary structure in different mRNA regions.

As already mentioned, the presence of a special three-dimensional structure at the UGA codon in the coding sequence determines the incorporation of selenocysteine, instead of inducing the regular stop signal (see Section 10.2.2 for more detail). Also some special folds within coding sequences may provoke frameshifting, or even jumping of translating ribosomes over a section of mRNA during elongation (see Sections 12.4.2 and 12.4.3).

Particular attention should be paid to noncoding mRNA sequences. Specifically, their function may be to create the specialized three-dimensional secondary and tertiary structures that control initiation, elongation, sometimes termination, and reinitiation. In prokaryotes the noncoding intercistronic spacers affect the transition of ribosomes from one cistron to another. In eukaryotes the 5′- and 3′-untranslated regions determine the binding to mRNA of special recognition proteins that affect translation and also may govern conservation and degradation of mRNA, its intracellular transport, and specific intracellular localization.

2.6. Messenger Ribonucleoproteins of Higher Eukaryotes

The presence of complexes between mRNA and proteins (messenger ribonucleoprotein particles) was first discovered in the cytoplasm of animal embryonic cells (Spirin, Belitsina, and Ajtkhozhin, 1964; Spirin and Nemer, 1965). These were called informosomes. Soon after it became clear that all mRNA in the eukaryotic cytoplasm of all cell types, at least in animals and higher plants, exists in the form of messenger ribonucleoproteins, or mRNPs.

Now several classes of mRNA–protein complexes in the cytoplasm may be distinguished: (1) polyribosomal mRNPs, the mRNA–protein complexes within translating polyribosomes; (2) free mRNP particles which are principally translatable, but either are in transit to polyribosomes, or represent a pool of excess mRNA for translation, or are not capable of efficiently competing with other, stronger mRNAs for initiation factors ("weak" mRNAs); (3) nontranslatable mRNP particles where initiation of translation is blocked by specific 5′-UTR-bound repressors (see Section 17.5); and (4) masked mRNP particles which are inactive in translation, stable, and stored in the cytoplasm until they receive a signal for unmasking (Section 17.6); they are typical of germ cells and other dormant states.

All these cytoplasmic mRNPs have certain features in common. First, they always have a relatively high proportion of protein: the protein to RNA ratio is universally about 3:1 to 4:1 in the free mRNPs and somewhat lower, down to 2:1, in the polyribosomal mRNPs. For comparison, ribosomes have a protein to RNA ratio from 1:2 in prokaryotic particles to 1:1 in eukaryotic ribosomes. Second, at least two major families of proteins are present in stoichiometry over one protein per RNA. One is represented by a basic protein (or a couple of closely related proteins) with a molecular mass of about 35 kDa, which is usually designated as "p50", or "Y-box protein(s)" (see Section 17.2.2); this protein (or proteins) possesses a high affinity for various heterologous mRNA sequences, and much lower affinity for poly(A) tails. The other is a protein with a molecular mass of about 70 to 80 kDa (p70, or PABP, poly(A)-binding protein) having a predominant affinity for poly(A) sequences. A great variety of minor protein species are also bound within the mRNP particles. Third, the mRNP particles are found to be rather resistant to removal of Mg^{2+}, in contrast to ribosomal particles.

The protein(s) designated as p50, or Y-box protein(s), seems to be the major mRNP protein component of all cytoplasmic mRNPs, both in dormant germ cells and in actively translating somatic cells. The same major mRNA-binding protein(s) can be detected both in free mRNP particles and in polyribosomal mRNPs. It seems likely that the p50 mentioned is the main protein component (mRNP core protein) physically forming the cytoplasmic mRNPs of eukaryotic cells, like histones form DNP. The role of the protein may be some kind of structural organization and sequence-nonspecific packaging of eukaryotic mRNA into mRNP particles. This universal form eukaryotic mRNA is available for intracellular transport, translation, masking, and degradation, depending on other protein components involved. Under certain circumstances, with participation of a specific masking protein (see Section 17.6), the protein may be responsible for some conformational rearrangements of mRNPs, for example, their condensation into inactive (masked) particles.

Among minor protein components of mRNPs, protein kinases may govern the composition and the activity of mRNPs by inducible phosphorylation of other mRNP proteins. Other enzymatic activities and proteins serving translation, including some initiation factors, can also be found associated with mRNPs. A schematic representation of the distribution of mRNA-binding proteins among different functional regions of eukaryotic mRNA is given in Fig. 2.6.

Generally, the massive loading of eukaryotic mRNA with proteins suggests that the following points may be very important in mRNA interactions with the translation machinery: (1) The binding of proteins may modify, melt, induce, or switch structural elements in mRNA, thus affecting its translational activity. (2) Specifically in eukaryotes, the mRNA-binding components involved in translation, such as ribosomes, translation initiation factors, and translational repressors and activators, must interact with mRNPs, rather than with mRNAs. The prebound

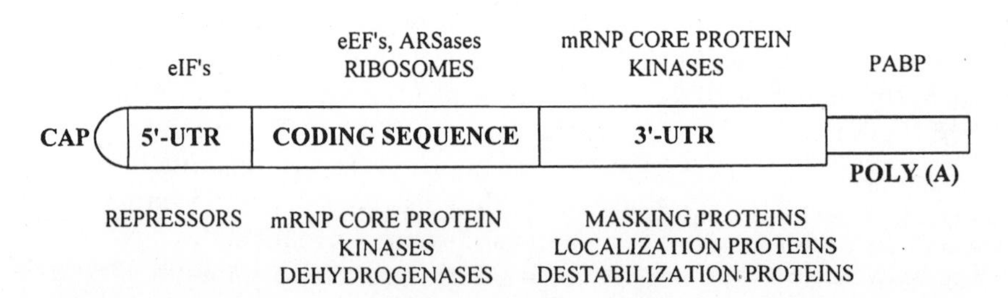

FIG. 2.6. Schematic representation of the distribution of mRNA-binding proteins among different regions of eukaryotic mRNA. Reproduced from Spirin, A. S. (1996) *in Translational Control* (J. W. B. Hershey, M. B. Mathews, and N. Sonenberg, eds.), pp. 319–334, CSHL Press, with permission.

mRNP proteins may exert either a competing (antagonistic) or attracting (synergistic) effect on the binding of the translation components to mRNA. (3) Numerous protein–protein interactions within mRNPs are likely. This can create additional possibilities for three-dimensional folding and packaging of mRNPs, thus controlling the accessibility of mRNA for translation, degradation, transporting systems, intracellular localization "anchors," and possible association with cytoskeleton structures.

References

Avery, O. T., MacLeod, C. M., and McCarty, M. (1944) Studies in the chemical nature of the substance inducing transformation of pneumococcal types, *J. Exp. Med.* **78**:137–158.

Barrell, B. G., Bankier, A. T., and Drouin, J. (1979) A different genetic code in human mitochondria, *Nature* **282**:189–194.

Belozersky, A. N., and Spirin, A. S. (1958) A correlation between the compositions of deoxyribonucleic and ribonucleic acids, *Nature* **182**:111–112.

Brachet, J. (1941–1942) La detection histochimique et le microdosage des acides pentosenucleiques, *Enzymologia* **10**:87–96.

Brenner, S., Jacob, F., and Meselson, M. (1961) An unstable intermediate carrying information from genes to ribosomes for protein synthesis, *Nature* **190**:576–581.

Casperson, T., Landström-Hyden, H., and Aquilonius, L. (1941) Cytoplasmanukleotide in eiweissproduzierenden Drüsenzellen, *Chromosoma* **2**:111–131.

Chambers, I., Frampton, J., Goldfarb, P., Affara, N., McBain, W., and Harrison, P. R. (1986) The structure of the mouse glutathione peroxidase gene: the selenocysteine in the active site is encoded by the "termination" codon TGA, *EMBO J.* **5**:1221–1227.

Crick, F. H. C. (1959) The present position of the coding problem, *Brookhaven Symp. Biol.* **12**:35–38.

Crick, F. H. C. (1966) The genetic code: Yesterday, today and tomorrow, *Cold Spring Harbor Symp. Quant. Biol.* **31**:1–9.

Crick, F. H. C., Barnett, L., Brenner, S., and Watts-Tobin, R. J. (1961) General nature of the genetic code for proteins, *Nature* **192**:1227–1232.

Furuichi, Y., and Miura, K.-I. (1975) A blocked structure at the 5′-terminus of mRNA from cytoplasmic polyhedrosis virus, *Nature* **253**:374–375.

Furuichi, Y., Morgan, M., Muthukrishnan, S., and Shatkin, A. J. (1975) Reovirus messenger RNA contains a methylated blocked 5′-terminus structure: $m^7G(5')ppp(5')G^mpCp$, *Proc. Natl. Acad. Sci. U.S.A.* **72**:362–366.

Gamov, G., Rich, A., and Ycas, M. (1956) The problem of information transfer from the nucleic acids to proteins, *Advan. Biol. Med. Phys.* **4**:23–68.

Gros, F., Hiatt, H., Gilbert, W., Kurland, C. G., Riseborough, R. W., and Watson, J. D. (1961) Unstable ribonucleic acid revealed by pulse labeling of *Escherichia coli, Nature* **190**:581–585.

Grunberg-Manago, M., and Ochoa, S. (1955) Enzymatic synthesis and breakdown of polynucleotides: Polynucleotide phosphorylase, *J. Amer. Chem. Soc.* **77**:3165–3166.

Hershey, A. D., and Chase, M. (1952) Independent functions of viral protein and nucleic acid in growth of bacteriophage, *J. Gen. Physiol.* **36**:39–56.

Jacob, F., and Monod, J. (1961) Genetic regulatory mechanisms in the synthesis of proteins, *J. Mol. Biol.* **3**:318–356.

Khorana, H. G., Büchi, H., Ghosh, H., Gupta, N., Jacob, T. M., Kössel, H., Morgan, R., Narang, S. A., Ohtsuka, E., and Wells, R. D. (1966) Polynucleotide synthesis and the genetic code, *Cold Spring Harbor Symp. Quant. Biol.* **31**:39–49.

Nirenberg, M. W., Jones, O. W., Leder, P., Clark, B. F. C., Sly, W. S., and Pestka, S. (1963) On the coding of genetic information, *Cold Spring Harbor Symp. Quant. Biol.* **28**:549–557.

Nirenberg, M. W., and Leder, P. (1964) RNA codewords and protein synthesis: The effect of trinucleotides upon the binding of sRNA to ribosomes, *Science* **145**:1399–1407.

Nirenberg, M. W., and Matthaei, J. H. (1961) The dependence of cell-free protein synthesis in *E. coli* upon naturally occurring or synthetic polyribonucleotides, *Proc. Natl. Acad. Sci. U.S.A.* **47**:1588–1602.

Speyer, J. F., Lengyel, P., Basilio, C., Wahba, A. J., Gardner, R. S., and Ochoa, S. (1963) Synthetic polynucleotides and the amino acid code, *Cold Spring Harbor Symp. Quant. Biol.* **28**:559–567.

Spiegelman, S. (1961) The relation of informational RNA to DNA, *Cold Spring Harbor Symp. Quant. Biol.* **26**:75–90.

Spirin, A. S., Belitsina, N. V., and Ajtkhozhin, M. A. (1964) Messenger RNA in early embryogenesis, *J. Gen. Biol.* (Russian) **25**:321–338. English translation (1965) *Fed. Proc.* **24**:T907–T922.

Spirin, A. S., and Nemer, M. (1965) Messenger RNA in early sea-urchin embryos: Cytoplasmic particles, *Science* **150**:214–217.

Volkin, E., and Astrachan, L. (1956) Phosphorus incorporation in *Escherichia coli* ribonucleic acid after infection with bacteriophage T2, *Virology* **2**:149–161.

Watanabe, K., and Osawa, S. (1995) tRNA sequences and variations in the genetic code. In *tRNA: Structure, Biosynthesis, and Function* (D. Soell and U. L. RajBhandary, eds.), pp. 225–250, ASM Press, Washington, DC.

Watson, J. D., and Crick, F. H. C. (1953a) Molecular structure of nucleic acids, *Nature* **171**:738–740.

Watson, J. D., and Crick, F. H. C. (1953b) Genetical implications of the structure of deoxyribose nucleic acid, *Nature* **171**:964–967.

Yamao, F., Muto, A., Kawauchi, Y., Iwami, M., Iwagami, S., Azumi, Y., and Osawa, S. (1985) UGA is read as tryptophan in *Mycoplasma capricolum, Proc. Natl. Acad. Sci. U.S.A.* **82**:2306–2309.

Zinoni, F., Birkmann, A., Leinfelder, W., and Boeck, A. (1987) Cotranslational insertion of selenocysteine into formate dehydrogenase from *Escherichia coli* directed by a UGA codon, *Proc. Natl. Acad. Sci. U.S.A.* **84**:3156–3160.

3

Transfer RNA and Aminoacyl-tRNA Synthetases

3.1. Discovery

Information on the amino acid sequences of proteins is written down as nucleotide sequences of the mRNA. The template triplet codon should determine unambiguously the position of a corresponding amino acid. However, there is no apparent steric fit between the structure of amino acids and their respective codons. In other words, codons cannot serve as direct template surfaces for amino acids. In order to solve this problem, in 1955 Francis Crick put forward his "adaptor hypothesis" in which he proposed the existence of special small adaptor RNA species and of specialized enzymes covalently attaching the amino acid residues to these RNAs (see Hoagland, 1960). According to this hypothesis each of the amino acids has its own species of adaptor RNA, and the corresponding enzyme attaches this amino acid only to a given adaptor. On the other hand, the adaptor RNA possesses a nucleotide triplet (subsequently termed the *anticodon)* that is complementary to the appropriate codon of the template RNA. Hence, the recognition of a codon by the amino acid is indirect and is mediated through a system consisting of the adaptor RNA and the enzyme: a specific enzyme concomitantly recognizes an amino acid and the corresponding adaptor molecule, so that they become ligated to each other; in turn, the adaptor recognizes an mRNA codon, and thus the amino acid attached becomes assigned specifically to this codon. In addition, this mechanism implied that the energy supply for amino acid polymerization was at the expense of chemical bond energy between the amino acid residues and the adaptor molecules.

This model was soon fully confirmed experimentally. In 1957 Hoagland, Zamecnik, and Stephenson, and simultaneously Ogata and Nohara, reported the discovery of a relatively low-molecular-weight RNA ("soluble RNA") and a special enzyme fraction ("pH 5 enzyme") that attached amino acids to this RNA. It was demonstrated that the aminoacyl-tRNA formed was indeed an intermediate in the transfer of amino acids into a polypeptide chain. Subsequently, this RNA was termed *transfer RNA* (tRNA); the enzymes were called *aminoacyl-tRNA synthetases* (ARSases).

The cell contains a specific ARSase for each of the 20 amino acids participating in protein synthesis (the individual amino acid-specific ARSases will be designated below as AlaRS, ArgRS, AspRS, etc.). Therefore, prokaryotic cells contain

20 different ARSases. The situation with eukaryotic cells is more complex, particularly because, in addition to the main cytoplasmic synthetases, there are special sets of ARSases for chloroplasts and mitochondria.

The number of different tRNA species is always greater than the number of amino acids and ARSases. For example, in *E. coli* 49 tRNA species encoded by different genes have been discovered (some tRNA species are encoded by multiple genes, so that the total number of tRNA genes approaches 80). This implies that several different tRNAs may be recognized by the same ARSase and, correspondingly, can be ligated to the same amino acid; such tRNAs are called *isoacceptor* tRNAs. Some isoacceptor tRNAs differ only in a few nucleotides and possess the same anticodon (thus recognizing the same codons), but in most cases different isoacceptor tRNA species have different anticodons and therefore recognize different codons for a given amino acid. In *E. coli* there are about 40 tRNA species carrying different anticodons, including tRNA for selenocysteine (recognizing UGA) and a special initiator tRNA (having the same anticodon as methionine tRNA). There are five different leucine tRNA species in *E. coli*, with anticodons CAG, GAG, U*AG (U* is modified uridine), CAA, and U*AA, recognizing six leucine codons; among them, $tRNA_1^{Leu}$ recognizes the leucine codon CUG (anticodon CAG), and $tRNA_5^{Leu}$ recognizes the leucine codons UUA and UUG (anticodon U*AA). The situation is similar in the cytoplasm of eukaryotic cells.

Cellular organelles (mitochondria and chloroplasts) of eukaryotic cells contain their own sets of tRNA species which are simpler than those of the cytoplasm, and also, as a rule, they have their own ARSases. Only 22 to 23 tRNA species encoded by the organelle genome can be found in animal mitochondria, and they are sufficient to recognize all 62 sense codons of mitochondrial mRNA. Thus, there usually exists just a single species of tRNA which corresponds to each amino acid and to all codons of a given amino acid. The exceptions are $tRNA^{Leu}$ and $tRNA^{Ser}$, where two species correspond to two different codon boxes.

3.2. Structure of tRNA

3.2.1. Primary Structure

In 1965 Holley and co-workers reported the nucleotide sequence of the first tRNA molecule. This molecule was yeast alanine tRNA (Fig. 3.1). Since then, hundreds of sequences of different tRNA from various sources have been determined. All of these structures have several common features.

The length of tRNA chains varies from 74 to 95 nucleotide residues (though in animal mitochondria it may be reduced to 60 or even 50 nucleotides). At the 3′-end all tRNA species contain a universal trinucleotide sequence, CCA_{OH}; it is the terminal invariant adenosine that accepts the amino acid residue when the aminoacyl-tRNA is being formed.

The anticodon triplet is located approximately in the middle of the tRNA chain (IGC in positions 34 to 36 in Fig. 3.1). As a rule, the 5′-side of the anticodon are two pyrimidine residues, whereas the 3′-side most often contains two purine residues, although the second residue on the 3′-side may be a pyrimidine, as in the case of the $tRNA^{Ala}$ (Fig. 3.1). These seven nucleotide residues together form the so-called anticodon loop (AC loop) which interacts with the mRNA and possesses a characteristic three-dimensional structure.

Approximately one-third of the way along the tRNA chain from its 3′-end there is a region common to most tRNA species; this region contains a sequence GTΨC, where Ψ is pseudouridine, or, much less frequently, GUΨC (or $Gm^1ΨΨC$ in archaebacteria), and is flanked on both sides by purine residues. In the eukaryotic initiator $tRNA_i^{Met}$, this sequence is substituted by GAΨC or GAUC. This sequence is the principal conservative sequence of tRNA. In mitochondrial tRNAs, however, the corresponding sequence region varies strongly and may even be absent.

Some other conservative parts of the sequence in the region of nucleotide residues 8 to 25 should be mentioned. Several invariants and semi-invariants are present here: U or its thio-derivative (s^4U) in position 8, G or its methyl-derivative (m^2G) in position 10, AG or AA in positions 14 or 15, GG in positions 17 to 21, and AG in positions 21 to 24 of different tRNAs.

In addition to the four main types of nucleotide residues (i.e., A, G, C, and U), the tRNA polynucleotide chain is characterized by a variety of modified nucleosides frequently referred to as "minor" nucleosides. These nucleosides are the result of posttranscriptional enzymatic modification of the usual nucleotide residues at specific positions of the tRNA polynucleotide chain. To date, several dozen modified nucleosides have been identified. Ribothymidine (5-methyluridine, abbreviated T or m^5U) and pseudouridine (5-ribofuranosyl-uracil, Ψ) are found in nearly all tRNAs and are particularly characteristic of the universal sequence GTΨC (Fig. 3.2). 5,6-Dihydrouridine (D or hU) is also an almost universal minor residue, especially in the region of residues 15 to 24. Bacterial tRNAs typically contain 4-thiouridine (s^4U) in position 8. The most common minor residues are methylated derivatives of the usual nucleosides, such as 1-methylguanosine (m^1G), N^2-methylguanosine (m^2G), N^2,N^2-dimethylguanosine (mG), 7-methylguanosine (m^7G), 2′-O-methylguanosine (Gm), 1-methyladenosine (m^1A), 2-methyladenosine (m^2A), N^6-methyladenosine (m^6A), 2′-O-methyladenosine (Am), 3-methylcytidine (m^3C), 5-methylcytidine (m^5C), and 2′-O-methylcytidine (Cm).

The first position of the anticodon may contain nonmodified G and C, but A

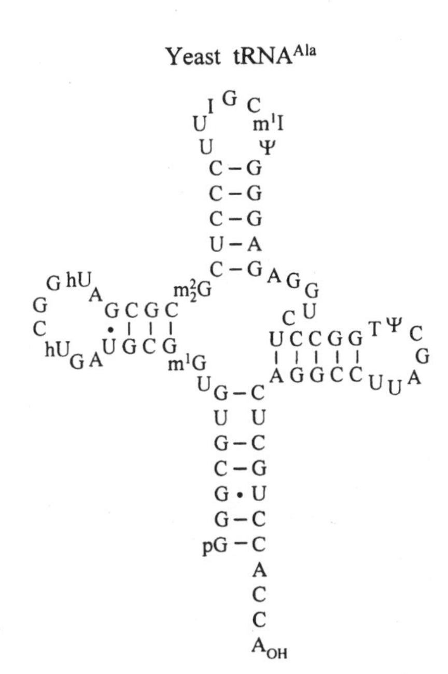

FIG. 3.1. Alanine tRNA of yeast ($tRNA_1^{Ala}$): nucleotide sequence and secondary structure in the form of the "cloverleaf." (Holley, R. W., Apgar, J., Everett, G. A., Madison, J. T., Marquisee, M., Merrill, S. H., Penswick, J. R., and Zamir, A. (1965) *Science* **147**:1462–1465). Ψ, hU and T are pseudouridine, dihydrouridine and ribothymidine, respectively (Fig. 3.2), I is inosine (see Fig. 3.3), and m^1G, $m_2{}^2G$, and m^1I are methylated guanosine and inosine (see text).

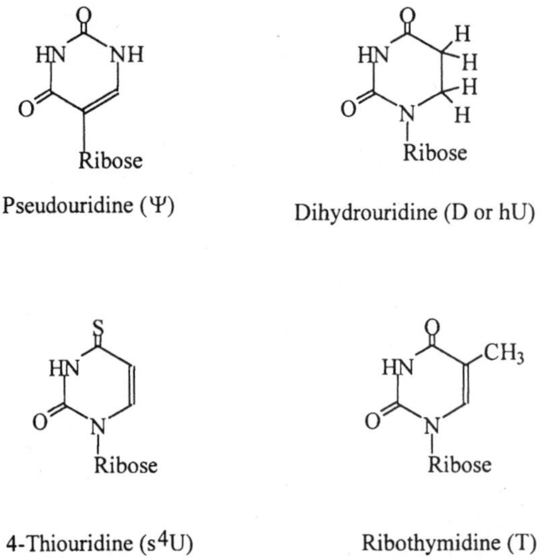

Pseudouridine (Ψ)

Dihydrouridine (D or hU)

4-Thiouridine (s^4U)

Ribothymidine (T)

FIG. 3.2. Modified uridine derivatives widely occurring in tRNAs. The full list of modified nucleosides in tRNAs and their formulas can be found in: Limbach, P. A., Crain, P. F., and McCloskey, J. A. (1995) *tRNA: Structure, Biosynthesis, and Function* (D. Soell and U. L. RajBhandary, eds.), pp. 551–555. ASM Press, Washington, DC.

and U are almost always derivatized. The exceptions are mitochondrial and a few special tRNA species (see below). The A in the first position of the anticodon is usually deaminated into inosine (I) (Fig. 3.3). I in this position is particularly characteristic of eukaryotic tRNAs, such as $tRNA^{Ile}$, $tRNA^{Val}$, $tRNA^{Ser}$, $tRNA^{Pro}$, $tRNA^{Thr}$, $tRNA^{Ala}$, and $tRNA^{Arg}$. The U derivatives present in the first anticodon position are 5-methoxyuridine (mo^5U) or 5-carboxymethoxyuridine (cmo^5U or V) in $tRNA^{Ala}$, $tRNA^{Ser}$, and $tRNA^{Val}$ of bacteria; 5-methylaminomethyl-2-thiouridine (mnm^5s^2U) in bacterial $tRNA^{Glu}$ and $tRNA^{Lys}$; 5-(methoxycarbonylmethyl)-2-thiouridine (mcm^5s^2U) in $tRNA^{Glu}$ and $tRNA^{Lys}$ of fungi; or 5-(methoxycarbonylmethyl)uridine (mcm^5U) in $tRNA^{Arg}$ of fungi (see Fig. 3.3). The presence of unmodified U has been demonstrated for one species of $tRNA^{Gly}$ in several bacteria and for one of the yeast $tRNA^{Leu}$. The unmodified U, however, is typical of mitochondrial tRNAs.

In some tRNAs, such as $tRNA^{Asp}$, $tRNA^{Asn}$, $tRNA^{His}$, and $tRNA^{Tyr}$ of bacteria and animals, the first position of the anticodon contains a hypermodified G derivative, the so-called queuosine (Quo or Q), the chemical name of which is 7-{[(cis-4,5-dioxy-2-cyclopenten-1-yl)amino]methyl}-7-deazaguanosine (see Fig. 3.3).

Hypermodifications are found to be typical of the position of the purine nucleoside adjacent to the anticodon on the 3′-side. For example, the residue flanking the anticodon at the 3′-side is N^6-isopentenyl adenosine (i^6A) in eukaryotic $tRNA^{Cys}$, $tRNA^{Ser}$, and $tRNA^{Tyr}$; 2-methylthio-N^6-isopentenyladenosine (N^6-isopentenyl-2-methylthioadenosine, ms^2i^6A) in the analogous bacterial tRNAs; and N^6-(threoninocarbonyl)-adenosine (t^6A) in $tRNA^{Ile}$, $tRNA^{Thr}$, $tRNA^{Lys}$, and $tRNA^{Met}$ of both eukaryotes and bacteria (Fig. 3.4). This position is even more hypermodified in the $tRNA^{Phe}$ of all eukaryotes, where it is represented by the so-called wybutosine (yW or Y) or its hydroxy-derivative (oyW) (see Fig. 3.4).

3.2.2. Secondary Structure

An analysis of even the first tRNA primary structure (i.e., $tRNA^{Ala}$ of yeast) revealed a number of interesting features concerning possible chain folding into the secondary structure. First, the 5′-terminal section (positions 1 to 7) has a marked complementarity with the 3′-end-adjacent section (positions 66 to 72) if the sections are arranged in an antiparallel fashion. In addition, three inner sections of the tRNA chain display self-complementarity when folded upon themselves; because of this they are capable of forming hairpin-like structures. Pairing these complementary sequences results in the structure schematically presented in Fig. 3.1, commonly called a *cloverleaf* structure. It is remarkable that without exception the nucleotide sequences of all the tRNA species studied so far reveal similar self-complementarity features and correspondingly can be folded into very similar cloverleaves.

The parts of the cloverleaf structure have been designated as follows: the *acceptor stem* (AA stem), with the universal 3′-terminal sequence CCA which accepts an amino acid residue; the *dihydrouridylic arm* (D arm), with the corresponding loop varying somewhat in length and containing, as a rule, between one and five dihydrouridylic acid residues; the *anticodon arm* (AC arm), with an anticodon loop of constant length equal to seven nucleotides; and the *thymidylpseudouridylic arm* (TΨ arm), which has a loop with the universal GTΨCGA or GTΨCAA sequence. In addition, the cloverleaf contains a *variable loop* (V loop) between the anticodon and TΨ arms; in $tRNA^{Ala}$ this loop is only five nucleotides long whereas in other tRNA species it may reach 15 to 20 nucleotide residues in length (the latter is the case for $tRNA^{Leu}$, $tRNA^{Ser}$, and bacterial $tRNA^{Tyr}$).

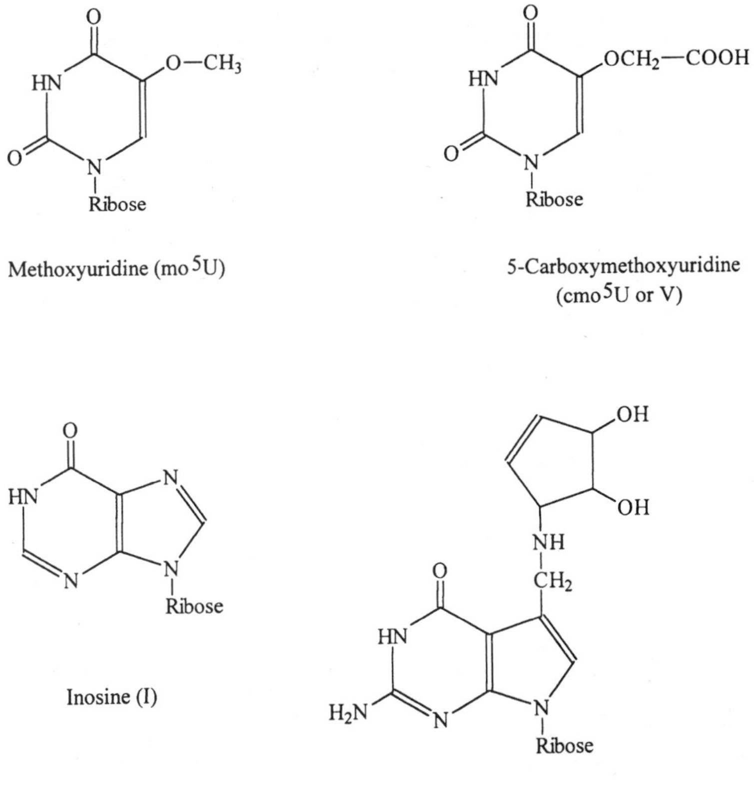
O
HN
O—CH₃
O
N
Ribose
Methoxyuridine (mo⁵U)
O
HN
OCH₂—COOH
O
N
Ribose
5-Carboxymethoxyuridine
(cmo⁵U or V)
O
HN
N
N
N
Ribose
Inosine (I)
OH
OH
NH
CH₂
O
HN
H₂N
N
N
Ribose

Queuosine (Quo or Q)

FIG. 3.3. Some modified nucleosides occurring in the first position of the tRNA anticodon.

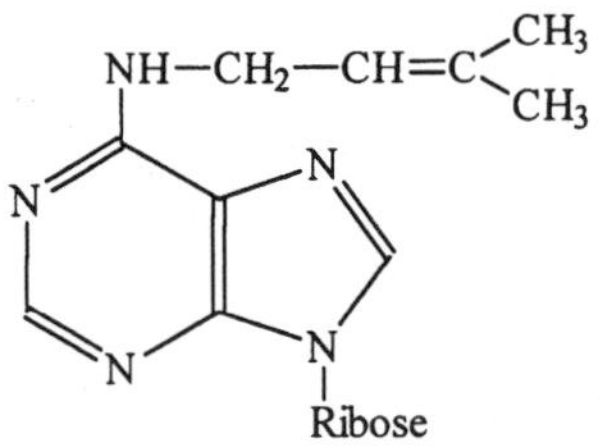

N6- Isopentenyl-adenosine (i 6A)

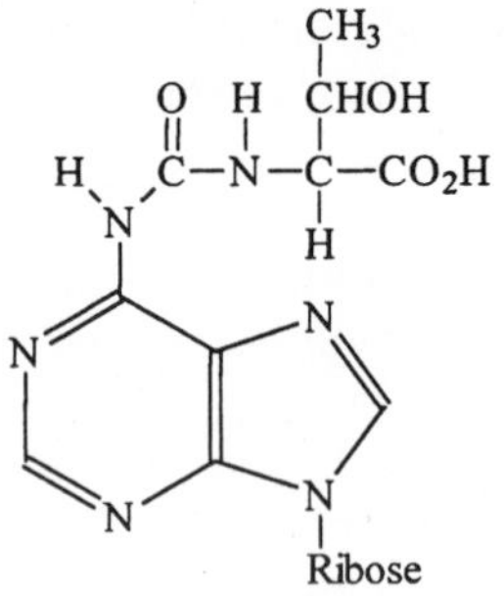

N6- (threoninocarbonyl) adenosine (t 6A)

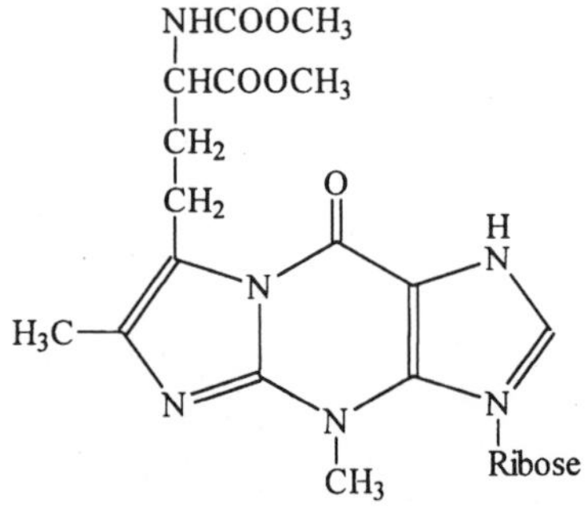

Wybutosine (yW or Y)

FIG. 3.4. Some hypermodified nucleosides occurring in the position adjacent to the anticodon at its 3′-side.

In animal mitochondrial tRNAs the D-arm or T-arm may be reduced or absent.

Structurally, the paired (double-stranded) part of each arm of tRNA is a double helix. The RNA double helix contains 11 pairs of nucleotide residues per turn. The parameters of this helix are similar to those of the A form of DNA. The double helix is the main element of tRNA secondary structure. In addition to the canonical Watson–Crick base pairs G:C and A:U, the double-stranded regions of tRNA often contain the G:U pair, which is close by its steric parameters to the canonical pairs (Fig. 3.5).

The secondary structure of the unpaired regions, such as loops and the acceptor ACCA or GCCA terminus, is of a different type. A single-helical arrangement of several residues maintained by base-stacking interactions can occur here. The structure of the anticodon loop is particularly interesting (Fig. 3.6); three anticodon bases and two subsequent bases adjacent to the anticodon from the 3′-side are stacked with each other and form a single-stranded, right-handed helix;

the first base of the anticodon is located at the top of the helix, and the groups capable of forming hydrogen bonds of all three anticodon bases are exposed outward. Such an orientation of the anticodon bases is extremely important for interaction with the mRNA codon. The features of the primary structure of the anticodon loop contribute specifically to the maintenance of the spatial arrangement described: the hypermodified purine base directly adjacent to the anticodon from the 3′-side as well as the next base, usually also a purine, provides for stable stacking interactions in the single-stranded helix, while the two "small" pyrimidine bases at the 5′-side of the anticodon, and particularly the adjacent invariant U, make a sharp bend in the chain (between the anticodon and U) and maintain the loop conformation, particularly at the expense of a hydrogen bond between the invariant U and the phosphate group of the third residue of the anticodon.

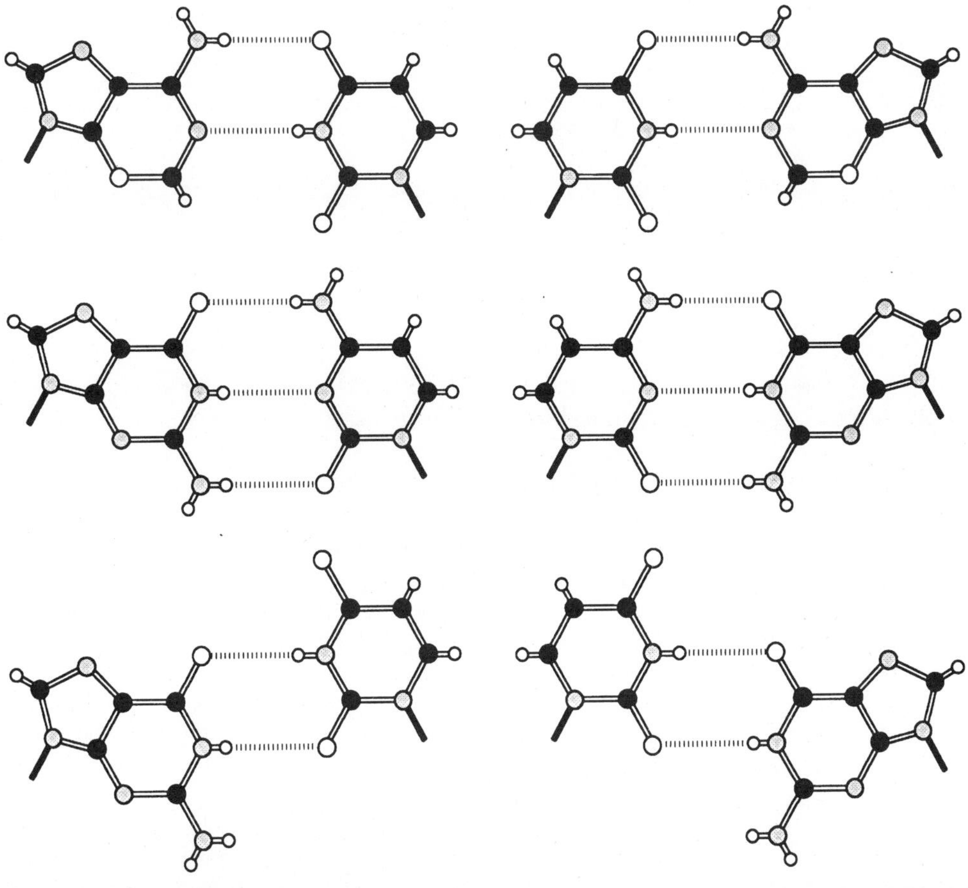

FIG. 3.5. Base pairing in RNA double helices: ball-and-stick drawing. (*Top to bottom*) A:U and U:A; G:C and C:G; G:U and U:G. Solid circles are carbons, shaded circles are nitrogens, large open circles are oxygens, and small open circles are hydrogens; solid sticks are N-glycosidic bonds between the base and ribose.

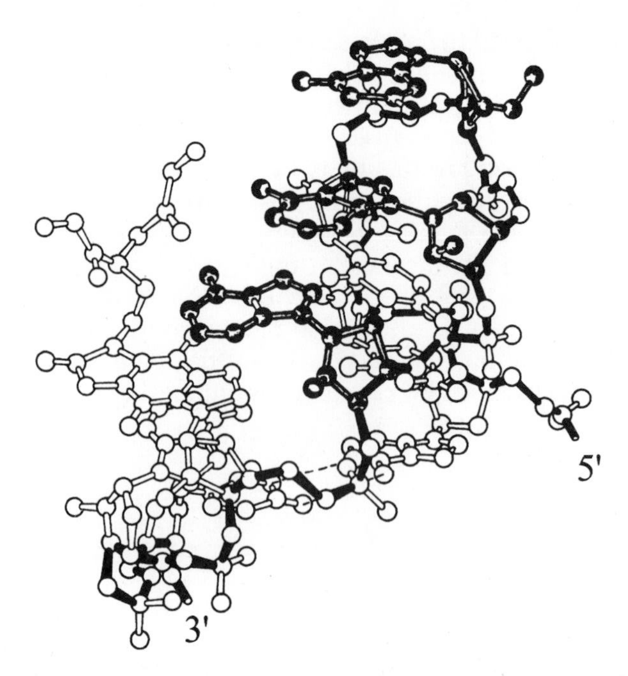

FIG. 3.6. Anticodon loop of yeast phenylalanine tRNA: ball-and-stick skeletal model without hydrogens. The path of the backbone is given in solid black; three anticodon residues are shaded. The three-dimensional structure of the yeast $tRNA^{Phe}$ is determined by X-ray crystallography. See Kim, S.-H., Suddath, F. L., Quigley, G. J., McPherson, A., Sussman, J. L., Wang, A. H.-J., Seeman, N. C., and Rich, A. (1974) *Science* **185**:435–440, 1974; Robertus, J. D., Ladner, J. E., Finch, J. T., Rhodes, D., Brown, R. S., Clark, B. F. C., and Klug, A., (1974) *Nature* **250**:546–551.

3.2.3. Tertiary Structure

The three-dimensional structure of tRNA was first reported for yeast $tRNA^{Phe}$. This structure was determined independently by the groups of Alexander Rich and Aaron Klug in 1974 through the use of X-ray analysis of $tRNA^{Phe}$ crystals (Kim *et al.,* 1974; Robertus *et al.,* 1974). A great deal of indirect evidence as well as direct determinations of the three-dimensional structures of several other tRNA species has demonstrated that the main pattern of tRNA chain folding into the tertiary structure is universal. Schematically, this folding may be represented as follows. The acceptor stem and the T-arm are arranged along a common axis, forming a continuous double helix 12 nucleotide pairs in length; the anticodon arm and the dihydrouridylic arm are also arranged along a common axis and yield another double helix, this one nine nucleotide pairs long. These two helices are oriented toward each other at approximately a right angle so that the dihydrouridylic loop is brought close to the T-loop, and the interaction between the GG invariant and the ΨC invariant fastens them together (Fig. 3.7). As a result, the structure looks like the letter L with the tops of its two limbs corresponding to the anticodon and the acceptor 3′-end. The short, single-stranded bridge between the acceptor stem and dihydrouridylic helix (residues 8 and 9), part of the dihydrouridylic loop, and the additional variable loop are superimposed on the dihydrouridylic helix in the region of the inner corner of the L-shaped molecule, resulting in the formation of the core of the molecule with a number of tertiary interactions. In a schematic drawing of the model of the yeast $tRNA^{Phe}$, the core can be seen as a concentration and intertwining of the chain sections in the region of the corner, especially at its inner side (Fig. 3.8).

Each limb of the L-shaped tRNA molecule is about 70 Å long, and the molecule has a "thickness" of approximately 20 Å. The distance between the anticodon and the acceptor end is 76 to 78 Å. All three bases of the anticodon on top of one of the limbs are turned toward the inner side of the corner in the L-shaped molecule.

There are a large number of noncanonical (non-Watson–Crick) interactions between chain bases in the tRNA tertiary structure. First, the corner of the L-shaped molecule is stabilized by both the stacking interactions and the hydrogen bonding between the dihydrouridylic loop and the T loop. The interaction between the invariant G19 and C56 is of the Watson–Crick type, whereas the interaction between the invariant G18 and Ψ55 is unusual, including the hydrogen bonding of O at C4 of the pyrimidine ring both with N1 and with the nitrogen atom at C2 of the purine ring of the G. In addition, there is an unusually strong stacking interaction between three guanosine residues in the same corner: G57 is found to be intercalated between G18 and G19. Moreover, G57 through N at C2 seems to form hydrogen bonds with the ribose residues of G18 and G19, whereas through its N7 it forms a hydrogen bond with the ribose of Ψ55.

Even more complex tertiary interactions are observed in the core. As has already been mentioned, different sections of the polynucleotide chain are interwound here. A characteristic feature is the noncanonical purine–purine G:A or A:G pairing (depending on the RNA species) between residues 26 and 44. G:C pairing, or A:U pairing in other tRNA species, between residues 15 and 48 is unusual for double helices; in this case the orientation of chains is parallel. Even more unusual is A:U pairing between residues 14 and 8 where N7 of the purine ring par-

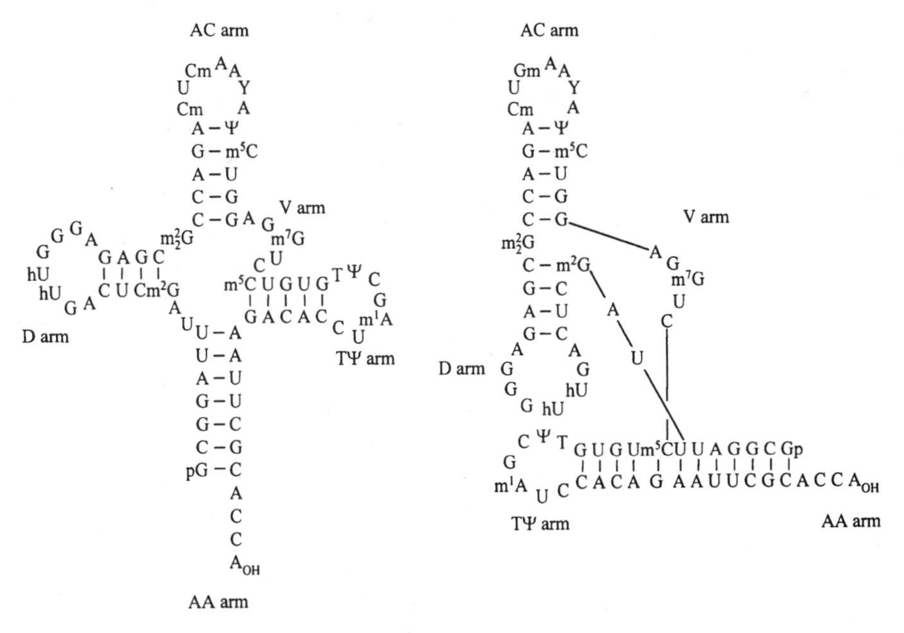

FIG. 3.7. Scheme illustrating the folding of the tRNA helical regions into the tertiary structure (yeast tRNA^{Phe}). From Kim, S.-H., Quigley, G. J., Suddath, F. L., McPherson, A., Sneden, D., Kim, J. J., Weinzierl, J., and Rich, A. (1973) *Science* **179**:285–288; **185**:435–440.

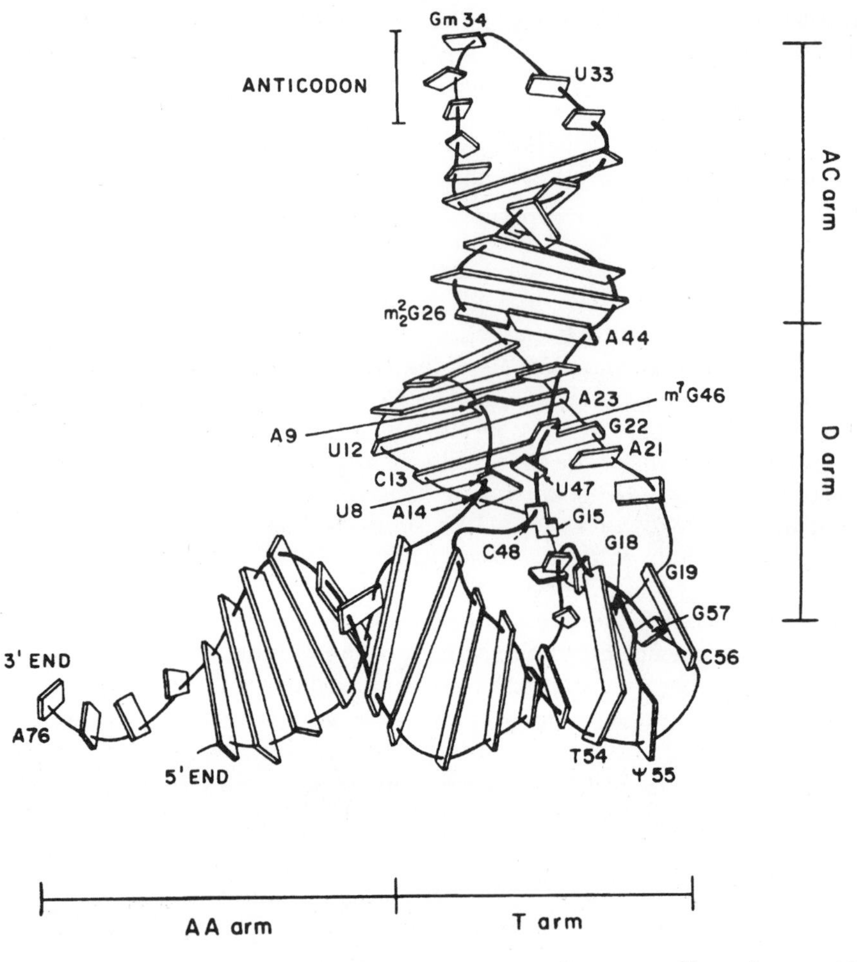

FIG. 3.8. Schematic drawing of the three-dimensional structure of yeast $tRNA^{Phe}$. Redrawn, with minor modifications, from Kim, S.-H. (1975) *Nature* **256**:679–681 with permission; see also Rich, A., and Kim, S.-H. (1978) *Scientific American* **238**:52–62.

ticipates in the formation of the hydrogen bond. Triple hydrogen bond interactions, such as U:A:A, or the equivalent G:C:G or U:A:G in other tRNAs, between residues 12, 23, and 9, respectively, are characteristic of this part of the molecule (Fig. 3.9, bottom). The triple hydrogen bond interaction is found also for C:G:G, or the equivalent U:A:A in other tRNAs, between residues 13, 22, and 46, respectively (Fig. 3.9, top).

Computer images of the atomic space-filling model and the skeletal model of full yeast $tRNA^{Phe}$ molecule (viewed from the side of its TΨ loop) are given in Fig. 3.10 (compare with the schematic representation in Fig. 3.8).

3.3. Aminoacyl-tRNA Synthetases

Despite the universality of the main features of the three-dimensional structure of tRNAs, ARSases show marked differences depending on their amino acid specificity. As a rule, ARSases are relatively large proteins with a molecular mass around 100,000 daltons, although both smaller (about 50 kDa for bacterial CysRS and GluRS) and larger (more than 200 kDa for GlyRS, AlaRS, and PheRS) enzymes also occur. One third of ARSases are monomeric, half of ARSases are homodimers (α_2 type), and the three large ARSases mentioned above are tetramers of the α_4 or $\alpha_2\beta_2$ type (see Table 3.1). The molecular masses of subunits of dimeric and tetrameric enzymes range from 35,000 to 90,000 daltons. In the case of large monomeric enzymes, such as bacterial and fungal ValRS, LeuRS, and IleRS with a molecular mass of 100,000 to 120,000 daltons, it appears that their single polypeptide chain consists of two homologous regions forming two similar domains, each with a molecular mass of about 50,000 to 60,000 daltons. At the same time, ArgRS, CysRS, GluRS, and GlnRS of bacteria consist of a single polypeptide chain (also α_1 type) with a molecular mass of 50,000 to 60,000 daltons, and do not appear to be subdivided into two homologous regions.

From the analysis of subunit and domain structure of ARSases it is tempting to suggest a generalized pattern of their principal organization. Indeed, most synthetases have a molecular mass around 100,000 daltons and consist either of two subunits or two similar halves (superdomains). Therefore, the principal building unit (i.e., subunit or superdomain) has a molecular mass ranging from 40,000 to 60,000 daltons, and many of the ARSases would be considered dimers or pseudodimers of the building unit, or $(40{,}000\text{—}60{,}000)_2$. The synthetases with greater molecular masses, around 200,000 daltons, may be "duplicated" enzymes of this

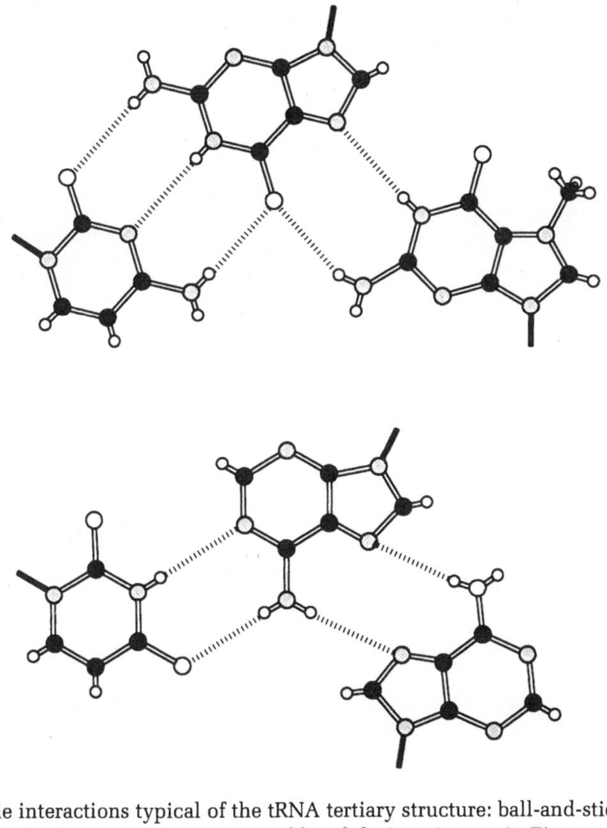

FIG. 3.9. Base triple interactions typical of the tRNA tertiary structure: ball-and-stick drawings. (*Top*) C13:G22:m^{7}G46. (*Bottom*) U12:A23:A9. Atom and bond designations as in Fig. 3.5.

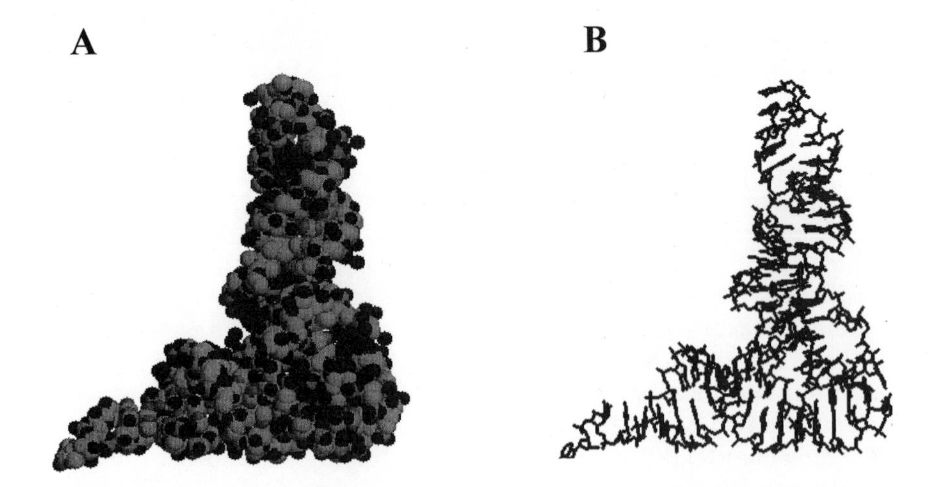

FIG. 3.10. Space-filling (*left,* **A**) and skeletal (*right,* **B**) models of yeast phenylalanine tRNA (From Westhof, E., Dumas, P., and Moras, D. (1989) *Protein Data Bank of Brookhaven National Laboratory,* PDB ID code 4TRA.

type. In reality, however, the nonrepeating unit in some cases can be markedly larger; bacterial AlaRS, for example, consists of four identical subunits, each with a molecular mass of about 100,000 daltons, and shows no evidence of any repeats in its amino acid sequence. On the other hand, such relatively small, one-subunit enzymes as CysRS and ArgRS (molecular mass 52,000 and 64,500 daltons, respectively) do not display two homologous superdomains in their structure.

In any case, according to functional tests the molecules of some (but not all) ARSases possess two sets of substrate-binding sites; in other words, they are dimers in the functional sense as well. The active sites, however, are not independent and can markedly affect each other in the dimeric or two-superdomain enzymes, thus displaying a certain cooperativity (see below).

Despite the apparent structural diversity of ARSases, they have been found to possess structural motifs that provide the basis for revealing homology between

TABLE 3.1. Classification of *E. coli* Aminoacyl-tRNA Synthetases

	Class I (ATP binding site: Rossmann fold)			Class II (ATP binding site: antiparallel β-sheet)		
Subclass	Amino acid	Oligomeric state	Aminoacylation site	Amino acid	Oligomeric state	Aminoacylation site
a	Leu	α	2′-OH	His	α_2	3′-OH
	Ile	α	2′-OH	Pro	α_2	3′-OH
	Val	α	2′-OH	Ser	α_2	3′-OH
	Cys	α	2′-OH or 3′-OH	Thr	α_2	3′-OH
	Met	α_2	2′-OH			
b	Glu	α	2′-OH	Asp	α_2	3′-OH
	Gln	α	2′-OH	Asn	α_2	3′-OH
	Arg	α	2′-OH	Lys	α_2	3′-OH
c	Tyr	α_2	2′-OH or 3′-OH	Gly	$\alpha_2\beta_2$	3′-OH
	Trp	α_2	2′-OH	Ala	α_2	3′-OH
				Phe	$\alpha_2\beta_2$	2′-OH

From Cavarelli, J., and Moras, D. (1995) in *tRNA: Structure, Biosynthesis, and Function* (D. Soell and U. RajBhandary, eds.), p. 412, ASM Press, Washington DC.

some of them and for unification of homologous species into classes (Eriani *et al.*, 1990). There are two distinct classes of ARSases, each consisting of ten enzymes (Table 3.1). Class I consists of Arg-, Cys-, Gln-, Glu-, Ile-, Leu-, Met-, Trp-, Tyr-, and Val-RSases. They are predominantly monomers, with the exception of Met-, Trp- and Tyr-RSases, which are homodimers. The monomeric globule is subdivided into different domains. The N-terminal region of the molecule is responsible for the binding of all three substrates, namely ATP, amino acid, and the acceptor stem of tRNA, and for the catalysis of the reactions between them. This region is characterized by the presence of the so-called Rossmann dinucleotide-binding fold (earlier found in dehydrogenases, kinases, and many other proteins utilizing ATP or other high-energy nucleotides), as well as by two conserved sequences (His-Ile-Gly-His and Lys-Met-Ser-Lys-Ser) localized within or in the vicinity of the catalytic center. The classical Rossmann fold is formed by a repeating $\beta\alpha\beta$ motif such that a six-stranded parallel β-sheet is sandwiched between two pairs of α-helices (see, e.g., domain 1 in Fig. 10.4 and G-domain in Fig. 12.2). The C-terminal half of the enzyme participates in the specific recognition of the tRNA molecule by interacting with its anticodon loop and stem. The enzyme as a whole approaches the tRNA molecule from the side of its D loop and from inside the L, contacting with the minor groove of the acceptor helix (Fig. 3.11).

Class II includes Ala-, Asn-, Asp-, Gly-, His-, Lys-, Phe-, Pro-, Ser-, and Thr-RSases, all of which are composed of two or four subunits (Table 3.1). Their folding pattern is quite different from that of the class I ARSases. First, their ATP-binding and catalytic domain is constructed as a seven-stranded antiparallel β-sheet (see, e.g., domain 2 in Fig. 10.4 and domain II in Fig. 12.2), in contrast to the Rossmann fold of class I ARSases with its parallel β-sheet between two layers of helices. The catalytic domain also contains relatively conserved, class-defining sequence motifs, but with nothing in common with the motifs in class I enzymes. This large

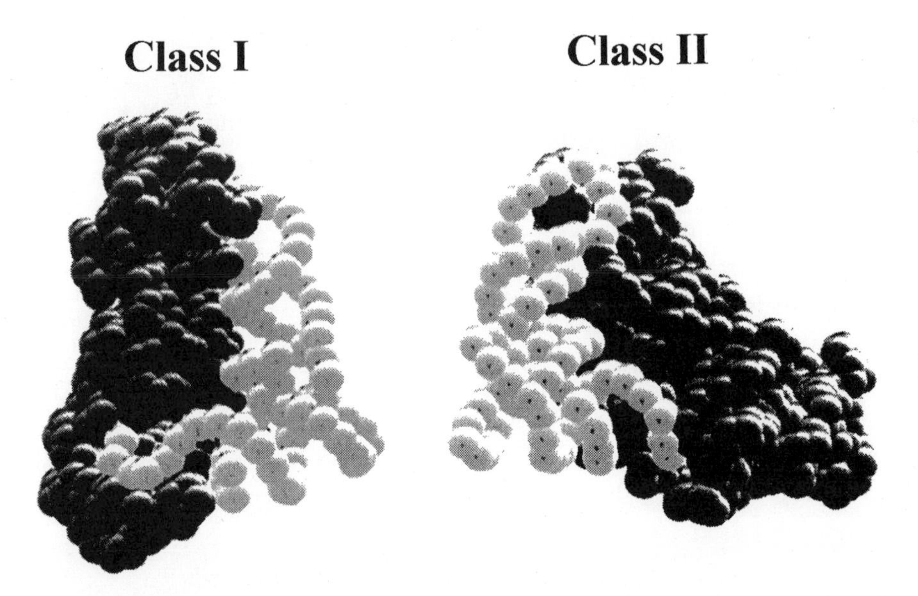

FIG. 3.11. Space-filling models of the complexes of tRNAs with ARSases of the two classes. **Class I:** GlnRS:tRNA^{Gln} complex. **Class II:** AspRS:tRNA^{Asp} complex. The opposite sites of approach of the tRNA molecules by the enzymes are demonstrated. Reproduced from Ruff, M., Krishnaswamy, S., Boeglin, M., Poterszman, A., Mitschler, A., Podjarny, A., Rees, B., Thierry, J. C., and Moras, D. (1991) *Science* **252:**1682–1689, with permission.

domain involved in ATP-binding, amino acid binding, tRNA acceptor end binding, and catalysis may constitute, as in the cases of Asp-, Asn-, Lys-, and Ser-RSases, the C-terminal part of the enzyme, whereas the N-terminal part participates in interactions with the distal region of tRNA such as the anticodon. In other class II synthetases, however, the catalytic domain may be at the N-terminal end (e.g., in Thr-, Pro-, His-, Gly-, and Ala-RSases). Each subunit of the enzyme binds one tRNA molecule. On the whole, the enzyme approaches the tRNA molecule from the side of its variable and TΨ loops and contacts with the major groove of the acceptor stem helix (Fig. 3.11), unlike the interaction in the case of class I synthetases.

Several characteristic features of eukaryotic ARSases primarily in animals, should be noted. Like many other proteins that play a role in the protein-synthesizing machinery of the eukaryotic cell (see, for example, Section 10.2.1), some eukaryotic, and particularly mammalian, synthetases are organized in large multienzyme complexes (Dang, Johnson, and Yang, 1982; Mirande, Cirakoglu, and Waller, 1983). For example, a complex with a molecular mass of about 10^6 daltons and containing nine ARSase activities, namely Arg-, Asp-, Glu-, Gln-, Ile-, Leu-, Lys-, Met-, and Pro-RSases, can be isolated from mammalian cells. Two of these activities, GluRS and ProRS, are carried by a single polypeptide. The existence of these enzymes in the form of a complex is not, however, vital to their activity: on one hand, the enzymes may be present in cells individually and not in the aggregate; on the other hand, they appear to function independently of each other in the complex. A smaller complex including ValRS and the elongation factors eEF1A and eEF1B (the so-called "heavy form" of elongation factor 1, or eEF-1_H, see Section 10.2.1), was also demonstrated in mammalian cells.

In addition, the eukaryotic ARSases can be directly associated with polyribosomes. Although the association is quite loose and reversible, at each given moment a large proportion of cellular ARSases are in a labile association with the functional ribosomes. These characteristic features of eukaryotic systems appear to be due to the molecular attributes of the enzymes. It has been demonstrated that in contrast to prokaryotic ARSases, their eukaryotic analogs possess an affinity for high-molecular-mass RNA, including mRNA and ribosomal RNA (Alzhanova *et al.*, 1980). This additional property of eukaryotic ARSases correlates with the facts that, compared with prokaryotic enzymes, the synthetases of eukaryotic cells are characterized by somewhat larger subunits, and polypeptide chains of eukaryotic ARSases have extensions at the N-terminus that are especially rich in basic amino acids. The nonspecific affinity for RNA (i.e., RNA-binding capacity) of eukaryotic ARSases may be responsible for their concentration and partial compartmentation on the protein-synthesizing particles.

3.4. Aminoacylation of tRNA

The ligation of an amino acid to the tRNA 3′-end catalyzed by ARSase is coupled with ATP cleavage. The overall equation of the process may be written as follows:

$$\text{Aa} + \text{ATP} + \text{tRNA} \overset{\text{E}}{\rightleftharpoons} \text{Aa-tRNA} + \text{AMP} + \text{PP}_i$$

where Aa is the amino acid, Aa-tRNA is the aminoacyl-tRNA, and PP_i is the inorganic pyrophosphate. It has been demonstrated that the enzyme catalyzes two different reactions comprising two consecutive steps of this process.

The first stage reaction, catalyzed by ARSase, is amino acid *activation*, where the carboxyl group of the amino acid attacks the bond between the α- and β-phosphates of ATP, resulting in the formation of a mixed anhydride aminoacyl adenylate and inorganic pyrophosphate (Fig. 3.12):

$$\text{Aa} + \text{ATP} \overset{E}{\rightleftharpoons} \text{Aa-AMP} + \text{PP}_i$$

This reaction is reversible and may be conveniently traced by pyrophosphate exchange: if [^{32}P]-pyrophosphate is added to the reaction mixture, the label is soon detected in [^{32}P]-ATP. Aminoacyl adenylate formed in the reaction remains bound to the enzyme and is not released into solution.

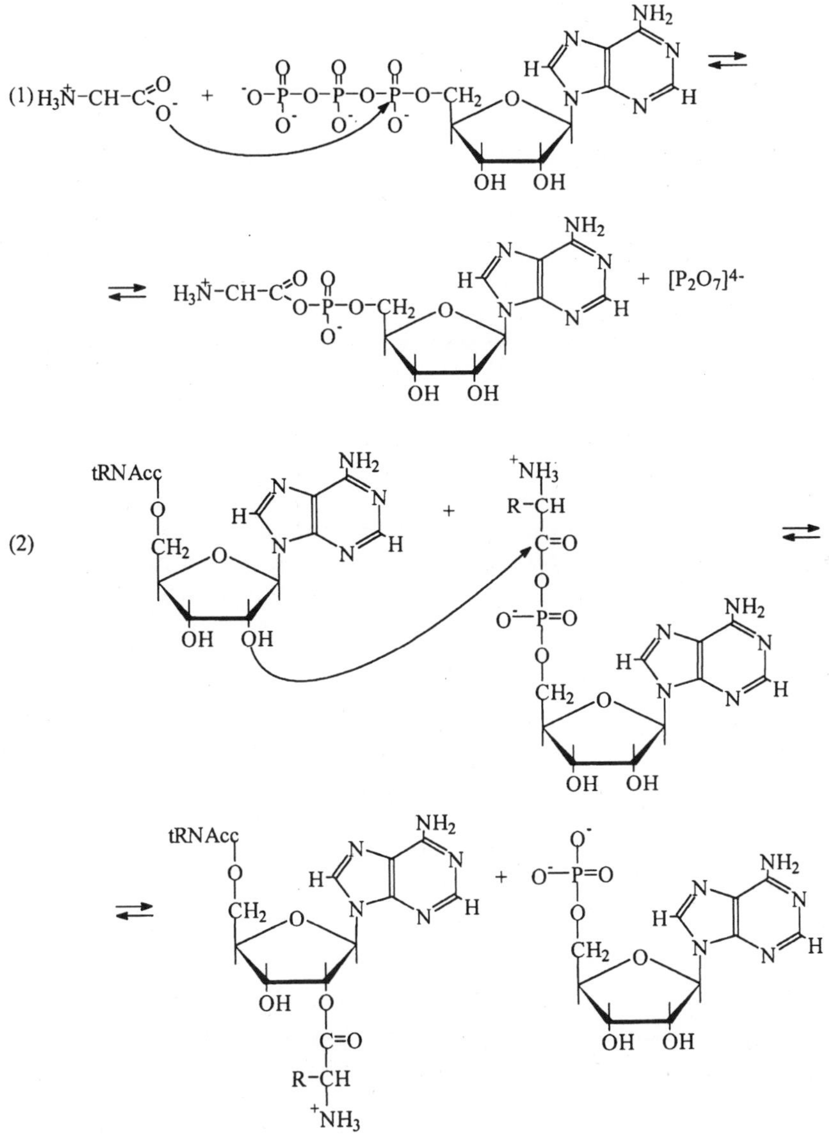

FIG. 3.12. Reactions of amino acid activation (1) and acceptance of aminoacyl residue by the tRNA molecule (2), catalyzed by ARSase.

The reaction, and consequently the overall reaction, is markedly shifted in the direction of aminoacyl adenylate and aminoacyl-tRNA formation due to the hydrolysis of the inorganic pyrophosphate which is catalyzed by pyrophosphatase. Therefore, the production of pyrophosphate in the amino acid activation step and the subsequent hydrolysis of the pyrophosphate to the inorganic orthophosphate play an important part in providing the energy that ensures the direction of the entire process.

A reaction catalyzed at the second stage by the same ARSase involves the *accepting* of the amino acid, where the 2′- or 3′-hydroxyl of the ribose residue of the tRNA 3′-terminal adenosine attacks the anhydride group of the aminoacyl adenylate, resulting in the formation of an ester bond between the aminoacyl residue and the tRNA, with the accompanying release of AMP (Fig. 3.12):

$$\text{Aa-ATP} + \text{tRNA} \overset{\text{E}}{\rightleftharpoons} \text{Aa-tRNA} + \text{AMP}$$

It is noteworthy that different ARSases possess different specificity with regard to the position of the ribose hydroxyl participating in the transacylation reaction (Table 3.1). All class I ARSases catalyze the coupling of amino acids to the 2′-position of the ribose of the 3′-terminal adenosine residue. TyrRS and CysRS, however, may catalyze the reaction with both the 2′- and the 3′-hydroxyl groups. At same time class II synthetases catalyze the reaction of the 3′-hydroxyl with the amino acid residue; the only exception among them is PheRS which ligates the amino acid to the 2′-position of tRNA. This is of no great importance to the subsequent fate of the aminoacyl-tRNA formed because in aqueous solution the aminoacyl residue spontaneously migrates between the 2′- and 3′-positions (through the formation of a 2′, 3′-derivative), and eventually the two forms are in equilibrium.

Thus, an ARSase uses three substrates of a different chemical nature: ATP, an amino acid, and tRNA. Correspondingly, it must possess three different substrate-binding sites. ATP is the universal substrate for all ARSases, whereas for the amino acid and tRNA, each ARSase displays high specificity.

As has already been mentioned, in many cases ARSases are dimers or pseudodimers, and, correspondingly, they possess two sets of substrate-binding sites. The substrate-binding sites within each subunit (or the equivalent domain) and on different subunits (or domains) are interdependent. Frequently synergism is observed: the binding of one substrate molecule facilitates the binding of the other. On the other hand, there is a negative cooperativity in the binding of two tRNA molecules: the binding of one tRNA molecule makes the binding of the other one less tight.

An example sequence of substrate addition to the dimeric enzyme is schematically presented in Fig. 3.13. Beginning with the enzyme free of substrates (upper part of the scheme), the first stages often involve the binding of small substrates, such as ATP and the amino acid; and the binding of one of these may stimulate the binding of the other (synergism). The substrates bound to the enzyme interact to yield aminoacyl adenylate, and the resulting pyrophosphate is released into solution. The binding of the small substrates and the formation of aminoacyl adenylate stimulate tRNA binding, resulting in the aminoacylation of tRNA by the enzyme and the release of AMP into solution. The aminoacyl-tRNA, when present in a single copy per dimeric enzyme molecule, may dissociate from the enzyme rather slowly, but the binding of the second tRNA molecule stimulates the dissociation.

This leads to the cycle shown in the lower part of Fig. 3.13, where one of the tRNA-binding sites is permanently occupied and the enzyme displays the reactivity of only half of its substrate-binding sites (half-of-the-sites-reactivity).

Under conditions where the enzyme works in substrate excess, the pathway shown in the lower part of Fig. 3.13 is the route that appears to occur. State 1 is exhibited when the active site of one subunit (or domain) is occupied by aminoacyl-tRNA while the other one is vacant. Therefore, only the substrate-binding sites of the other active center of the enzyme are capable of binding substrate ligands. The consecutive or independent binding of the small substrates, ATP, and the amino acid (states 2 and 3) results in the formation of the enzyme-bound aminoacyl adenylate (state 4), which in turn stimulates tRNA association with the second active center of the enzyme (state 5). Because of the negative cooperativity mentioned above, the binding of tRNA with the second active center weakens the holding of aminoacyl-tRNA in the first binding center; as a result, this aminoacyl-tRNA dissociates into solution, leaving the enzyme with one active center occupied and the other vacant (state 6). Thus, the two active centers of a dimeric (or two-domain) enzyme appear to work alternately. The final product aminoacyl-tRNA is not released into solution immediately after its synthesis has been completed, but "waits" until the second substrate tRNA enters its binding site. It should be pointed out again that the above model is just an example of a possible reaction pathway and cannot be regarded as general.

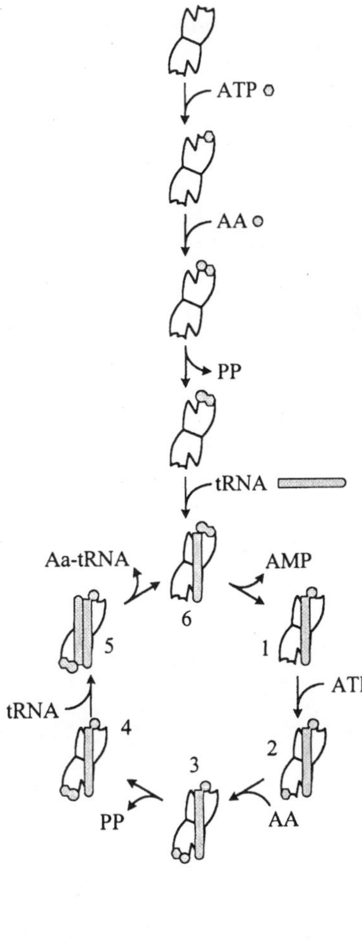

FIG. 3.13. Possible sequence of events in the functioning of the two-domain (or dimeric) ARSase.

3.5. Specificity of tRNA Aminoacylation

3.5.1. Specificity for Amino Acids

To provide unambiguous mRNA decoding during translation, ARSases should possess an extremely high specificity when selecting amino acids and tRNAs as substrates. In the case of amino acid selection the enzyme has to discriminate between substrates, which sometimes possess very similar structures, such as isoleucine and valine. The error rate in tRNA aminoacylation is indeed extremely low, and even for related amino acids (e.g., isoleucine and valine) it does not appear to exceed 1 per 10,000 under normal physiological conditions.

However, analysis of the stages of amino acid binding and the subsequent reversible formation of aminoacyl adenylate measured by ATP–pyrophosphate exchange has shown that the enzyme cannot provide such high specificity in the discrimination of related amino acids at these stages. For example, IleRS synthetase can effectively bind valine and form valyl adenylate. Similarly, ValRS can bind and activate isoleucine as well as a number of other amino acids, such as alanine, serine, cysteine, and threonine. The phenylalanine enzyme activates methionine, leucine, and tyrosine. Nevertheless, none of the listed misactivated amino acids is accepted by the tRNA. Thus, a number of ARSases may possess, in addition to amino acid discrimination at the binding stage, a special error-correcting mechanism which acts *after* the aminoacyl adenylate has been formed (Baldwin and Berg, 1966). Basically, the binding of the cognate tRNA by the enzyme results in a hydrolytic release of the free amino acid if the amino acid residue was noncognate for the enzyme. It seems that at least in some cases misactivated amino acid bound to the enzyme as aminoacyl adenylate is transferred to tRNA, but the ester bond between the noncognate amino acid and tRNA is hydrolyzed immediately by the enzyme:

$$\text{Val} + \text{ATP} + \text{IleRS} \rightleftharpoons \text{Val-AMP:IleRS} + \text{PP}_i$$

$$\text{Val-AMP:IleRS} + \text{tRNA}^{\text{Ile}} \rightleftharpoons \text{Val-tRNA}^{\text{Ile}}\text{:IleRS} + \text{AMP}$$

$$\text{Val-tRNA}^{\text{Ile}}\text{:IleRS} + \text{H}_2\text{O} \rightarrow \text{Val} + \text{tRNA}^{\text{Ile}} + \text{IleRS}$$

This implies that the enzyme has a second chance to discriminate between the aminoacyl residues, now in the form of their ester derivatives; if the residue is noncognate, then the water molecule is activated and the ester bond is attacked. In the hydrolysis of valyl-tRNA^{Ile} by IleRS, the free hydroxyl of the tRNA terminal ribose plays an important part.

In some other cases a different correction (proofreading) mechanism may be in operation when a noncognate aminoacyl adenylate is hydrolyzed by the enzyme prior to the transfer of the aminoacyl residue to tRNA.

3.5.2. Specificity for tRNA

It has already been stated that the binding of tRNA with ARSase is a multistep process. The initial tRNA binding is not very specific, so the enzyme may interact with a number of noncognate tRNAs. The IleRS, for example, can bind tRNA^{Val}, and its binding is only one-fifth the strength of the binding of the cognate tRNA^{Ile}. The enzyme interacts with tRNA^{Glu} as well; this latter binding, however, is 10,000 times weaker than the binding of the cognate $tRNA^{Ile}$. Generally, very different affinities are found for various combinations of ARSases with noncognate tRNA species, from an almost complete absence of affinity to an affinity close to that of the cognate tRNA. The affinity of the enzymes for tRNA usually increases with the decrease in pH and ionic strength and is stimulated by organic solvents; this suggests that ionic interactions contribute considerably toward binding. Correspondingly, the same factors stimulate the nonspecific binding of tRNA by ARSases. Magnesium ions, however, frequently have the opposite effect: they may decrease the binding of noncognate tRNA species to an ARSase, that is, increase binding specificity. The latter effect is usually considered to be the result of the action of magnesium ions upon the conformation of both the enzyme and tRNA.

The initial binding of tRNA to the enzyme is a fast step—the rates of both forward and reverse reactions (association and dissociation) are high. This fast step of initial recombination may be followed by a slower step, when the complex somehow rearranges. Such a rearrangement takes place only if the enzyme has bound the cognate tRNA. This is the recognition stage during which the main discrimination between the cognate and noncognate tRNA species is accomplished. Thus, the first binding step involves only a rough selection of tRNAs, and the main function of this step is the rapid scanning of the various tRNA species. If the bound tRNA is noncognate, it will be inactive in the induction of the structural rearrangement of the enzyme complex and, hence, is incapable of entering the next stage; as a result, it will be dissociated easily from the fast reversible initial complex. Only if the cognate tRNA is bound, is the next phase, involving the *rearrangement* of the complex and proper *fitting* of tRNA and the enzyme, required for the subsequent aminoacylation reaction initiated:

$$\text{ARSase:Aa-AMP} + \text{tRNA} \xrightleftharpoons[]{\substack{\text{fast,}\\ \text{not very}\\ \text{specific}}} (\text{ARSase:Aa-AMP:tRNA})' \xrightarrow{\substack{\text{slow,}\\ \text{specific}}}$$

$$\longrightarrow (\text{ARSase:Aa-AMP:tRNA})'' \longrightarrow \text{ARSase:Aa-tRNA} + \text{AMP}$$

This mechanism, however, is not applicable to all ARSases. For example, the TyrRS from *E. coli* and the SerRS from yeast, as well as the ArgRS, show a very high specificity even at the stage of the initial complex; they bind little of the noncognate tRNA species.

Regardless of which mechanism is realized, the final result is a very high specificity of the selection of tRNA by the enzyme. This raises the problem of the specific tRNA–protein recognition. It is apparent that certain specific regions of the tRNA molecule are involved in this recognition. There are two major regions of the tRNA molecule that are in most cases directly involved in the recognition by the enzyme: the acceptor stem and the anticodon. This is not a general rule, however. For example, tRNAs such as $tRNA^{Ala}$ and $tRNA^{Ser}$ are recognized by their cognate synthetases without participation of their anticodons, mainly by specific binding of the acceptor stem. On the other hand, $tRNA^{Met}$ seems to be specifically recognized by its synthetase predominantly at the anticodon.

The acceptor stem is involved in the recognition in most cases of tRNA–ARSase interactions. The base of the fourth nucleotide from the 3'-end (position 73) is often designated a "discriminator base" (Crothers, Seno, and Soell, 1972) because it divides tRNAs into four recognition groups: A73 must be present in tRNAs specific for Ala, Arg, Ile, Leu, Lys, Pro, Tyr, Val; G73 in tRNAs for Asn, Asp, Gln, Glu, Trp; C73 in $tRNA^{His}$; and U73 in $tRNA^{Cys}$ and $tRNA^{Gly}$. Further discrimination may be determined by the end-proximal base pairs. For example, the presence of the pair G3:U70 in the acceptor stem determines the recognition of the tRNA by AlaRS (provided A73 is present). For the recognition of $tRNA^{Gln}$ by the cognate synthetase the stem should include the pairs U1:A72, G2:C71, and G3:C70, in addition to the unpaired discriminator G73. The recognition at the acceptor stem is considered to be the earliest in the evolution of ARSases.

The anticodon is the second major site of recognition of tRNAs by ARSases. In some tRNAs, all three nucleotides of the anticodon seem to be important for the recognition (e.g., $tRNA^{Asp}$, $tRNA^{Cys}$, $tRNA^{Gln}$, $tRNA^{Met}$, $tRNA^{Phe}$, $tRNA^{Trp}$), whereas in others just one (C35 in $tRNA^{Arg}$) or two anticodon residues (C35C36 in $tRNA^{Gly}$, A35C36 in $tRNA^{Val}$) are known to be essential for recognition.

The tight interaction of the ARSase with the acceptor stem and the anticodon may induce serious conformational distortions in these regions of the tRNA molecule. The $tRNA^{Gln}$, when interacting with its cognate GlnRS (a representative of class I synthetases), changes the "classical" conformation of the 3' single-stranded terminal sequence (Fig. 3.14): now the strand makes a hairpin turn toward the inside of the L, with the disruption of the adjacent base pair of the acceptor stem. The anticodon loop of $tRNA^{Gln}$ in the complex with the synthetase also adopts an

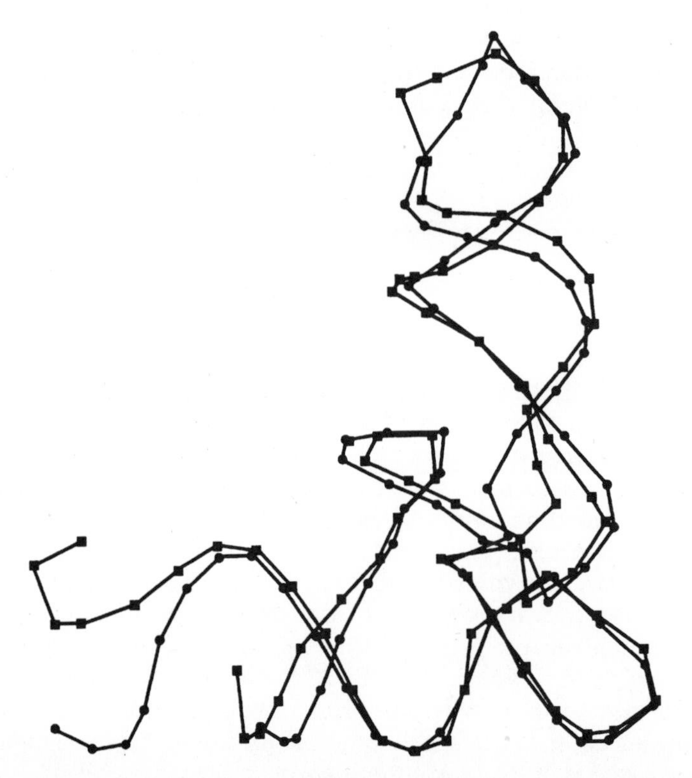

FIG. 3.14. Superposition of the phosphate backbone of uncomplexed yeast $tRNA^{Phe}$ ("classical" tRNA structure, filled circles; see Figs. 3.8 and 3.10) on that of the $tRNA^{Gln}$ (filled squares) complexed with the GlnRS. Major differences are seen in the acceptor strand conformation, the anticodon loop, and the width of the grooves of the acceptor stem and anticodon stem. Reproduced from Rould, M. A., Perona, J. J., Söll, D., and Steitz, T. A. (1989) *Science* **246**:1135–1142, with permission.

unusual structure: the middle anticodon base U35 becomes stacked with the anticodon adjacent base A37, whereas the C34 and G36 of the anticodon are unstacked and project outward to interact with the corresponding protein groups (Rould, Perona, and Steitz, 1991). It seems likely that the structure distortions of this type are typical of interaction of class I synthetases with their tRNAs.

On the contrary, the interaction of a representative of the class II synthetases, AspRS, with its $tRNA^{Asp}$ stabilizes the stacked helical conformation of the GCCA single-stranded portion of the acceptor end; as in the crystal structure of $tRNA^{Phe}$ (see Fig. 3.8), it continues the acceptor stem helix. The anticodon loop, however, undergoes a large conformational change in order to bind with the contacting protein groups: the U-turn conformation is disrupted with a concomitant unstacking of all the anticodon bases, and the loop as a whole is moved toward the inside of the L (Cavarelli *et al.*, 1993).

3.5.3. Specific Modifications of Aminoacyl Residues after Aminoacylation

In addition to the ordinary tRNA species which accept amino acids and then immediately transfer them to ribosomes, there are peculiar tRNAs that present their aminoacyl residues for enzymatic modification, prior to participation in translation.

The first, well-known example is the formylation of the amino group of one of the two Met-tRNA species in bacteria (Marker and Sanger, 1964). Formylmethionyl-tRNA (F-Met-tRNA) is formed and serves as the *initiator tRNA:* it enters the ribosome first and starts translation at the initiation codon AUG or GUG, so that every polypeptide chain synthesized by bacterial ribosomes begins with formylmethionine at the N-terminus. Both $tRNA^{Met}_{m}$ and $tRNA^{Met}_{f}$ are aminoacylated by the same MetRS, but Met-$tRNA^{Met}_{f}$ is found to be a substrate for methionyl-tRNA transformylase, resulting in the formation of N-formylmethionyl-$tRNA^{Met}_{f}$:

$$\text{Met-tRNA}^{\text{Met}}_{\text{f}} + \text{N}^{10}\text{-Formyltetrahydrofolate} \rightarrow \text{F-Met-tRNA}^{\text{Met}}_{\text{f}} + \text{Tetrahydrofolate}$$

The enzyme is highly specific and attacks only Met-$tRNA^{Met}_{f}$ (for review, see RajBhandary and Chow, 1995).

Archaebacteria, many gram-positive eubacteria, cyanobacteria, plant chloroplasts, and plant and animal mitochondria lack glutaminyl-tRNA synthetase (GlnRS). At the same time they have a special tRNA which is aminoacylated by GluRS. The Glu-tRNA formed serves as a substrate for a specific amidotransferase converting Glu-tRNA into Gln-tRNA:

$$\text{Glu-tRNA} + \text{Glutamine} + \text{ATP} \rightarrow \text{Gln-tRNA} + \text{ADP} + \text{Glutamic acid} + \text{P}_{\text{i}}$$

Since this tRNA species recognizes glutamine codons, it is designated as $tRNA^{Gln}$, despite the fact that it is aminoacylated by GluRS (for review, see Verkamp *et al.*, 1995).

The 21st amino acid of natural proteins, selenocysteine, is also formed on tRNA. Special $tRNA^{Sec}$ with anticodon UCA is aminoacylated by regular serine tRNA synthetase (SerRS). The resultant Ser-$tRNA^{Sec}$ is a specific substrate for selenocysteine synthase. The enzyme binds to Ser-$tRNA^{Sec}$ in such a way that, in addition to specific recognition of the tRNA moiety, the serine residue becomes covalently attached via its amino group to the pyridoxal phosphate prosthetic group in the active center. Then the dehydration reaction takes place, removing the serine hydroxy group and thus converting the seryl residue into an aminoacryloyl residue. This intermediate reacts with a selenophosphate resulting in the formation of selenocysteinyl residue still linked to the pyridoxal phosphate by its amino group. The final stage is the release of the selenocysteinyl-tRNA from this covalent bond and from the complex with the enzyme. The sequence of the reactions catalyzed by selenocysteine synthase (SCSase) is presented below (for review, see Baron and Boeck, 1995):

$$\text{Ser-tRNA}^{\text{Sec}} + \text{SCSase:Pyridoxal-P} \rightarrow \text{SCSase:(Pyridoxal-P)-Ser-tRNA}^{\text{Sec}} \rightarrow$$

$$\rightarrow \text{SCSase:(Pyridoxal-P)-Aminoacryloyl-tRNA}^{\text{Sec}} + H_2O$$

$$\text{SCSase:(Pyridoxal-P)-Aminoacryloyl-tRNA}^{\text{Sec}} + \text{Se-P} \rightarrow$$

$$\rightarrow \text{SCSase:(Pyridoxal-P)-Selenocysteinyl-tRNA}^{\text{Sec}} + P_i$$

$$\text{SCSase:(Pyridoxal-P)-Selenocysteinyl-tRNA}^{\text{Sec}} \rightarrow$$

$$\rightarrow \text{SCSase:Pyridoxal-P} + \text{Selenocysteinyl-tRNA}^{\text{Sec}}$$

References

Alzhanova, A. T., Fedorov, A. N., Ovchinnikov, L. P., and Spirin, A. S. (1980) Eukaryotic aminoacyl-tRNA synthetases are RNA-binding proteins whereas prokaryotic ones are not, *FEBS Letters* **120**:225–229.

Baldwin, A. N., and Berg, P. (1966) tRNA induced hydrolysis of valyl adenylate bound to isoleucyl-tRNA synthetase, *J. Biol. Chem.* **241**:839–845.

Baron, C., and Boeck, A. (1995) The selenocysteine-inserting tRNA species: structue and function, in *tRNA: Structure, Biosynthesis, and Function* (D. Soell and U. L. RajBhandary, eds.), pp. 529–544, ASM Press, Washington, DC.

Cavarelli, J., Rees, B., Ruff, M., Thierry, J. C., and Moras, D. (1993) Yeast tRNA^{Asp} recognition by its cognate class II aminoacyl-tRNA synthetase, *Nature* **362**:181–184.

Crick, F. H. C. (1957) Discussion in "The structure of nucleic acids and their role in protein synthesis, *Biochemical Society Symposium* (E. M. Crook, ed.), No. 14, pp. 25–26, Cambridge University Press, Cambridge.

Crothers, D. M., Seno, T., and Soell, D. G. (1972) Is there a discriminator site in transfer RNA?, *Proc. Natl. Acad. Sci. USA* **69**:3063–3067.

Dang, C. V., Johnson, D. L., and Yang, D. C. H. (1982) High molecular mass amino acyl-tRNA synthetase complexes in eukaryotes, *FEBS Letters* **142**:1–6.

Eriani, G., Delarue, M., Poch, O., Gangloff, J., and Moras, D. (1990) Partition of tRNA synthetases into two classes based on mutually exclusive sets of sequence motifs, *Nature* **347**:203–206.

Hoagland, M. B. (1960) The relationship of nucleic acid and protein synthesis as revealed by studies in cell-free systems, *Nucleic Acids* (E. Chargaff and J. N. Davidson, eds.), vol. 3, pp. 349–408, New York, Academic Press.

Hoagland, M. B., Zamecnik, P. C., and Stephenson, M. L. (1957) Intermediate reactions in protein biosynthesis, *Biochim. Biophys. Acta* **24**:215–216.

Holley, R. W., Apgar, J., Everett, G. A., Madison, J. T., Marquisee, M., Merrill, S. H., Penswick, J. R., and Zamir, A. (1965) Structure of a ribonucleic acid, *Science* **147**:1462–1465.

Kim, S. H., Suddath, F. L., Quigley, G. J., McPherson, A., Sussman, J. L., Wang, A. H.-J., Seeman, N. C., and Rich, A. (1974) Three-dimensional tertiary structure of yeast phenylalanine transfer RNA, *Science* **185**:435–440.

Marker, K., and Sanger, F. (1964) N-formyl-methionyl-s-RNA, *J. Mol. Biol.* **8**:835–840.

Mirande, M., Cirakoglu, B., and J.-P. Waller (1983) Seven mammalian aminoacyl-tRNA synthetases associated within the same complex are functionally independent, *Eur. J. Biochem.* **131**:163–170.

Ogata, K., and Nohara, H. (1957) The possible role of the ribonucleic acid (RNA) of the pH 5 enzyme in amino acid activation, *Biochim. Biophys. Acta* **25**:659–660.

RajBhandary, U. L., and Chow, C. M. (1995) Initiator tRNAs and initiation of protein synthesis, in *tRNA: Structure, Biosynthesis, and Function* (D. Soell and U. L. RajBhandary, eds.), pp. 511–528, ASM Press, Washington, DC.

Robertus, J. D., Ladner, J. E., Finch, J. T., Rhodes, D., Brown, R. S., Clark, B. F. C., and Klug, A. (1974) Structure of yeast phenylalanine tRNA at 3 Å resolution, *Nature* **250**:546–551.

Rould, M. A., Perona, J. J., and Steitz, T. A. (1991) Structural basis of anticodon loop recognition by glutaminyl-tRNA synthetase, *Nature* **352**:213–218.

Verkamp, E., Kumar, A. M., Lloyd, A., Martins, O., Stange-Thomann, N., and Soell, D. (1995) Glutamyl-tRNA as an intermediate in glutamate conversions, in *tRNA: Structure, Biosynthesis, and Function* (D. Soell and U. L. RajBhandary, eds.), pp. 545–550, ASM Press, Washington, DC.

4

Ribosomes and Translation

4.1. First Observations

By 1940 Albert Claude had succeeded in isolating from animal cells cytoplasmic RNA-containing granules that were smaller than mitochondria. These granules varied from 50 to 200 μ in diameter and later Claude began calling them *microsomes.* Chemical analyses indicated that Claude's microsomes were "phospholipid-ribonucleoprotein complexes."

On the other hand, cytochemical studies by Casperson (1941) and Brachet (1942) demonstrated the preferential localization of RNA in the cytoplasm and the existence of a correlation between the amount of RNA in the cytoplasm and the intensity of protein synthesis. Later, a number of scientists reported on the isolation of RNA-containing particles, which were much smaller than microsomes, from the cytoplasm of animal and plant cells as well as from bacteria. Electron microscopy and sedimentation analysis in the ultracentrifuge indicated that these particles were compact; had a more or less spherical shape; were homogeneous in size, with a diameter of 100 to 200 Å; and exhibited sharp sedimentation boundaries corresponding to sedimentation coefficients ranging from 30S to 100S. The first unambiguous evidence that such particles from bacteria were ribonucleoproteins was probably obtained by Schachman, Pardee, and Stanier in 1952.

Improved techniques of microtomy and electron microscopy of ultrathin sections of animal cells resulted in the detection of uniform dense granules, with a diameter of about 150 Å, directly in the cell. Palade's electron microscopic studies (1955) demonstrated that small dense granules are abundant in animal cell cytoplasm. These granules were seen either attached to the membrane of the endoplasmic reticulum or freely dispersed throughout the cytoplasm. Claude's microsomes were identified as fragments of the endoplasmic reticulum with these granules attached. It became clear that the Palade granules were ribonucleoprotein particles and that they accounted for most of the cytoplasmic RNA involved in protein synthesis (Palade and Siekevitz, 1956a,b).

Purified preparations of ribonucleoprotein particles were isolated and studied in several laboratories between 1956 and 1958; these investigations included isolating 80S particles from yeast, accomplished by Chao and Schachman (1956); from plants, by Ts'o, Bonner, and Vinograd (1956); and from animals, by Petermann and Hamilton (1957); and isolating 70S particles from bacteria (*E. coli*) by Tissieres and Watson (1958). In 1958, in the first symposium devoted to these particles and their role in protein biosynthesis it was suggested that ribonucleoprotein particles be called *ribosomes* (Roberts, 1958).

Studies of the functional role played by ribosomes proceeded hand-in-hand with their structural description. The experiments of Zamecnik and co-workers provided the first convincing demonstration that ribonucleoprotein particles of microsomes are responsible for the incorporation of radioactive amino acids into newly synthesized proteins (Littlefield *et al.*, 1955). This was followed by other experiments conducted at the same laboratory which demonstrated that the free ribosomes unattached to the endoplasmic reticulum membranes also incorporate amino acids (Littlefield and Keller, 1957). The functions of bacterial ribosomes were the subject of intense studies conducted by Roberts' group; the 1959 publication of McQuillen, Roberts, and Britten finally established that proteins are synthesized on ribosomes and then distributed throughout the bacterial cell.

4.2. Localization of Ribosomes in the Cell

Ribosomes are abundant in cells involved in intense protein synthesis. In the bacterial cell they are dispersed throughout the protoplasm and account for about 30%, sometimes even more, of its dry weight. In electron micrographs all the intracellular space, except nucleoid (DNA) regions, is packed with ribosomes (Fig. 4.1). About 10^4 ribosomes, on average, are present in one bacterial cell.

The relative content (concentration) of ribosomes in eukaryotic cells is lower; here, the number of ribosomes varies considerably depending on the protein-synthesizing activity of the corresponding tissue or individual cell. Most of the ribosomes are found in the cytoplasm. In the cells with an active protein secretion and

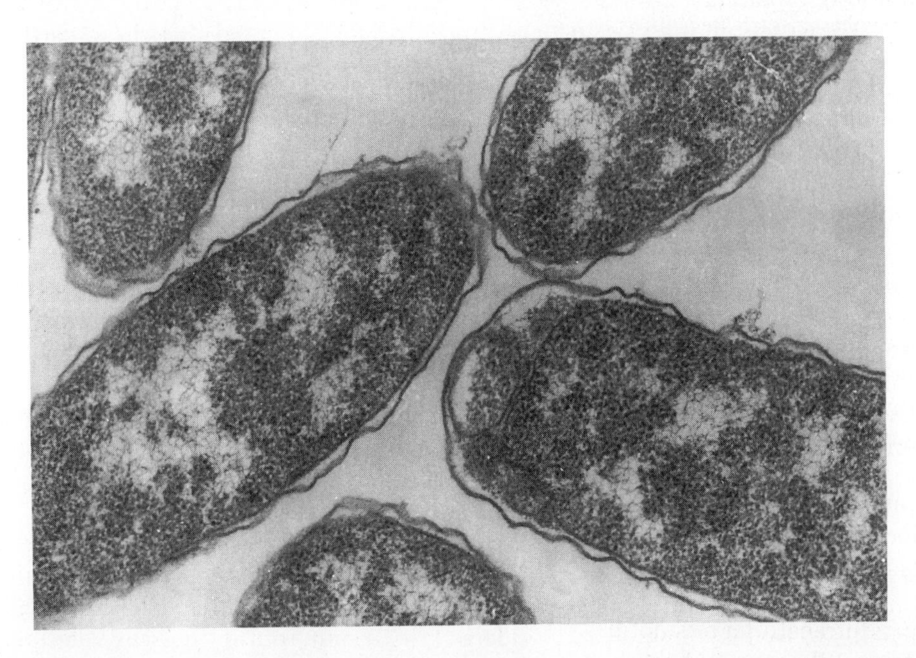

FIG. 4.1. Electron micrograph of ribosomes on an ultrathin section of the bacterium *Vibrio alginolyticus*. The cells are fixed with osmium tetraoxide. Ribosomes appear as the abundant granular material filling in the cytoplasm. (Courtesy of L. Ye. Bakeyeva, Moscow State University).

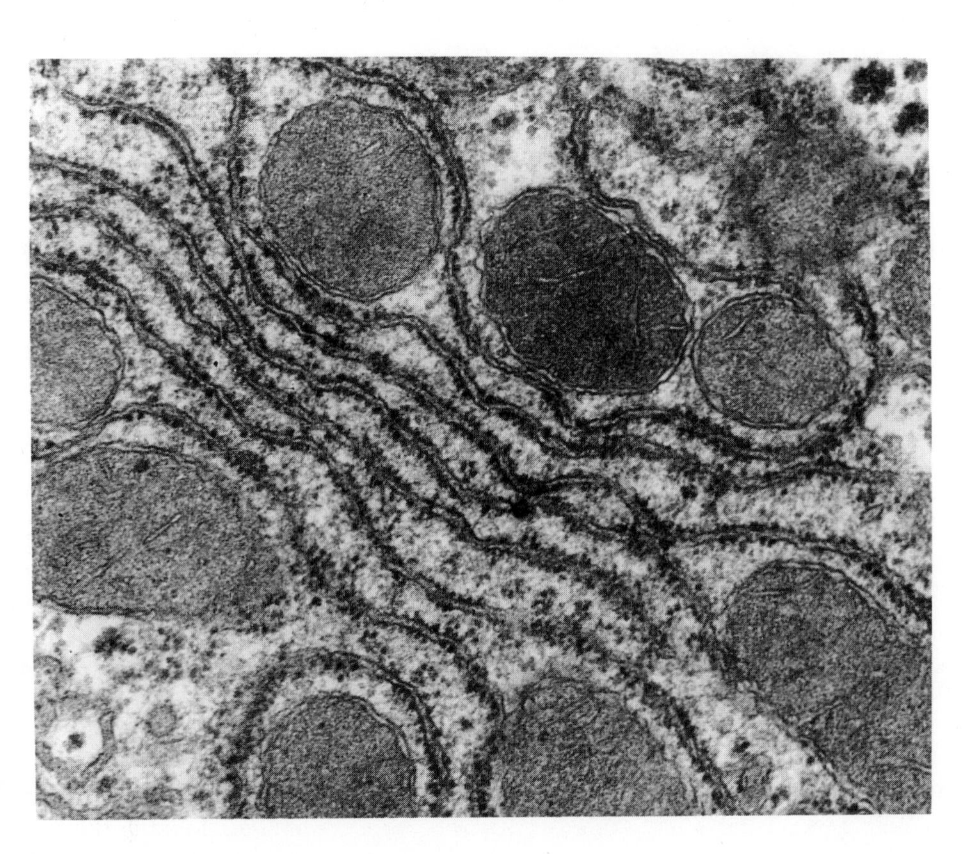

FIG. 4.2. Electron micrograph of ribosomes on an ultrathin section of a rat liver cell. Fixation with glutaraldehyde. Ribosomes on the membranes of rough endoplasmic reticulum, as well as some clusters of free ribosomes, are seen. (Courtesy of Yu.S. Chentsov, Moscow State University).

a developed network of endoplasmic reticulum, a marked proportion of cytoplasmic ribosomes are attached to the endoplasmic reticulum membrane, specifically to the surface facing the cytoplasmic matrix (Fig. 4.2). The ribosomes are distributed unevenly on the reticulum: they may be abundant in one part and virtually nonexistent in others. These ribosomes synthesize proteins that are directly transported into the membrane lumen for subsequent secretion (see Chapter 18). Protein synthesis for "housekeeping" purposes inside the cell takes place primarily on the free cytoplasmic ribosomes that are not associated with the membrane but are scattered in the cytoplasmic matrix. This is why the cytoplasm of embryonic, nondifferentiated, rapidly growing or proliferating cells contains mainly free ribosomes.

The formation of all ribosomes present in the cytoplasmic matrix, both membrane-bound and free ones, takes place in the nucleolus of the eukaryotic cell, and ribosomes can naturally also be detected in this compartment of the cell nucleus; it is thought, however, that nucleolar ribosomes are not active in protein synthesis.

In addition, the eukaryotic cell contains different populations of ribosomes in such intracellular organelles as mitochondria and, in the case of plant cells, chloroplasts. Ribosomes of these organelles differ from cytoplasmic ribosomes in that they are slightly smaller and have a different chemical composition and different functional characteristics. These ribosomes are formed directly in the organelles.

4.3. Prokaryotic and Eukaryotic Ribosomes

Two main types of ribosomes can be found in nature (Fig. 4.3). All prokaryotic organisms, including gram-positive and gram-negative eubacteria, actinomycetes, and blue-green algae (cyanobacteria), as well as archaebacteria (archaea), contain 70S ribosomes. These ribosomes exhibit a sedimentation coefficient of about 70S; their molecular mass is approximately 2.5×10^6 daltons, and their mean diameter in a lyophilized state is about 200 to 250 Å; in chemical composition they are pure ribonucleoproteins, that is, they consist of only RNA and protein. The RNA-to-protein weight ratio in them is about 2:1; correspondingly, the partial specific volume of 70S ribosomes is about 0.60 cm^3/g, and buoyant density in CsCl is 1.64 g/cm^3. RNA is present in the ribosomes mainly as a Mg^{2+}, and perhaps partially as a Ca^{2+} salt; magnesium may account for up to 2% of the ribosomes' dry weight. Furthermore, ribosomes may contain various amounts (up to 2.5% of the dry weight) of such organic cations as spermine, spermidine, cadaverine, and putrescine. The amount of water bound in 70S ribosomes is about 1 g/g; in other words, ribosomes are rather compact unswollen particles in an aqueous medium.

The morphology of 70S ribosomes of prokaryotic organisms is almost universal, and only ribosomes of archaebacteria have been shown to possess some differences from their eubacterial counterparts (see Chapter 5).

The cytoplasm of all eukaryotic organisms, including animals, fungi, plants, and protozoans, contains the somewhat larger 80S ribosomes. The molecular mass of these ribosomes is about 4×10^6 daltons, and the mean diameter is about 250

	PROKARYOTIC TYPE RIBOSOMES:	EUKARYOTIC TYPE RIBOSOMES:
	70S RIBOSOMES OF EUBACTERIA, BLUE-GREEN ALGAE AND CHLOROPLASTS	CYTOPLASMIC 80S RIBOSOMES OF ANIMALS, FUNGI AND PLANTS
MOL. MASS, DALTONS	2.5×10^6	4×10^6
SIZE	200 - 250 Å	250 - 300 Å
RNA : PROTEIN, W/W	2 : 1	1 : 1
	70S RIBOSOMES OF ARHAEBACTERIA MITOCHONDRIAL 75S RIBOSOMES OF FUNGI MITOCHONDRIAL 55S RIBOSOMES ("MINIRIBOSOMES") OF MAMMALS	

FIG. 4.3. Prokaryotic and eukaryotic types of ribosomes.

to 300 Å. Like prokaryotic ribosomes they contain only two types of biopolymers—RNA and protein—but the protein content is markedly greater; the RNA-to-protein ratio in 80S ribosomes is about 1:1 by weight, the partial specific volume is about 0.65 cm^3/g, and the buoyant density in CsCl is about 1.55 to 1.59 g/cm^3. It is important to point out that the absolute content of both RNA and protein per particle in 80S ribosomes is markedly greater than in 70S ribosomes. The ribosomal RNA of 80S ribosomes is also bound with divalent cations, Mg^{2+} and Ca^{2+}, as well as with small amounts of polyamines and diamines (e.g., spermine, spermidine, and putrescine).

Again, the morphological characteristics of all 80S ribosomes, whether they have been obtained from animals, plants, or lower eukaryotes, are universal. The chloroplasts and mitochondria of eukaryotic cells, however, contain ribosomes that differ from the 80S type. The chloroplast ribosomes of higher plants belong to the true 70S type and are difficult to distinguish from the ribosomes of eubacteria and blue-green algae by the above characteristics or by more subtle molecular features. Mitochondrial ribosomes are more diverse; their properties depend on the taxonomic position of the organism from which they originate. Mitochondrial ribosomes of fungi and mammals have been studied in some detail. Mitochondrial ribosomes from fungi *(Saccharomyces* or *Neurospora)* resemble prokaryotic 70S ribosomes but are slightly larger (about 75S) and contain relatively more protein; the absolute content of ribosomal RNA seems almost identical to that found for typical 70S ribosomes. Mitochondrial ribosomes of mammals, however, are, significantly lighter than typical 70S ribosomes. The absolute content of ribosomal RNA per particle is also significantly lower. That is why they are sometimes called "mini-ribosomes." The sedimentation coefficient of mini-ribosomes from mammalian mitochondria is only about 55S, and the total mass of ribosomal RNA per particle is about two-thirds of that in typical 70S ribosomes. At the same time, mammalian mitochondrial ribosomes contain a high proportion of protein, so their total size does not seem to differ greatly from that of prokaryotic ribosomes. Despite some unusual features, mammalian mitochondrial ribosomes are similar to prokaryotic 70S ribosomes in their characteristics, including functional ones.

4.4. Sequential Readout of mRNA by Ribosomes; Polyribosomes

Throughout the course of protein synthesis, the ribosome is associated with a limited section of the template polyribonucleotide. Since the ribosome-bound sections of the template are protected from nuclease action, they may be isolated after nuclease treatment of the ribosome–template complexes. Such sections have been found to have a length of 40 to 60 nucleotide residues. It must be noted again that the length of the mRNA coding sequence usually exceeds 300 nucleotides. Therefore, in order to read the entire mRNA coding sequence, the ribosome should *sequentially run over* the template (or thread through itself) from the 5′-terminal part of the coding sequence to the 3′-terminal part. In other words, the ribosome should work as a tape-driving mechanism.

At what rate, then, would the ribosome move along the mRNA? In a bacterial cell (e.g., *E. coli*) a polypeptide with a length of 300 amino acids is synthesized for about 20 seconds at 37°C, (i.e., one ribosome runs over about 40 to 50 nucleotides per second). The rate of mRNA readout in eukaryotes can approach 30 nucleotides per second, but regulatory effects may reduce it to 5 to 10 nucleotides per second (see Chapter 13).

While moving along the template polynucleotide from the 5′-end to the 3′-end, the ribosome, after some time, moves away from the 5′-terminal section of the template. As a result, this section becomes exposed and is capable of binding with another free ribosome. The second ribosome will start the readout, and move away from the 5′-terminus, giving the third ribosome an opportunity to bind and start reading, and so on. In this way, moving along the template one after another, a number of ribosomes simultaneously perform a readout of the same information and, hence, synthesize identical polypeptide chains (of course, at any given moment the chains on different ribosomes are at different stages of completion). This process is schematically presented in Fig. 4.4, where the ribosomes at the 3′-end of the template contain an almost completed polypeptide, the ribosomes located in the middle of the mRNA carry the polypeptide only half the length of the complete one, and the ribosomes near the 5′-end contain only short peptides which have just started elongation. A structure in which the template polynucleotide is associated with many translating ribosomes is called the *polyribosome.*

Early electron microscopic observations in the mid-1950s demonstrated that ribonucleoprotein granules (ribosomes) are not dispersed uniformly in animal cell cytoplasm or in the preparations of microsome-derived particles but are clustered in groups (Palade and Siekevitz, 1956a,b). Evidence that such aggregates of ribosomes consist of particles that are connected by the mRNA chain and are engaged in translation was provided simultaneously by several groups in 1963 (Gierer; Warner, Knopf, and Rich; Wettstein, Staehelin, and Noll; Penman, *et al.*; Watson). Polyribosomes were shown to be a form of actively translating ribosomes both in eukaryotes and prokaryotes.

The eukaryotic polyribosomes often appear to be ordered structures, rather than like beads on a randomly flexible thread. For example, most membrane-free polyribosomes from sea urchin eggs and embryos were visualized in the form of zig-zags (Martin and Miller, 1983), although linear forms were also present. The endoplasmic reticulum membrane-bound polyribosomes of protein-secreting cells are represented mostly by circular ("O-like") and spiral ("G-like") forms (Fig. 4.5). The functional significance of these distinctive arrangement patterns is not clear.

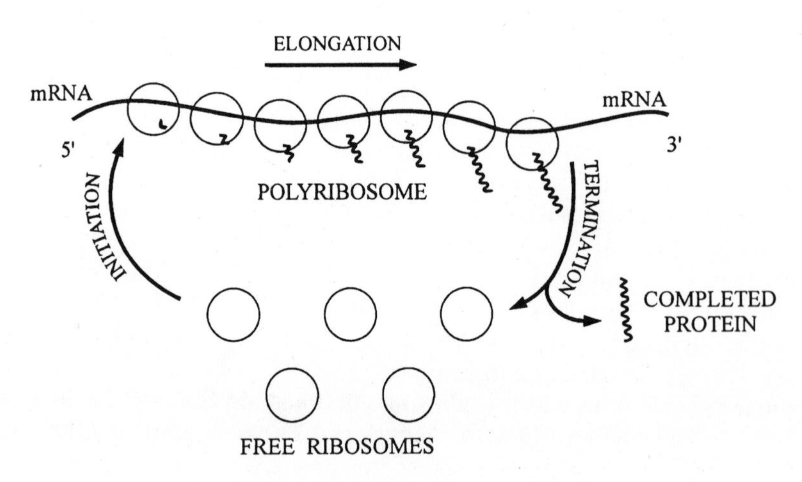

FIG. 4.4. Schematic representation of a polyribosome.

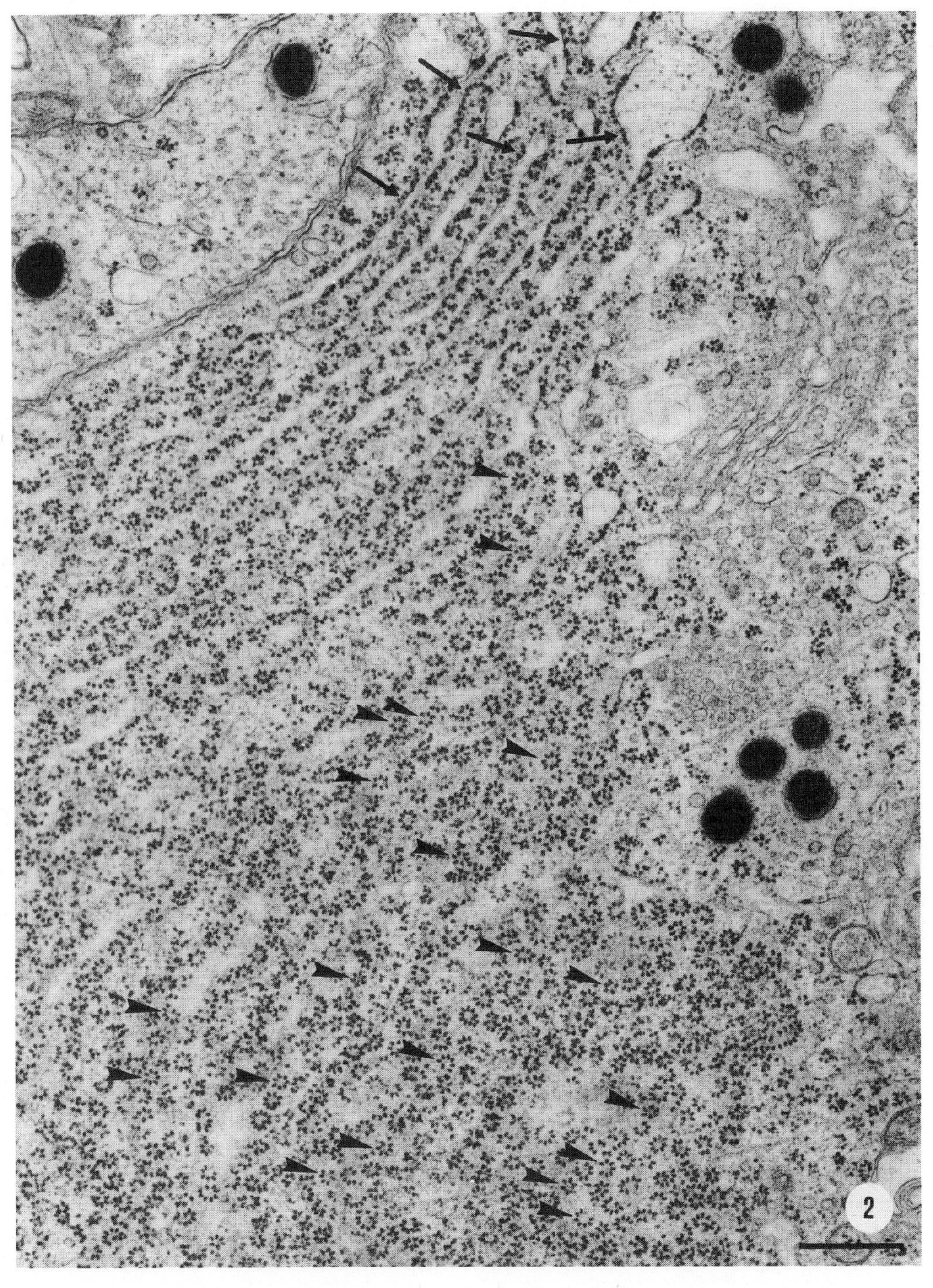

FIG. 4.5. Electron micrograph showing predominantly circular, and sometimes spiral "G-like", polyribosomes on the rough endoplasmic reticulum of somatotrope cytoplasm from the rat pituitary. Reproduced from Christensen, A. K., Kahn, L. E., and Bourne, C. M. (1987) *Amer. J. Anat.* **178**:1–10. (Courtesy of A. K. Christensen, Department of Anatomy and Cell Biology, University of Michigan Medical School, Ann Arbor, MI).

In the case of circular polyribosomes it can be speculated that such a shape provides an efficient reinitiation of translation due to the proximity of the 5′-terminus to the termination codon. Instead of the release of terminating ribosomes from mRNA they can recycle directly onto the 5′-end of the message to begin a new round of translation (see Chapters 15 and 17, and specifically Figs. 15.14 and 17.12).

Under conditions of intensive protein synthesis, the distance between ribosomes along the mRNA chain within the polyribosome may be extremely short, so the ribosomes may be packed together very tightly. This means that there may be roughly 50 nucleotide residues of the template per ribosome in the polyribosome. The implication here is that every 1 to 3 seconds, a ribosome finishes synthesizing the protein molecule near the 3′-end of the mRNA coding section and then jumps off the template or reinitiates; correspondingly, one new ribosome will become associated with the template at its 5′-end and will start moving toward the 3′-terminus. In more common cases about 100 nucleotide residues of mRNA per ribosome have been estimated. Also gaps and tails that are not covered by ribosomes sometimes can be visualized along mRNA in polyribosomes; they may reflect the existence of some barriers inducing temporary stops (pauses) during elongation and the presence of untranslatable terminal sequences.

The existence of polyribosomes as a form of translating ribosomes in the cell explains the observation that ribosomes are abundant in the cell while the amount of mRNA is low. Indeed, ribosomal RNA accounts for about 80% of the total cellular RNA, whereas the mRNA content does not, as a rule, exceed 5%. This is easily understandable if one takes into account that the translation machinery of the cell is organized on the basis of polyribosomes: one mRNA is translated by many ribosomes, and one part of the translatable mRNA corresponds to 100 to 200 parts of ribosomal RNA by weight.

4.5. Stages of Translation: Initiation, Elongation, and Termination

A ribosome begins to read mRNA from a strictly definite point in its sequence, that is, from the beginning of its coding region. It has already been noted that this point generally does not coincide with the 5′-terminal mRNA nucleotide and as a rule is located at a certain, sometimes significant, distance from the 5′-end of the polynucleotide chain. The ribosome should in some way identify the readout origin, bind to it, and then begin translation. The series of events that provide for the beginning of translation is called *initiation*. Initiation requires a special initiation codon, initiator tRNA, and proteins, which are referred to as initiation factors.

After initiation, the ribosome consecutively reads mRNA codons in the direction of its 3′-end. The mRNA readout implies concomitant synthesis of the polypeptide chain coded by the mRNA. Synthesis takes place on the ribosome by the sequential addition of amino acid residues to the nascent polypeptide chain; it is in this way that the peptide elongation is accomplished. Each new amino acid residue is added to the carboxyl terminus (C-terminus) of the peptide; in other words, the C-terminus of the peptide is the end that grows. The addition of one amino acid residue corresponds to the readout of one nucleotide triplet. This whole process involving the actual translation of mRNA coding region is termed *elongation*.

When a ribosome reaches the mRNA termination codon, synthesis of the polypeptide stops. In the presence of the termination codon the ribosome does not bind any aminoacyl-tRNA; instead, specialized proteins called termination factors come into play. These factors induce the release of the synthesized polypeptide from the ribosome. This stage is designated as *termination.* After termination the ribosome may either jump off the mRNA or continue to slip along it, without translating. When a ribosome comes across a new initiation codon either on a new mRNA chain or on the same chain downstream from the termination codon, a new initiation takes place. Thus each ribosome passes through the whole translation cycle including initiation, elongation, and termination; such an epicycle results in the readout of the whole mRNA coding sequence and synthesis of a complete polypeptide. Thereafter, a given ribosome may repeat the cycle with the same mRNA chain, another mRNA chain, or another coding sequence in the same chain (Fig. 4.4).

4.6. Chemical Reactions and Overall Energy Balance of Protein Biosynthesis

Thus, proteins are synthesized from amino acids in *three consecutive chemical reactions.* The first two reactions are catalyzed by ARSases, and the third and final one is accomplished by the ribosome (RS):

$$\text{Aa} + \text{ATP} \xrightarrow{\text{ARSase}} \text{Aa-AMP} + \text{PP}_\text{i}$$

$$\text{Aa-AMP} + \text{tRNA} \xrightarrow{\text{ARSase}} \text{Aa-tRNA} + \text{AMP}$$

$$\text{Aa-tRNA} + \text{X-Aa}'\text{-tRNA}' \xrightarrow{\text{RS}} \text{X-Aa}'\text{-Aa-tRNA} + \text{tRNA}'$$

In the first reaction the amino acid carboxyl group reacts with the polyphosphate group of ATP, resulting in the replacement of a pyrophosphate residue by the aminoacyl residue; a mixed *anhydride,* the aminoacyl adenylate, is formed. In the second reaction the adenylate residue is exchanged for tRNA, an *ester* bond being formed between the carboxyl group of the aminoacyl residue and the ribose hydroxyl of the tRNA terminal nucleoside. The third reaction, catalyzed by the ribosome, is the substitution of the tRNA residue (tRNA′) by the aminoacyl-tRNA; this results in the formation of an *amide (peptide)* bond between the amino group of the aminoacyl-tRNA and the carboxyl group of the other aminoacyl residue (Aa′). If a complete protein molecule consists of *n* aminoacyl residues, the overall balance of the reactions may be written as follows:

$$n\,\text{Aa} + n\,\text{ATP} \xrightarrow{\text{ARSases, tRNAs, RS}} \text{protein} + n\,\text{AMP} + n\,\text{PP}_\text{i}$$

Furthermore, pyrophosphate is hydrolyzed in the cell by pyrophosphatase to orthophosphate:

$$n\,\text{PP}_\text{i} + n\,\text{H}_2\text{O} \rightarrow 2n\,\text{P}_\text{i}$$

The free energy of the hydrolysis of ATP pyrophosphate bonds under standard conditions ($\Delta G^{0\prime}$) is about −7 to −8 kcal/mole. The anhydride bond of the aminoacyl adenylate and the ester bond of the aminoacyl-tRNA possess similar values of free energy of hydrolysis under standard conditions. The free energy of the hydrolysis of the peptide bond in an infinitely long polypeptide (protein) under standard conditions is equal to −0.5 kcal/mole. It can therefore be seen that the whole process of protein synthesis involves releasing a considerable amount of free energy; in other words, protein synthesis is a thermodynamically spontaneous and energetically ensured process:

$$n\ \text{Aa} + n\ \text{ATP} \rightarrow \text{protein} + n\ \text{AMP} + n\ \text{PP}_i - n \times 7\ \text{kcal}$$

If pyrophosphate hydrolysis is added, the overall energy balance will be $-n \times 15$ kcal per mole of protein. Thus, the free-energy gain under standard conditions for a protein of about 200 aminoacyl residues will be roughly 3,000 kcal/mole.

An analysis of the energy balance of each of the three reactions shows that the first two reactions do not by themselves achieve any gain in free energy (under standard conditions), and therefore the preribosomal stages should not be shifted markedly toward the synthetic side; the shift, however, will be generated provided pyrophosphate is hydrolyzed in a parallel reaction. The main difference in the free-energy levels between substrates and products is found in the third reaction. This implies that the shift of the overall reaction toward synthesis is provided mainly by the ribosomal stage.

It is surprising that despite the full energy support of protein biosynthesis at the expense of ATP (or the ester bond energy of aminoacyl-tRNA), the ribosomal stage still requires two GTP molecules per amino acid residue:

$$2n\ \text{GTP} + 2n\ \text{H}_2\text{O} \rightarrow 2n\ \text{GDP} + 2n\ \text{P}_i$$

This gives an additional free-energy gain of about 15 kcal per mole of amino acid (under standard conditions).

Thus, the sum total of all the chemical reactions in protein synthesis may be written as follows:

$$n\ \text{Aa} + n\ \text{ATP} + 2n\ \text{GTP} + 3n\ \text{H}_2\text{O} \xrightarrow{\text{ARSases, PPase, tRNAs, RS}} \text{protein} + n\ \text{AMP} + 2n\ \text{GDP} + 4n\ \text{P}_i$$

The total energy balance of the overall reaction $\Delta G^{0\prime}$ is equal to about −30 kcal per mole of amino acid or −6,000 kcal per mole of protein with a length of 200 amino acid residues.

Here, only the chemical aspect of the process has been taken into account. It is important to analyze to what extent this estimate may be changed if we take into account entropy loss due to the ordered arrangement of the amino acid residues along the chain of synthesized protein, and due to the fixed three-dimensional protein structure. It seems that the entropy loss due to amino acid ordering in the polypeptide chain may introduce only a small correction, around 2.5 kcal per mole of amino acid. In the three-dimensional ordering of the chain in the protein molecule, the entropy loss (decrease) is significant, but it is compensated by the enthalpy gain resulting from noncovalent interactions of amino acid residues. Thus, in any case, the protein synthesis is accompanied by dissipation of a large amount of free energy.

The meaning of the release of such a tremendous excess of energy is an enigma and an extremely interesting problem in molecular biology. Energy excess which is dissipated into heat and not used for any accumulated useful work (in the form of chemical bonds or nonrandom arrangement of residues) should play an important part in the functioning of the protein-synthesizing system. It is likely that this energy excess is necessary to support the high rates and high fidelity of protein synthesis.

4.7. Cell-free Translation Systems

One of the most remarkable discoveries of the 1950s was the understanding that protein synthesis does not require the integrity of the cell and can be performed after cell disruption. This laid the basis for the creation of the so-called *cell-free translation systems.* The incorporation of amino acids into proteins in cell homogenates, in cell extracts, and in cell-free fractions containing microsomes was demonstrated long ago; the first examples were the cell-free systems from animal tissues, specifically from rat liver, described by Siekevitz in 1952 and by Zamecnik in 1953. It was shown soon thereafter that the incorporation of amino acids corresponding to protein synthesis in a cell-free system proceeds on ribonucleoprotein particles or ribosomes (Littlefield *et al.,* 1955). Cell-free protein-synthesizing systems with bacterial *(E. coli)* ribosomes were developed almost simultaneously in Zillig's, Zamecnik's, and Tissieres' groups in 1959 and 1960 (Schachtschabel and Zillig, 1959; Lamborg and Zamecnik, 1960; Tissieres, Schlessinger, and Gros, 1960). All of these systems were programmed by endogenous mRNAs; in these systems ribosomes simply continued to synthesize polypeptides upon the mRNA molecules to which they were attached at the time of cell disruption. In 1961 Nirenberg and Matthaei improved the system, separated ribosomes from endogenous messages, and introduced the exogenous template for polypeptide synthesis (Matthaei and Nirenberg, 1961; Nirenberg and Matthaei, 1961). One of their main achievements was the use of synthetic polynucleotide templates prepared by polynucleotide phosphorylase, including the simple templates, such as poly(U) and poly(A). It is this innovation that made it possible to break the genetic code.

Today, cell-free protein-synthesizing systems may be reconstituted from well-characterized, highly purified components, including ribosomes, template polynucleotides, and a set of aminoacyl-tRNAs or a system of tRNA aminoacylation, namely, tRNA, amino acids, ATP, and ARSases. In addition, the system should be supplied with a set of special proteins called elongation factors, as well as with GTP. The simplest cell-free ribosomal system of polypeptide synthesis, which can be used to study the fundamental mechanisms of translation, includes only six high-molecular-mass components plus GTP; for example, the poly(U)-directed system may be reconstituted from the following ingredients: *E. coli* 70S ribosomes; Poly(U); Phe-tRNA; EF-T_u (protein with a molecular mass of 47,000 daltons); EF-T_s (protein with a molecular mass of 34,000 daltons); EF-G (protein with a molecular mass of 83,000 daltons); GTP. As a result of poly(U) translation, polyphenylalanine is synthesized.

For translating natural cellular mRNA and viral RNA, the prokaryotic cell-free system should be supplemented by a complete set of aminoacyl-tRNAs, three proteins necessary for initiating translation (IF1, IF2, and IF3), and three proteins necessary for terminating translation (RF1, RF2, and RF3). When eukaryotic 80S ribosomes are used for cell-free translation, all corresponding protein factors should be of eukaryotic origin. These include the elongation factors, namely eEF1 which is equivalent to bacterial EF-T_u plus EF-T_s, and eEF2 equivalent to bacterial EF-G, numerous initiation factors (eIF1, eIF2, eIF3, eIF4A, eIF4B, eIF4C, eIF5, etc.), and one high-molecular-mass termination factor (eRF). In addition, initiation in eukaryotic systems requires ATP.

Usually, however, crude cell extracts comprising all these endogenous components and factors are used in a routine laboratory practice. Pre-incubation of the cell extract at physiological temperature is often sufficient to remove the endogenous mRNA from the ribosomes, due to the digestion of it by endogenous nucleases. The vacant ribosomes of the extract accept either exogenous natural mRNA or synthetic polynucleotides as templates. The treated extract (including ribosomes, tRNAs, ARSases, and translation factors), in addition to an exogenous message for polypeptide synthesis, should be also supplemented with amino acids, ATP, GTP, and an ATP/GTP regenerating system (either phosphoenol pyruvate and pyruvate kinase, or creatine phosphate and creatine kinase, or acetyl phosphate and acetyl kinase).

An alternative strategy is the use of partially fractionated cell extract. Thus, ribosomes and all RNA are removed from the extract by ultracentrifugation with subsequent DEAE cellulose treatment, and the remaining extract fraction (the so-called S100 fraction which means "supernatant prepared at 100,000 g") is combined with purified ribosomes, total tRNA, and mRNA. In this case the S100 fraction contains all necessary protein translation factors and ARSases. Again, amino acids, ATP, GTP, and an ATP/GTP regenerating system should be added.

Sometimes it is expedient to produce mRNA immediately in the translation system rather than to add an isolated mRNA (DeVries and Zubay, 1967; Gold and Schweiger, 1969). This is easy in the case of prokaryotic systems, since prokaryotic cell extracts contain RNA polymerase. Then a corresponding DNA species, such as plasmid, isolated gene, synthetic DNA fragment, or viral DNA, is added to the DNA-free extract instead of mRNA, and the proper mRNA is synthesized by the endogenous RNA polymerase *in situ*. In this case ribosomes start to translate the nascent chains of mRNA, even prior to the completion of their synthesis. That is why such systems are called *coupled transcription–translation systems*. Of course, the coupled systems should be supplemented with all four nucleoside triphosphates for RNA synthesis, rather than with just ATP and GTP which are required for translation alone.

The eukaryotic extracts are prepared from the cytoplasmic fraction, so they lack endogenous RNA polymerase activity. This limitation can be overcome by addition of a prokaryotic RNA polymerase–usually bacteriophage T7 or SP6 RNA polymerase–to the eukaryotic extract, in order to produce mRNA *insitu* using DNA species with corresponding T7 or SP6 promoters. In this case, however, no real coupling between transcription and translation takes place, since the bacteriophage RNA polymerases work much faster than the translation system. Nevertheless, the eukaryotic transcription–translation systems of this type are found to be practical and productive.

The ionic strength and specifically the Mg^{2+} concentration are important factors for the cell-free systems. The usual range of Mg^{2+} concentrations, within which ribosomes are active in the cell-free system, is from 3 to 20 mM; the optimum is somewhere between these values and depends on the ribosome origin and monovalent cation (K^+ or NH_4^t) concentration, as well as on the concentration of di- and polyamines; it also depends on the incubation temperature. As a rule, thiol compounds, such as mercaptoethanol, dithiothreitol, and glutathione, should be present in the translation mixture in order to maintain the reduced state of translation factors.

One principal shortcoming of all cell-free translation and transcription–translation systems should be mentioned: in contrast to *in vivo* protein synthesis, they have short lifetimes and, as a consequence, give a low yield of the protein synthesized. This makes them useful only for analytical purposes and inappropriate for preparative syntheses of polypeptides and proteins. Indeed, the bacterial (*E. coli*) cell-free systems are usually active for 10 to 60 min at 37°C. The systems based on rabbit reticulocyte lysate or wheat germ extract are capable of working for 1 hour, although in some cases the lifetime may be prolonged up to 3 or 4 hours.

Spirin and co-workers (1988) have found that this shortcoming can be conquered if the incubation is performed under conditions of continuous removal of the products (synthesized polypeptide, AMP, GDP, inorganic phosphates, etc.) and continuous supply with the consumable substrates (amino acids, ATP, and GTP). This can be achieved with the use of a porous barrier limiting the reaction mixture. One way is to pass the flow of the substrate-containing solution through the reactor (*continuous-flow cell-free system*); the outflow will remove the products through the barrier, including the protein synthesized, provided the proper membrane is selected. It is interesting that the components involved in translation (or transcription–translation) are retained in the reactor volume under these conditions, even when some of them (in an individual state) are smaller than the barrier (membrane) pores. This may imply that the components of the protein-synthesizing system in a functional state are present as large dynamic complexes with each other. Another way, but still based on the same principle, is to put the reaction mixture into a dialysis bag or other dialysis device against a large volume of the substrate-containing solution; during incubation the low-molecular-mass products will be removed and new portions of the consumable substrates will be provided through the dialysis membrane (*continuous-exchange cell-free system*). In this case, however, the protein synthesized is retained in the reactor. The lifetimes of the systems described, especially of the flow system, increases up to 50 hours at least, both for prokaryotic and eukaryotic ones. The yields of proteins synthesized are typically around 100 to 200 μg, and may be up to 4 mg in some cases, from a 1 ml reactor.

The most important information regarding translation and its molecular mechanisms has been obtained with the aid of cell-free systems of different types. A historical review on ribosome preparation and cell-free protein synthesis was presented by A. S. Spirin in 1990.

References

Brachet, J. (1942) La localisation des acides pentosenucleiques dans les tissues animaux et les oeufs d'amphibiens en voie de developement, *Arch. Biol.* **53**:207–257.

Casperson, T. (1941) Studien uber den Eiweissumsatz der Zelle, *Naturwissenschaften* **29**:33–43.

Chao, F.-C., and Schachman, H. K. (1956) The isolation and characterization of a macromolecular ribonucleoprotein from yeast, *Arch. Biochem. Biophys.* **61**:220–230.

Claude, A. (1940) Particulate components of normal and tumor cells, *Science* **91**:77–78.

DeVries, J. K., and Zubay, G. (1967) DNA-directed peptide synthesis, II. The synthesis of the α-fragment of the enzyme β-galactosidase, *Proc. Natl. Acad. Sci. USA* **57**:1010–1012.

Gierer, A. (1963) Function of aggregated reticulocyte ribosomes in protein synthesis, *J. Mol. Biol.* **6**:148–157.

Gold, L. M., and Schweiger, M. (1969) Synthesis of phage-specific α- and β-glucosyl transferases directed by T-even DNA *in vitro, Proc. Natl. Acad. Sci. USA* **62**:892–898.

Lamborg, M., and Zamecnik, P. C. (1960) Amino acid incorporation by extracts of *E. coli, Biochim. Biophys. Acta.* **42**:206–211.

Littlefield, J. W., and Keller, E. B. (1957) Incorporation of C^{14} amino acids into ribonucleoprotein particles from the Ehrlich mouse ascites tumor, *J. Biol. Chem.* **224**:13–30.

Littlefield, J. W., Keller, E. B., Gross, J., and Zamecnik, P. C. (1955) Studies on cytoplasmic ribonucleoprotein particles from the liver of the rat, *J. Biol. Chem.* **217**:111–123.

Martin, K. A., and Miller, O. L. (1983) Polysome structure in sea urchin eggs and embryos: An electron microscopic analysis, *Dev. Biol.* **98**:338–348.

Matthaei, H., and Nirenberg, M. W. (1961) The dependence of cell-free protein synthesis in *E. coli* upon RNA prepared from ribosomes, *Biochem. Biophys. Res. Commun.* **4**:184–189.

McQuillen, K., Roberts, R. B., and Britten, R. J. (1959) Synthesis of nascent protein by ribosomes in *E. coli, Proc. Natl. Acad. Sci. USA* **45**:1437–1447.

Nirenberg, M. W., and Matthaei, J. H. (1961) The dependence of cell-free protein synthesis in *E. coli* upon naturally occurring or synthetic polynucleotides, *Proc. Natl. Acad. Sci. USA* **47**:1588–1602.

Palade, G. E. (1955) A small particulate component of the cytoplasm, *J. Biophys. Biochem. Cytol.* **1**:59–68.

Palade, G. E., and Siekevitz, P. (1956a) Liver microsomes: An integrated morphological and biochemical study. *J. Biophys. Biochem. Cytol.* **2**:171–200.

Palade, G. E., and Siekevitz, P. (1956b) Pancreatic microsomes: An integrated morphological and biochemical study, *J. Biophys. Biochem. Cytol.* **2**:671–691.

Penman, S., Scherrer, K., Becker, Y., and Darnell, J. (1963) Polyribosomes in normal and poliovirus-infected HeLa cells and their relationship to messenger-RNA, *Proc. Natl. Acad. Sci. USA* **49**:654–662.

Petermann, M. L., and Hamilton, M. G. (1957) The purification and properties of cytoplasmic ribonucleoprotein from rat liver, *J. Biol. Chem.* **224**:725–736.

Roberts, R. B., ed. (1958) *Microsomal particles and protein synthesis,* First Symposium of the Biophysical Society, at The Massachusetts Institute of Technology, Cambridge, Mass., February 5, 6, and 8, 1958 (London, New York, Paris, and Los Angeles: Pergamon Press).

Schachman, H. K., Pardee, A. B., and Stanier, R. Y. (1952) Studies on the macromolecular organization of microbial cells, *Arch. Biochem. Biophys.* **38**:245–260.

Schachtschabel, D., and Zillig, W. (1959) Untersuchungen zur Biosynthese der Proteine. I. Uber den Einbau ^{14}C-markierter Aminosauren ins Protein zellfreier Nucleoproteid-Enzym-Systeme aus *E. coli* B, *Hoppe-Seyler's Z. Physiol. Chem.* **314**:262–275.

Spirin, A. S. (1990) Ribosome preparation and cell-free protein synthesis, in *The Ribosome: Structure, Function, and Evolution* (W. E. Hill, A. Dahlberg, R. A. Garrett, P. B. Moore, D. Schlesinger, and J. R. Warner, eds.) pp. 56–70, ASM Press, Washington, DC.

Spirin, A. S., Baranov, V. I., Ryabova, L. A., Ovodov, S. Yu., and Alakhov, and Yu. B. (1988) A continuous cell-free translation system capable of producing polypeptides in high yield, *Science* **242**:1162–1164.

Tissieres, A., and Watson, J. D. (1958) Ribonucleoprotein particles from *E. coli, Nature* **182**:778–780.

Tissieres, A., Schlessinger, D., and Gros, F. (1960) Amino acid incorporation into proteins by *E. coli* ribosomes, *Proc. Natl. Acad. Sci. USA* **46**:1450–1463.

Ts'o, P. O. P., Bonner, J., and Vinograd, J. (1956) Microsomal nucleoprotein particles from pea seedlings, *J. Biophys. Biochem. Cytol.* **2**: 451–465.

Warner, J. R., Knopf, P. M., and Rich, A. (1963) A multiple ribosomal structure in protein synthesis, *Proc. Natl. Acad. Sci. USA* **49**:122–129.

Watson, J. D. (1963) The involvement of RNA in the synthesis of proteins, *Science* **140**:17–26.

Wettstein, F. O., Staehelin, T., and Noll, H. (1963) Ribosomal aggregate engaged in protein synthesis: Characterization of the ergosome, *Nature* **197**:430–435.

Zamecnik, P. C. (1953) Incorporation of radioactivity from D,L-leucine-1-^{14}C into proteins of rat liver homogenates, *Fed. Proc.* **12**:295.

Zamecnik, P. C., and Keller, E. B. (1954) Relation between phosphate energy donors and incorporation of labeled amino acids into proteins, *J. Biol. Chem.* **209**:337–354.

II

Structure of the Ribosome

5

Morphology of the Ribosome

with Victor D. Vasiliev

5.1. Size, Appearance, and Subdivision into Subunits

When examined by electron microscopy the isolated bacterial ribosomes at first approximation look like compact rounded particles with linear sizes of about 200 to 250 Å (Fig. 5.1), and somewhat larger, from 200 to 300 Å, in the case of eukaryotic ribosomes. Ribosomes from different organisms and cells, whether prokaryotic or eukaryotic, have a strikingly similar appearance.

A characteristic feature of one of the visible ribosomal projections is a groove dividing the ribosome into two unequal parts (Fig. 5.2). This subdivision reflects the fact that the ribosome consists of two separable subparticles, or *ribosomal subunits.* Under certain conditions, for example, if the concentration of magnesium ions in the medium is sufficiently low, the ribosome dissociates into two subunits with a mass ratio of about 2:1 (Fig. 5.3). The prokaryotic 70S ribosome dissociates into subunits with the sedimentation coefficients 50S (molecular mass 1.65×10^6 daltons) and 30S (molecular mass 0.85×10^6 daltons):

$$70S \rightarrow 50S + 30S$$

The eukaryotic 80S ribosome dissociates into 60S and 40S subunits:

$$80S \rightarrow 60S + 40S$$

The dissociation can be also induced by Na^+, Li^+, and urea, as well as by high concentrations (above 0.5 M) of such "physiological" monovalent cations as K^+ and NH_4^+. The dissociation of *E. coli* 70S ribosomes following a decrease of the Mg^{2+} concentration is illustrated by the sedimentation patterns shown in 5.3.

The dissociation is reversible. The restoration of a proper Mg^{2+} concentration and the removal of dissociating agents result in reassociation of ribosomes. The reassociation is also promoted by Ca^{2+}, diamines and polyamines, and alcohols. Some factors contributing to and counteracting the dissociation of ribosomes are indicated in Fig. 5.4.

After dissociation, the ribosomal subunits can be separated in the preparative

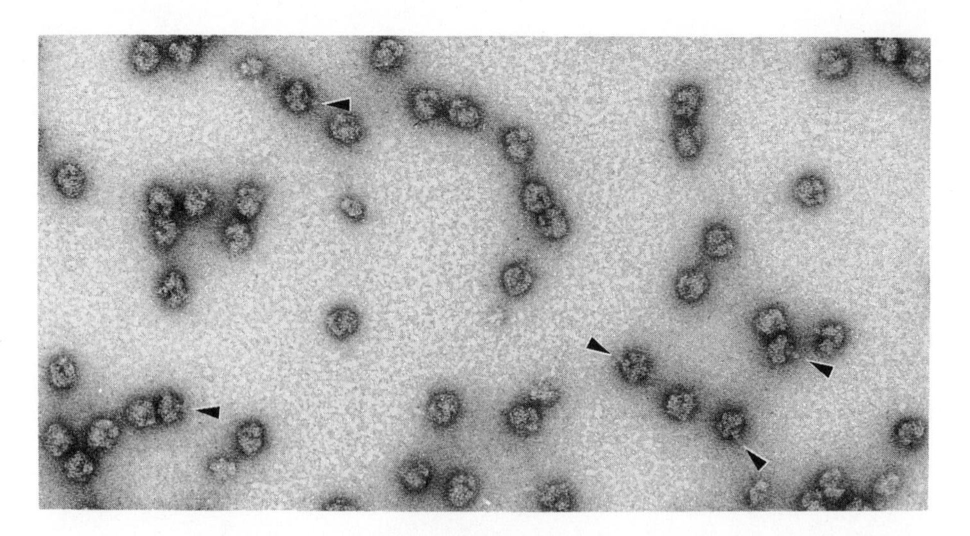

FIG. 5.1. Electron micrograph of the 70S ribosomes isolated from *E. coli.* To achieve the contrast necessary for the particles to be seen in the electron microscope, the isolated 70S ribosomes are applied on an ultrathin carbon film; the film with attached particles is treated by uranyl acetate solution and dried in air. The particles become embedded in the uranyl acetate that fills the cavities and grooves. The ribosomal particles that have lower electron density than uranyl acetate appear *negatively stained* against the background of uranyl acetate. The arrows indicate the L7/L12 stalk described in the text. Original photo by V. D. Vasiliev.

ultracentrifuge, and then studied individually. A unique asymmetrical shape for each of them has been detected (Vasiliev, 1974; Lake, 1976; Kastner, Stöffler-Meilicke, and Stöffler, 1981; Vasiliev, Selivanova, and Ryazantsev, 1983; see also Nonomura, Blobel, Sabatini, 1971 for 80S ribosomes and their subunits) and is described below.

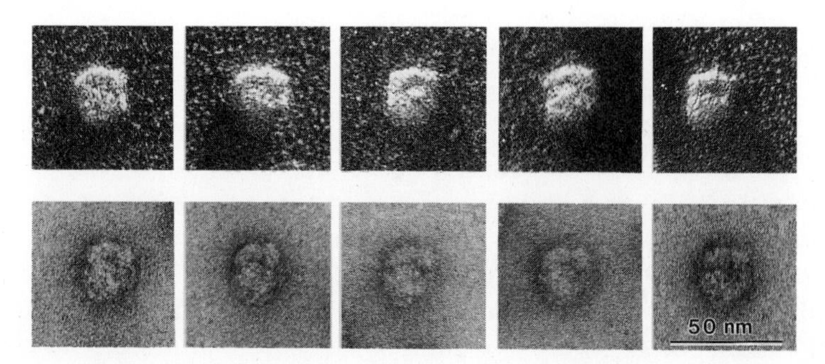

FIG. 5.2. Electron micrographs of individual *E. coli* 70S ribosomes illustrating their subdivision into two unequal subunits; the images are oriented such that the small subunit is at the top and the large subunit is at the bottom. Original photos by V. D. Vasiliev; see also Hall, C. E., and Slater, H. S. (1959) *J. Mol. Biol.* **1**:329–332; Huxley, H. E., and Zubay, G. (1960) *J. Mol. Biol.* **2**:10–18; Vasiliev, V. D. (1971) *FEBS Lett.* **14**:203–205. (**A**) Ribosomes contrasted by metal shadowing. In this case, to achieve the necessary contrast the suspension of isolated 70S ribosomes is applied to a carbon film surface and freeze-dried; the particles are shadowed by metal (tungsten or tungsten-rhenium alloy) using vacuum evaporation at an angle of about 75° to film surface; this yields shadow-cast particles. (**B**) Negatively stained ribosomes, prepared as described in the legend to Fig. 5.1.

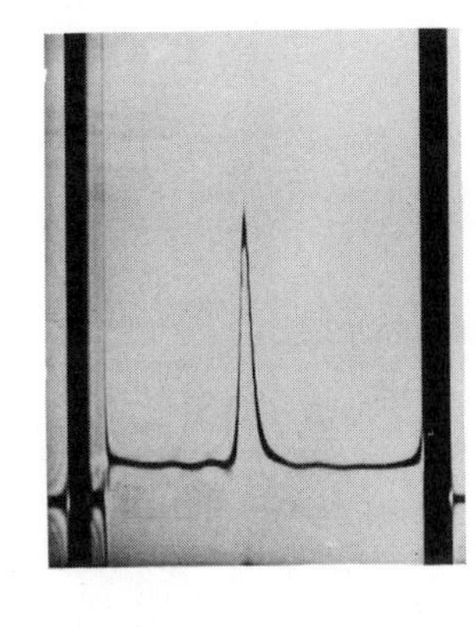

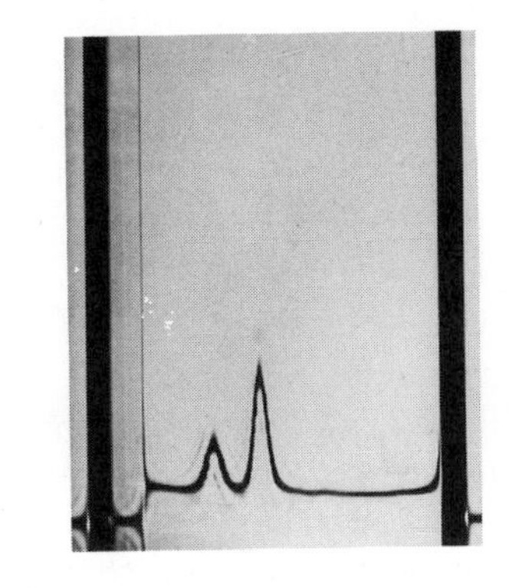

FIG. 5.3. Sedimentation pattern (analytical ultracentrifugation with schlieren optic) of the *E. coli* 70S ribosomes and the products of their dissociation achieved by lowering the Mg^{2+} concentration in the medium. (**A**) 70S ribosomes in 10 mM $MgCl_2$, 100 mM NH_4Cl. (**B**) 30S and 50S subunits in 1 mM $MgCl_2$, 100 mM NH_4Cl.

5.2. Small Subunit

Different electron microscopic projections of the bacterial (*E. coli*) ribosomal 30S subunit and the corresponding crude morphological model are shown in Fig. 5.5. The 30S subunit is somewhat elongated, and its length is about 230 Å. The subunit may be subdivided into lobes which are referred to as the "head" (H), "body" (B), and "side bulge" or "platform" (SB). The groove separating the head from the body is quite distinct.

The eukaryotic 40S subunit has a similar morphology, although two additional details of structure may be mentioned. The first is a protuberance, or "eukaryotic bill" on the head. Second, the end of the body distal to the head appears to be bifurcated due to the presence of some additional mass; this bifurcation is referred to as the "eukaryotic lobes" (Fig. 5.6).

More reliable information about ribosomal subunits can be derived from electron microphotographs if averaged images rather than individual ones are examined. Averaging allows the statistical noise on electron microphotographs to be eliminated. This contributes to a better visualization of the common features in the images of a given particle type. For such an averaging, a set of particle images in the same projection is digitized using a microdensitometer and processed by a computer. The images are aligned precisely with respect to each other and then summed together to give an "average" image. All nonreproducible details of original images such as

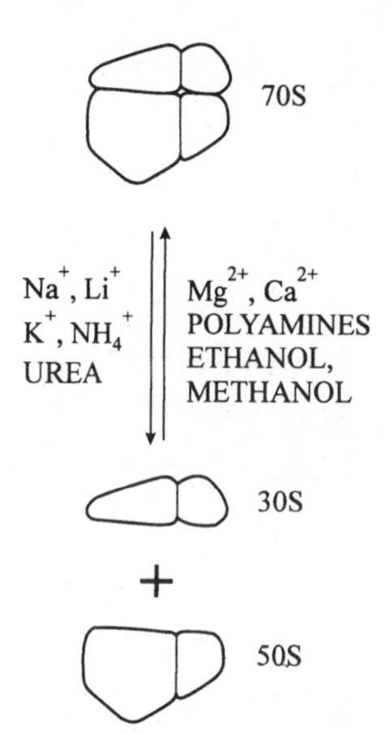

FIG. 5.4. Scheme of ribosome dissociation into subunits. Some factors inducing the dissociation and promoting the reassociation are indicated.

those resulting from variations of stain distribution around the particle, radiation-induced structural alterations, and variations in the background support film, are removed, leaving the common elements on the averaged image. Examples of such an averaging for negatively stained 30S subunits of *E. coli* are given in Fig. 5.7A. All three projections show that the head is separated from the remainder of the subunit by a distinct deep groove, the "neck" being rather thin. An example of the averaging for negatively stained 40S subunits of rat liver ribosomes is presented in Fig. 5.7B. Again, the "neck" is thin, and the bill of the head can be seen clearly in two of the projections; the bifurcated tail, or eukaryotic lobes, are also prominent.

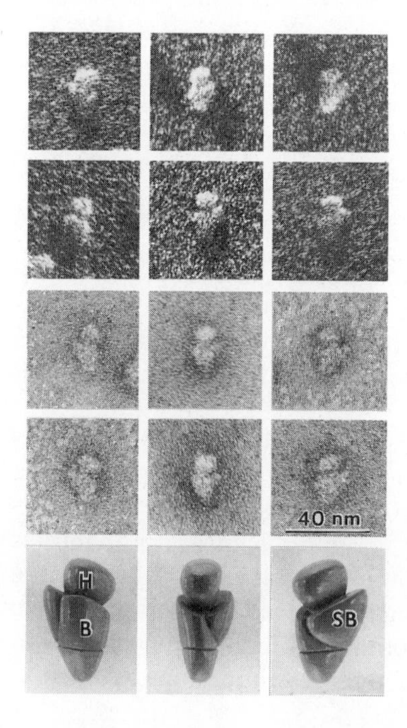

FIG. 5.5. Electron micrographs of individual 30S ribosomal subunits of *E. coli* and a model of them in three projections. See Vasiliev, V. D. (1974) *Acta Biol. Med. Germ.* **33**:779–793). The upper two rows show metal-shadowed particles, prepared as described in the legend to Fig 5.2A. The next two rows show uranyl acetate-stained particles, prepared as described in the legend to Fig. 5.1. The lower row is the model. The left column shows the images of the 30S subunit and its model in the projection when it is viewed from the side opposite to that facing the 50S subunit in the complete ribosome. The middle column contains the images of the 30S subunit and its model in the narrow side (frontal) projection. The right column has the images of the 30S subunit and its model in the projection when it is viewed from the side facing the 50S subunit in the ribosome. Original photos by V. D. Vasiliev.

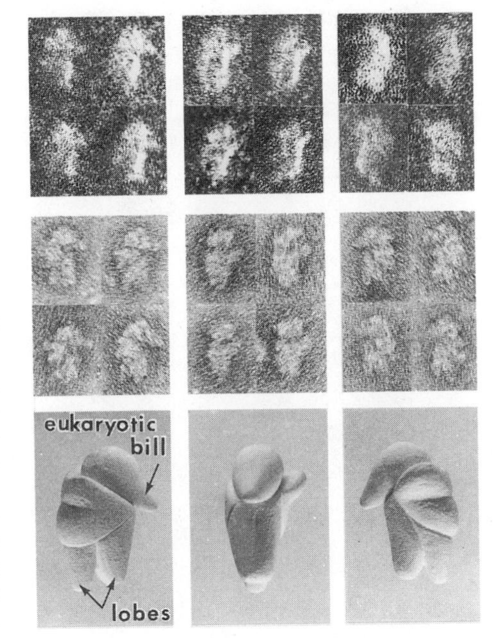

FIG. 5.6. Electron micrographs of the individual 40S subunits of rat liver ribosomes and a model of them in three projections. See Vasiliev, V. D., Selivanova, O. M., Lutsch, G., Westermann, P., and Bielka, H. (1989) *FEBS Letters* **248**:92–96. The upper two rows show metal-shadowed particles, prepared as described in the legend to Fig. 5.2A. The next two rows show uranyl acetate-stained particles, prepared as described in the legend to Fig. 5.1. The lower row is the model. The three columns of images are of the 40S subunit and its model in the same projections as those for the 30S subunit shown in Fig. 5.3. Original photos by V. D. Vasiliev.

It should be noted that the small ribosomal 30S subunit of archaebacteria has a morphology which is intermediate between that of the eubacterial 30S subunit and the eukaryotic 40S subunit: the archaebacterial subunit has a characteristic bill on the head but does not possess the eukaryotic lobes at the end of the body.

5.3. Large Subunit

Different projections of the bacterial 50S subunit and its crude model are shown in Fig. 5.8. This subunit is more isometric than the small one, with a linear

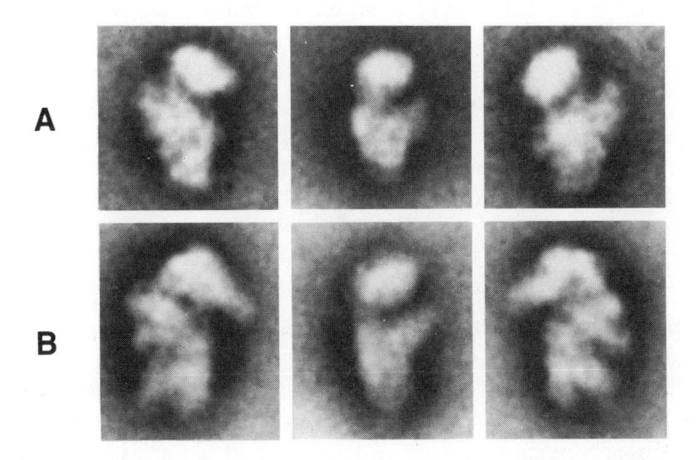

FIG. 5.7. Averaged images of negatively stained 30S (*upper row*) and 40S (*lower row*) ribosomal subunits in the same three projections as shown in Figs. 5.5 and 5.6. Original photos by V. D. Vasiliev.

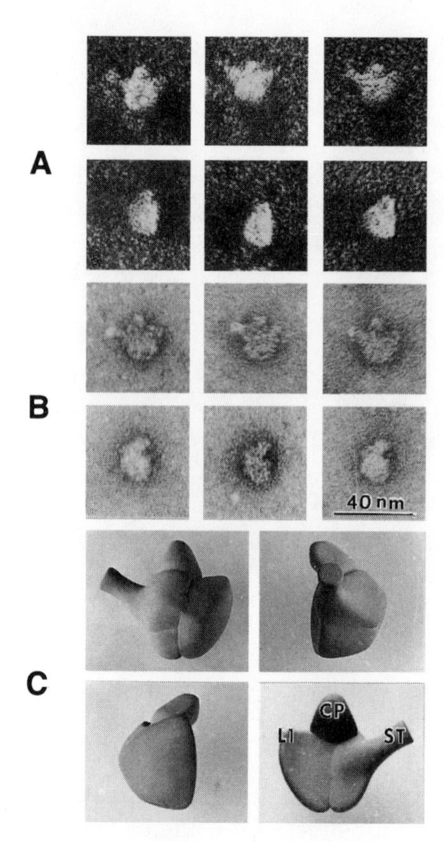

FIG. 5.8. Electron micrographs of the individual 50S subunits of *E. coli* ribosomes and a model of them. See Vasiliev, V. D., Selivanova, O. M., and Ryazantsev, S. N. (1983) *J. Mol. Biol.* **171**:561–569. (**A**) Metal-shadowed particles prepared as described in the legend to Fig. 5.2A. (*Upper row*) the so-called crown-like projection when the particle is viewed from its "back" convex side turned away from the 30S subunit in the ribosome. (*Lower row*) the lateral projection when the particle is viewed from the side of its L1-ridge. (**B**) Negatively stained subunits prepared as described in the legend to Fig. 5.1. (*Upper row*) the crown-like projection. (*Lower row*) the lateral projection. (**C**) The model of the 50S subunit viewed at different angles when rotated around the vertical axis. Original photos by V. D. Vasiliev.

size of 200 to 230 Å in all directions. Three peripheral protuberances can be distinguished: the central one (CP) is termed the head, the lateral finger-like protuberance is called the L7/L12 stalk, and the other lateral protuberance, located on the other side of the central protuberance, is referred to as the side lobe or L1 ridge (in the case of *E. coli* 50S ribosomal subunit, the two lateral protuberances contain ribosomal proteins L7/L12 and L1, respectively; see Chapter 7 for more details).

Averaged images of the large ribosomal subunit of *E. coli* (Fig. 5.9) reveal all the details in the individual particle images. Again, the deep groove separates the head (CP) from the remainder of the subunit, the groove being deeper on the side of the L1 ridge than on the side of the L7/L12 stalk. The body of the subunit is bifurcated on the side opposite the head. In addition, the averaged images demonstrate that many of the smaller details are not random features but are reproducible in numerous images. The eukaryotic 60S subunit has all the same main morphological features.

5.4. Association of Subunits into the Complete Ribosome

In an intact ribosome the two ribosomal subunits are joined in a very specific manner. The flattened (or concave) side of the 50S subunit is involved in the contact between the subunits; if the subunit is viewed from this surface, the head of the subunit is up, and the stalk is on the right (Fig. 5.10A). The subunits are associated in the "head-to-head" and "side lobe-to-side lobe" manner. The head-to-head association may be seen clearly on an electron microphotograph of the other projection of 70S ribosomes (Fig. 5.10B). An electron microscopic image of the "overlap" projection demonstrates that the 30S subunit covers only a part of the flattened side of the 50S subunit (Fig. 5.10A). The region at the base of the stalk remains exposed. This region appears to accommodate functionally important ribosomal sites (see Chapter 9). A photograph of the low-resolution model of the 70S ribosome with the coupled 30S and 50S subunits in head-to-head and side lobe-to-side lobe association is presented at the bottom of Fig. 5.10.

5.5. Morphology of the Ribosome at 25-Å Resolution

Recently, further significant progress in electron microscopy of ribosomes has been achieved. It is based on several developments of the ribosome imaging and image processing techniques. The development of cryo-electron microscopy has allowed for the visualization of ribosomes embedded in a thin film of vitreous ice without any staining. The images are recorded under low-dose conditions and optimal use of the phase contrast. The vitrification preserves the particles in a fully hydrated state, and investigation of the native structure of the ribosomes becomes possible. Special computer programs making use of several thousands of images have been created for three-dimensional reconstruction of the ribosome. The three-dimensional reconstruction is based on numerous projections that show the particles from different directions. In electron microscopy, in contrast to medical X-ray computerized tomography, the necessary projections cannot be obtained by consecutive image recording of the same particles under different angles to the electron beam. To prevent radiation damage each particle under investigation must be illuminated only once. Therefore, a full data set necessary for three-dimensional reconstruction is collected simultaneously from different particle projections of the same microphotograph. The projections arise by chance and the problem is to assign an orientation to each of them in a common coordinate system. Two differ-

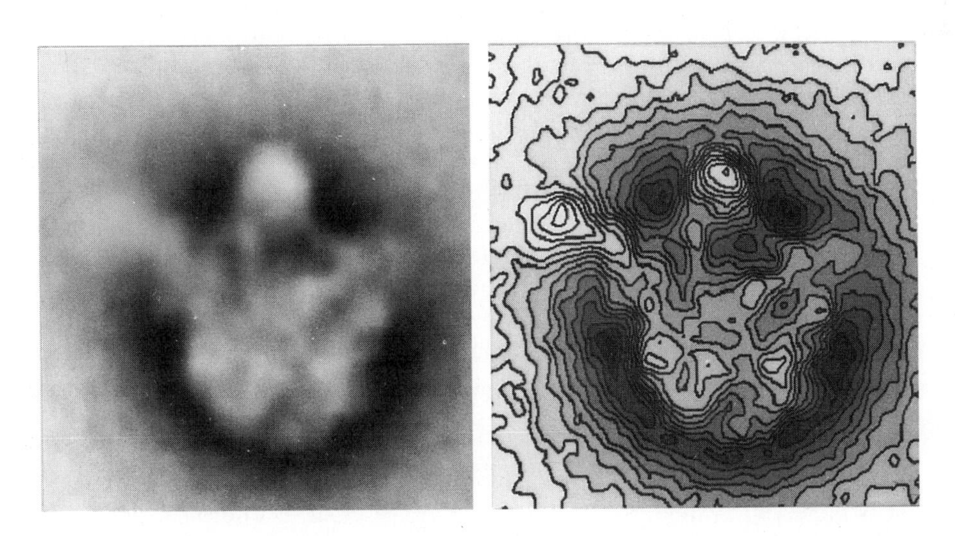

FIG. 5.9. Averaged images of the negatively stained 50S subunit in the crown-like projection. Original photos by V. D. Vasiliev.

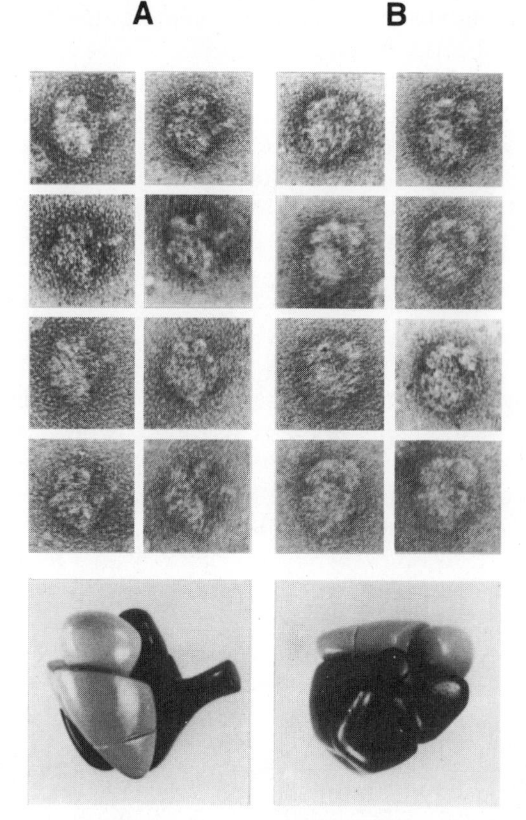

FIG. 5.10. Electron micrographs of the individual 70S ribosomes from *E. coli* and the model in two different projections. See Vasiliev, V. D., Selivanova, O. M., Baranov, V. I., and Spirin, A. S. (1983) *FEBS Lett.* **155**:167–172. (**A**) The so-called overlap projection when the 30S subunit faces the viewer and covers a part of the 50S subunit. (**B**) The nonoverlap or lateral projection viewed from the side of the L7/L12 stalk. The particles were stained with uranyl acetate as described in the legend to Fig. 5.1. Original photos by V. D. Vasiliev.

ent approaches were applied to solve this problem resulting in a three-dimensional reconstruction of the bacterial ribosome with a resolution of about 25 Å (Frank *et al.,* 1995; van Heel et al., 1995; see Figs. 5.11 and 5.12, respectively).

The overall shape of the new high-resolution models of the 70S ribosome (Figs. 5.11 and 5.12) is very close to that of the aforementioned low-resolution model derived by visual interpretation of the ribosome images (Fig. 5.10). Although the surface of the ribosome is much more irregular at high resolution, the main characteristic morphological features of both the subunits are well recognized. The small subunit consists of a head, a side lobe, and a body, and the large subunit is roughly hemispherical with three protuberances.

The new models are similar, although they have somewhat different details. The Frank model (Fig. 5.11) appears rather solid, whereas the van Heel model (Fig. 5.12) is full of internal cavities and channels. This seems to result from different contouring of their density maps (different density threshold values). In the model of van Heel and co-workers, approximately half the total volume of the ribosome is found to consist of solvent regions, including the intersubunit space and intrasubunit channels and cavities that give the appearance of "Swiss cheese pieces" to the ribosomal particles. The density threshold chosen by Frank and co-workers was evidently lower, and one bifurcated channel in the 50S subunit part of their model is observed instead of the extensive network of channels of the van Heel model.

The most important new feature visible on the reconstructed, unstained, ice-embedded ribosome (both 70S and 80S, see also Fig. 5.13) is a large cavity between the ribosomal subunits, in the region of their necks. This intersubunit space is sufficient to accommodate tRNA molecules, so it is likely that the cavity serves as a tRNA-binding pocket of the ribosome.

New specific details can also be seen on each subunit. Generally, the shape and morphological features of the coupled ice-embedded subunits are somewhat different from those of the isolated particles. The most marked differences are the structural changes of the small subunit in the neck and side lobe ("platform") regions. An additional thin connection between the head and the body appears, so that a channel penetrating the neck is formed instead of an open gap. The side lobe of the 30S subunit now is further separated from the head when compared to the crude model of the isolated particle. It appears to be spade-shaped and forms a well-defined cavity with the head. Other structural elements are the "beak" in the head region and the "toe" or "spur" extending from the end of the body distal to the head.

The large subunit seems to be less changed upon association into the full ribosome. The most apparent difference with the isolated subunit is that the rod-like (L7/L12) stalk is truncated and rather directed perpendicular to the subunit interface, instead of being stuck out from the subunit body in the plane of the contacting surface of the isolated subunit.

It is noteworthy that the subunits in the full ribosome (at least in the "empty," nontranslating ribosome) are seen as drawn apart, especially at the side of the rod-

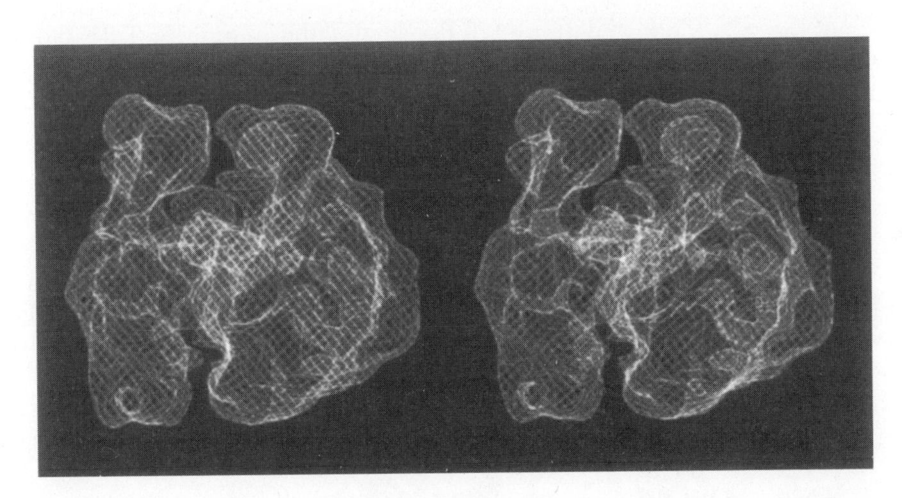

FIG. 5.11. Stereo representation of the three-dimensional density map of the *E. coli* 70S ribosome in the nonoverlap projection viewed from the L7/L12 stalk side from Frank, J., Zhu, J., Penczek, P., Li, Y., Srivastava, S., Verschoor, A., Radermacher, M., Grassucci, R., Lata, R. K., and Agrawal, R. K. (1995) *Nature* **376:**441–444, with permission. The three-dimensional reconstruction is based on so-called random-conical-tilt-series approach. This elegant approach exploits the random azimuthal orientations of asymmetrical particles lined in one (or several) preferred orientations relative to the support film. It starts from a pair of micrographs showing the same field both from a high tilt angle and without tilt. The tilted-field image is recorded first and only particle images from this field enter the three-dimensional reconstruction. The untilted-specimen images are used as references to convert the random planar particle orientations in the real definite projections. The first three-dimensional reconstruction is then improved by iterative procedures. Finally, the orientation of each projection is determined individually by matching it to computed projections of the previous model. This 25Å three-dimensional reconstruction of the 70S ribosome is obtained by combining 4,300 individual images. Courtesy of J. Frank, New York State Department of Health, Albany.

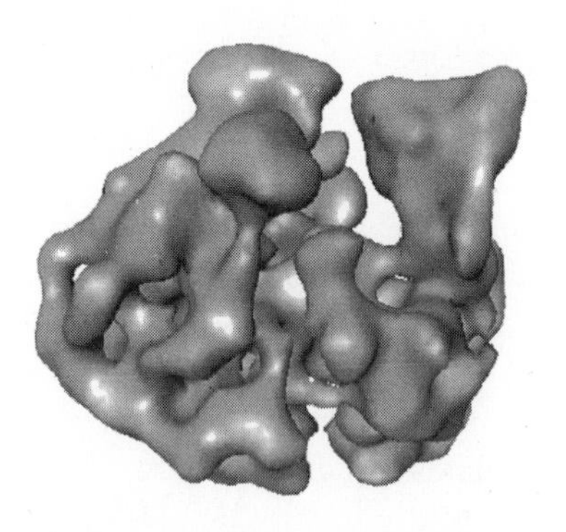
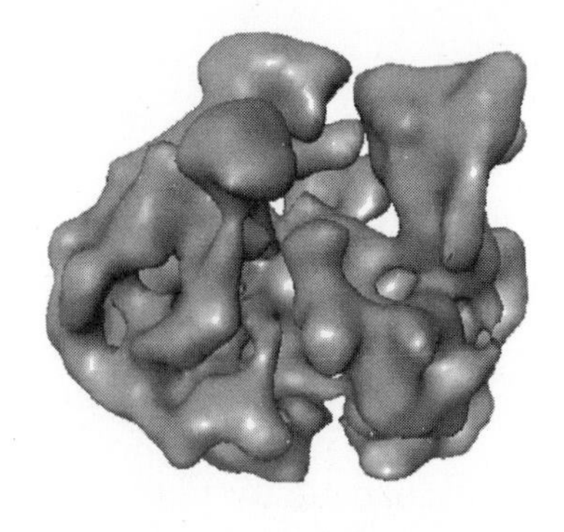

FIG. 5.12. Stereo representation of the three-dimensional reconstruction of the *E.coli* 70S ribosome in the nonoverlap projection viewed from the side opposite to the L7/L12 stalk from Stark, H., Müller, F., Orlova, E. V., Schatz, M., Dube, P., Erdemir, T., Zemlin, F., Brimacombe, R., and van Heel, M. (1995) *Structure* **3**:815–821 with permission. The three-dimensional reconstruction is based on the "angular reconstitution" approach which allows one to determine the relative angular orientations of the particles arbitrarily arranged within a vitreous ice matrix. This 23Å three-dimensional reconstruction is derived from 2447 individual images. Courtesy of M. van Heel and H. Stark, Fritz Haber Institute, the Max Planck Society, Berlin.

like stalk, thus leaving free access from outside to the intersubunit space. The numerous tight contacts between the two subunits are clearly visible mainly in the region of the side lobes ("platform" of the small subunit and the L1 ridge of the large subunit). The contact between the heads of the two subunits looks less prominent and, possibly, nonpermanent. If the subunits are capable of swinging in and out (Section 9.6), the side lobe contact may serve as a main hinge of the pulsating ribosome.

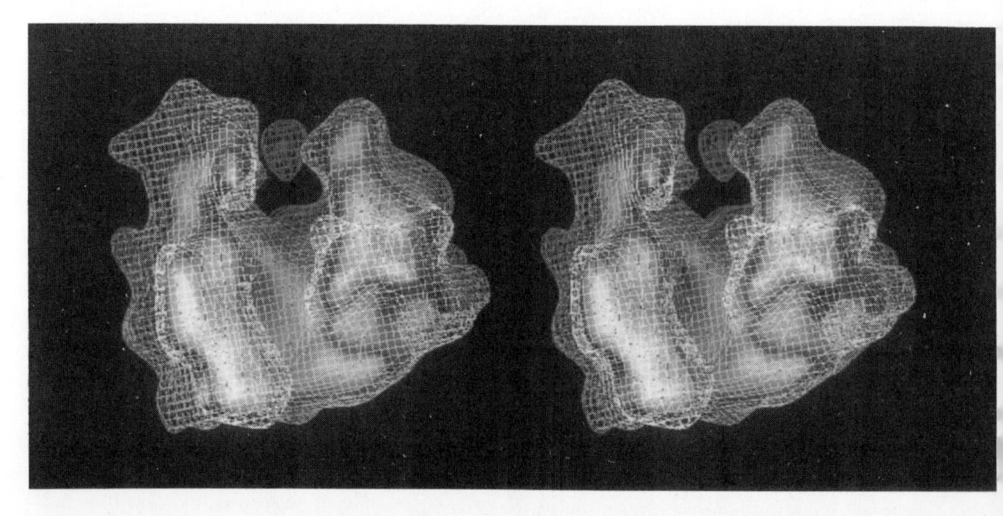

FIG. 5.13. Stereo representation of the three-dimensional reconstruction of the 80S ribosome in the nonoverlap projection. See Verschoor, A., Srivastava, S., Grassucci, R., and Frank, J. (1996) *J. Cell Biol.* **133**:495–505. Courtesy of J. Frank, New York State Department of Health, Albany.

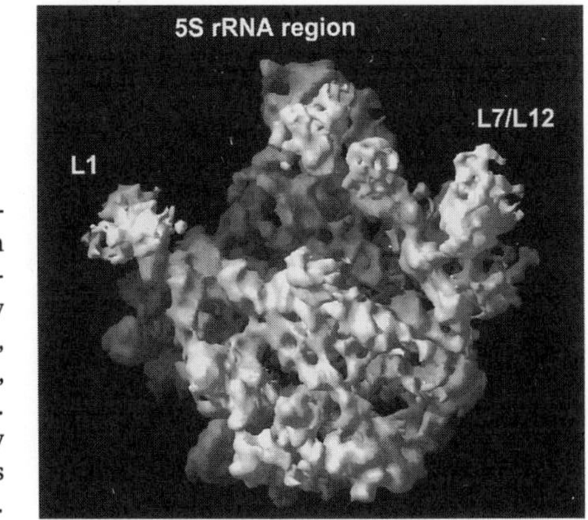

FIG. 5.14. The 50S ribosomal subunit of archaebacterium *Haloarcula marismurtui* shown in the crown view from the side that interacts with the 30S subunit: a surface rendition of a 9 Å resolution X-ray electron density map. Reproduced from Ban, N., Freeborn, B., Nissen, P., Penczek, P., Grassucci, R. A., Sweet, R., Frank, J., Moore, P. B., and Steitz, T. A. (1998) *Cell* **93**:1105–1115 with permission (courtesy of T. A. Steitz, Department of Molecular Biophysics and Biochemistry, Yale University, New Haven, CT).

5.6. X-Ray Crystallography of the Ribosome

It is remarkable that the ribosomal particles, including full ribosomes and their isolated subunits, can be crystallized, and in some cases the crystals are well ordered and diffract X-rays up to about 3 Å resolution. Generally, the crystallographic studies of ribosomal particles have demonstrated the similarity of the X-ray electron density maps to the electron microscopy maps. Figure 5.14 shows the result of the first X-ray crystallographic study of a ribosomal particle where the electron density map at 9 Å resolution was obtained for the 50S subunit of the archaean ribosome. All the morphological features revealed earlier by electron microscopy are confirmed. At the same time, several important details of the ribosome structure have been defined more precisely. Further work with X-ray crystallography of ribosomes is aimed at the solution of the problem of detailed quaternary structure of the particles.

References

Hall, C. E., and Slayter, H. S. (1959) Electron microscopy of ribonucleoprotein particles from *E. coli, J. Mol. Biol.* **1**:329–332.

Huxley, H. E., and Zubay, G. (1960) Electron microscope observations on the structure of microsomal particles from *Escherichia coli, J. Mol. Biol.* **2**:10–18.

Kastner, B., Stoeffler–Meilicke, M., and Stoeffler, G. (1981) Arrangement of the subunits in the ribosome of *Escherichia coli:* Demonstration by immunoelectron microscopy, *Proc. Natl. Acad. Sci. USA* **78**:6652–6656.

Lake, J. A. (1976) Ribosome structure determined by electron microscopy of *Escherichia coli* small subunits, large subunits and monomeric ribosomes, *J. Mol. Biol.* **105**:131–159.

Nonomura, Y., Blobel, G., and Sabatini, D. (1971) Structure of liver ribosomes studied by negative staining, *J. Mol. Biol.* **60**:303–324.

Vasiliev, V. D. (1971) Electron microscopy study of 70S ribosomes of *Escherichia coli, FEBS Lett.* **14**:203–205.

Vasiliev, V. D. (1974) Morphology of the ribosomal 30S subparticle according to electron microscopy data, *Acta Biol. Med. Germ.* **33**:779–793.

Vasiliev, V. D., Selivanova, O. M., and Ryazantsev, S. N. (1983) Structure of the *Escherichia coli* 50S ribosomal subunit, *J. Mol. Biol.* **171**:561–569.

6

Ribosomal RNA

with Alexey A. Bogdanov

6.1. Introduction

Ribosomal RNAs (rRNAs) comprise 50 to 70% of the mass of the ribosomal particles. They principally determine the size and the shape of the ribosomal subunits. The rRNA molecules form a framework for specific positioning of ribosomal proteins in the ribosome. rRNA plays the leading role in ribosome function, participating in every aspect of protein synthesis. Therefore, it can be said that the ribosome is its RNA.

It should be noted that this concept was not generally accepted among molecular biologists until the early 1980s. In the 1970s, despite the growing numbers of findings that indicated the participation of rRNA in the organization of ribosomal functional centers (such as the discovery of mRNA–rRNA interactions during translation initiation by Shine and Dalgarno [1974] or the identification of rRNA mutations that affected ribosome activities), it was generally believed that active centers of the ribosome were formed mainly by ribosomal proteins, whereas rRNA in the ribosome served only as a scaffold for specific protein binding. The shift from the protein to the RNA paradigm came after the realization that the primary, secondary, and tertiary structure of rRNA is highly conserved throughout evolution. The discovery of catalytic RNAs also strongly strengthened the concept of the function-defining role of rRNA in translation. Now there is strong belief that the ancient ribosome was composed entirely of RNA, and the modern ribosome proteins only help to organize the intraribosomal structure of rRNA and fine-tune its activity.

6.2. Types of Ribosomal RNAs and their Primary Structures

Like all other single-stranded polynucleotides, rRNAs respond to changes in ionic strength and temperature by altering their structure from completely unfolded to rather compact. At the same time, it is customary to characterize and designate rRNAs (as well as ribosomal subunits) by their sedimentation coefficients (Kurland, 1960), that are a function of macromolecular size and shape. In this connection, it should be emphasized that the values of sedimentation coefficients generally used to mark different types of rRNA are valid only within a limited range of conditions and practically obtained at ionic strength 0.1, 20°C, and in the absence of Mg^{2+} and other divalent cations.

The small ribosomal subunit (30S or 40S) contains one molecule of high-molecular-weight rRNA that is designated 16S rRNA in the case of ribosomes of *E. coli* and other bacteria, and 16S-like rRNA in other cases (after the 16S rRNA of *E. coli* ribosomes). After 1978, when the first complete nucleotide sequence of 16S rRNA was determined in Ebel's and Noller's laboratories (Carbon *et al.*, 1978; Brosius *et al.*, 1978), many different rRNA genes were sequenced. The shortest 16S-like rRNA, only 610 nucleotides long, was found in mitochondrial small ribosomal subunits from the homoflagellate *Leishmania tarentolae.* Relatively short 16S-like rRNAs (10–12S rRNA, 960–970 nucleotides long) were also discovered in mitochondrial ribosomes of higher eukaryotes; interestingly, cytoplasmic ribosomes of the same organisms contain the longest 16S-like rRNA (18S rRNA, about 1,880 nucleotides long). The best studied 16S rRNA of *E. coli* ribosomes consists of 1,542 nucleotide residues.

The vast majority of 16S and 16S-like rRNAs are continuous (uninterrupted) polynucleotide chains. However, several examples of split (fragmented) 16S-like rRNA were found in mitochondria of some species. An example is the mitochondrial 16S-like rRNA of *Chlamydomonas reinhardtii* which consists of four separate polynucleotides (Boer and Gray, 1988).

The large ribosomal subunit (50S or 60S) contains a high-molecular-mass rRNA called 23S rRNA in the case of bacteria and 23S-like rRNA in other cases, and a low-molecular-mass rRNA designated 5S RNA. The bacterial 23S rRNAs are covalently continuous polynucleotide chains, as are the 16S RNAs. At the same time the molecules of 23S-like rRNA of the large ribosomal subunits of cytoplasmic ribosomes of all eukaryotes are discontinuous. They consist of two tightly associated polynucleotide chains: a high-molecular-mass 28S rRNA with a length of 4,700–4,800 nucleotide residues, and a low-molecular-weight 5.8S rRNA that is about 160 nucleotides long. In corresponding rRNA genes, 5.8S and 28S rRNA sequences are separated by an internal transcribed spacer (ITS), and the corresponding rRNAs are formed as a result of removal of the ITS during the processing of a common precursor. The 5.8S rRNA appears to be the structural equivalent of the 5′-terminal 160-nucleotide segment of the prokaryotic 23S rRNA. In other words, the 5′-end of 23S rRNA was split during evolution to form the eukaryotic 5.8S rRNA.

Another example of discontinuity among large-subunit rRNAs is the 23S-like rRNA of plant chloroplast ribosomes. It contains a 4.5S fragment (about 110 nucleotides long) that is the structural counterpart of the 3′-terminal segment of *E. coli* 23S rRNA; the fragment is also tightly associated with the high-molecular-mass rRNA. The mitochondrial 23S-like rRNA of *Chlamydomonas reinhardtii* consists of eight separate RNA pieces (Boer and Gray, 1988). The most striking example of discontinuous rRNA is *Euglena gracilis* cytoplasmic 23S-like rRNA, which consists of 14 fragments (Schnare and Gray, 1990).

As in the case of rRNA from small ribosomal subunits, large subunit rRNAs vary greatly in length. For example, 23S rRNA from the 50S subunit of *E. coli* ribosomes consists of 2,904 nucleotide residues, whereas human cytoplasmic 28S rRNA is 5,025 nucleotides long. It is interesting that both cytoplasmic and mitochondrial yeast 23S-like rRNAs (26S rRNAs) have the same size (3,392 and 3,273 nucleotide residues, respectively, in the case of *Saccharomyces cerevisiae*). Large ribosomal subunits of mitochondrial ribosomes of higher eukaryotes contain relatively short 23S-like rRNAs (1,560–1,590 nucleotide residues).

As mentioned above, in addition to one molecule of 23S rRNA or one 28S:5.8S rRNA complex the large ribosomal subunits of cytoplasmic ribosomes of all prokaryotes and eukaryotes contain one 5S rRNA molecule that is about 120 nucleotides long. The 5S rRNA of *E. coli* was the first ribosomal RNA species whose primary structure was determined (Brownlee, Sanger, and Barell, 1967). The 5S rRNA forms a separate domain of the large subunit. In contrast to the 5.8S rRNA it is not tightly associated with 23S-like rRNA and therefore cannot be considered a component of this rRNA. 5S rRNA was also found in chloroplast ribosomes but is apparently absent from mitochondrial ribosomes except land plant mitochondria.

In addition to "normal" G, A, U, and C residues the high-molecular-mass rRNAs contain modified nucleosides. They are mostly represented with pseudouridine (Ψ) and methylated (both at the base and at the 2′-OH of ribose) nucleoside residues. Although some modification sites in rRNAs are extremely conserved in evolution (such as m^6_2 A1518/m^6_2 A1519 and m^7 G527 in the 16S-like rRNA; see Fig. 6.1) the number of modified residues in rRNA differs strongly in different organisms and increases dramatically from eubacteria to multicellular eukaryotes. For instance, *E. coli* rRNA has nine Ψ residues per 70S ribosome (one in 16S rRNA and eight in 23S rRNA), whereas vertebrate rRNAs contain about 95 Ψ residues. The latter ones contain also about 100 2′-methylated ribose residues and ten methylated bases. The distribution of modified residues through the rRNA molecules and their possible role in the organization of rRNA structure will be discussed in the following sections. It should be noted that several sequence-specific methylases and pseudouridylases have been found in *E. coli*. In eukaryotic cells, small nucleolar RNAs (snoRNAs) participate in rRNA modifications forming complementary rRNA–snoRNA complexes at modification sites.

6.3. Secondary Structure of Ribosomal RNAs

6.3.1. General Principles

The current view of macromolecular structure of rRNA (as well as all other single-stranded RNAs) is based on ideas developed in the late 1950s and early 1960s (Fresco, Alberts, and Doty, 1960; Spirin, 1960; Cox, 1966). It was postulated that rRNA is built of numerous, rather short double-stranded regions, in which base pairing occurs between neighboring RNA segments connected by single-stranded sequences. Early secondary-structure models of the *E. coli* 16S and 23S rRNA were generated by maximizing Watson–Crick base pairing within the putative helical regions. It was also considered that double-stranded regions in rRNA can be formed not only between neighboring sections, but also between quite distant regions of the polynucleotide chain. The models satisfied the physicochemical data accumulated by that time. It was clear, however, that more information was needed to get realistic models of rRNA secondary structure.

The breakthrough in rRNA secondary structure study occurred when several rRNA sequences became known. This allowed Woese and co-workers (1980) to propose and use successfully the comparative sequence analysis of rRNA structure based on a very simple principle. This principle suggests that "the functionally equivalent RNA molecules should form a comparable three-dimensional structure no matter how similar or divergent their sequences are." In other words, if this principle is correct, rRNAs from different organisms have to form isomorphic sec-

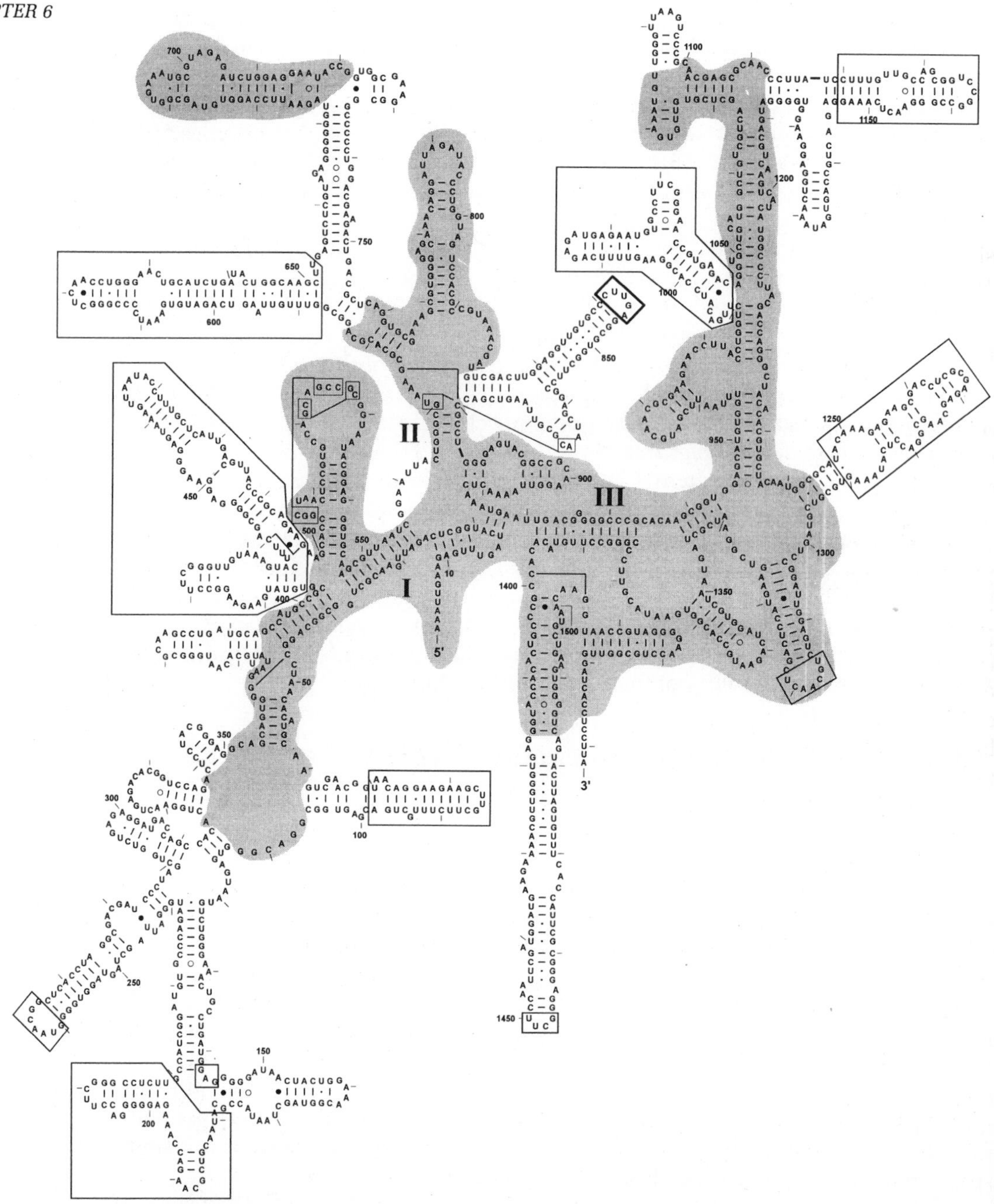

FIG. 6.1. Secondary-structure model for *E.coli* 16S rRNA. Watson–Crick base pairs are connected with short lines. Dots and open and closed circles show noncanonical base paring. Every tenth nucleotide position is marked, and every 50th position is numbered. Tertiary interactions are shown with long solid lines. The universal core sequences are shaded, and variable regions are shown in boxes. See Gutell, R. R., (1993) *Nucleic Acid Res.* **21**:3051–3054; Gerbi, S. A. (1996) in *Ribosomal RNA: Structure, Evolution, Processing, and Function in Protein Biosynthesis* (R. A. Zimmermann and A. Dahlberg, eds.), pp. 71–87, CRC Press, Boca Raton.

ondary and tertiary structures. This means that the coordinated compensatory base changes should be observed in homologous rRNA double-stranded regions (e.g., A:U ↔ G:C). The crucial point of the comparative approach is therefore searching for positional covariance in proposed secondary-structure elements (double helices).

Thus, the realistic secondary-structure models for archaebacterial, eubacterial, eukaryotic, mitochondrial, and chloroplast rRNA have been created and corroborated by the comparative method, in combination with experimental approaches (e.g., chemical modifications, enzymatic probing, RNA–RNA cross-linking, complementary oligonucleotide binding, etc.) that allow one to distinguish between single-stranded and double-stranded regions in RNA molecules (Woese *et al.*, 1983).

6.3.2. Secondary Structure of the Small-Subunit rRNA

The current versions of secondary structures for prokaryotic (*E. coli*), eukaryotic (*Saccharomyces cerevisiae*), and mitochondrial (*Caenorhabditis elegans*) 16S or 16S-like rRNAs are presented in Figs. 6.1, 6.2, and 6.3. The most important features of these structures are explained here.

The rRNA chain folds back into a series of structural motifs: (1) ideal hairpins with external (end) loops (e.g., the hairpin 1506–1529; here and below, all examples are taken from the *E. coli* 16S rRNA secondary structure model, Fig. 6.1); (2) helices with a single bulged nucleotide (e.g., the helix 27–37:547–556), or with a pair of bulged nucleotides, like in the helix 61–106; (3) helices with larger bulge (side) loops, such as the hairpin 289–311; (4) the so-called compound hairpins or helices with interior loops where double-helical regions alternate with noncomplementary regions (e.g., the hairpin 1241–1296), and (5) different sorts of branched and bifurcated helices, such as the structures in the regions 122–239 and 997–1044.

Although rRNA helices are formed predominantly due to antiparallel Watson–Crick base pairing (more than 80% of the base pairs in the known rRNAs), G:U and U:G pairs also occur with relatively high frequency (13%). Among other possible noncanonical base pairs, A:G (G:A) and U:U are observed rather frequently (3% and 1%, respectively).

One can expect that rRNA helices are in the A-type conformation (see also Section 3.2.2). When synthetic or nucleolytic fragments of rRNA were studied by NMR spectroscopy or X-ray crystallography it was proved that indeed they adopted conformations very close to the classical A-form.

NMR studies of short RNAs representing different elements of rRNA secondary structure demonstrate that both external (end) and internal (side) loops may have quite complicated and well-ordered structure. For example, the structure of the UUCG tetraloop closed by the C:G pair (it is present in the *E. coli* 16S rRNA at positions 420–423, 1029–1032, and 1450–1453) is characterized by an additional wobble G:U base pairing in the double-helix stem, with guanosine in the *syn* conformation, so that the loop proper includes only two nucleotides (UC). A sharp turn in the phosphodiester backbone is stabilized by the hydrogen bond between the amino group of the cytidine and the oxygen of the UpU phosphate and the extensive base stacking (Fig. 6.4).

Another example of an ordered three-dimensional structure of an external loop region in rRNA is the conformation of GNRA tetraloops, one of the most com-

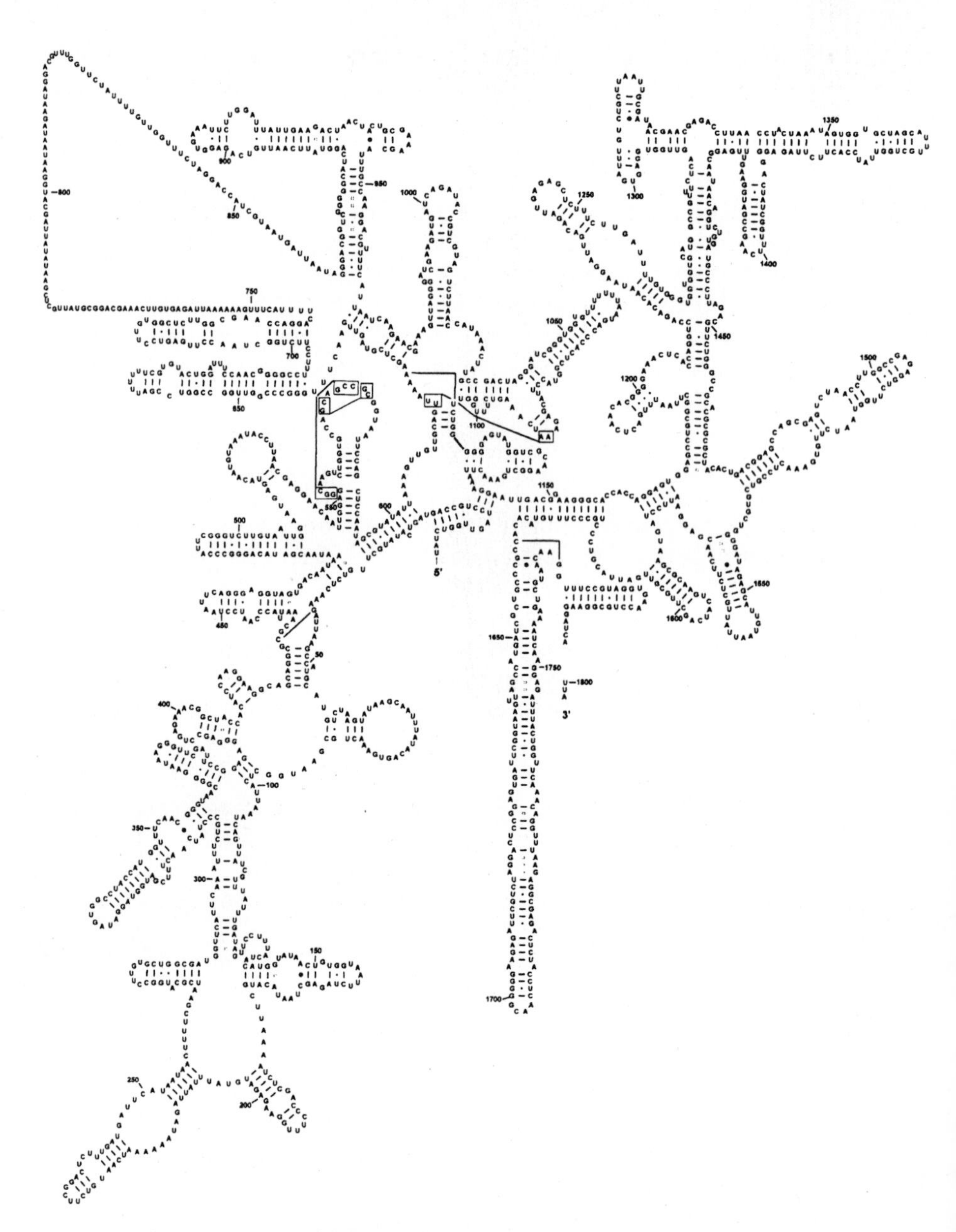

FIG. 6.2. Secondary-structure model for *Saccharomyces cerevisiae* cytoplasmic 16S-like rRNA (see the legend to Fig. 6.1 for details and references).

mon tetraloop families in rRNAs (see positions 159–162, 187–190, 297–300, 380–383, 898–901, 1013–1016, 1077–1080, 1266–1269, and 1516–1519 in Fig. 6.1). It was shown by NMR studies that they include a G:A base pair in the double-helix stem and, hence, are also characterized by a two-nucleotide loop with a sharp turn (Heus and Pardi, 1991). The turn is stabilized by a hydrogen bond between the G base and the RpA phosphate, a hydrogen bond between the R (A or G) base and the guanosine 2′-OH, and extensive base stacking. The "U-turn" motif typical of the anticodon loop structure of tRNAs (see Section 3.2.2 and Fig. 3.6) can be identified in the structure under consideration. The nucleotide N (C, A, U, or G) is on top, and its base is not engaged in any intraloop interactions and therefore seems to be prepared for tertiary base pairings.

The analysis of NMR structure of the internal loop in the *E. coli* 16S rRNA fragment formed with sequences 1404–1412/1488–1498 (see Fig. 6.1) has shown that it adopts a fully helical conformation with base pairs U1406:U1495 and C1407:G1494 and the three adenines (A1408, A1492, and A1493) stacked within the helix (Fourmi *et al.*, 1996). On the other hand, it was noted that the bulges and the mismatches can considerably alter the conformation (e.g., groove dimensions) of neighboring helical regions, thereby bending the structure.

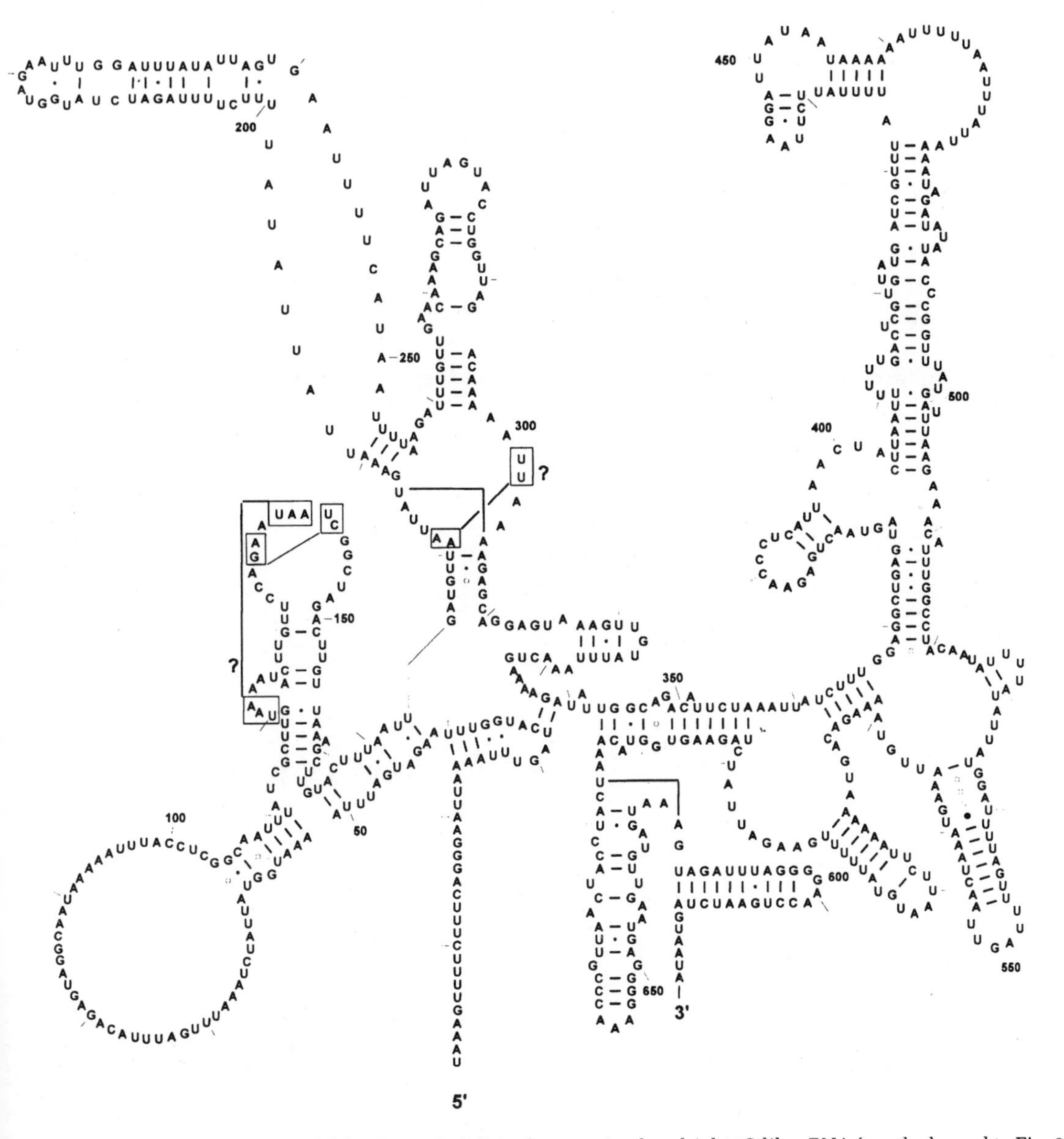

FIG. 6.3. Secondary-structure model for *Caenorhabditis elegans* mitochondrial 16S-like rRNA (see the legend to Fig. 6.1 for details and references).

There are three pseudoknot helices within the secondary structure of 16S rRNA shown in Fig. 6.1: the helices 17–19/916–918, 505–507/524–526, and 570–571/865–866. They are highly conserved in the 16S-like rRNAs and may play an important role in the organization of ribosome functional centers.

It is customary to divide the 16S rRNA secondary structure into four parts: three major domains, namely the 5′-domain, central domain, and 3′-major domain, and the 3′-end minor domain. In many aspects these domains behave like autonomous structural units. The major domains of 16S rRNA are enclosed by long-range double helices: the helix 27–37/547–556 encloses, as a stem, the 5′-domain, the helix 921–933/1384–1396 confines the 3′-major domain, and the central domain is between these two helices. The sequence 912–920 and the pseudoknot helix 17–19/916–918 connect the three major domains. Interestingly, the two other 16S rRNA pseudoknot helices are positioned near the interdomain junctions.

These long-range base-pair interactions define also a core secondary structure that is universal among the 16S and 16S-like rRNAs (see Fig. 6.1). The universal core seems to comprise the most basic structural and functional part of the 16S-like rRNA molecules. Indeed, the vast majority of mutations altering ribosome activities are localized in the 16S rRNA universal core. Most of the modified nucleotide residues are also clustered in the universal core (Brimacombe *et al.*, 1993). The primordial ribosome is thought to consist of its rRNA core.

Comparative analysis of secondary structures of different 16S and 16S-like rRNAs reveals regions of variable size that interrupt the universal core (see Fig. 6.1). They are not evolutionarily conserved and have been termed "variable" or "divergent" regions (or "expansion or contraction segments"). The positions in the *E. coli* 16S rRNA secondary structure where variable regions occur (Fig. 6.1) can be expanded or contracted in rRNAs of other organisms (compare with structures in Figs. 6.2 and 6.3). The role of variable regions in ribosome structure and function is unknown. Up to now no functionally meaningful mutations have been found in the variable regions of 16S-like rRNAs. It is worth noting that the breaks

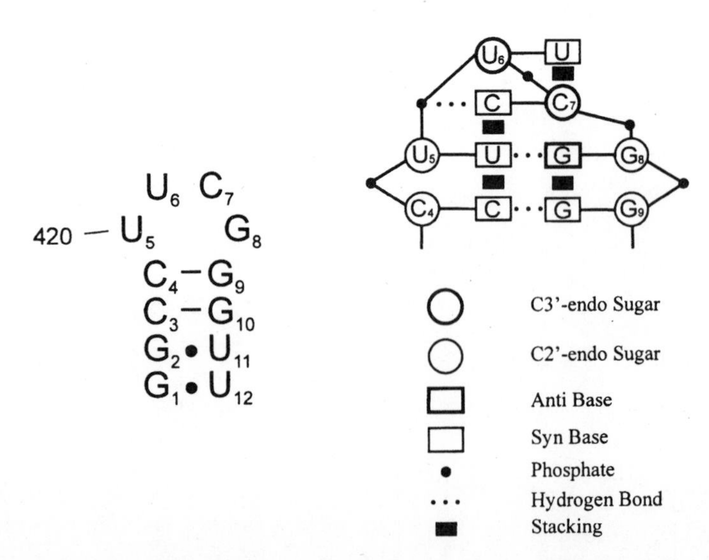

FIG. 6.4. Traditional secondary structure of *E. coli* 16S rRNA hairpin 416–427, and schematic diagram of its UUCG tetraloop conformation. See Varani, G., Cheong, C., and Tinoco, I., Jr. (1991) *Biochemistry* **30**:3280–3289; Allain, F. H.-T. and Varani, G. (1995) *J. Mol. Biol.* **250**:333–353.

in the polynucleotide chain of the discontinuous rRNA molecules (see above, Section 6.2) have been found to occur only in the variable regions.

6.3.3. Secondary Structure of the Large-Subunit rRNA

As one can see from Fig. 6.5, eubacterial 23S rRNA, the general principles of organization of the small-subunit and large-subunit rRNA secondary structures are the same. The relative frequencies of Watson–Crick and non-Watson–Crick base pairs in the 16S-like and 23S-like rRNAs are equal. The structures of several hairpin-loop fragments of bacterial 23S rRNA have been resolved with atomic resolution, demonstrating that their single-stranded regions, just as in the case of the 16S rRNA, are well ordered.

The major difference between these two classes of molecules is that the 3′- and 5′-terminal sequences of 23S and 23S-like rRNAs, in contrast to the 16S-like rRNAs, are mutually complementary and form a long stable stem. In eukaryotic large-subunit rRNAs whose 5′-terminal region is represented with 5.8S rRNA, the 3′-end sequence of 23S-like (25S or 28S) rRNA forms a double-helical structure with the 5′-end sequence of the 5.8S rRNA, the latter being associated with the large rRNA due to formation of two more double helices. In chloroplasts, the 5′-end sequence of the 23S rRNA forms a double-helical complex with the 3′-end region of the 4.5S rRNA.

The secondary structure of prokaryotic, eukaryotic, chloroplast, and some mitochondrial 23S-like rRNAs consists of six domains (I–VI) enclosed by long-range double helices. In the case of *E. coli* 23S rRNA, they are helix 15–24/516–525 (domain I); helix 579–584/1256–1261 (domain II); helix 1295–1298/1642–1645 (domain III); helix 1648–1667/1979–1988 (domain IV); helix 2043–2057/2611–2625 (domain V); and helix 2630–2644/2771–2788 (domain VI). There are approximately 15 pseudoknots in the *E. coli* 23S rRNA secondary structure (see Fig. 6.5); most of them are localized near the interdomain regions, as in the case of the 16S-like rRNAs.

Domains I and III have not been found in the small mitochondrial 23S-like rRNAs. That is why the nucleotide sequences of these domains are not included in the universal core of 23S and 23S-like rRNAs presented in Fig. 6.5. All main functional sites of the large ribosomal subunit determined by genetic studies were localized in the universal core of 23S-like rRNAs, and no mutations that affected the ribosome activities were found within the variable regions.

Interestingly, the modified nucleotides in both prokaryotic and eukaryotic 23S-like rRNAs are clustered mainly in domains II, IV, and V, which are the most functionally important domains of 23S-like rRNAs (see Sections 9.3 and 9.4).

6.3.4. Secondary Structure of 5S rRNA

In contrast to the large ribosomal rRNA, the length of 5S rRNA polynucleotide chain is highly conserved in ribosome evolution. It varies from 115 to 125 nucleotides in length. In addition, all 5S rRNA molecules known to date have very similar secondary structures. The 5S RNA forms an independent structural domain in the large ribosomal subunit. However, it seems to be premature to consider the 5S rRNA as part of the rRNA universal core of the large subunit since 5S rRNA sequences are absent from mitochondrial genomes of fungi, protozoa, algae, and animals, and the question of whether mitochondrial ribosomes in these organisms

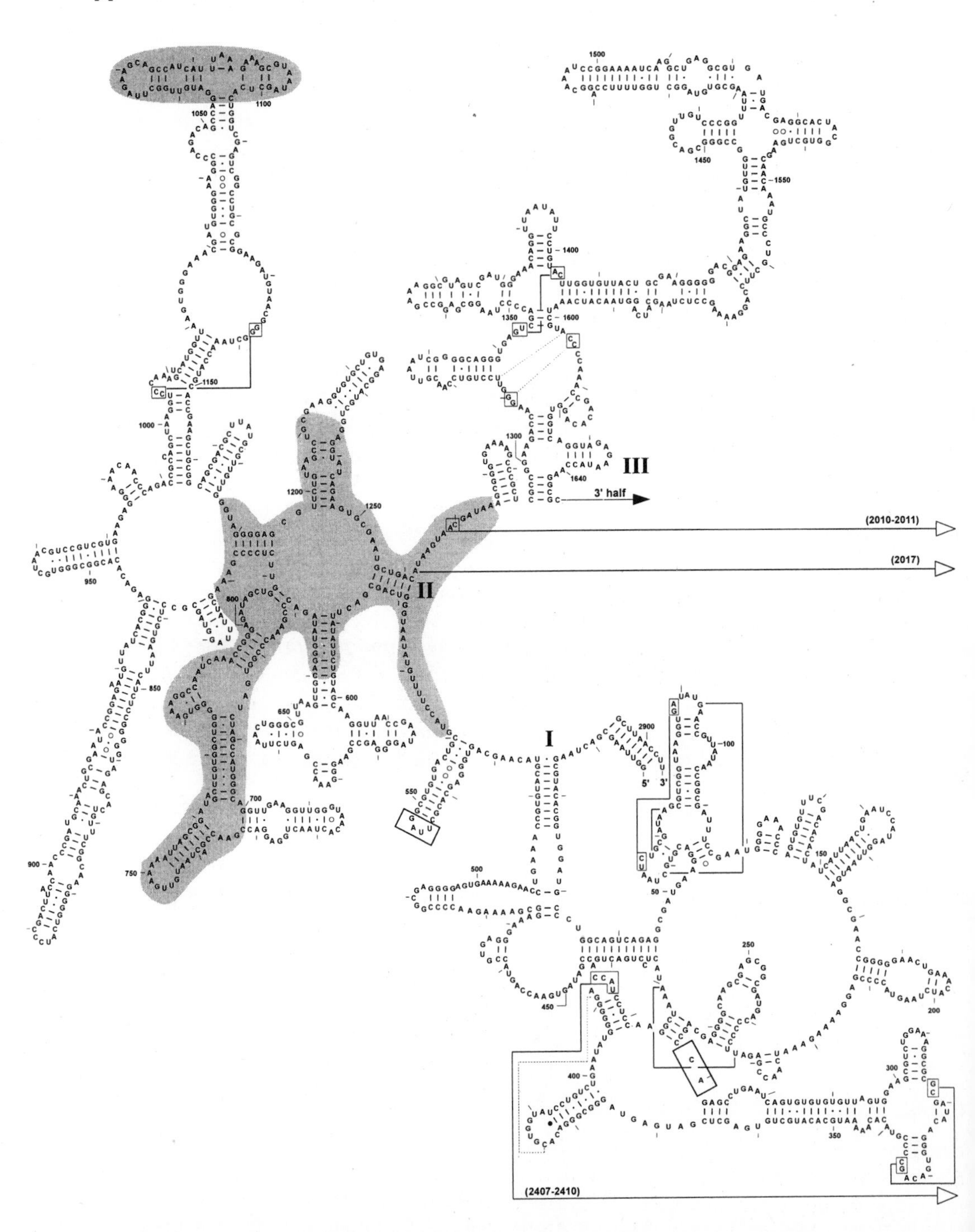

FIG. 6.5. Secondary-structure model for *E. coli* 23S rRNA. The universal core is shaded, and the most variable regions are shown in boxes. See Gutell, R. R., Gray, M. W., and Schnare, M. N. (1993) *Nucleic Acid Res.* **21**:3055–3074. See also the legend to Figure 6.1 for details and references. **(A)** 5′-half. **(B)** 3′-half.

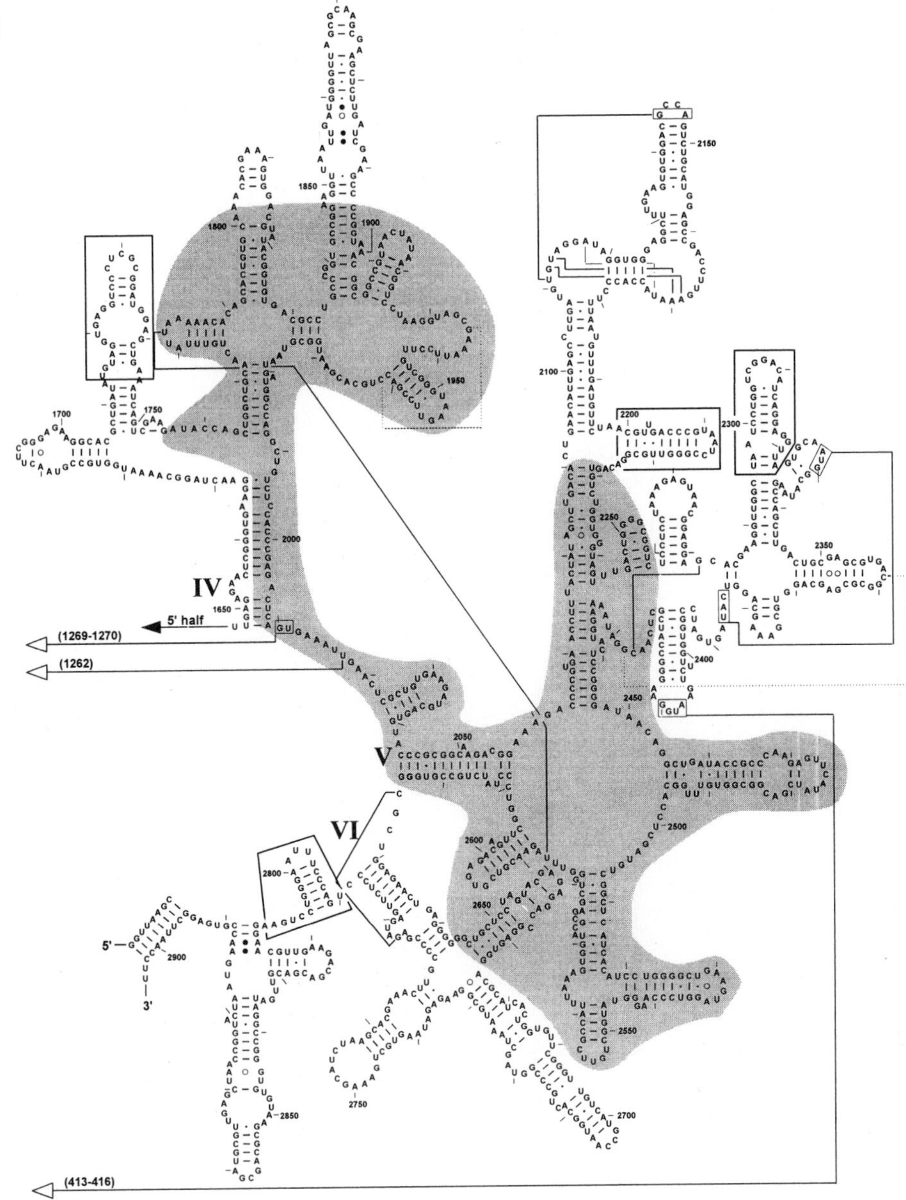

FIG. 6.5. (*continued*)

can function without 5S rRNA or whether they somehow use nuclear-transcribed 5S rRNA remains open.

The model for the secondary structure of 5S rRNA was proposed by Fox and Woese (1975) entirely on the basis of the comparative phylogenetic analysis (see Section 6.3.1). Since then the model has been verified in numerous studies, mainly by chemical or enzymatic probing and NMR spectroscopy, and has not undergone any serious alterations. A recent generalized version of this model (the so-called three-stem model; Fig. 6.6) contains five helices (I–V) and five single-stranded ele-

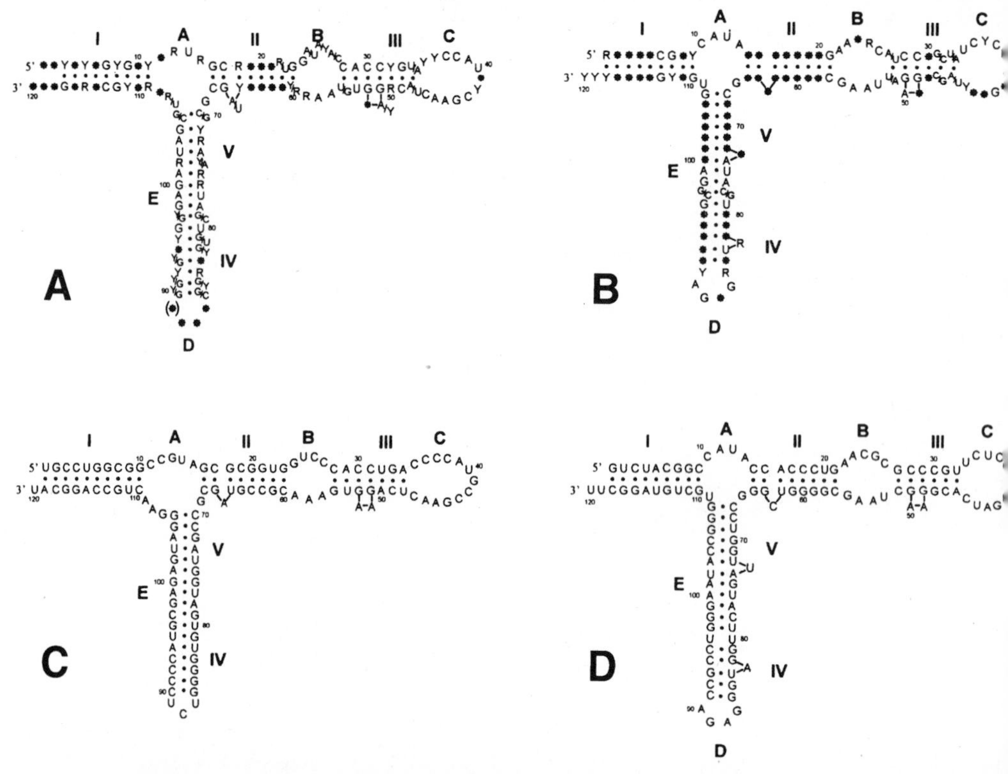

FIG. 6.6. Secondary structure models for 5S rRNA. See Delihas, N., Andersen, J., and Singhal, R. P. (1984) *Progr. Nucleic Acid Res. Mol. Biol.* **31**:160–190. See also the text for details. (**A**) a generic prokaryotic 5S rRNA; positions that can be occupied by any nucleotide residue are indicated by stars. (**B**) a generic eukaryotic 5S rRNA; positions that can be occupied by any nucleotide residue are indicated by stars. (**C**) *E. coli* 5S rRNA. (**D**) human 5S rRNA.

ments consisting of two internal loops (B and E), two external loops (C and D), and one joint loop (A) that serves to connect helices I, II, and V. The 3′- and 5′-terminal sections of 5S rRNA are base-paired, forming a stem. Helices II and III have conservative one-nucleotide and two-nucleotide bulges, respectively, that are believed to participate in RNA–protein interactions. As seen from Fig. 6.6, several base pairs and some nucleotide residues at the equivalent positions are also highly conserved throughout the evolution. The major part of the universal 5S rRNA secondary structure, however, is organized from quite divergent sequences. On the whole, the 5S rRNA gives us a very impressive example of the correctness of the general principle of organization of rRNA three-dimensional structure that has been formulated in Section 6.3.1.

The recent X-ray crystallographic analysis of the 3′/5′ terminal stem of *T. flavus* 5S rRNA has proved that helix I has the classical A-conformation. NMR studies of the *E. coli* 5S rRNA fragment containing loop E have shown once again that single-stranded regions in RNA molecules may have rather well-ordered structure. In particular, loop E appears to resemble a double helix but formed by non-Watson–Crick base pairs such as G:G, G:A, reversed Hoogsteen A:U, and G:U stabilized by a single hydrogen bond (Fig. 6.7).

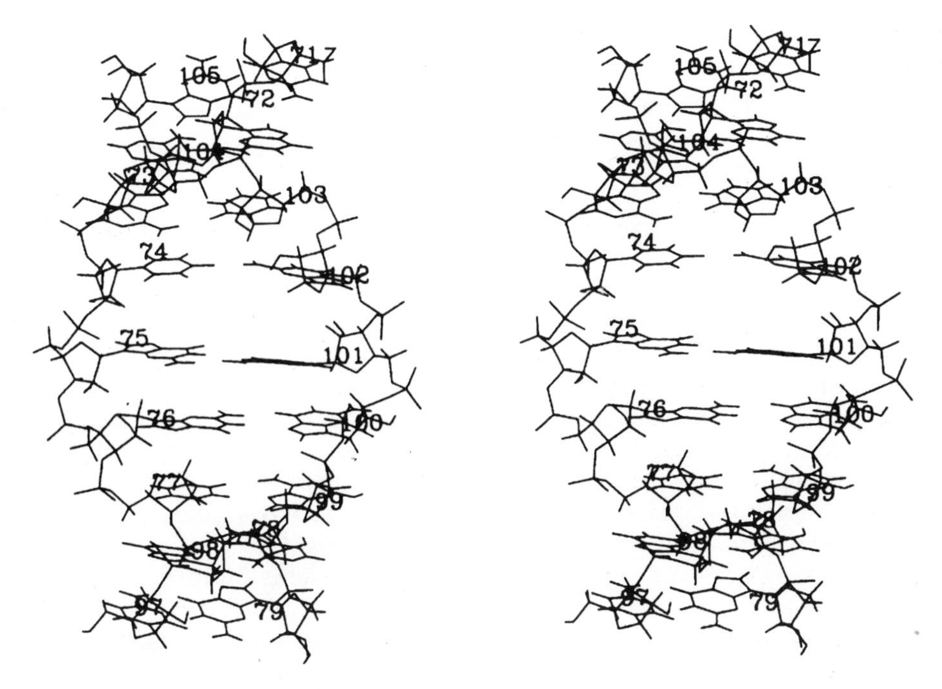

FIG. 6.7. A model for the three-dimensional structure of loop E of *E. coli* 5S rRNA as determined by NMR spectroscopy: a stereoscopic view (see the text for details). From Dallas, A., Rycyna, R., and Moore, P. (1995) *Biochem. Cell Biol.* **73**:887–897.

6.4. Tertiary Structure and Compact Folding of Ribosomal RNAs

6.4.1. General Principles and Properties

Combination of chemical, enzymatic, and physical approaches led to the conclusion that much of the double-helical structure in rRNAs of all types is similar in solution and within ribosomal particles. In other words, although some alterations in conformations of rRNA regions involved in interactions with ribosomal proteins were detected (mainly by CD spectroscopy), one can argue that the intraribosomal secondary structure of rRNA is known in all major details.

On the contrary, the tertiary structures of rRNAs, namely a mutual arrangement of the rRNA secondary-structure elements, both in solution and in the ribosome, are unknown. Nevertheless, serious efforts have been made to solve this problem and at least three fundamental features of rRNA spatial structure have been established:

1. Within the ribosome, rRNAs are folded in a compact way, and the internal part of ribosomal subunits is represented almost entirely with RNAs that form the so-called rRNA core.
2. The overall shape and size of the isolated (naked) rRNA in solution can be similar, under certain conditions, to those of the corresponding ribosomal subunit.
3. Three-dimensional structure of rRNA in the ribosome can undergo alterations (conformational changes) probably related to ribosome function.

In addition, several well-conserved tertiary Watson–Crick base pairs have been found by the comparative analysis of rRNA sequences (shown in secondary structure models for both 16S-like and 23S-like rRNAs; see Figs 6.1–6.3 and 6.5).

6.4.2. Compact Folding of rRNAs

The macromolecular size and shape of rRNA, as well as any other single-stranded RNA, strongly depend on ionic strength, mono- and divalent ion concentrations, pH, and temperature (Spirin, 1960, 1964). rRNA can adopt conformations ranging from completely unfolded threads (at zero ionic strength, high temperature, or low pH) to tightly folded coils (at high ionic strength and in the presence of Mg^{2+}). Electron microscopy studies show that at low, but not zero, ionic strength rRNAs can acquire an intermediate, strongly elongated (rod-like) conformation that still retains a substantial fraction of their secondary structure.

The systematic comparison of the size and shape of isolated high-molecular-mass rRNAs and ribosomal subunits by sedimentation analysis; light, X-ray or neutron scattering; and diffusion coefficient measurements (see Vasiliev *et al.,* 1986) showed that at Mg^{2+} and monovalent ion concentrations optimal for ribosome function *in vitro* (about 5 mM $MgCl_2$ at 100 mM NH_4Cl or KCl) the isolated rRNAs are much less compact than within the ribosome. Neutron small-angle scattering measurements with variation of contrast (allowing one to estimate rRNA parameters within the ribosome, see Section 8.1) have demonstrated that under these conditions the radius of gyration (R_g) for the RNA core of the *E. coli* 50S subunit (6.5 nm) is almost half that for the isolated 23S rRNA in the same solvent. In accordance with this, under "physiological" ionic conditions the *E. coli* ribosomal subunits, despite their higher molecular weight and lower net negative charge, have a greater electrophoretic mobility in polyacrylamide gels than corresponding rRNAs. This is expected because a mass of basic ribosomal protein interacting with rRNA in the ribosome should strongly contribute to the ionic atmosphere of the RNA.

Yet, the external ionic conditions can be selected to maintain a compactly folded state of isolated rRNA in solution (Vasiliev *et al.,* 1986). These conditions include relatively high Mg^{2+} concentration (about 20 mM), elevated ionic strength (0.3 to 0.5), and sometimes the presence of di- and polyamines and alcohol. It is noteworthy that the same conditions are optimal for *in vitro* reconstitution of ribosomal particles from rRNA and ribosomal protein (see Section 7.6.2). Under these conditions the isolated rRNA acquires a compact conformation approaching that in the ribosome. Nevertheless, the compactness of the isolated rRNA does not fully attain that of the rRNA in the ribosome: the R_g of the isolated rRNA in the compact form in solution is still one-fourth larger than the R_g of the rRNA *in situ.* The diffusion and viscosity measurements also point to a more compact rRNA folding within the ribosome than in the free state. Thus, ribosomal proteins not only exert ionic effects on rRNA, but probably also impose additional constrains on rRNA folding.

6.4.3. Specific Shapes of the Folded rRNAs

In the late 1970s, Vasiliev and co-workers provided evidence that the isolated rRNA, under the conditions when it acquires the compact state, folds into particles with a shape similar to that of the corresponding ribosomal subunit (Fig. 6.8). Both 16S and 23S rRNAs of *E. coli* prepared in a buffer containing selected salt proportions were studied by electron microscopy. The shapes of these rRNAs were found to differ sharply from one another. A substantial fraction of the 16S rRNA molecules had a specific Y- or V-like configuration and was similar in the overall shape to that of the 30S subunit (Fig. 6.8A). On the other hand, the 23S rRNA images resembled those of the original 50S subunits: they had a characteristic central protuberance with smaller protuberances on either side (Fig. 6.8B). The con-

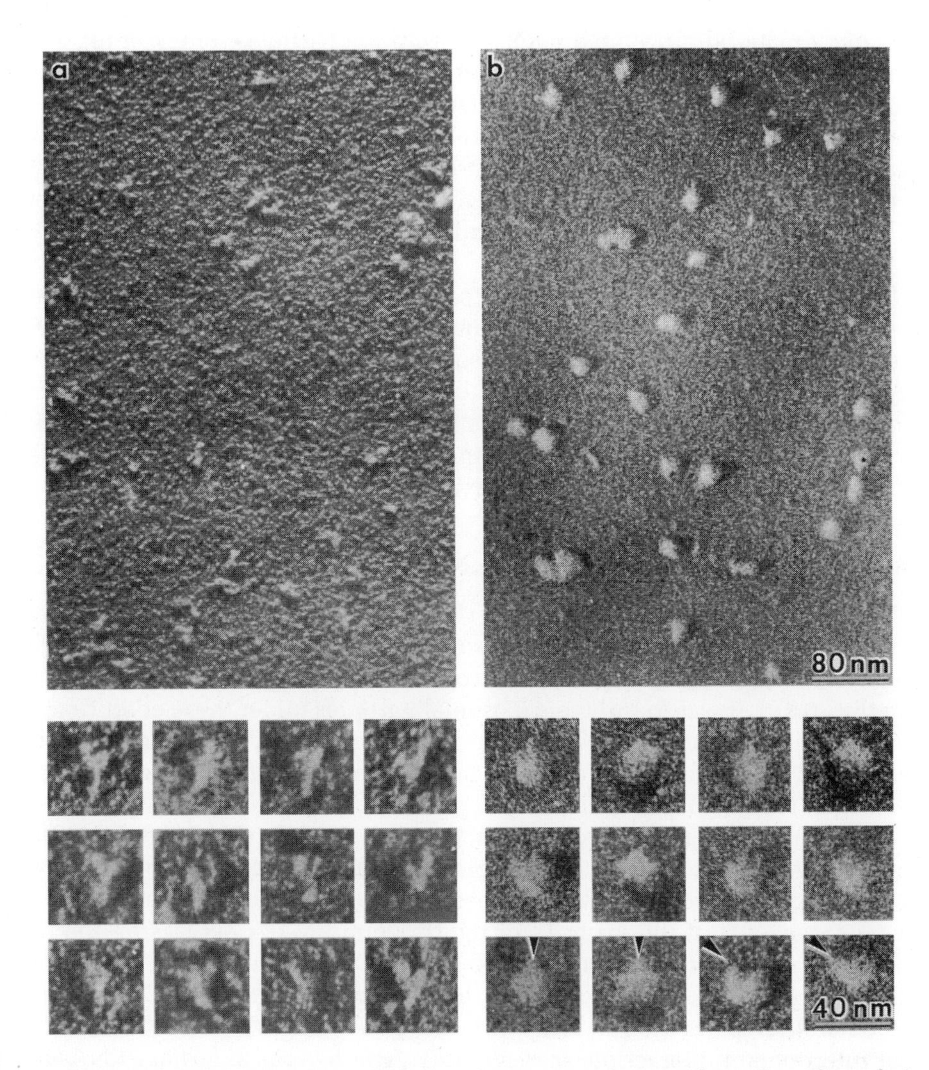

FIG. 6.8.–Electron micrographs of *E. coli* ribosomal RNAs in compact conformations: General views and galleries of images. Specimens from proper solutions were freeze-dried and shadow-cast. (**A**) 16S RNA. Y-like and V-like particles are seen. From Vasiliev, V. D., Selivanova, O. M., and Koteliansky, V. E. (1978) *FEBS Lett.* **95**:273–276. (**B**) 23S rRNA of *E. coli*. The arrowheads indicate the location of the central protuberance in the crown-like region of roughly hemispherical particles. From Vasiliev, V. D., and Zalite, O. M. (1980) *FEBS Lett.* **95**:273–276. Original photos of V. D. Vasiliev.

clusion was made that the general patterns of compact rRNA folding observed by electron microscopy with isolated rRNA samples anticipate the major morphological features of the ribosomal particles.

Subsequent electron microscopic studies of RNA distribution within the ribosomal particles demonstrated a surprisingly close resemblance between the shapes of the isolated and intraribosomal rRNAs. First, the use of element-specific (spectroscopic) electron microscopy revealed the distribution of phosphorus, and therefore of RNA, in the ribosomal subunits *in situ* (Korn, Spitnik-Elson, and Elson, 1983). Second, the method of cryo-electron microscopy in conjunction with the random-conical image reconstruction technique (see Section 5.5) was applied to study *E. coli* 70S ribosomes and showed the spatial distribution of high-electron-density material, presumably RNA, within the particles (Frank *et al.,* 1991). In both cases, the RNA component within the 30S subunit was found to have an asymmetric Y-like (or V-like) shape similar to the electron microscopic images of the isolated 16S rRNA in a compact form, as earlier described by Vasiliev and coworkers (1986) and fit the subunit contours well (Fig. 6.9A). In the case of the 50S subunit, the shape of the RNA component was much more isometric, characterized by three reduced protuberances, and also inscribed into the subunit shape (Fig. 6.9B).

Thus, rRNAs have been shown to be capable of specific self-folding into compact particles of unique shapes. The resemblance of the specific shapes of rRNAs in the isolated state and within the ribosomal particles suggests that the specific rRNA self-folding mainly determines the tertiary structure of rRNA in the ribosome as well. From the resemblance of the shapes of rRNAs and the corresponding ribosomal subunits it follows that the compact folding pattern of rRNA principally sets the morphology of the ribosomal particles.

6.4.4. Conformational Changes of rRNAs

The first evidence of the key role of rRNA in the determination of the specific structure of ribosomal subunits came from the demonstration of unfolding of the ribosome compact structure in response to decreasing Mg^{2+} concentration in solution (Fig. 6.10). It was found that ribosomal particles under these conditions behave like typical polyelectrolytes, similar to isolated RNAs, and their dimensions are expanded due to repulsion of the negatively charged phosphate groups of the rRNAs. It is remarkable that ribosomal proteins remain bound to rRNA in the course of the unfolding. At the same time, their presence on rRNA certainly affects the unfolding process. In particular, in contrast to the unfolding of isolated rRNAs, the unfolding of ribosomal particles proceeds as a stepwise cooperative process. Thus, after removal of the majority of tightly bound Mg^{2+}, decreasing the ionic strength causes abrupt transformation of 30S and 50S subunits into 26S and 35S particles, respectively. The next step is the appearance of 15S and 22S particles. These discrete steps in the ribosome unfolding are thought to reflect the destruction of interdomain interactions and possibly certain types of tertiary RNA–RNA interactions stabilized by ribosomal proteins. These steps are not accompanied by a significant melting of the secondary structure of rRNA. Further unfolding, however, proceeds more gradually; the transition from 15S and 22S particles down to strongly unfolded 5S particles upon depletion of Mg^{2+} and other counter-ions reflects predominantly a successive independent melting of numerous helical regions of rRNA.

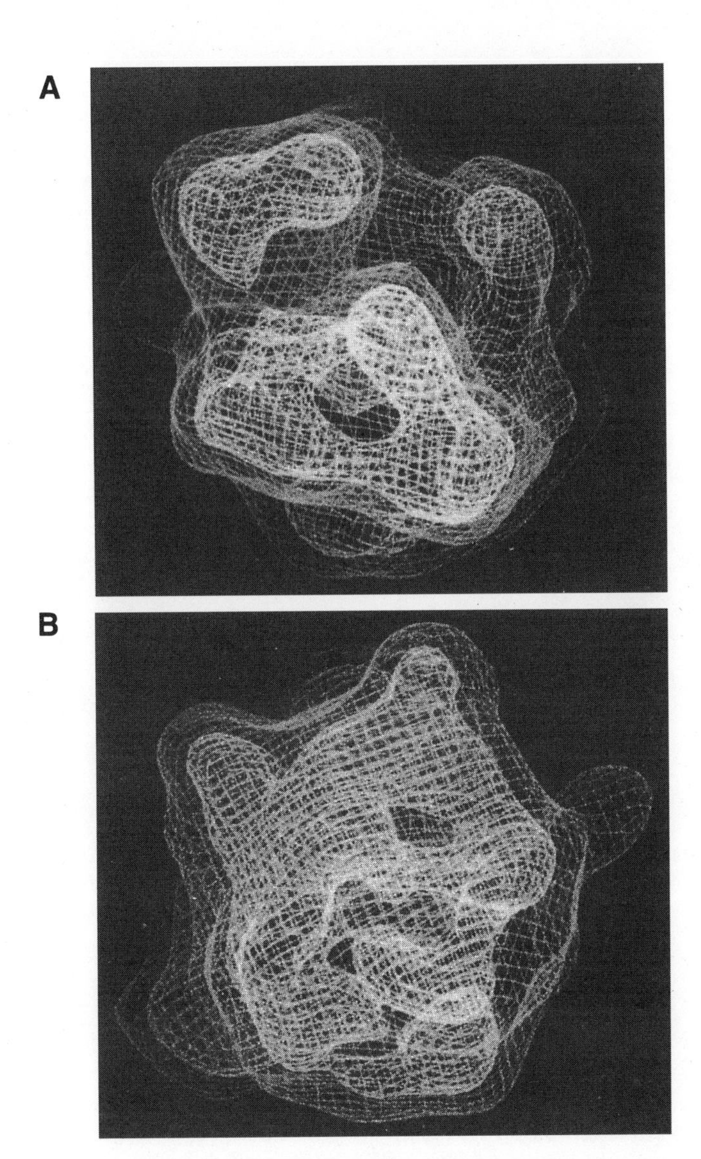

FIG. 6.9. Computer-graphical representation of the three-dimensional reconstruction of the 70S ribosomes. (**A**) View from the solvent side of the 30S subunit. (**B**) View from the solvent side of 50S subunit. Lighter parts of subunit images represent the distribution of rRNA electron densities. Adapted from Frank, J., Penczek, P., Grassucci, R., and Srivastava, S. (1991) *J. Cell Biol.* **115**:597–605.

An interesting conformational change of the *E. coli* 30S subunit was observed at the early stage of depletion of tightly bound Mg^{2+}. Although the small subunits retained a rather compact form (their sedimentation coefficient value was close to 30S) they lost their ability to bind tRNAs and 50S subunits. The inactivation of the 30S subunits was reversible: after heating in solution with appropriate Mg^{2+} concentration they were converted into fully active particles. It was shown by chemical probing that the spatial 16S rRNA structure undergoes specific local alterations within its 3′-domain during the interconversion of active and inactive particles, and several nucleotide residues located at 16S rRNA functional sites change their intramolecular contacts (see, e.g., Ericson and Wollenzein, 1989).

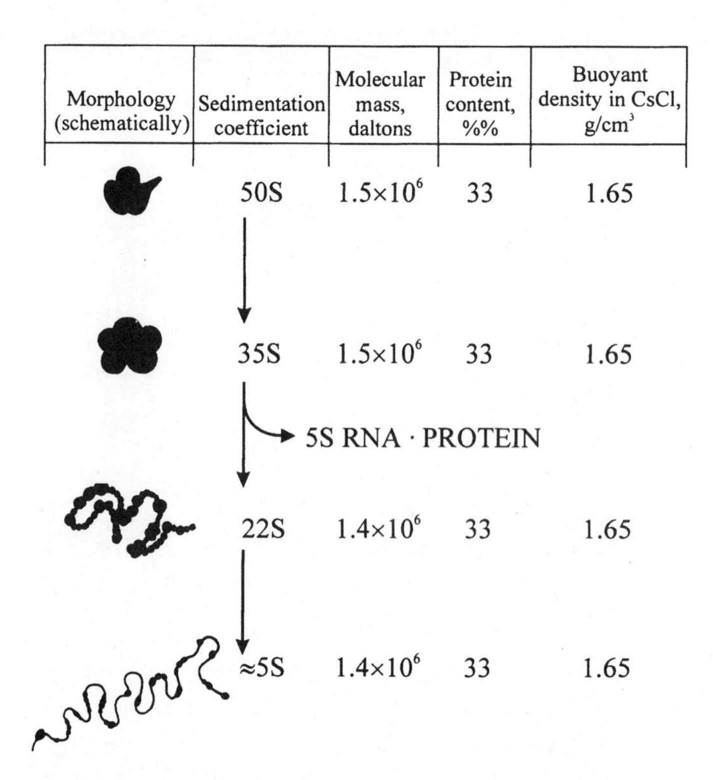

FIG. 6.10. Schematic representation of the process of stepwise unfolding of ribosomal particles (50S subunits) in response to Mg^{2+} depletion and ionic strength reduction. From Spirin, A. S., Kisselev, N. A., Shakulov, R. S., and Bogdanov, A. A. (1963) *Biokhimiya* **28**:920–930; Gavrilova, L. P., Ivanov, D. A., and Spirin, A. S. (1966) *J. Mol. Biol.* **16**:473–489; Gesteland, R. (1966) *J. Mol. Biol.* **18**:356–371.

It is worth noting in this connection that several potential switches in rRNA structure based on formation of alternative tertiary contacts have been discussed in the theoretical plane. In particular, it was suggested that 16S rRNA pseudoknot elements (e.g., the pseudoknot between positions 17–19 and 916–918; see Fig. 6.1) could be disrupted and then formed again in the elongation cycle. From genetic studies it follows that opening of the 16S rRNA pseudoknots strongly affects some ribosome activities. Neither of the alternative pseudoknot structures, however, was supported by direct experimental evidence.

The direct evidence for a conformational switch in the 16S rRNA that affects the decoding process has been obtained recently (Lodmell and Dahlberg, 1997). Using the combination of site-directed mutagenesis and chemical probing it was shown that the interdomain compound hairpin 885–912 of the 16S rRNA in the *E. coli* 30S subunit can exist in two alternative conformations. In one of these conformations the trinucleotide segment C912-U911-C910 is paired with the segment G885-G886-G887 (see Fig. 6.1). In the alternative conformation, positions 912–910 are paired with G888-A889-G890. These short duplexes are located in a very important region of the 16S rRNA molecule where all three major domains meet together. (As mentioned previously, all the 16S rRNA pseudoknots are also located near this region). Switches between the two alternative structures were shown to cause the alterations in conformation of several functional sites of the 30S subunit (rather distant in primary and secondary structure of the 16S RNA) involved in tRNA and mRNA binding.

Generally, switches between alternative conformations in rRNA tertiary structure may take an important part in ribosome functions. They may be relevant to functional switches during the elongation cycle (Section 9.1) and to the dynamic character of the ribosome's work (Section 9.6).

6.4.5. Model of Three-Dimensional Folding of rRNA

There is a vast amount of biochemical data that describe RNA–RNA and RNA–protein contacts in the ribosome (see Sections 7.5 and 7.6), as well as the topography of rRNA-bound proteins and certain nucleotide residues of rRNA on the ribosome surface (Sections 8.2 and 8.3). Numerous nucleotide residues that interact directly with ribosome ligands, such as tRNA, mRNA, protein factors, and antibiotics, have also been identified (Chapter 9). On the other hand, the remarkable progress in cryo-electron microscopy of ribosomes combined with image processing techniques have made it possible to analyze, at 20 to 25 Å resolution, morphological details of their structure comparable in their dimensions with RNA helices (Section 5.5). This encouraged several groups to suggest provisional models of tertiary structure of rRNAs within ribosomal subunits by fitting elements of well-established secondary structure of rRNA into certain elements of electron microscopy ribosome structures. As an example, the recent model of three-dimensional folding of the 16S rRNA within the 30S subunit of *E. coli* ribosome proposed by Brimacombe and co-workers (Fig. 6.11) is considered here.

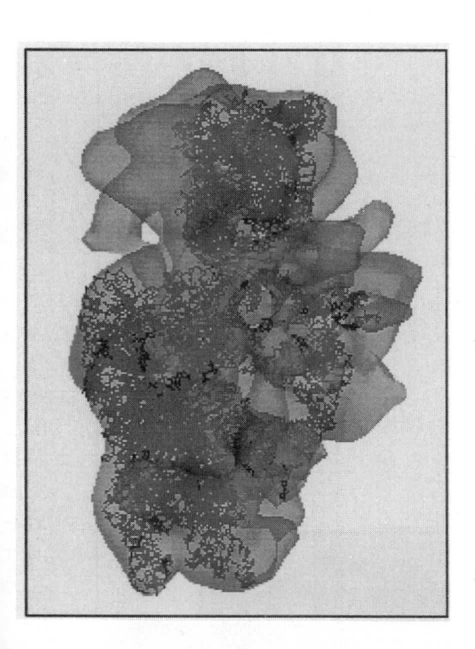
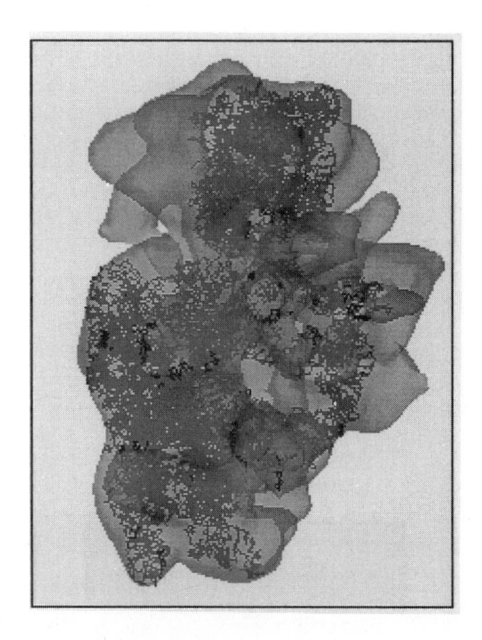

FIG. 6.11. Model of tertiary structure of 16S rRNA in the 30S subunit of the *E. coli* ribosome: Stereo view of the complete 16S rRNA molecule in the 30S subunit, viewed from the interface side of the latter. Adapted from Mueller, F., and Brimacombe, R. (1997) *J. Mol. Biol.* **271**:524–544 (Courtesy of R. Brimacombe and F. Müller, Max Planck Institüt für Moleculare Genetik, Berlin).

In the model, an attempt to solve the problem of matching the biochemical data (obtained at single-nucleotide resolution) with the electron microscopy data (obtained at about ten-nucleotide resolution) was undertaken. The starting point of the modeling process was to fit the elements of the 16S rRNA secondary structure which are involved in the organization of the decoding center (the biochemical data; see Section 9.2) into the "neck" region (the groove separating the 30S subunit head from its body; see Section 5.2) of the high-resolution electron microscopic image of the 30S subunit (Section 5.5) known to accommodate mRNA. The 16S rRNA regions adjoining the decoding area were selected on the basis of intra-RNA and RNA–protein cross-linking data. Then the structure derived was extrapolated to other regions of the 16S rRNA molecule. The electron microscopy contour of the 30S subunit was used as an important set of constrains at each stage of the modeling. This step-by-step process resulted in an arrangement of all helices and single-stranded regions of the 16S rRNA in the ribosomal subunit (Fig. 6.11). The model agrees with most of the biochemical data on the interactions of the 16S rRNA with ribosomal proteins. It looks like an atomic-resolution model, although the authors selected this type of presentation "purely for visual purposes."

References

Boer, P. H., and Gray, M. W. (1988) Scrambled ribosomal gene pieces in *Chlamydomonas reinhardtii* mitochondrial DNA, *Cell* **55**:399–411.

Brimacombe, R., Mitchell, P., Osswald, M., Stade, K., and Bochkariov, D. (1993) Clustering of modified nucleotides at the functional center of bacterial ribosomal RNA, *FASEB J.* **7**:161–167.

Brosius, J., Palmer, M., Kennedy, P. J., and Noller, H. F. (1978) Complete nucleotide sequence of a 16S ribosomal RNA gene from *E. coli, Proc. Natl. Acad. Sci. USA* **75**:4801–4805.

Brownlee, G. G., Sanger, F., and Barell, B. G. (1967) Nucleotide sequence of 5S-ribosomal RNA from *Escherichia coli, Nature* **215**:735–737.

Carbon, P., Ehresmann, C., Ehresmann, B., and Ebel, J. P. (1978) The sequence of *E. coli* ribosomal 16S RNA determined by new rapid gel methods, *FEBS Lett.* **94**:152–156.

Cox, R. A. (1966) The secondary structure of ribosomal RNA in solution, *Biochem. J.* **98**:841–857.

Ericson, G., and Wollenzein, P. (1989) An RNA secondary structure switch between the inactive and active conformations of the *Escherichia coli* 30S ribosomal subunit, *J. Biol. Chem.* **264**:540–545.

Fourmy, D., Recht, M. I., Blanchard, S. C., and Puglisi, J. D. (1996) Structure of the A site of *E. coli* 16S ribosomal RNA complexed with an aminoglycoside antibiotic, *Science* **274**:1367–1371.

Fox, J. W., and Woese, K. (1975) 5-S RNA secondary structure, *Nature* **256**:505–507.

Frank, J., Penczek, P., Grassucci, R., and Srivastava, S. (1991) Three-dimensional reconstruction of the 70S *Escherichia coli* ribosome in ice: The distribution of ribosomal RNA, *J. Cell Biol.* **115**:597–605.

Fresco J. R., Alberts, B. M., and Doty, P. (1960) Some molecular details of secondary structure of ribonucleic acids, *Nature* **188**:98–104.

Heus, H. A., and Pardi, A. (1991) Structural features that give rise to the unusual stability of RNA hairpins containing GNRA loops, *Science* **253**:191–194.

Korn, A. P., Spitnik–Elson, P., and Elson, D. (1983) Specific visualization of ribosomal RNA in the intact ribosome by electron spectroscopic imaging, *Eur. J. Cell Biol.* **31**:334–340.

Kurland, C. G. (1960) Molecular characterization of ribonucleic acids from E. coli ribosomes. I. Isolation and molecular weights, *J. Mol. Biol.* **2**:83–91.

Lodmell, J. S., and Dahlberg, A. E. (1997) A conformational switch in *Escherichia coli* 16S ribosomal RNA during decoding of messenger RNA, *Science* **277**:1262–1267.

Schnare, M. N., and Gray, M. W. (1990) Sixteen discrete RNA components in the cytoplasmic ribosome of *Euglena gracilis, J. Mol. Biol.* **215**:73–83.

Shine, J., and Dalgarno, L. (1974) The 3′-terminal sequence of *Escherichia coli* 16S ribosomal RNA: Complementarity to nonsense triplets and ribosome binding sites, *Proc. Natl. Acad. Sci. USA* **71**:1342–1346.

Spirin, A. S. (1960) On macromolecular structure of native high-polymer ribonucleic acid in solution *J. Mol. Biol.* **2**:436–446.

Spirin, A. S. (1964) *Macromolecular Structure of Ribonucleic Acids.* Reinhold, New York.

Vasiliev, V. D., Serdyuk, I. N., Gudkov, A. T., and Spirin, A. S. (1986) Self-organization of ribosomal RNA, in *Structure, Function, and Genetics of Ribosomes* (Hardesty, B. and Kramer, G., eds.), pp. 129–142, Springer-Verlag, New York.

Woese, C. R., Guttell, R., Gupta, R., and Noller, H. F. (1983) Detailed analysis of high-order structure of 16S-like ribosomal ribonucleic acids, *Microbiol. Rev.* **47**:621–669.

Woese, C. R., Magrum, L. J., Gupta, R., Siegel, R. B., Stahl, D. A., Kop, J., Crawford, N., Brosius, J., Gutell, R., Hogan, J. J., and Noller, H. F. (1980) Secondary structure model for bacterial 16S ribosomal RNA: phylogenetic, enzymatic and chemical evidence, *Nucleic Acids Res.* **8**:2275–2293.

7

Ribosomal Proteins

7.1. Diversity and Nomenclature

Each of the two ribosomal subunits contains many different proteins, most of which are represented by only one copy per ribosome. This is a fundamental difference between the structurally asymmetric ribosomal ribonucleoprotein and the symmetric viral nucleoproteins which are formed by the ordered packaging of many identical protein subunits. The discovery in the pioneering studies of Waller (Waller and Harris 1961; Waller 1964) that the ribosome contains many nonidentical protein molecules established an important principle of the structural organization of ribosomes.

The best technique for analytically separating ribosomal proteins is gel electrophoresis. Even one-dimensional gel electrophoresis under denaturing conditions gives a considerable fractionation of ribosomal proteins by charge and molecular size. Moderately basic polypeptides predominate among ribosomal proteins from most organisms, although several neutral and acidic proteins are always present as well. The molecular masses of ribosomal proteins are usually in the range of 10,000 to 30,000 daltons. Only a few proteins have a greater size, up to about 50,000 or 60,000 daltons (these are two proteins of the large subunit of mammalian ribosomes and one protein of the small subunit of *E. coli* ribosomes, respectively). On the other hand, the large subunit of both prokaryotic and eukaryotic ribosomes contains several (three to six) low-molecular-mass proteins, or polypeptides, of only about 50–60 amino acid residues in length, or even less.

The small (30S) subunit of prokaryotic ribosomes contains about 20 proteins, while there are around 30 in the large (50S) ribosomal subunit. Eukaryotic ribosomes contain a broader spectrum of proteins: the small (40S) subunit contains about 30 proteins, and the large (60S) about 50. Nearly all of these proteins are present as a single copy per ribosome.

A complete analytical resolution of all ribosomal proteins may be achieved by two-dimensional gel electrophoresis under denaturing conditions. For example, 8% polyacrylamide gel at pH 8.6 can be used for electrophoresis in the first direction, and 18% gel at pH 4.6 in the second direction. These conditions lead to almost complete separation of all proteins present in the 30S (Fig. 7.1) or 50S (Fig. 7.2) ribosomal subunits of *E. coli*. The first electrophoretic separation in a less concentrated gel at neutral or slightly alkaline pH results in the migration of acidic and neutral proteins toward the anode (left), while the basic proteins migrate toward the cathode (right); here the separation is largely on the basis of charge. Electrophoresis in the second direction is conducted in a highly cross-linked gel a acidic pH, and all the proteins migrate toward the cathode (downward); in thi case the separation occurs largely on the basis of the molecular size of the com ponents (the smaller the size the greater the mobility).

Separation in the above system provides the basis for the nomenclature of ri bosomal proteins (Kaltschmidt and Wittmann, 1970). It has been proposed that ri bosomal proteins be designated by numbering in a downward direction, as seen from the two-dimensional electrophoretic separation patterns (Figs. 7.1 and 7.2). Proteins of the small ribosomal subunit (30S or 40S) are denoted by the letter S (S1, S2, S3, etc.), while proteins of the large subunit (50S or 60S) are designated by the letter L (L1, L2, L3, etc.). The small *E. coli* ribosomal subunit contains 21 proteins, from S1 to S21. The large ribosomal subunit contains 32 different pro teins, from L1 to L34; the spot initially referred to as L8 is not an individual pro tein but a complex between proteins L10 and L12; the spots designated as L7 and L12 correspond to the same protein, L7 being the N-acetylated derivative of L12. Protein S20 of the small ribosomal subunit is identical to protein L26 of the large subunit. Therefore, there are 52 different ribosomal proteins in the *E. coli* 70S ri bosome.

The acidic L7/L12 protein present in *E. coli* ribosomes (120 amino acid residues, molecular mass 12,200 daltons) is unique in the sense that there are four molecules of this protein per ribosome; it appears to form a tetramer with a mole cular mass of about 50,000 daltons. With this exception, all other proteins in the *E. coli* ribosome seem to be present in a single copy per ribosome. The sizes of *E. coli* ribosomal proteins are given in Table 7.1.

Originally ribosomal proteins of each species had their own nomenclature, ac cording to their own electrophoretic patterns. Thus, the *E. coli* ribosomal proteins

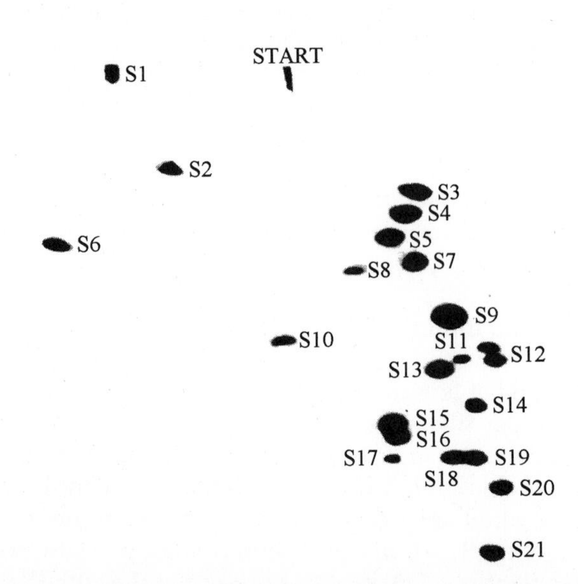

FIG. 7.1.–Two-dimensional electrophoretic separation and nomenclature of proteins from *E. coli* 30S ribosomal subunit. According to Kaltschmidt, E., and Wittmann, H. G. (1970) *Anal. Biochem.* **36**:401–412; *Proc. Natl. Acad. Sci. USA* **67**:1276–1282. First direction (horizontal): 4% polyacrylamide gel containing 8 M urea, pH 8.6; second direction (vertical, downward): 18% polyacrylamide gel contain ing 6 M urea, pH 4.6.

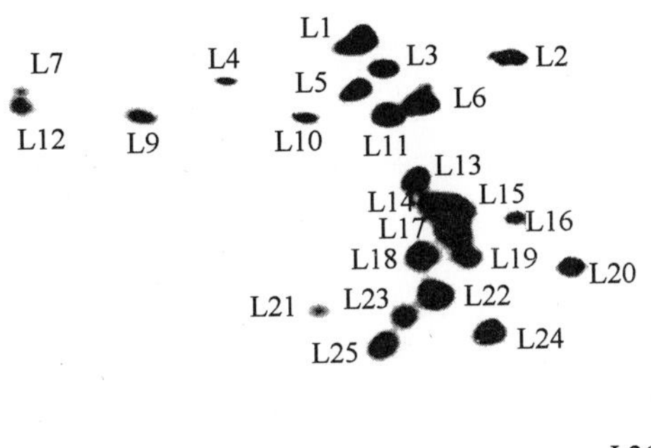

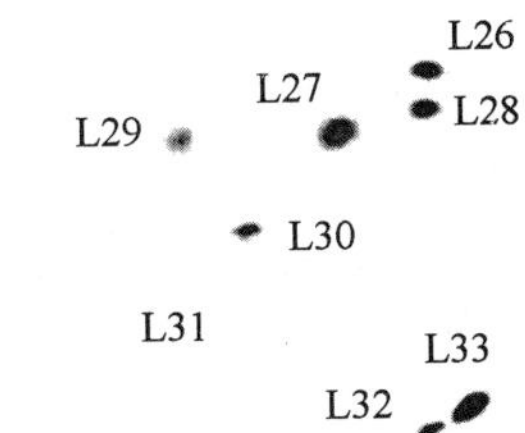

'IG. 7.2. Two-dimensional electrophoretic sepa-ation and nomenclature of proteins from *E. coli* 0S ribosomal subunit (L34 spot is not seen on this lectrophoregram). Reference and separation con-litions are the same as in Fig. 7.1.

vere designated as EcS1, EcS2, . . . EcL1, EcL2, and so on; the ribosomal proteins)f *Bacillus stearothermophilus* as BsS1, BsS2, and so on. Similarly, the ribosomal)roteins of yeast (*Saccharomyces cerevisiae*), the rat, and the human have the pre-ix Y, R, and H, respectively. Naturally, the same protein number may refer to non-analogous proteins of different species. Now, when it becomes clear that most ri-)osomal proteins are evolutionarily conserved, the homology between proteins of lifferent species can be found and thus the relationship of proteins can be estab-ished. This creates the basis for a universal nomenclature. The universal nomen-:lature for ribosomal proteins of bacteria, including eubacteria and archaea, based)n the protein numbers of *E. coli* ribosomes, is already in operation. The univer-al nomenclature for eukaryotic ribosomal proteins based on the protein numbers)f rat or human ribosomes has been also recently introduced. The correlation be-ween some ribosomal proteins of prokaryotes and eukaryotes is given in Table 7.2.

Rat and human ribosomes have 33 proteins in the 40S subunit (from Sa, S2, 33, S3a to S30) and 47 proteins in the 60S subunit (from P0, P1, P2, L3 to L41). As n the case of *E. coli* ribosomes, the acidic proteins of the large subunit, P1 and P2, which are analogs of the bacterial L7/L12, are present in more than one copy per ibosome: both form homodimers, and the two dimers are combined together. It should be mentioned that at least 30 proteins of mammalian ribosomes display some sequence homology with eubacterial ribosomal proteins and thus can be cor-related with corresponding *E. coli* proteins (Table 7.2). RP1 and RP2 are related to EcL12 indeed. Homologs of these acidic proteins of the large subunit are present n ribosomes of all studied organisms, both prokaryotic and eukaryotic ones.

7.2. Primary Structures

The primary structures of all the *E. coli* ribosomal proteins have been deter-mined (Wittmann-Liebold, 1984). A comparison of the primary structures showed

TABLE 7.1. Size of *Escherichia coli* Ribosomal Proteins

Protein	Number of amino acid residues	Molecular mass (kDa)	Protein	Number of amino acid residues	Molecular mass (kDa)
S1	557	61.2	L1	233	24.6
S2	240	26.6	L2	272	29.7
S3	232	25.9	L3	209	22.3
S4	203	23.1	L4	201	22.1
S5	166	17.5	L5	178	20.2
S6	135	15.7	L6	176	18.8
S7	157 (177)	17.1 (19.8)	L7/L12	120	12.2 (four copies
S8	129	14.0	L9	148	15.7
S9	129	14.7	L10	164	17.6
S10	103	11.7	L11	141	14.9
S11	128	13.7	L13	142	16.0
S12	123	13.6	L14	123	13.5
S13	117	13.0	L15	144	15.0
S14	98	11.2	L16	136	15.3
S15	88	10.1	L17	127	14.4
S16	82	9.2	L18	117	12.8
S17	83	9.6	L19	114	13.0
S18	74	8.9	L20	117	13.4
S19	91	10.3	L21	103	11.6
S20	86	9.6	L22	110	12.3
S21	70	8.4	L23	100	11.2
			L24	103	11.2
			L25	94	10.7
			L26 = S20	86	9.6
			L27	84	9.0
			L28	77	8.9
			L29	63	7.3
			L30	58	6.4
			L31	62	7.0
			L32	56	6.3
			L33	54	6.3
			L34	46	5.4

From Wittmann-Liebold, B. (1986) in *Structure, Function, and Genetics of Ribosomes* (B. Hardesty and G. Kramer eds.), p. 331, Springer-Verlag, New York.

a lack of any sequence similarities: each ribosomal protein is unique by the criterion of its amino acid sequence, and no homologies between different ribosomal proteins have been detected. Thus the *E. coli* ribosome contains 52 different amino acid sequences possessing neither common blocks nor homologous regions. A similar conclusion can be made from analyses of amino acid sequences of ribosomal proteins of other eubacteria, archaebacteria, and eukaryotes.

No peculiarities of the primary structures of ribosomal proteins, compared with normal soluble globular proteins, have been found. Most sequences, however, contain a large number of lysine and arginine residues which are sometimes clustered in lysine/arginine-rich blocks; this fact appears to be directly related to the RNA-binding properties of ribosomal proteins and results in the net positive charge of their molecules. Many ribosomal proteins do not contain tryptophan.

Several eukaryotic ribosomal proteins are phosphorylated at serine residues These include the acidic proteins P0, P1, and P2. The phosphorylation is stable that is, unaffected by physiological changes, and necessary for the ribosome assembly and function. Another phosphorylated protein is eukaryotic S6 where several seryl residues are clustered at the C-terminus and can be targets for different kinases in response to different physiological stimuli. This is the case of changeable phosphorylation. The phosphorylation of protein S6 does not seem to be a prerequisite for ribosome assembly or function. As in the case of acetylation of the N-terminal serine of prokaryotic protein L12, the functional significance of this variable modification of protein S6 is unknown.

As already mentioned, the primary structures of ribosomal proteins are very conservative in evolution. Proteins of rat and human ribosomes are either fully identical (32 proteins), or differ just in several amino acid residues per chain, the identity varying from 90 to 100%. This suggests that ribosomal proteins of all mammals are almost identical. There is an extensive homology of amino acid sequences between the equivalent ribosomal proteins of two taxonomically distant groups of eubacteria, Gram-negative *(Escherichia)* and Gram-positive *(Bacillus)*; as a rule, at least 50% of the amino acid residues in the polypeptide chains of corresponding proteins are identical. Similarly, considerable homology of amino acid sequences has been found for ribosomal proteins of various evolutionarily distant eukaryotic organisms; animals, higher plants, and fungi display 40 to 80% sequence identity between the equivalent ribosomal proteins. It is noteworthy that archaebacteria (or archeae) also show certain homology (from 20 to 50% identity) between their ribosomal proteins and the equivalent ribosomal proteins of eukaryotes, and somewhat less homology with the ribosomal proteins of eubacteria. Finally, some homology (from 20 to 30% identity) can be detected between sequences of eubacterial (*E. coli*) and eukaryotic ribosomal proteins, including 15 proteins of the 30S and 16 proteins of the 50S subunit (Table 7.2). Returning to eukaryotic ribosomes, it can be said that at least half of their ribosomal proteins have equivalents or homologues among prokaryotic ribosomal proteins, whereas others may be unique to eukaryotes (see Wool, Chan, and Glueck, 1996).

TABLE 7.2. Correlation (Homology) between Prokaryotic (Eubacteria and Archaea) and Eukaryotic (Fungi and Mammals) Ribosomal Proteins

Escherichia coli *Halobacterium marismortui*	Yeast Rat	*Escherichia coli* *Halobacterium marismortui*	Yeast Rat
Ec or Hm S2	Y or RSa	Ec or Hm L2	Y or RL8
S3	S3	L3	L3
S4	S9	L5	L11
S5	S2	L6	L9
S7	S5	L7/L12	P1, P2
S8	S15	L10	P0
S9	S16	L11	L12
S10	S20	L13	L3
S11	S14	L14	L23
S12	S23	L15	L27a
S13	S18	L18	L5
S14	S29	L22	L17
S15	S13	L23	L23a
S17	S11	L24	L26
S19	S15	L29	L35
		L30	L7

From Wool, I., Chan, Y.-L., and Glück, A. (1996) In *Translational Control* (J.W.B. Hershey, M.B. Mathews, and N. Sonenberg, eds.), pp. 685–732, CSHL Press.

7.3. Three-dimensional Structures

Generally, ribosomal proteins have compact, typical globular conformations with well-developed secondary and tertiary structures (Serdyuk, Zaccai, and Spirin 1978; Ramakrishnan *et al.*, 1981; Nierhaus *et al.*, 1983). The conformations of ribosomal proteins in the ribosome are stabilized by interactions with RNA and other ribosomal proteins. When isolated from the ribosome, many ribosomal proteins are not stable and can be easily denatured. That is why physical studies of isolated ribosomal proteins, and especially their crystallization and X-ray analyses, are performed mostly with the proteins of thermophilic organisms, such as *Bacillus stearothermophilus* and *Thermus thermophilus,* that are characterized by very stable three-dimensional structures.

At the same time, at least some ribosomal proteins may possess noncompact "tails." For example, the *E. coli* protein S6 possesses a strongly acidic C-terminal sequence containing several glutamic acid residues added at the end posttranslationally; it is unlikely that this sequence is included in the globular part of this protein. Protein S7 is also a typical compact globular protein with developed secondary and tertiary structures, but in *E. coli* strain K it has an additional sequence at the C-end, which does not seem to be an indispensible part of the globular structure.

The acidic protein L7/L12 has a number of distinctive features. It has already been pointed out that this protein appears to form a tetramer in the ribosome. In solution it is stable in the dimeric form (Moeller *et al.,* 1972). The dimers appear to be packaged in the tetramer which participates in the formation of a rod-like stalk of the large ribosomal subunit (see Chapter 5). The monomeric subunit of the L7/L12 protein consists of two domains: the globular C-terminal one with about 70 to 80 amino acid residues, and the nonglobular, purely helical N-terminal one with approximately 40 amino acid residues; they are connected by an easily cleavable hinge.

The globular domain of the *E. coli* L7/L12 protein (fragment 47–120) was the first protein element of the ribosome to be crystallized and studied by X-ray analysis. Its three-dimensional structure has been solved at 1.7 Å resolution. The secondary structure of the domain includes α-helices and β-sheets; their sequences along the polypeptide chain is as folows: βααβαβ. The general pattern of folding can be presented as a globule composed of two layers: three α-helices are arranged into one sheet while the antiparallel β-structure consisting of three strands forms the other sheet (Fig. 7.3A).

More recently the whole structure of the L7/L12 dimer was resolved by using NMR. The arrangement of monomers is parallel. Two globular domains are in a weak contact with each other, if any. The monomers are firmly joined together by their N-terminal parts. The two α-helical N-terminal parts form a four-helix antiparallel bundle, as shown in Fig. 7.3B. The N-terminal part of the dimer is responsible for interaction with protein L10 and integration with the ribosome, whereas the globular domains possess a mobility relative to the rest of the ribosome, hanging on flexible hinges. It is not clear yet how the two dimers are arranged in the tetramer.

The two-layer pattern of domain formation similar to that found in the globular domain of protein L7/L12 (Fig. 7.3A) is very typical of other ribosomal proteins (Fig. 7.4). Protein S6 of bacterial ribosomes is a one-domain protein (βαββαβ) where two α-helices lie on a four-strand β-sheet. It is interesting that this folding pattern is identical to those of domain V of EF-G (see Chapter 12, Fig. 12.2) and the nuclear spliceosomal protein U1A of eukaryotes. Protein S5 is a two-domain protein with tightly associated halves where the C-terminal domain (ββαβα) has an exposed β-sheet lying on two α-helices. The folding pattern of S5 has some similarities to that of domain IV of EF-G (see Fig. 12.2). Protein L1 is also a two-domain protein; its RNA-binding domain is composed of two layers where two α-helices lie on four-strand β-sheet. It is believed that the β-sheet, and specifically its middle strands with their basic and aromatic amino acid residues, plays an important part in RNA recognition and binding. Another two-domain protein—S8—has the N-terminal domain (αβαββ) where the three-strand β-sheet lies on the two α-helices and is exposed for interaction with RNA; the C-terminal domain of protein S8 has an unusual, mainly β-sheet conformation.

The five-β-strand barrel structure is another folding motif found among ribosomal proteins. The representative is bacterial protein S17 (Fig. 7.5, left). Protein S17 is an RNA-binding protein and extensively interacts with ribosomal RNA. The same folding motif is known to occur among cytoplasmic mRNA-binding and nuclear hnRNA-binding proteins in eukaryotes. The β-barrel proteins can also be presented as two-layer globules where one β-sheet lies on the other, with an RNA-binding site at one of them. Long flexible loops between the strands are likely participants of RNA binding as well.

On the other hand, fully α-helical proteins, such as protein S15 (Fig. 7.5, right), can occur among RNA-binding ribosomal proteins.

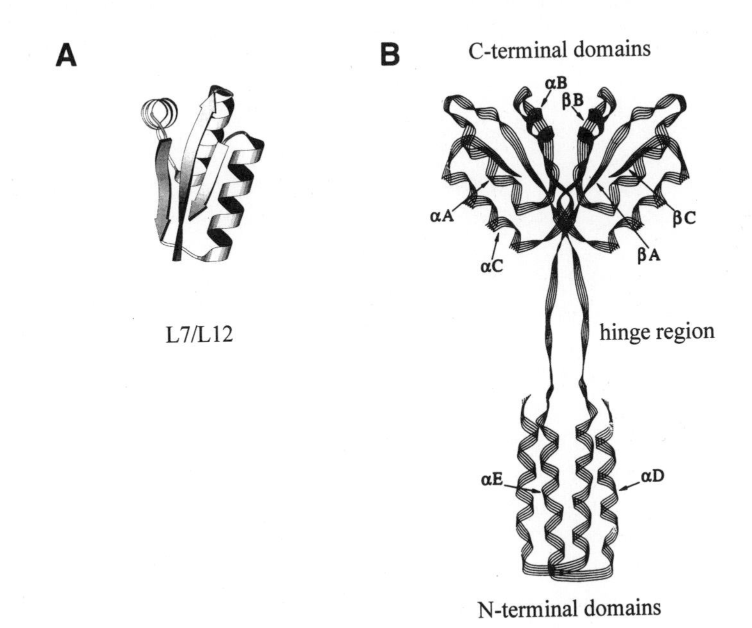

FIG. 7.3. Ribbon diagram of the structure of ribosomal protein L7/L12. (**A**) The structure of the globular C-terminal domain of the protein, as determined by X-ray crystallography. Adapted from Leijonmarck, M., Eriksson, S., and Liljas, A. (1980) *Nature* **286**:824–826; Liljas, A. (1982) *Prog. Biophys. Molec. Biol.* **40**:161–228. (**B**) The L7/L12 dimer structure, as determined by NMR spectroscopy analysis in solution. Reproduced from Bocharov, E. V., Gudkov, A. T., and Arseniev, A. S. (1996) *FEBS Lett.* **379**:291–294 with permission.

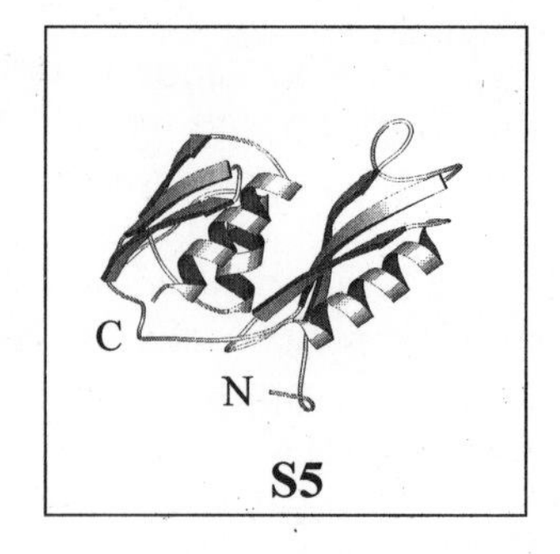

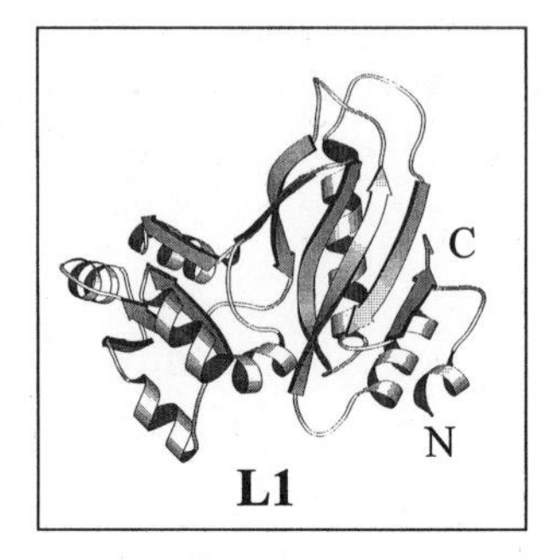

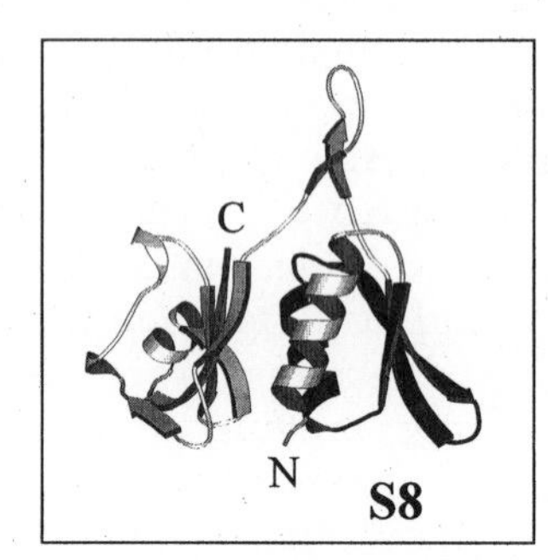

FIG. 7.4. Ribbon diagrams of several ribosomal proteins with globular domains characterized by a two-layer structure: α-helices lying on a β-sheet. **Protein S6** from *Thermus thermophilus.* From Lindahl, M., Svensson, L. A., Liljas, A., Sedelnikiva, S. E., Eliseikina, I. A., Fomenkova, N. P., Nevskaya, N., Nikonov, S. N., Garber, M. B., Muranova, T. A., Rykonova, A. I., and Amons, R. (1994) *EMBO J.* **13**:1249–1254. **Protein S5** from *Bacillus stearothermophilus.* Ramakrishnan, V., and White, S. W. (1992) *Nature* **358**:768–771. **Protein L1** from *Thermus thermophilus.* Nikonov, S., Nevskaya, N., Eliseikina, I., Fomenkova, N., Nikulin, A., Ossina, N., Garber, M., Jonsson, B.-H., Briand, C., Al-Karadaghi, S., Svensson, A., Aevarsson, A., and Liljas, A. *EMBO J.* (1996) **15**:1350–1359. **Protein S8** from *Thermus thermophilus.* From Nevskaya, N., Tishchenko, S., Al-Karadaghi, S., Liljas, A., Ehresmann, B., Ehresmann, C., Garber, M., and Nikonov, S. (1998) *J. Mol. Biol.* **279**:233–244. (Protein S8 from *Bacillus stearothermophilus* has a very similar structure: Davies, C., Ramakrishnan, V., and White, S. W. (1996) *Structure* **4**:1093–1104.

7.4. Protein Complexes

A relatively mild technique for dissociating ribosomal proteins from the ribosome involves treatment with monovalent salts such as CsCl, LiCl, or NH_4Cl at high concentrations. As a result of such treatment, many proteins dissociate in groups rather than as individual molecules. This reflects a certain cooperativity of protein retention within ribosomal subunits. In a number of cases such groups of proteins may be removed from the particles as stable complexes. A pentamer formed by one molecule of protein L10 and the tetramer of protein L7/L12 is an example of such a stable complex in bacterial (*E. coli*) ribosomes. It can be selectively removed from the 50S ribosomal subunit by treatment with a 1:1 mixture of 1 M NH_4Cl and ethanol. As a result, the 50S subunit loses its stalk. Globular structures of both protein L10 and protein L7/L12 within the complex are markedly more stable than in the individual state. The pentameric complex may be loosely associated with yet another protein, L11.

Similarly, the pentameric complex $P0:P1_2:P2_2$ of acidic ribosomal proteins is revealed in eukaryotic ribosomes and can be selectively removed from the 60S subunit resulting in the loss of its stalk (rod-like protuberance).

Another example of the complex is the pair of proteins S6:S18 in *E. coli* ribosomes. Individually, protein S18 is especially unstable and can hardly be isolated in a globular (nondenatured) state. When these proteins form a complex, however, they stabilize each other.

Protein–protein complexes may play an important part in the structure and function of the ribosome, participating in the formation of protein quaternary structures on the ribosomal surface. Generally, ribosomal proteins are not dispersed over the compactly folded ribosomal RNA but rather clustered. For example, four clusters of ribosomal proteins can be mapped on the bacterial small (30S) subunit: S4-S5-S12-S16-S20 at the central part ("body"), S8-S15-S6-S18-S21-S11 on the side bulge, and S7-S9-S13-S19 and S3-S10-S14 on the head. Such clusters may be protein complexes with quaternary structure. At the same time, proteins of different clusters may interact with each other; for example, proteins S5 and S8 have complementary hydrophobic areas on their surfaces that may be responsible for their contact within the ribosomal particle.

7.5. Interactions with Ribosomal RNA

The quaternary structures of ribosomal domains and subdomains are formed with the participation of both ribosomal RNA and ribosomal proteins. Ribosomal proteins are arranged on rRNA and, at the same time, may contribute to the compact folding of rRNA in the ribosome. They are thought to stabilize RNA–RNA tertiary contacts in rRNAs and sometimes induce local alterations in rRNA secondary and tertiary structure. The major role in stabilization and induction of alterations in rRNA structure is attributed to the so-called *primary,* or core rRNA binding proteins. By definition, these proteins interact with rRNAs directly and independently of other ribosomal proteins. Most other ribosomal proteins (if not all of them) also interact with rRNAs, but their binding to rRNA depends on the presence of the primary binding proteins. Thus, in the course of ribosome assembly (see Section 7.6) these proteins stabilize and reorganize local rRNA structures in

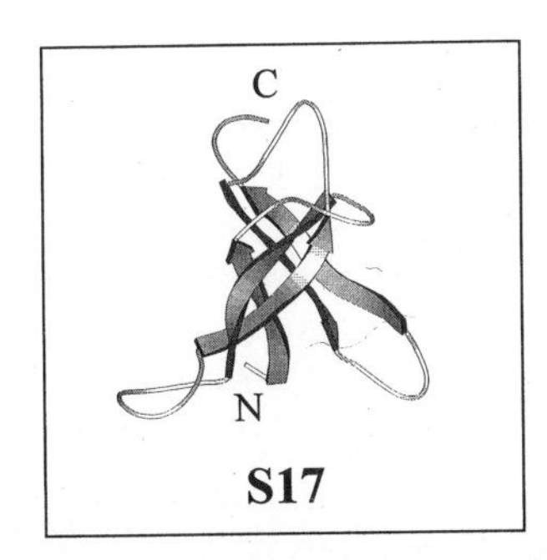

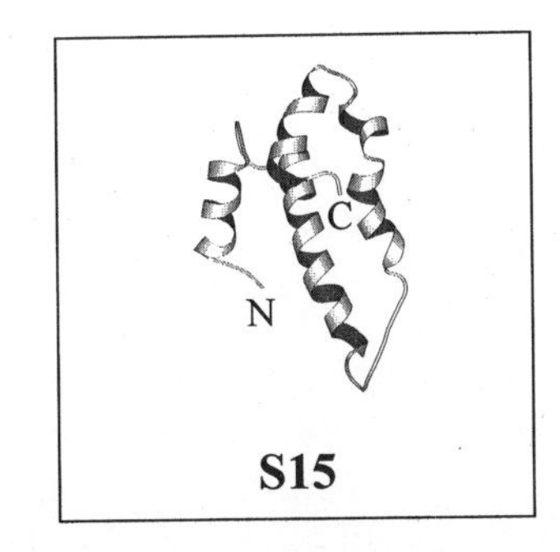

FIG. 7.5. Ribbon diagrams of representatives of β-barrel and α-helical ribosomal proteins: β-Barrel **protein S17** from *Bacillus stearothermophilus.* From Jaishree, T. N., Ramakrishnan, V., and White, S. W. (1996) *Biochemistry* **35**:2845–2853. α-Helical **protein S15** from *Thermus thermophilus.* From Berglund, H., Rak, A., Serganov, A., Garger, M., and Härd, T. (1997) *Nature Struct. Biol.* **4**:20–23.

such a way that new protein binding sites become available (at the same time, the role of protein–protein interactions in this process should not be neglected).

The low-molecular-mass 5S ribosomal RNA may interact with several proteins in the large subunit (Horne and Erdmann, 1972), forming a complex located in the region of the central protuberance (see Chapters 5 and 8). In *E. coli* ribosomes, three proteins—L5, L18, and L25—form a rather stable nucleoprotein complex with the 5S RNA. In *Thermus thermophilus* ribosomes, however, protein L25 is replaced by another protein, TtL5, which is twice as large as EcL25 and has a low sequence homology with EcL25. In eukaryotes only one protein (RL5 or YL3) is shown to form a stable complex with 5S RNA; it seems that the N-terminal part of the protein is homologous to EcL18, and the C-terminal part may be equivalent to EcL5. In any case, the 5S RNA–protein complex looks less conservative than other elements of the ribosome structure.

Of particular interest, of course, are the interactions between the ribosomal proteins and the high-molecular-mass ribosomal RNA (16S and 23S prokaryotic RNAs or 18S and 28S eukaryotic RNAs), since these RNA species serve as the main covalent backbone and the structural core of the ribosomal subunits. As mentioned above, the primary (core) rRNA-binding proteins interact with the corresponding RNA sites on the high-molecular-mass rRNAs, more or less independently of other proteins. In the case of the *E. coli* 30S ribosomal subunit, proteins S4, S7, S8, S15, S17, and S20 are the core proteins that can independently interact with the 16S RNA (Mizushima and Nomura, 1970). Each of these proteins binds only to a specific site on the 16S RNA, recognizing the corresponding nucleotide sequence and three-dimensional structure. The location of the binding sites of the six abovementioned proteins along the 16S chain is shown schematically in Fig. 7.6A. Proteins S4, S17, and S20 are complexed in the region of the 5′-terminal third of the 16S RNA (domain I), proteins S8 and S15 interact with the middle part of the 16S RNA (domain II), while protein S7 has its binding site in the region of the 3′-terminal third of this RNA (domain III).

In the *E. coli* 50S ribosomal subunit, proteins L1, L2, L3, L6, L9, L11, L23, and L24 bind directly and specifically to the 23S rRNA (Fig. 7.6B). The pentameric complex $(L7/L12)_4$:L10 should be added to this list. The latter protects the entire three-way helical structure (positions 1030–1125) of domain II of the 23S rRNA (see Fig. 6.5A), suggesting that this peripheral element of the rRNA together with the protein complex form the lateral protuberance, called the L7/L12 stalk, of the 50S subunit.

For most of the primary rRNA–binding proteins their binding sites on rRNAs have been defined and characterized in detail. At the first stage footprinting approaches (based on both rRNA chemical modification and RNAase limited digestion; see Section 9.2) and cross-linking techniques (based on the application of bifunctional reagents; see Section 8.2.1) were used to define the points of contact between rRNA and proteins (Fig. 7.6). This information helped to construct minimal rRNA fragments containing protein-binding sites and to reconstitute corresponding RNP complexes that were used successfully in physical studies. As an example, the results of studies of the interaction of proteins S8 and S15 with a respective *E. coli* 16S rRNA fragment are considered below.

As mentioned, in the 30S subunit proteins S8 and S15 interact with the central domain of the 16S rRNA. Their rRNA binding sites were located in the adjacent double-helical segments, namely helix 21 (588–604/634–651) for protein S8, and helix 22 (655–672/734–751) for protein S15 (see Figs. 6.1 and 7.7). The pro-

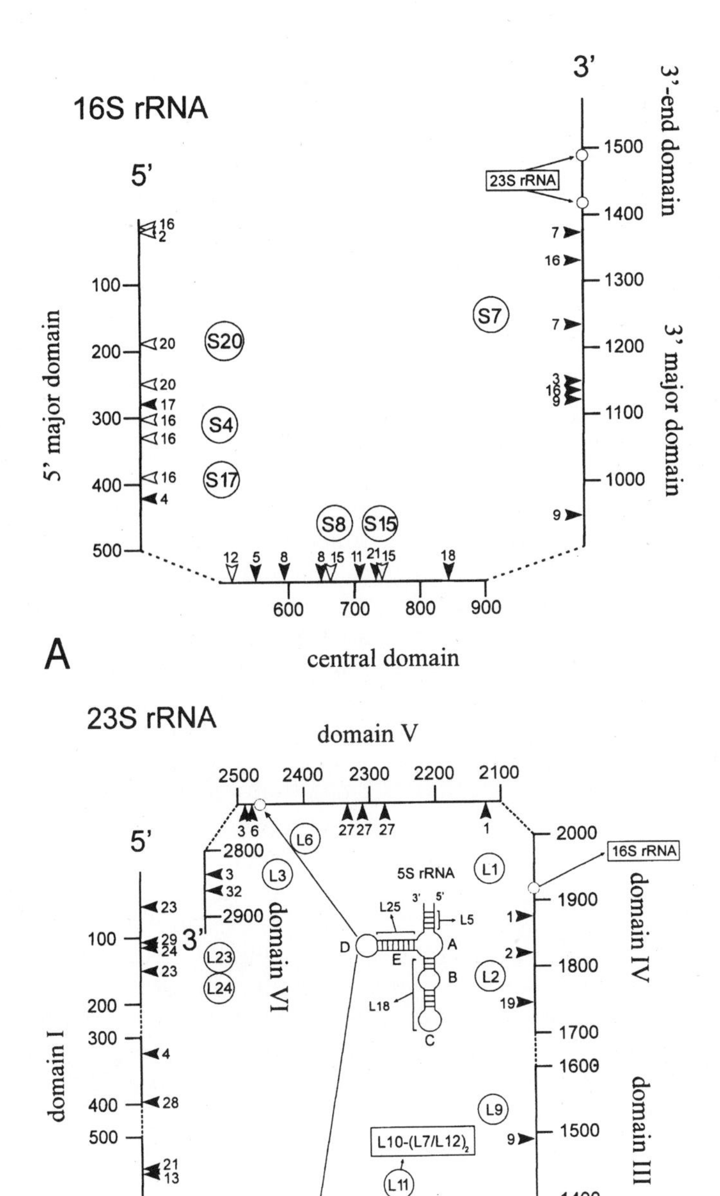

FIG. 7.6. Diagram showing distribution of ribosomal protein binding sites along rRNA chains. (**A**) Diagram of the binding sites of the 30S subunit proteins on the 16S rRNA. Primary binding proteins are circled. Cross-linking sites are indicated by filled arrowheads; centers of protection sites for proteins defined in footprinting experiments are indicated by open arrowheads. Sites of cross-linking of the 23S rRNA to 3′-terminal region of the 16S rRNA are indicated by arrows. Based on the data summarized by Mueller, F. & Brimacombe, R. (1997) *J. Mol. Biol.* **271:**545–565. (**B**) Diagram of the binding sites of the 50S subunit proteins on the 23S rRNA and in the 5S rRNA–protein complex. Primary binding proteins are circled. Protein cross-linking sites are indicated by filled arrowheads. Protein binding sites on the 5S rRNA are marked by lines. RNA–RNA cross-links (23S rRNA to 16S rRNA and 5S rRNA to 23S rRNA) are indicated by long arrows. Based on the data summarized by Brimacombe, R. (1995) *Eur. J. Biochem.* **230:365–383.**

tein S8 binding site contains the highly conservative irregular element 595–598/640–644 flanked with double-helical regions. Protein S8 was cross-linked to A595 in the conservative element, as well as to U653 (Fig. 7.7A). NMR spectroscopy study of the rRNA-binding site for protein S8 has shown that the base triplet A595:A596:U644 is present in the core element (Fig. 7.7B). It has been also proved that U598 is base-paired with A640 in the protein S8–rRNA fragment complex, as shown in Fig 7.7A, and the formation of this base pair is promoted by the protein. The bulged A residues (A595 and A642; see Fig. 7.7A) seem to be directly involved in protein S8 recognition. Indeed, the deletion of A642 strongly decreases the stability of protein S8 complex with 16S rRNA. On the whole, the NMR data suggest that S8–RNA interaction is accomplished without significant changes in the RNA. At the same time, some base pairs (e.g., A596:U644 and G597:C643, as well as U598:A640) are found to be more stable in the RNA–protein complex than in the isolated rRNA fragment.

The studies of the 16S rRNA fragments that represent the protein S15 binding site led to the conclusion that the nucleotide residues responsible for the RNA-protein recognition are located in two regions (marked by heavy lines in Fig. 7.7A): at and near the three-way helical junction, and close to the internal loop A663-G664-A665/G741-G742 (which probably also adapts a double-helical conformation due to formation of two A:G base pairs). The nucleotide residues that are directly involved in the 16S rRNA–protein S15 interaction were revealed by means of site-directed mutagenesis and chemical modification interference. In this approach, selected nucleotide residues in an RNA fragment are either replaced with other nucleotides or modified, for instance, ethylated at their phosphate groups, and then mutations and modifications that interfere with protein binding are determined. It was found that protein S15 recognizes two centers located on the same face of helix 22: the phylogenetically conserved G666:U740 base pair, and the noncanonical G654:G752 and U653:A753 base pairs (see Fig. 7.7A). Structural analysis suggests that protein S15 forms several specific contacts with the hydrogen bond donor and acceptor groups of nucleotide bases situated in the minor groove of the RNA helix. Several electrostatic contacts of protein S15 with phosphate groups surrounding these base pairs enhance the binding of the protein with the 16S rRNA.

Protein S15 causes alterations in the tertiary structure of the central domain of 16S rRNA. As seen from the scheme presented in Fig. 7.7C, helices 22 and 20 in the RNA–protein complex adopt a nearly parallel orientation, and the overall structure of the complex becomes more compact than that of free RNA. The compacting of the RNA fragment structure upon protein S15 binding was convincingly demonstrated by electrophoresis in polyacrylamide gel under nondenaturing conditions: in spite of higher molecular mass and lower net negative charge, the complex displayed a higher electrophoretic mobility than the free RNA fragment.

The mechanism of interactions of other primary rRNA-binding proteins with specific rRNA sites seems to have a great deal in common with that of protein S8–16S rRNA and protein S15–16S rRNA interactions. As a rule, proteins recognize unusual elements of rRNA secondary structure (e.g., noncanonical base pairs or bulged nucleotides) and stabilize them. Protein–RNA interactions can induce a more compact folding of rRNAs and stabilize their tertiary structure. The interactions with rRNA may also affect the ribosomal protein conformation: some elements of protein three-dimensional structure, such as flexible loops, can aquire a specific fixed conformation in the RNA–protein complexes.

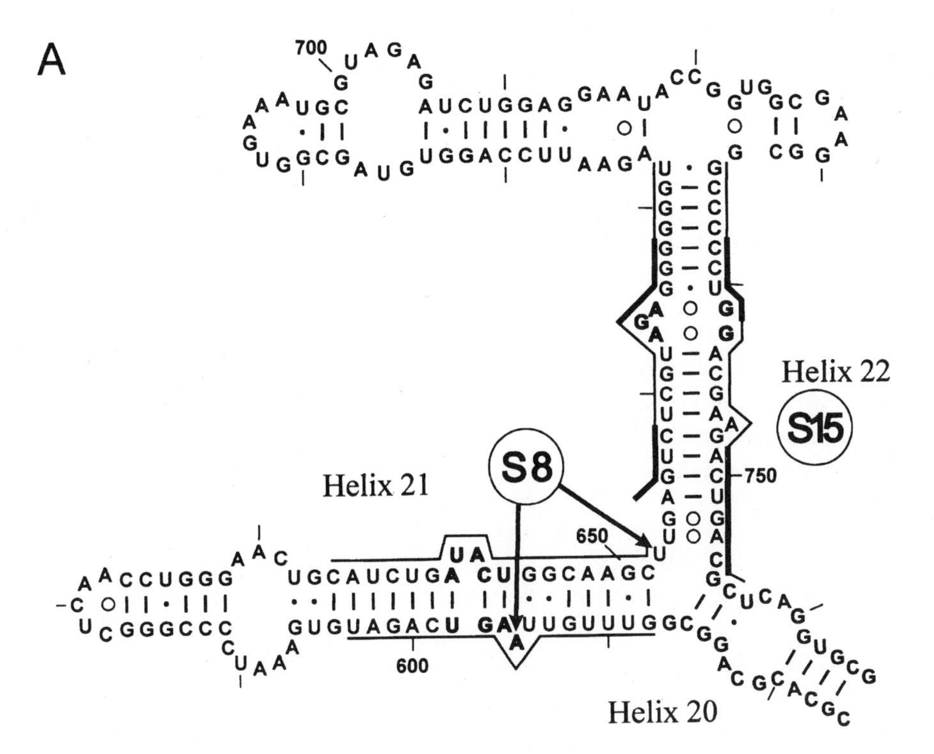

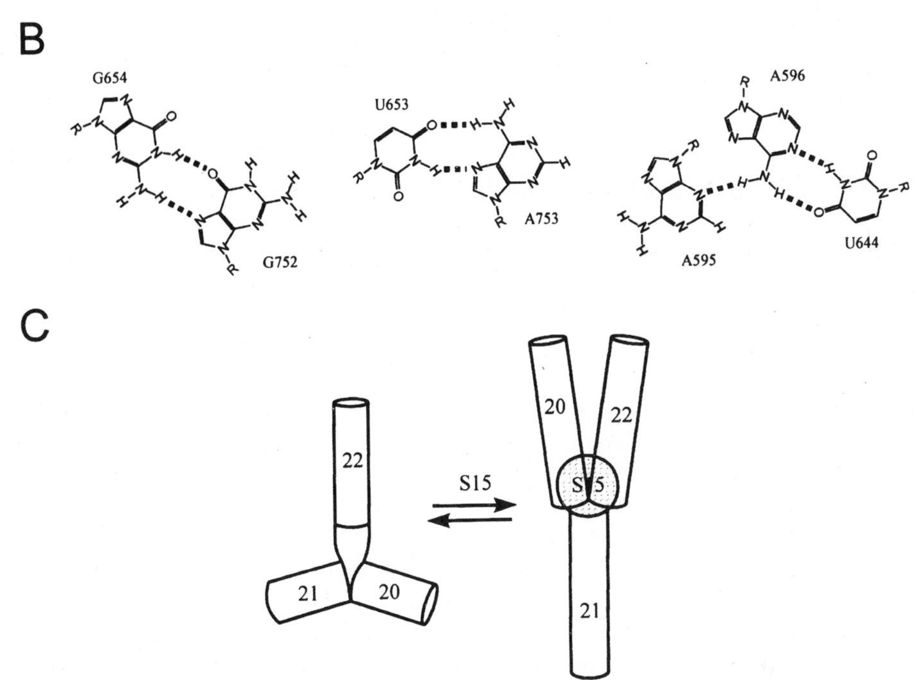

FIG. 7.7. Interaction of *E.coli* ribosomal proteins S8 and S15 with 16S rRNA. (**A**) Secondary structure of the segment of the 16S rRNA central domain that contains the binding sites for proteins S8 and S15. The helices are numbered as in Brimacombe, R. (1995) *Eur. J. Biochem.* **230**:65–87. The sequences protected from chemical modification by proteins are marked by lines. The nucleotide residues most important for the RNA–protein recognition are shown in bold letters. Based on the data summarized by Batley, T. R., and Williamson, J. R. (1996) *J. Mol. Biol.* **261**:536–549. (**B**) Non-Watson–Crick base pairs recognized by proteins S15 and S8. See Batley, R. T., and Williamson, J. R. 1996) *J. Mol. Biol.* **261**:550–567; Kalurachchi, K., Uma, K., Zimmermann, R. A., and Nikonowicz, E. P. (1997). *Proc. Natl. Acad. Sci. USA* **94**:2139–2144. (**C**) Diagram showing a possible rearrangement of the 16S rRNA segment induced by protein S15 binding. Adapted from Batley, R. T., and Williamson, J. R. (1996) *J. Mol. Biol.* **261**:550–567.

Other ("nonprimary") ribosomal proteins interacting with ribosomal RNA require the presence of at least one, even several, core RNA-binding proteins for the formation of sufficiently firm complexes. Two patterns are possible: either an intrinsic interaction between a given protein and RNA is insufficient for the stable complex to be formed, and should therefore be supported by protein–protein interaction with the already bound protein; or, alternatively, the protein bound earlier induces (or stabilizes) the local conformation of the RNA required for binding a given protein. For example, the binding of protein S7 to the *E. coli* 16S RNA contributes to a tighter binding of proteins S9, S13, and S19, as well as of S10 and S14 in the 3′-proximal 16S RNA region (domain III); it may well be that proteins S9 and S13:S19 directly interact with protein S7, while the effect of S7 on the binding of S10 and S14 is less direct.

Thus, both in the small and in the large ribosomal subunits, certain cooperative groups of proteins assigned to definite sites of the three-dimensional ribosomal RNA structure are present. In the case of the *E. coli* 30S ribosomal subunit one such group is formed by proteins of the 3′-proximal domain of 16S RNA (domain III): S7, S9, S13, S19, S10, and S14. It will be demonstrated later that they are all located on the head of the 30S subunit and may contribute to the formation of its tRNA-binding site. Another cooperative group of 30S subunit proteins is associated with the middle domain of the 16S RNA (domain II) and includes the RNA-binding proteins S8 and S15, the S6:S18 pair, as well as S11 and S21; these proteins are found mainly on the side bulge (platform) of the 30S ribosomal subunit. This group, through proteins S5 and S12, is connected with the cooperative group of the 16S RNA 5′-terminal domain (domain I); the latter group includes RNA-binding proteins S4, S17, and S20, as well as proteins S5, S12, and S16, which are constituents of the central body of the 30S ribosomal subunit.

7.6. Disassembly and Reassembly of Ribosomal Subunits

7.6.1. Disassembly

If ribosomal subunits are incubated at a high ionic strength with a sufficiently high Mg^{2+} concentration, the compactness of the subunits is retained, but ribosomal proteins partly dissociate from them. This dissociation is primarily the result of a weaker holding of proteins on the RNA scaffold due to the suppression of their electrostatic interactions. Both during incubation at a high salt concentration and upon a stepwise increase in ionic strength, groups of proteins sequentially split from the particles, resulting in the formation of a series of protein-deficient derivatives. This is a stepwise disassembly of ribosomal particles (Spirin, Belitsina, and Lerman, 1965; Itoh, Otaka, and Osawa, 1968).

The stepwise dissociation of proteins from the *E. coli* 30S subunit by increasing the LiCl or CsCl concentrations is schematically shown in Fig. 7.8. Initially, incubation with high salt results in the release of such relatively loosely bound proteins as S1, S2, S3, S14, and S21; the removal of these proteins yields 28S particles with a protein content of 30% and a buoyant density in CsCl equal to 1.67 g/cm^3. Incubation in a higher salt concentration leads to the splitting off of the next portion of proteins, S5, S9, S10, S12, S13, and S20; the resulting 25S particles contain almost half the initial proteins (20%), and their buoyant density in CsCl is equal to 1.74 g/cm^3. In the range from 3 to 3.5 M LiCl, or during long-term centrifuga-

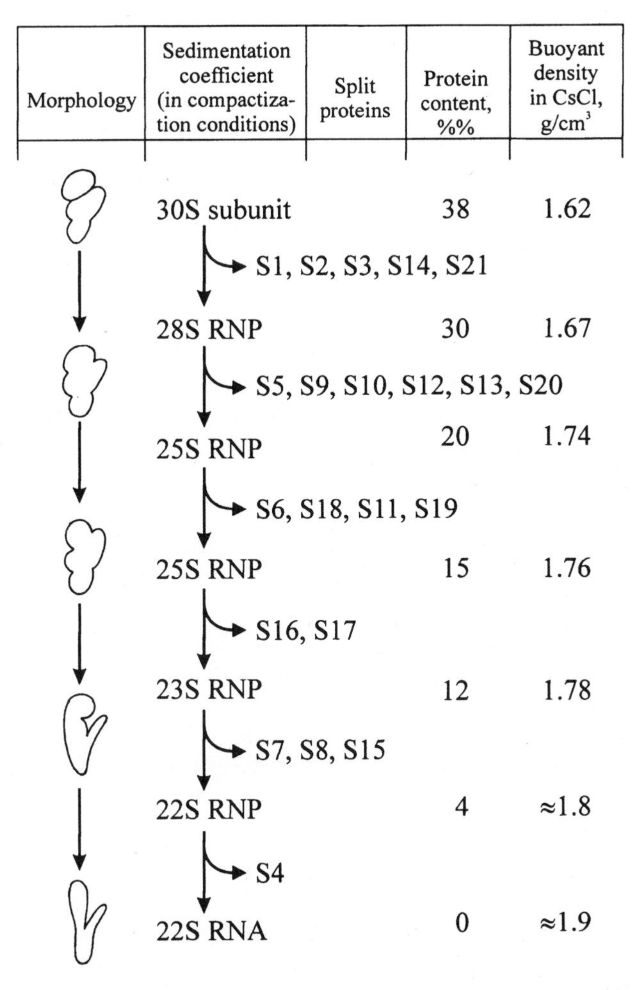

FIG. 7.8. Scheme of the disassembly of the 30S ribosomal subunit achieved by high salt concentration (e.g., by an increased concentration of LiCl in the presence of 5 mM $MgCl_2$; see the text).

tion in 5 M CsCl proteins S6, S18, S11, S19, and then S16 and S17, are released; the residual 23S particles contain just four ribosomal RNA-binding proteins: S4, S7, S8, and S15. The removal of the latter group of proteins from RNA requires more drastic treatment, that is, a combination of high salt and urea.

The dissociation of most of the proteins from the ribosomal particles does not appear to induce disruption of the overall tertiary structure and compactness of the ribosomal RNA. Electron microscopic observations in the course of stripping *E. coli* 30S ribosomal subunits have demonstrated that removing half of all the proteins does not lead to significant morphological changes in the particles; they retain the same size, axial ratio (2:1), and their characteristic subdivision into a head, body, and side bulge. Moreover, morphologically similar particles can be seen after 15 of the 21 ribosomal proteins have been removed. Measuring the compactness of the ribosomal RNA in particles with different protein content using X-ray and neutron scattering confirms the electron microscopic observations; the 16S RNA retaining only six proteins, specifically S4, S7, S8, S15, S16, and S17, maintains the compactness and shape characteristic of RNA within the 30S ribosomal subunit (Vasiliev *et al.*, 1986).

Removal of those core RNA-binding proteins, however, affects the stability of RNA conformation more drastically: as follows from the measurement of the radius of gyration, the compactness of RNA decreases somewhat, corresponding to an increase in the linear size of about one quarter. Nevertheless, free 16S RNA at a sufficient Mg^{2+} concentration and ionic strength, like the 16S RNA carrying the four proteins S4, S7, SS, and S15, is still quite compact and retains its specific overall folding pattern; it can be visualized as a characteristic Y-shaped particle, the contours of which can be inscribed in those of the 30S ribosomal subunit (see Section 6.4.3). This implies that the general pattern of 16S RNA folding is governed and maintained by its internal intramolecular interactions, although the stabilization of the eventual completely folded conformation requires the set of six core RNA-binding proteins.

Other ribosomal proteins may, of course, contribute to the folding and stabilization of ribosomal RNA, but they affect its local structures.

Similar trends may be noted during the stripping of the *E. coli* 50S ribosomal subunit. The 23S RNA retains its initial compactness until the stage when just 9 of the 32 proteins, namely, L2, L3, L4, L13, L17, L20, L21, L22, and L23, remain in the particles. The further removal of proteins leads to a reduction in compactness which, nevertheless, remains reasonably high, and the overall shape of the molecule does not undergo any marked changes.

Thus, the stepwise stripping or disassembly of ribosomal particles clearly demonstrates that high-molecular-mass RNA plays the role of scaffold for the arrangement of ribosomal proteins. The phenomenon of unfolding (see Section 6.4.4) has shown that the RNA chain serves as a covalently continuous backbone of the particle, carrying all ribosomal proteins. The phenomenon of disassembly, during which the basic compactness and shape of RNA remain unchanged, suggests that the RNA tertiary structure forms a three-dimensional scaffold for the proper spatial arrangement of ribosomal proteins.

7.6.2. Reassembly (Reconstitution)

The disassembly is reversible, implying that under proper ionic conditions the ribosomal particles can be reassembled; this includes the recovery of their functional activities (Lerman *et al.*, 1966; Spirin and Belitsina, 1966; Hosokawa, Fujimura, and Nomura, 1966; Staehelin and Meselson, 1966). The reconstitution of bacterial ribosomal particles, both 30S and 50S, can be achieved from isolated ribosomal RNA and the complete set of individual ribosomal proteins (Traub and Nomura, 1968; Nomura and Erdmann, 1970).

Conditions for the reassembly of ribosomal particles of *E. coli* include (1) a moderate ionic strength (0.3 to 0.5); (2) a rather high Mg^{2+} concentration (10 to 30 mM); and (3) an increased temperature (about 40°C to 50°C). A higher ionic strength suppresses interactions between the proteins and RNA, while at a lower ionic strength the contribution of competing nonspecific interactions between basic proteins and the negatively charged polynucleotide increases markedly. The relatively high concentration of Mg^{2+} appears to be necessary primarily for the maintenance of the RNA tertiary and secondary structure which provides the scaffold for the arrangement of proteins. In general, the reconstitution buffer provides conditions under which ribosomal RNA is sufficiently compact in the isolated state and maintains its unique shape. Elevated temperature is believed to be necessary for facilitating the structural rearrangement of an intermediate ribonucleoprotein complex from a less compact to a more compact conformation.

The first proteins to bind to ribosomal RNA in the course of self-assembly are the core proteins which are capable of binding to RNA independently of each other. In the case of *E. coli* 30S subunits, these are proteins S4, S7, S8, S15, S17, and S20. Protein S16 binds along with these. The addition of the six proteins S4, S7, S8, S15, S16, and S17 (step I in the scheme of Fig. 7.9) is a prerequisite for the transition of the intermediate ribonucleoprotein from a less compact to a more compact state (step II in Fig. 7.9). Apparently, it is this transition that requires an elevated temperature during self-assembly. As a result of this transition, the 16S RNA almost reaches the maximal compact state of its overall folding which is characteristic of this RNA within the mature 30S ribosomal subunit.

Proteins S20, S6-S18, S5, S9, S11, S12, S13, and S19 may enter the complex concurrently with the aforementioned proteins, even before the transition of the complex to a more compact state. However, Fig. 7.9, which shows the sequence and interdependence of protein binding in the course of *E. coli* 30S subunit reconstitution, presents the incorporation of these proteins into the ribonucleoprotein as step III of self-assembly, since these proteins, even when bound, are not strictly necessary for the transition into a more compact state and can bind to the complex after this transition. Moreover, as shown in Fig. 7.9, the binding of most proteins at this stage of self-assembly depends on the presence of the set of previously bound proteins. For example, the binding of proteins S9 and S19 requires that protein S7 be bound with RNA. The attachment of proteins S6:S18 depends on the presence of protein S15 and, probably to a lesser extent, of protein S8. The binding of protein S11 requires the presence of proteins S6:S18. The attachment of protein S5 is induced by proteins S8 and S16. (It should be emphasized once again that steps II and III shown in the scheme of Fig. 7.9 are not strictly sequential but appear to proceed concurrently. In other words, the compacting and the binding of nine proteins do not greatly depend on each other and may proceed in parallel; *in vitro,* step II can be accomplished even after step III has been completed.)

Only after the ribonucleoprotein has undergone transition to a compact conformation can the last set of proteins, consisting of S3, S10, S14, S21, as well as S2 and S1, be added to the complex (Fig. 7.9, step IV); this step yields the completed biologically active 30S ribosomal subunit. The incorporation of each of these proteins into the complex requires the presence of proteins bound at previous stages, as well as the final overall folding of the 16S RNA. The binding of protein S10 requires the presence of protein S9, the addition of protein S14 depends on protein S19, protein S3 may become incorporated only if proteins S5 and S10 are present, and the binding of protein S21 is stimulated by the presence of protein S11. The binding of protein S2 is affected by protein S3 and probably by the whole local structure of the ribonucleoprotein. The binding of the largest acidic protein, S1, also requires the correct folding of the ribonucleoprotein; however, it is difficult to determine which specific proteins are necessary for its addition.

An analysis of the complete map of 30S ribosomal subunit reconstitution (Fig. 7.9) demonstrates that the assembly of each structural lobe of the particle proceeds on the corresponding domain of 16S RNA more or less independently. Thus, proteins S4, S16, S17, S20, as well as S12, are assembled on the 5′-terminal domain (I), forming the subunit body. The middle domain (II) binds proteins S8, S15,

S6:S18, as well as S11 and S21, yielding the assembled side bulge of the particle. The 3′-proximal domain (III) with protein S7 incorporates proteins S9, S13, and S19, followed by proteins S10 and S14, and forms the head of the 30S ribosomal subunit. The independence of the assembly of the structural lobes of the 30S subunit has been confirmed in experiments where the isolated RNA fragments representing all three main domains of the 16S rRNA are shown to form compact and specifically shaped ribonucleoprotein particles with corresponding cognate sets of ribosomal proteins (Weitzmann *et al.*, 1993; Samaha *et al.*, 1994; Agalarov *et al.*, 1998). The specific *in vitro* assembly of the 30S subunit fragments equivalent or similar to the main structural lobes of the integral ribosomal particle supports the idea of a large-block organization of ribosomal particles in general.

At the same time, interdomain and interlobe interactions should also receive some attention. The most characteristic cases are the addition of protein S5, which depends simultaneously on domains I and II with proteins contained therein; and

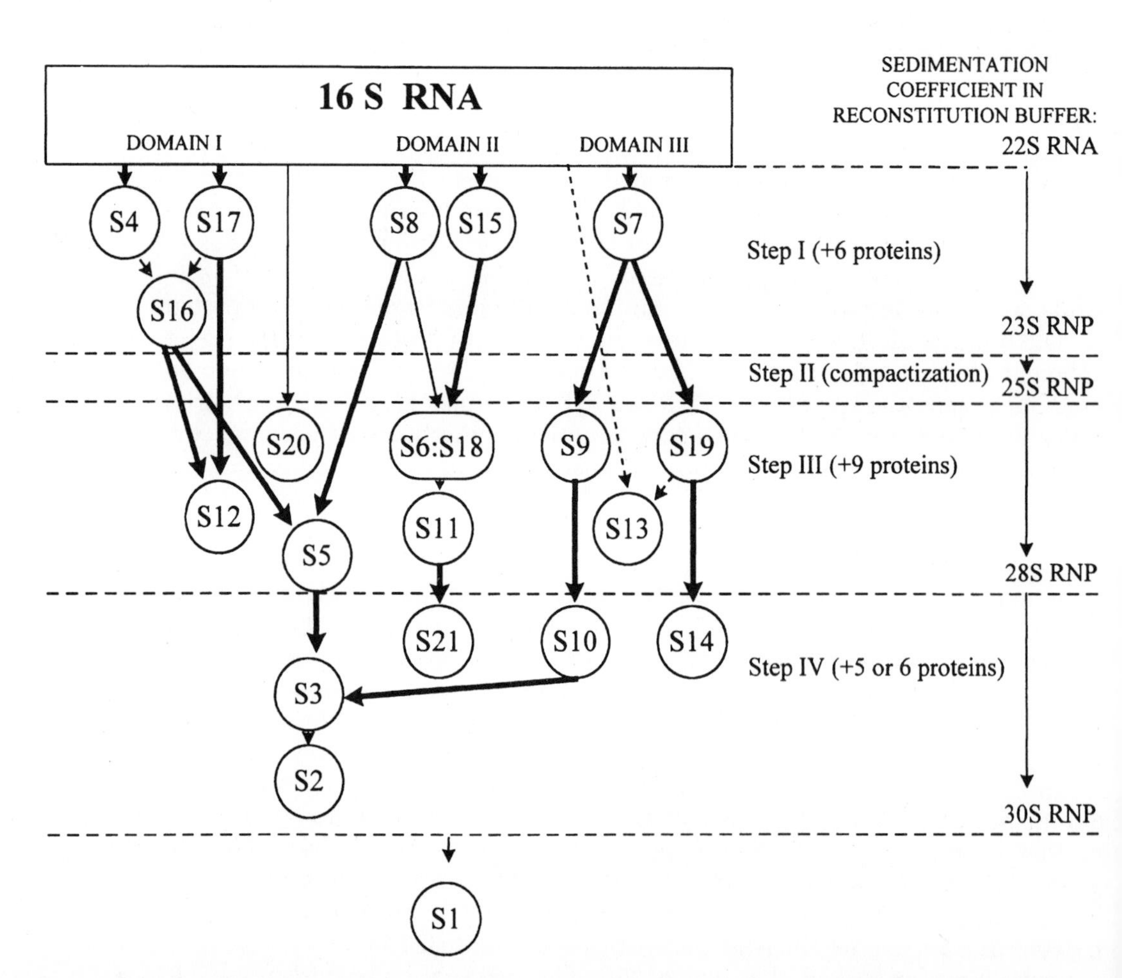

FIG. 7.9. Scheme of self-assembly (reconstitution) of the 30S ribosomal subunit from 16S ribosomal RNA and 21 proteins: "assembly map." Modified from Mizushima, S., and Nomura, M. (1990) *Nature* **226**:1214–1218; Held, W. A., Ballon, B., Mizushima, S., and Nomura, M. (1974) *J. Biol. Chem.* **249**:3103–3111. The thick arrows from the RNA to a protein or from one protein to another symbolize the great dependence of the binding of the subsequent partner on the previous one; the thin arrows indicate a weaker dependence. Some weak interactions are omitted for the sake of clarity.

the attachment of protein S3, which depends on all three domains of RNA and their corresponding proteins (Fig. 7.9). It is likely that protein S5 finds its place somewhere on the boundary between the subunit body and its side bulge, while protein S3 is located at the junction of the head, body, and side bulge of the 30S ribosomal subunit.

A similar analysis of the *E. coli* 50S ribosomal subunit assembly from 23S RNA, 5S RNA, and 32 proteins revealing the interdependence of protein binding and the sequence of stages can also be conducted on the basis of the experimental data available.

There is every reason to assume that the assembly of ribosomes *in vivo* proceeds mainly via the route demonstrated in the course of their reconstitution *in vitro*.

References

Agalarov, S. C., Zheleznyakova, E. N., Selivanova, O. M., Zheleznaya, L. A., Matvienko, N. I., Vasiliev, V. D., and Spirin, A. S. (1998) *In vitro* assembly of a ribonucleoprotein particle corresponding to the platform domain of the 30S ribosomal subunit, *Proc. Natl. Acad. Sci. USA* **95**:999–1005.

Horner, J. R., and Erdmann, V. A. (1972) Isolation and characterization of 5S RNA-protein complexes from *Bacillus stearothermophilus* and *Escherichia coli* ribosomes *Molec. Gen. Genetics* **119**:337–344.

Hosokawa, K., Fujimura, R., and Nomura, M. (1966) Reconstitution of functioning active ribosomes from inactive subparticles and proteins, *Proc. Natl. Acad. Sci. USA* **55**:198–204.

Itoh, T., Otaka, E., and Osawa, S. (1968) Release of ribosomal proteins from *Escherichia coli* ribosomes with high concentrations of lithium chloride, *J. Mol. Biol.* **33**:109–122.

Kaltschmidt, E., and Wittmann, H. G. (1970) Ribosomal proteins. XII. Number of proteins in small and large ribosomal subunits of *E. coli* as determined by two-dimensional gel electrophoresis, *Proc. Natl. Acad. Sci. USA* **67**:1276–1282.

Lerman, M. I., Spirin, A. S., Gavrilova, L. P., and Golov, V. F. (1966) Studies on the structure of ribosomes. II. Stepwise dissociation of protein from ribosomes by caesium chloride and the re-assembly of ribosome-like particles, *J. Mol. Biol.* **15**:268–281.

Mizushima, S., and Nomura, M. (1970) Assembly mapping of 30S ribosomal proteins from *E. coli, Nature* **226**:1214–1218.

Moeller, W., Groene, A., Terhorst, C., and Amons, R. (1972) 50S ribosomal proteins: Purification and partial characterization of two acidic proteins, A_1 and A_2, isolated from 50S ribosomes of *Escherichia coli, Eur. J. Biochem.* **25**:5–12.

Nierhaus, K. H., Lietzke, R., May, R. P., Nowotny, V., Schulze, H., Simpson, K., Wurmbach, P., and Stuhrmann, H. B. (1983) Shape determinations of ribosomal proteins *in situ, Proc. Natl. Acad. Sci. USA* **80**:2889–2893.

Nomura, M. and Erdmann, V. (1970) Reconstitution of 50S ribosomal subunits from dissociated molecular components, *Nature* **228**:744–748.

Ramakrishnan, V. R., Yabuki, S., Sillers, I.-Y., Schindler, D. G., Engelman, D. M., and Moore, P. B. (1981) Positions of proteins S6, S11, and S15 in the 30S ribosomal subunit of *Escherichia coli, J. Mol. Biol.* **153**:739–760.

Samaha, R. R., O'Brien, B., O'Brien, T. W., and Noller, H. F. (1994) Independent *in vitro* assembly of a ribonucleoprotein particle containing the 3′ domain of 16S rRNA, *Proc. Natl. Acad. Sci. USA* **91**:7884–7888.

Serdyuk, I. N., Zaccai, G., and Spirin, A. S. (1978) Globular conformation of some ribosomal proteins in solution, *FEBS Letters* **94**:349–352.

Spirin, A. S., and Belitsina, N. V. (1966) Biological activity of the re-assembled ribosome-like particles, *J. Mol. Biol.* **15**:282–283.

Spirin, A. S., Belitsina, N. V., and Lerman, M. I. (1965) Use of formaldehyde fixation for studies of ribonucleoprotein particles by caesium chloride density-gradient centrifugation, *J. Mol. Biol.* **14**:611–615.

Staehelin, T., and Meselson, M. (1966) *In vitro* recovery of ribosomes and of synthetic activity from synthetically inactive ribosomal subunits, *J. Mol. Biol.* **16**:245–249.

Traub, P., and Nomura M. (1968) Structure and function of *E. coli* ribosomes. V. Reconstitution of functionally active 30S ribosomal particles from RNA and proteins, *Proc. Natl. Acad. Sci. USA* **59**:777–784.

Vasiliev, V. D., Gudkov, A. T., Serdyuk, I. N., and Spirin, A. S. (1986) Self-organization of ribosomal RNA, in *Structure, Function, and Genetics of Ribosomes* (B. Hardesty and G. Kramer, eds.), pp. 128–142, Springer-Verlag, New York.

Waller, J.-P. (1964) Fractionation of the ribosomal protein from *Escherichia coli, J. Mol. Biol.* **10**:319–336.

Waller, J.-P., and Harris, J. I. (1961) Studies on the composition of the proteins from *Escherichia coli* ribosomes, *Proc. Natl. Acad. Sci. USA* **47**:18–23.

Weitzmann, C. J., Cunningham, P. R., Nurse, K., and Ofengand, J. (1993) Chemical evidence for domain assembly of the *Escherichia coli* 30S ribosome, *FASEB J.* **7**:177–180.

Wittmann-Liebold, B. (1984) Primary structure of *Escherichia coli* ribosomal proteins, *Adv. Protein Chem.* **36**:56–78.

Wool, I. G., Chan, Y.-L., and Glueck, A. (1996) Mammalian ribosomes: The structure and the evolution of the proteins, in *Translational Control* (J. W. B. Hershey, M. B. Mathews, and N. Sonenberg, eds.), pp. 685–732, CSHL Press.

8

Mutual Arrangement of Ribosomal RNA and Proteins (Quaternary Structure)

8.1. Peripheral Localization of Proteins on the RNA Core

In contrast to the RNA present in viral nucleoproteins, the RNA of ribosomal particles is not entirely covered by a protein envelope. As was demonstrated many years ago, extended regions of rRNA in the ribosome are exposed to the environment and are open to the action of various agents, such as nucleases. This fundamental difference compared to viral particles is understandable, since the ribosome is a functional structure in which RNA should actively participate in interactions with external factors, and is not used for storing genetic information.

At the same time, protein and rRNA are not just "scrambled" in the ribosome. The high-molecular-mass rRNA of each ribosomal subunit is self-folded into a compact structure with a unique shape (see Section 6.4.3), and it appears that proteins do not associate with the "inside" of this structure. Hence, ribosomal proteins are positioned mainly *on* the compactly folded high-molecular-mass rRNA. This implies that proteins preferentially occupy a position outside the rRNA core.

This principle of ribosomal organization was first deduced from experiments conducted to measure the R_g of ribosomal subunits. The radius of gyration measured by the diffuse small-angle X-ray scattering was found to be markedly lower than expected on the basis of the size of the subunit assuming that it was a uniformly dense body (Serdyuk *et al.*, 1970). It followed from this observation that a more electron-dense component of the particle (e.g., rRNA) lay closer to the center of gravity of the particle, while a less dense component (e.g., protein) tended to be closer to the periphery. Furthermore, measurements of the R_g of ribosomal subunits using different types of radiation (e.g., X-rays, neutrons, and light) demonstrated that the greater the contribution to the total scattering by the protein component compared to RNA (the relative scattering capacity of the protein increases in the series from X-rays to neutrons to light), the greater the value of the particle's R_g (Serdyuk and Grenader, 1975). Finally, neutron scattering experiments in solvents with a different scattering capacity for neutrons (i.e., with different proportions of H_2O and D_2O) allowed for direct measurement of the R_g of either the rRNA or the protein components *in situ* (Stuhrmann *et al.*, 1976). The basis is that H_2O and D_2O are known to differ greatly in their scattering capacity for neutrons, while the scattering capacities of biological macromolecules are intermediate between those of H_2O and D_2O. Because of this, a proportion between H_2O and D_2O in the medium can be selected when the scattering values of a given macromolecule, either protein or RNA, and the solvent are equal, that is, a given type of macromolecule is not "seen" by neutrons or is contrast-matched. Experiments have shown that the neutron scattering of protein is matched by 40 to 42% D_2O, whereas RNA is not "seen" by neutrons in 70% D_2O. Correspondingly, measurements of the R_g of ribosomal subunits in 42% D_2O yield values only for ribosomal RNA *in situ*, while measurements in 70% D_2O give the R_g of the total protein component of the particle. In the case of *E. coli* 50S ribosomal subunits, these values were found to be equal to 65 Å and 100 Å for RNA and protein, respectively (Fig. 8.1). In other words, the RNA is located preferentially at the center as a core, while protein, on average, occupies a more peripheral position. In the case of the 30S ribosomal subunit, this difference is less pronounced—65 Å and 80 Å for RNA and protein, respectively. This smaller difference is understandable since the 16S RNA, despite its lower mass, has a less compact shape or is less isometric than the 23S RNA of the 50S ribosomal subunit. Furthermore, some protein material may be located between the lobes (branches) of the 16S RNA in the 30S ribosomal subunit.

The RNA core in the ribosomal subunit seems to be dense, that is, the extent of RNA folding *in situ* is high. It follows from the value of the R_g and the scattering curve that the volume of RNA in the 50S subunit is equal to only 2×10^6 Å^3. This value is only twice as much as the "dry" volume of RNA. A similar conclusion has been made for 16S RNA in the 30S ribosomal subunit. Therefore, the density of RNA packaging in the ribosomal particle is approximately equal to that found for the crystalline packaging of hydrated RNA helices or tRNA.

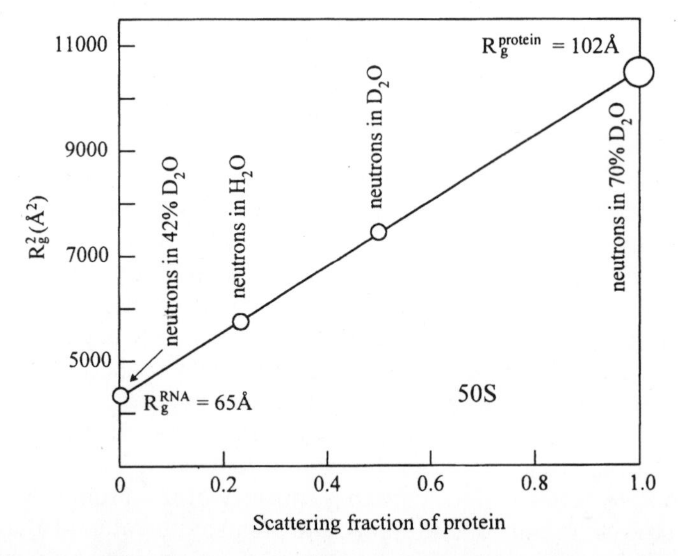

FIG. 8.1. Dependence of the radii of gyration of the *E.coli* 50S ribosomal particles measured by neutron scattering at different contrasts on the relative contribution of the protein component into the scattering. Adapted from Serdyuk, I. N., Grenader, A. K., and Zaccai, G. (1979) *J. Mol. Biol.* **135**:691–707

8.2. Topography of Proteins

After the core position of ribosomal RNA is determined, elucidation of pro-
ein distribution on the surface of the particle, or *protein topography*, becomes the
ext crucial step in determining the quaternary structure of the ribosome. A large
umber of experimental approaches to the study of protein topography have been
leveloped. These approaches will now be discussed using the *E. coli* 30S riboso-
nal subunit as an example.

8.2.1. Identification of Neighboring Proteins

Some information regarding protein neighbors can even be taken from the data
on protein binding sites on the primary and secondary structure of ribosomal RNA
see Section 7.5 and Fig. 7.6). Indeed, if the binding sites of proteins on rRNA are
ocated close to each other, it is clear that these proteins are neighbors in the ribosome. For example, the previously discussed proteins S8 and S15 recognize and
bind adjacent sections of the chain and adjacent hairpins in the secondary struc-
ure of 16S RNA (see Fig. 7.7); therefore it may be concluded that proteins S8 and
S15 are neighbors in the topographic sense as well. Their neighbors are proteins
S6 and S18, which for their binding require the initial binding of proteins S8 and
S15 (Fig. 7.9) and have the recognition sites in the same region of the RNA sequence (Fig. 7.6).

Another example of a group of neighboring proteins includes proteins S4, S16,
S17, and S20 which are located close to each other on the 16S RNA chain within
domain I (Fig. 7.6).

A more universal approach makes use of bifunctional chemical reagents
which are capable of cross-linking neighboring proteins with each other. After
treatment of ribosomal subunits with such reagents, the identification of proteins
in the cross-linked pairs provides the means of establishing that the corresponding proteins are neighbors in the ribosome. Diimidoesters of a different carbon
chain length have been widely used as bifunctional cross-linking agents:

$$\begin{array}{ccc} HN\!\!\searrow & & \nearrow\!\!NH \\ & C-(CH_2)_n-C & \\ H_3C-O\!\!\nearrow & & \searrow\!\!O-CH_3 \end{array}$$

The ester groups of such a reagent are effectively attacked by the ε-amino groups
of lysyl residues present in ribosomal proteins, resulting in the formation of ami-
dine bonds instead of ester bonds. Using reagents of a different length, for example, dimethylsuberimidate ($n = 6$) or dimethyladipimidate ($n = 4$), permits a rough
estimation of the distance between neighboring proteins.

The identification of proteins in cross-linked pairs may present some problems because the corresponding proteins are not in the individual state. One possible solution involves immunological identification of the partners within the
pair without separating them. Another approach makes use of cleavable cross-
links. For example, ribosomal particles may be treated by a sulfhydryl derivative
of the lysine-specific reagent, such as 2-iminothiolane (Traut *et al.*, 1980). It reacts
with protein amino groups, and the subsequent oxidation yields pairs of proteins
cross-linked by disulfide bridges:

$$\text{Protein}'-\text{NH}-\underset{\underset{\text{NH}}{\|}}{\text{C}}-(\text{CH}_2)_3-\text{S}-\text{S}-(\text{CH}_2)_3-\underset{\underset{\text{NH}}{\|}}{\text{C}}-\text{NH}-\text{Protein}''$$

Pairs of proteins cross-linked in this way are isolated, the disulfide bonds reduced, and individual proteins identified electrophoretically.

The summary of the results on protein cross-linking in the 30S ribosomal subunit of *E. coli* is schematically illustrated in Fig. 8.2. Circles connected with lines designate cross-linked proteins; groups of neighboring proteins on RNA chain are boxed.

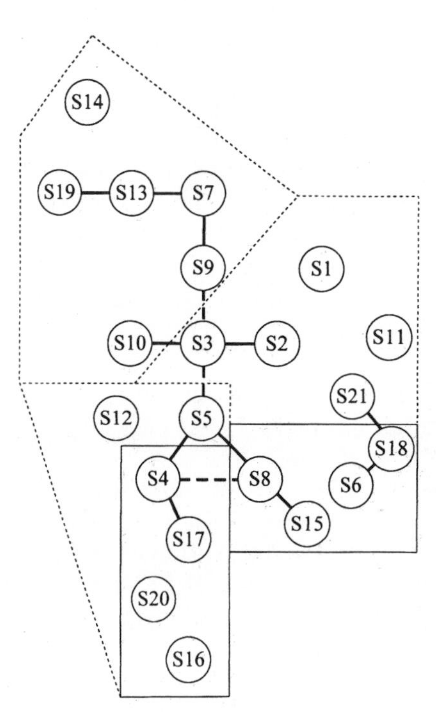

FIG. 8.2. Scheme showing proteins neighboring in the 30S ribosomal subunit. Proteins connected by a solid line are cross-linkable with a short reagent (most direct contact); those connected by a broken line are cross-linkable with longer reagents. The proteins in a solid-line box have adjacent binding sites on the ribosomal RNA sequence. The three groups of proteins within the broken-line boxes correspond to the three RNA domains and the three particle lobes.

8.2.2. Measuring Distances between Proteins and Triangulation

The problem of the mutual arrangement of proteins in the ribosome may be solved even more comprehensively by measuring the distances between the proteins. These approaches are not limited to determining the nearest neighbors. Technically, however, these approaches appear far more complex.

The most informative approach to measuring the distances between ribosomal proteins is based on the use of neutron scattering by ribosomal particles containing selectively deuterated pairs of proteins (Engelman, Moore, and Schoenborn, 1975). Since protonated and deuterated proteins exhibit different neutron scattering, comparing the scattering of correspondingly unlabeled and labeled ribosomal particles allows the contribution of the deuterated pair to be distinguished and used for estimating the distance between mass centers of the two proteins, as well as the degree of asymmetry (or compactness) of each of the proteins *in situ*. In selecting the solvent composition (proportion of H_2O and D_2O) in order to match the scattering of protonated proteins, one can further increase the apparent relative contribution of the deuterated pair. Using measured distances between mass centers of proteins in numerous deuterated pairs, the method of triangulation can be exploited in constructing a model of the three-dimensional arrangement of ribosomal proteins in the *E. coli* 30S subunit (Fig. 8.3). These results provide one of the most accurate and fundamental contributions to our knowledge of the arrangement of proteins in the ribosomal particle.

8.2.3. Immuno-Electron Microscopy

The above approaches provide evidence of the arrangement of proteins with respect to each other but without reference to the morphology of the ribosomal particle. The use of electron microscopy for visualizing proteins on the ribosome allows the location of a protein on a morphologically visible contour of the ribosomal particle to be determined (Wabl, 1974; Lake *et al.*, 1974; Tischendorf, Zeichhardt, and Stoeffler, 1975); combined with the above data, this provides an opportunity for superimposing the entire network of protein topography (Figs. 8.2 and 8.3) on vis-

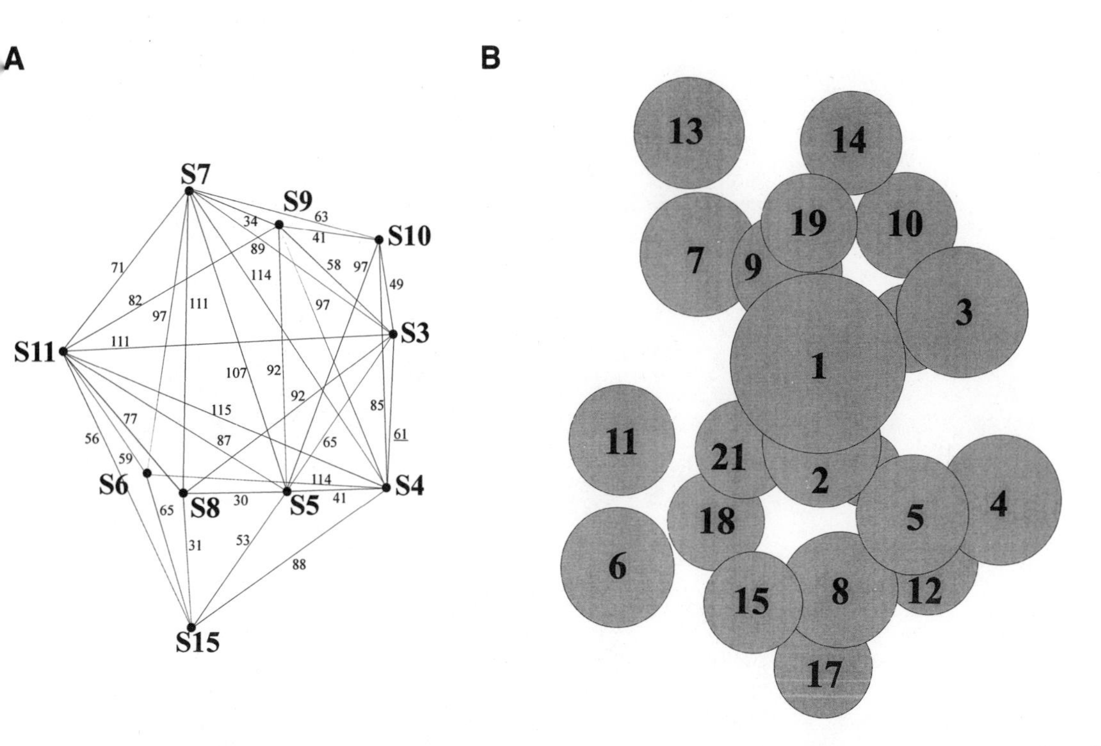

FIG. 8.3. Three-dimensional arrangement of proteins in the 30S ribosomal subunit of *Escherichia coli*, as determined by neutron scattering technique. (**A**) Triangulation of some proteins present in the 30S ribosomal subunit, on the basis of neutron-scattering data. Adapted from Moore, P. B., Langer, J. A., Schoenborn, B. P., and Engelman, D. M. (1977) *J. Mol. Biol.* **112:**199–234; Moore, P. B., Capel, M., Kjeldgaard, M., and Engelman, D. M. in *Structure, Function, and Genetics of Ribosomes* (B. Hardesty and G. Kramer, eds.), pp. 87–100, Springer-Verlag, New York. Figures at the lines connecting the protein positions indicate the distances measured between the protein mass centers, in angstroms (not all pairwise distances are given). (**B**) Map of three-dimensional disposition of proteins in the 30S subunit deduced from triangulation data. From Capel, M. S., Engelman, D. M., Freeborn, B. R., Kjeldgaard, M., Langer, J. A., Ramakrishnan, V., Schindler, D. G., Schneider, D. K., Schoenborn, B. P., Sillers, I. Y., Yabuki, S., and Moore, P. B. (1987) *Science* **238:**1403–1406. Proteins are approximated by spheres whose volumes correspond to the volume occupied by the corresponding anhydrous protein.

ible projections of the particle. Electron microscopic visualization of proteins on the ribosome makes use of specific antibodies against individual ribosomal proteins. The bivalent antibody bound to a given protein may interact with two identical ribosomal particles, yielding their dimer through the bridge of the antibody molecule. By observing dimers with an electron microscope, one may identify sites on the surface responsible for the joining; these sites correspond to the localization site of a given protein on the surface. In a number of cases, provided the resolution is sufficiently high, one can directly see the attachment of the Y-shaped antibody molecule to a certain region on the ribosomal surface. Using this approach, it has been established that protein L7/L12 forms the lateral rod-like stalk of the 50S ribosomal subunit, protein L1 is located in another lateral protuberance (side lobe) of the 50S subunit, and the 5S RNA–protein complex is in the central protuberance, or head, of the 50S subunit (Fig. 8.4).

Great efforts have been made to localize all of the proteins of the *E. coli* 30S ribosomal subunit. Despite the feasibility of obtaining specific antibodies against each of the 21 individual proteins, the task was far from simple and the technique yielded many false localizations. It should be pointed out that this method, which appears so direct and illustrative, may result in artifactual information due to the

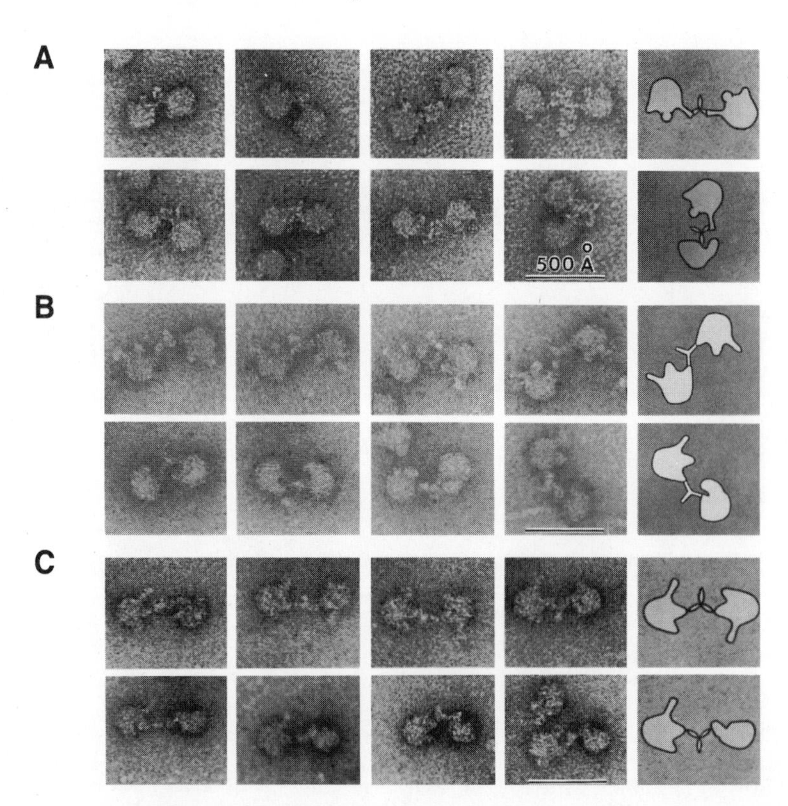

FIG. 8.4. Electron micrographs of 50S ribosomal subunits reacted with antibodies. (**A**) Antibodies against protein L7/L12. From Strycharz, W. A., Nomura, M., and Lake, J. A. (1978) *J. Mol. Biol.* **126:**123–140. Original photo was provided by J. A. Lake. (**B**) Antibodies against protein L1. From Dabbs, R. Ehrlich, R., Hasenbank, R., Schroeter, B. H., Stöffler-Meilicke, M., and Stöffler, G. (1981) *J. Mol. Biol.* **149:**553–578. Original photo provided by G. Stoffler. (**C**) Antibodies reacted with the 5S RNA–protein complex. The 50S particles are viewed from their convex ("back") side. From Shatsky, I. N., Evstafieva, A. G., Bystrova, T. F., Bogdanov, A. A., and Vasiliev, V. D. (1980) *FEBS Lett.* **121:**97–100. Original photo provided by V. D. Vasiliev.

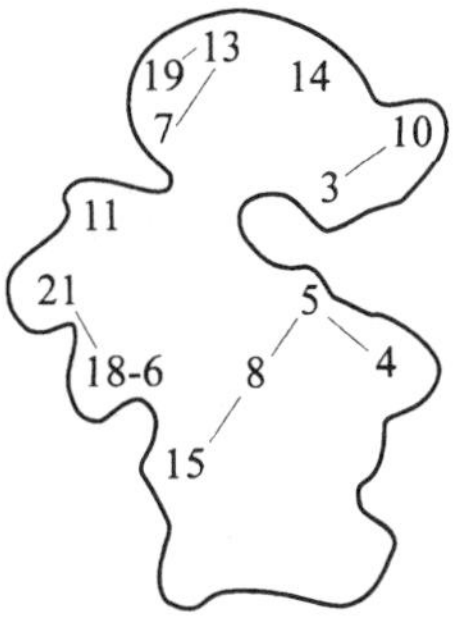

FIG. 8.5. Contour of the 30S ribosomal subunit, according to van Heel, M., and Stöffler-Meilicke, M. (1985) *EMBO J.* **4**:2389–2395, with positions of some proteins localized by immuno-electron microscopy. See Stöffler-Meilicke, M., and Stöffler, G. (1990) in *The Ribosome: Structure, Function and Evolution* (W. E. Hill, A. Dahlberg, R. A. Garrett, P. B. Moore, D. Schlessinger, and J. Warner, eds.), pp.123–133, ASM Press, Washington, DC. Cross-linkable proteins are connected by lines.

insufficient purity of antibodies, the nonspecific binding of antibodies to certain regions of the ribosomal surface, distortion of the specific position of the antibody molecule on the ribosome caused by the orientation of the ribosomal particle on the substrate, and other factors. Nevertheless, some reliable results have been obtained. They are schematically summarized in Fig. 8.5. Generally, it was demonstrated that proteins S3, S7, S10, S13, S14, and S19 are localized on the head of the 30S subunit. In more detail, proteins S13, S14 and S19 were detected at the top position of the head, whereas protein S3 and S7 were located below this group of proteins, near the groove separating the head from the body, but on two opposite sides. Protein S5 was localized even lower, also close to the groove but on the body of the subunit. Proteins S6 and S11 were localized on the other side of the 30S subunit, that is, on its side bulge or platform. Protein S8, according to the data provided by immuno-electron microscopy, is also located near the side bulge, somewhere between the bulge and the body.

8.2.4. Exposure of Proteins on the Ribosome Surface

The accessibility of antigenic determinants of many ribosomal proteins to antibodies does not mean that all the proteins are well exposed on the ribosome surface. An experimental approach to estimate the degree of the surface exposure of different ribosomal proteins can be based on a technique of labeling just the surface of a big molecular complex. Such a technique using thermally activated tritium atoms was developed (Shishkov *et al.*, 1976) and successfully applied for studies of the surfaces of multimeric protein complexes, viruses, membranes, and ribosomes. The principle of the technique is that high-energy tritium atoms are produced by dissociation of tritium gas ($^3H_2$) on a heated tungsten wire, and the bombardment of biological molecules by these atoms results in the replacement of surface hydrogens by tritium in covalent bonds, including C–H bonds. In this way only the surface of a molecule exposed to tritium atom flow becomes tritium-labeled.

The analysis of the surface of the *E. coli* ribosomes with the hot tritium bombardment technique (Yusupov and Spirin, 1988) demonstrates that ribosomal proteins can be divided into three groups: well exposed, fairly exposed, and buried or weakly exposed proteins. The best exposed proteins are S1, S4, S5, S7, S18, S20, and S21 on the 30S subunit, and L7/L12, L9, L10, L11, L16, L17, L24, and L27 on the 50S subunit (Fig. 8.6). The buried proteins are S8, S10, S12, S16, and S17 in the 30S subunit of the 70S ribosome, and L14, L20, L29, L30, L31, L32, L33, and L34 in the 50S subunit. It is interesting that the association of the two ribosomal

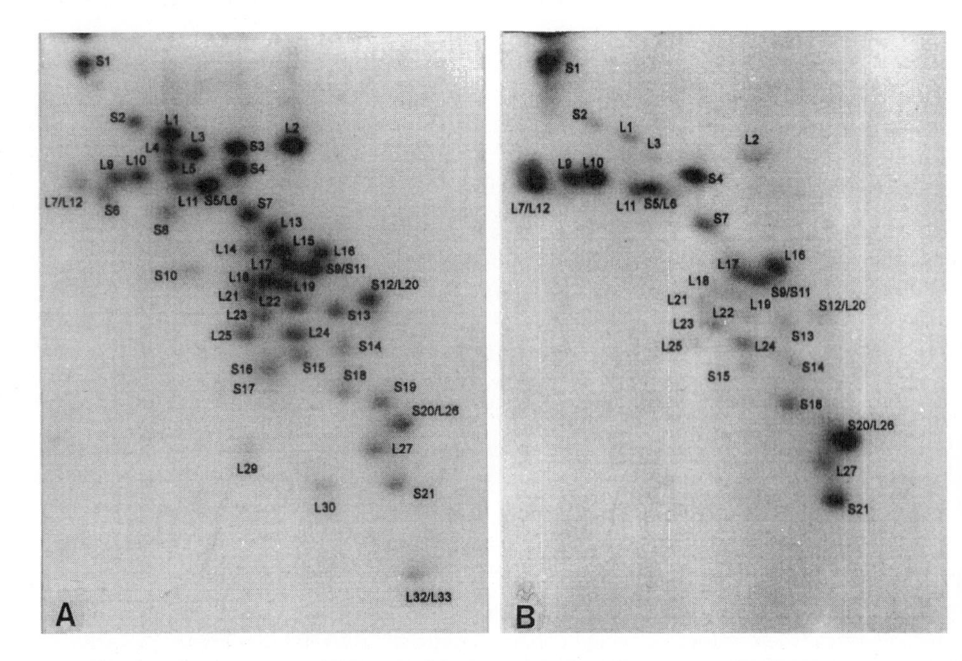

FIG. 8.6. Two-dimensional electrophoresis map of the ribosomal proteins exposed on the surface of th *E. coli* ribosome, as compared with the full set of the ribosomal proteins. From Agafonov, D. E., Kolb V. A., and Spirin, A. S. (1997) *Proc. Natl. Acad. Sci. USA* **94**:12892–12897. (**A**) Full set of the 70S ri bosome proteins separated by two-dimensional gel electrophoresis: Coomassie-stained gel. (**B**) The pro teins exposed to the tritium bombardment of the 70S ribosome surface: fluorogram of the ^{3}H-labele proteins on the same gel.

subunits into the 70S ribosome does not lead to shielding of any exposed proteins strongly suggesting that the subunits associate only by their RNA surfaces. Th topography of ribosomal proteins of the 30S subunit with the exposure data take into account is shown schematically in Fig. 8.7.

Considering the protein clusters of the ribosomal particles, some informatio about orientation of a cluster relative to the surface can be obtained. The protei clusters of the 30S subunit (see Section 8.2) with the well-exposed proteins writ ten in bold and buried proteins in italic are:
S4-S5-S20-*S12-S16* at the central part ("body"); **S18-S21**-S6-S11-S15-*S8* on th side bulge; and **S7**-S9-S13-S19 and S3-S14-*S10* on the head.

8.3. Topography of RNA

8.3.1. Assignment to Protein Topography

Data regarding protein topography and protein-binding sites on the primar and secondary structure of ribosomal RNA allow the approximate topography o the protein-binding regions of rRNA on the ribosomal particle to be deduced. Thi information has helped researchers to assign the rRNA domains to certain mor phological parts of ribosomal subunits.

In discussing general aspects of topography of the 16S RNA main domains and their correspondence to the main morphological lobes of the 30S ribosomal subunit (i.e., the body, side bulge, and head), one can use available data about mapping proteins on rRNA and on the 30S subunit. These data are as follows: (1) Proteins S4, S16, S17, and S20 are bound to the 5′-terminal domain, and at the same time are revealed on the body of the 30S subunit. (2) Proteins S8, S15, S6, and S18 interact with the middle domain of 16S RNA, and on the morphological image of the 30S subunit they are located either directly on the side bulge (platform) or on the line of contact between the side bulge and the body. (3) A group of proteins including S7, S9, S10, S13, S14, and S19 is attached to 16S RNA in the region of its 3′-proximal major domain, and all these proteins are found in the head of the 30S subunit (see Figs. 7.6A and 8.5). It can be deduced from this evidence that the three main structural domains of 16S RNA generally correspond to the three main morphologically visible lobes of the 30S ribosomal subunit. Thus, the 5′-terminal domain (I) forms the core of the subunit body, the middle domain (II) contributes to the formation of the side bulge or platform, and the 3′-proximal domain (III) fills the head of the subunit. The extreme 3′-terminal region (minor domain) of 16S RNA seems to protrude from the head base, or "neck," to the tip of the side bulge or platform, as evidenced by the immuno-electron microscopy data on the mapping of the 3′-end and 3′-terminal hairpin (see the next section).

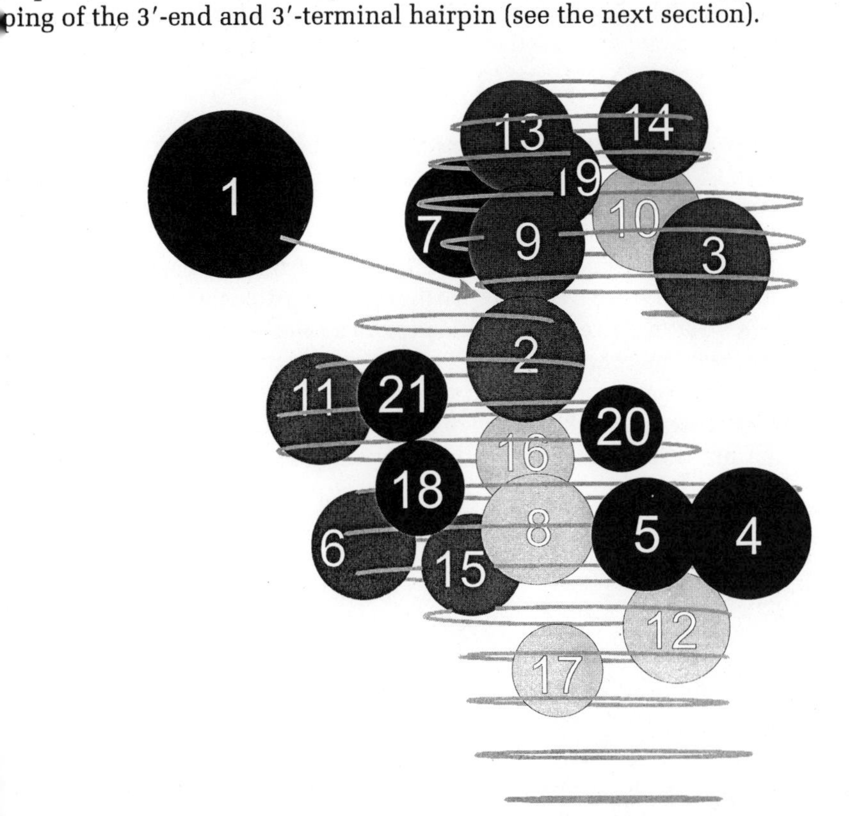

FIG. 8.7. Modified model of the three-dimensional arrangement of the 30S ribosomal subunit proteins by P.B. Moore *et al.* (see Fig. 8.3B) demonstrating the proteins well exposed on the surface (black spheres), moderately exposed (dark shaded spheres) and unexposed (light shaded spheres), as determined by hot tritium bombardment technique (see the text and Fig. 8.6). The model is fitted into the electron microscopic contour of the subunit (see Fig. 8.5). From Spirin, A. S., Agafonov, D. E., Kolb, V. A., and Kommer, A. (1996) *Biochemistry (Moscow)* **61**:1366–1368.

For the large subunit of the *E. coli* ribosome it has been inferred that domains I and III of the 23S rRNA occupy the lateral part of the subunit below the protein L1 site; domain II is located at the base of the L7/L12 stalk; and domains IV and V reside at the base of the central protuberance (see the morphological model of the 50S subunit in Fig. 5.8C and the domain structure of the 23S rRNA in Fig. 6.5). Taking into account the universal character of the ribosome three-dimensional structure one can suggest that all these conclusions are equally applicable to ribosomes from other sources.

More detailed assignments can also be made in some cases. For example, since protein S7 is known to bind to the region comprising the sequences 935–950, 1235–1255, and 1285–1380, the corresponding cluster of five helices (internal helices 938–943/1340–1345 and 945–955/1225–1236, and hairpins 1241–1296, 1303–1334, and 1350–1372 in Fig. 6.1) should be positioned in the head of the 30S subunit, near the "neck," on the side of the subunit bulge (see Figs. 5.5 and 8.5). According to the position of protein S8 which is bound to hairpin 588–651 (see Fig. 6.1), this helix is located on the border between the bulge and the body, near the central part of the 30S subunit (Fig. 8.5). At the same time, the end of hairpin 673–717 of the same central rRNA domain (see Fig. 6.1) known to bind protein S11 should be placed at the upper part of the side bulge of the subunit (Fig. 8.5). A characteristic multi-hairpin node 400–550 (see Fig. 6.1) that is involved in protein S4 binding has to be positioned on the extremity of the opposite side of the 30S subunit (Fig. 8.5).

8.3.2. Immuno-Electron Microscopy Data

By using either antibodies against naturally modified (minor) bases of rRNA, or against haptens such as dinitrophenyl or carbohydrate groups artificially linked to selected sites of rRNA, one may employ immuno-electron microscopy to study the topography of rRNA on the surface of the ribosomal particles. In the latter case, the approach consists of the chemical modification of a selected nucleotide residue in rRNA with a hapten, the reconstitution of ribosomal subunits from the modified rRNA and total ribosomal protein, and the localization of the modified site of rRNA by electron microscopy with the use of hapten-specific antibodies (Fig. 8.8).

This approach provided the successful localization of rRNA termini that are

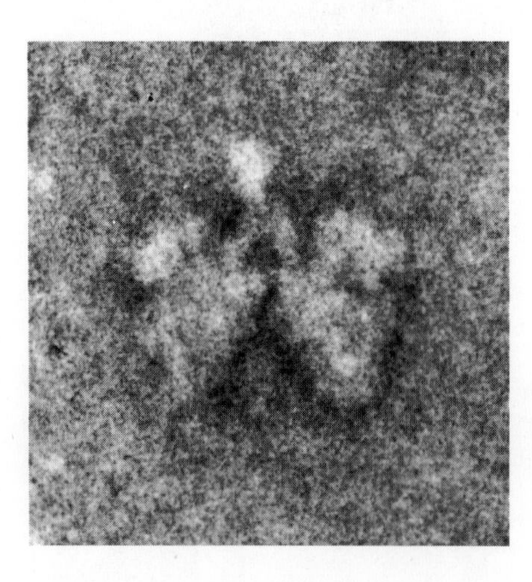

FIG. 8.8. Electron micrograph of a pair of the 30S ribosomal subunits with an antibody molecule connecting the hapten-modified 3′-ends of their 16S RNAs. From Shatsky, I. N., Mochalova, L. V., Kojouharova, M. S., Bogdanov, A. A., and Vasiliev, V. D. (1979) *J. Mol. Biol.* **133**:501–515. Original photo is provided by V. D. Vasiliev.

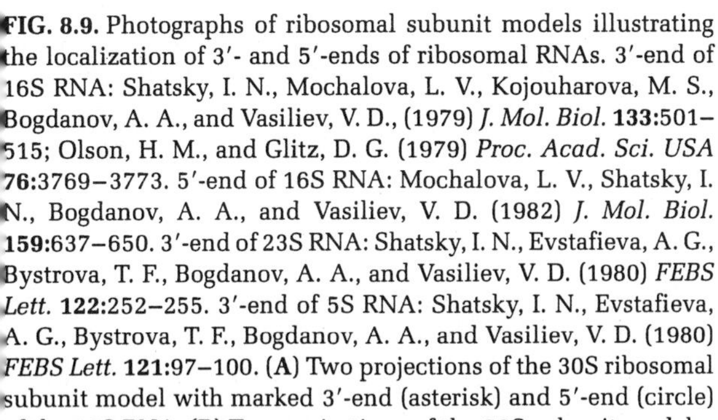

FIG. 8.9. Photographs of ribosomal subunit models illustrating the localization of 3′- and 5′-ends of ribosomal RNAs. 3′-end of 16S RNA: Shatsky, I. N., Mochalova, L. V., Kojouharova, M. S., Bogdanov, A. A., and Vasiliev, V. D., (1979) *J. Mol. Biol.* **133**:501–515; Olson, H. M., and Glitz, D. G. (1979) *Proc. Acad. Sci. USA* **76**:3769–3773. 5′-end of 16S RNA: Mochalova, L. V., Shatsky, I. N., Bogdanov, A. A., and Vasiliev, V. D. (1982) *J. Mol. Biol.* **159**:637–650. 3′-end of 23S RNA: Shatsky, I. N., Evstafieva, A. G., Bystrova, T. F., Bogdanov, A. A., and Vasiliev, V. D. (1980) *FEBS Lett.* **122**:252–255. 3′-end of 5S RNA: Shatsky, I. N., Evstafieva, A. G., Bystrova, T. F., Bogdanov, A. A., and Vasiliev, V. D. (1980) *FEBS Lett.* **121**:97–100. (**A**) Two projections of the 30S ribosomal subunit model with marked 3′-end (asterisk) and 5′-end (circle) of the 16S RNA. (**B**) Two projections of the 50S subunit model with marked 3′-end of 5S RNA (circle) and 3′-end of 23S RNA (asterisk). Courtesy of V. D. Vasiliev.

easily labeled selectively by hapten groups. The positions of the 3′- and 5′-terminal nucleotides of *E. coli* 16S rRNA on the 30S subunit, the 3′-terminal nucleotide of *E. coli* 5S rRNA, and the 3′-terminal nucleotide of 23S rRNA on the 50S subunit have been determined (Fig. 8.9). As seen, the 3′-end of the 16S RNA on the *E. coli* 30S ribosomal subunit is mapped in the region of the tip of the side bulge (platform), or somewhere between the bulge and the head. The 5′-end of the 16S RNA is localized on the body of the 30S subunit on the side opposite the side bulge. The 23S RNA 3′-end on the surface of the *E. coli* 50S subunit is mapped in the region of the L7/L12 stalk base, on the external side (the side turned away from the 30S subunit). The 3′-end of the 5S RNA is detected on the head or central protuberance of the 50S subunit; this defines the localization of the entire 5S RNA–protein complex including proteins L5, L18, and L25.

In addition, the position of nucleotide U40 of the 5S rRNA has been located on the central protuberance (head) of the 50S subunit (Evstafieva *et al.*, 1985). In the case of the U40 residue of the 5S rRNA, the hapten was attached to the 3′-end of the 5S rRNA fragment U1-U40, the modified fragment was associated with the 5S rRNA fragment G41-U120 to form the modified 5S rRNA molecule that was subsequently incorporated into the 50S subunit.

With antibodies against naturally modified rRNA bases, the locations of two

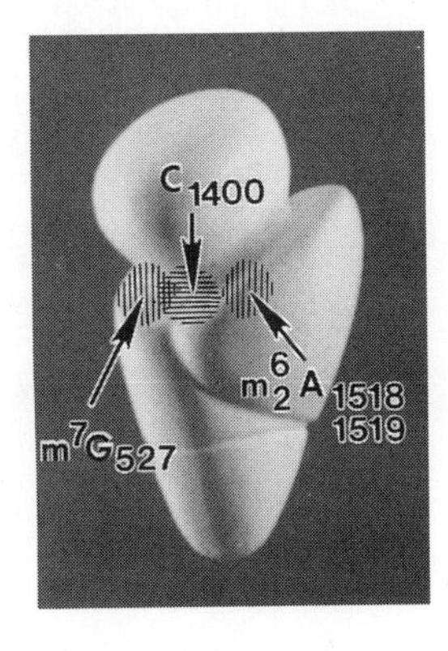

FIG. 8.10. Location of some modified nucleosides of 16S rRNA on the surface of the 30S subunit as determined by immuno-electron microscopy. $m^6{}_2$A1518–1519: Politz, S. M., and Glitz, D. G. (1977) *Proc. Natl. Acad. Sci. USA* **74**:1468–1472. m^7G527: Trempe, M. R., Ohgi, K., and Glitz, D. G. (1982) *J. Biol. Chem.* **257**:9822–9829. C1400: Gornicki, P., Nurse, K., Hellmann, W., Boublik, M., and Ofengand, J. (1984) *J. Biol. Chem.* **259**:10493–10498. Compiled on the basis of the 30S subunit model proposed by V. D. Vasiliev. Courtesy of V. D. Vasiliev.

neighboring N^6,N^6-dimethyladenosines (positions 1518 and 1519), N^7-methyl guanosine (position 527), and the cyclobutane dimer of 5′-anticodon base of tRNA with C1400 of 16S rRNA (that formed under UV-irradiation of the 30S subunit complex with $tRNA_1^{Val}$ at P site) in the *E. coli* 30S subunit have been determined (Fig. 8.10). They all are found to be localized in the groove separating the head and the body, that is, in the "neck" of the 30S subunit, mA1518–1519 being at the bulge side (near the 3′-terminal A1542), and m^7G527 and C1400 on the opposite side of the subunit. All the positions mentioned are known to belong to functionally important regions of the 16S rRNA.

8.4. Quaternary Structure

Determination of the precise mutual arrangement of all the structural elements of each ribosomal subunit, including proteins and their groups, the compact domains of rRNA, and individual rRNA helices depends on the progress in crystallographic studies of the particles. It is encouraging that the ribosomes and their isolated subunits can be crystallized, and the crystals diffract well. Crystallographic studies of bacterial ribosomal particles are being conducted by several groups.

At the same time, several preliminary models of the quaternary structure of ribosomal particles, and specifically the small (30S) subunit, have been proposed on the basis of numerous indirect data, such as protein and rRNA topography, chemical cross-linking, footprinting, neutron scattering, and stereochemical analyses fitted to electron microscopy models (e.g., Spirin *et al.*, 1979; Schueler and Brimacombe, 1988; Mueller and Brimacombe, 1997). The most recent 30S subunit model is based on the three-dimensional folding pattern of the 16S rRNA *in situ*, as deduced mainly from fitting individual elements of the well-known secondary structure of the rRNA to the fine structural elements of the 20 Å cryo-electron microscopy contours (see Fig. 6.11); the model is the combination of the proposed 16S rRNA folding with the protein map and the protein–RNA cross-linking and footprinting data.

References

Engelman, D. M., Moore, P. B., and Schoenborn, B. P. (1975) Neutron scattering measurements of separation and shape of proteins in 30S ribosomal subunits of *Escherichia coli:* S2-S5, S5-S8, S3-S7, *Proc. Natl. Acad. Sci. USA* **72**:3888–3892.

Evstafieva, A. G., Shatsky, I. N., Bogdanov, A. A. and Vasiliev, V. D. (1985) Topography of RNA in the ribosome; location of the 5S RNA residues A39 and U40 on the central protuberance of the 50S subunit, *FEBS Lett.* **185**:57–62.

Lake, J. A., Pendergast, M., Kahan, L., and Nomura, M. (1974) Localization of *Escherichia coli* ribosomal proteins S4 and S14 by electron microscopy of antibody-labeled subunits, *Proc. Natl. Acad. Sci. USA* **71**:4688–4692.

Mueller, F., and Brimacombbe, R. (1997) A new model for the three-dimensional folding of *Escherichia coli* 16S ribosomal RNA. II. The RNA-protein interaction data, *J. Mol. Biol.* **271**:545–565.

Schueler, D., and Brimacombe, R. (1988) The *E. coli* 30S ribosomal subunit; an optimized three-dimensional fit between the ribosomal proteins and the 16S RNA, *EMBO J.* **7**:1509–1513.

Serdyuk, I. N., and Grenader, A. K. (1975) Joint use of light, X-ray and neutron scattering for investigation of RNA and protein mutual distribution within the 50S particle of *E. coli* ribosomes, *FEBS Lett.* **59**:133–136.

Serdyuk, I. N., Smirnov, N. I., Ptitsyn, O. B., and Fedorov, B. A. (1970) On the presence of a dense internal region in the 50S subparticle of *E. coli* ribosomes, *FEBS Letters* **2**:324–326.

Shishkov, A. V., Filatov, E. S., Simonov, E. F., Unukovich, M. S., Goldanskii, V. I., and Nesmeyanov, A. N. (1976) Production of tritium-labeled biologically active compounds, *Dokl. Akad. Nauk SSSR* **228**:1237–1239.

Spirin, A. S., Serdyuk, I. N., Shpungin, J. L., and Vasiliev, V. D. (1979) Quaternary structure of the ribosomal 30S subunit: Model and its experimental testing, *Proc. Natl. Acad. Sci. USA* **76**:4867–4871.

Stuhrmann, H. B., Haas, J., Ibel, K., De Wolf, B., Koch, M. H. J., Parfait, R., and Crichton, R. R. (1976) New low resolution model for 50S subunit of *Escherichia coli* ribosomes, *Proc. Natl. Acad. Sci. USA* **73**:2379–2383.

Tischendorf, G. W., Zeichhardt, H., and Stoeffler, G. (1975) Architecture of the *Escherichia coli* ribosome as determined by immune electron microscopy, *Proc. Natl. Acad. Sci. USA* **72**:4820–4824.

Traut, R. R., Lambert, J. M., Boileau, G., and Kenny, J. W. (1980) Protein topography of *Escherichia coli* ribosomal subunits as inferred from protein crosslinking, in *Ribosomes: Structure, Functions, and Genetics* (G. Chambliss, G. R. Craven, J. Davies, K. Davis, L. Kahan, and M. Nomura, eds.), pp. 89–110, Baltimore, University Park Press.

Wabl, M. R. (1974) Electron microscopic localization of two proteins on the surface of the 50S ribosomal subunit of *Escherichia coli* using specific antibody markers, *J. Mol. Biol.* **84**:241–247.

Yusupov, M. M., and Spirin, A. S. (1988) Hot tritium bombardment technique for ribosome surface topography, *Methods Enzym.* **164**:426–439.

III

Function of the Ribosome

9

Functional Activities and Functional Sites of the Ribosome

9.1. Working Cycle of the Ribosome

At any given time in the course of polypeptide elongation, the ribosome is attached to the coding region of mRNA and retains the molecule of the peptidyl-tRNA (Fig. 9.1). The peptidyl-tRNA is a nascent peptide chain bound through its C-terminus to the tRNA that has donated the last amino acid residue to the peptide. Such a ribosome can bind or may become capable of *binding the aminoacyl-tRNA* determined by the next mRNA codon (Fig. 9.1 step I). The binding of the aminoacyl-tRNA results in the retained peptidyl-tRNA and the newly bound aminoacyl-tRNA being present on the ribosome simultaneously. Their side-by-side location and the catalytic activity of the ribosome are prerequisites of the *transpeptidation* reaction: the C-terminus of the peptidyl residue is transferred from the tRNA (to which it had previously been bound) to the amino group of the aminoacyl-tRNA (Fig. 9.1 step II). As a result, the formation of a new peptidyl-tRNA with the peptide elongated by one amino acid residue at the C-end takes place; the other product of the transpeptidation reaction is the deacylated tRNA. In order to make the ribosome competent to bind the next aminoacyl-tRNA, the intraribosomal ligands (tRNAs and mRNA) must be displaced, resulting in the vacation of a place for the aminoacyl-tRNA and in the positioning of the next mRNA codon (Fig. 9.1 step III); this step is called *translocation.*

Thus, the working cycle of the ribosome in the course of elongation consists of three principal steps: codon-dependent binding of aminoacyl-tRNA (step I), transpeptidation (step II), and translocation (step III). The binding of aminoacyl-tRNA requires the presence of a special protein called elongation factor 1 (EF1A); it is also called EF-Tu in the case of prokaryotes, and eEF1 in the case of eukaryotes. The binding is accompanied by the hydrolysis of a GTP molecule. Transpeptidation is catalyzed by the ribosome itself. Translocation requires another protein, elongation factor 2 (EF2), or EF-G in prokaryotes and eEF2 in eukaryotes, and is also accompanied by GTP hydrolysis.

The central chemical reaction of the elongation cycle is transpeptidation where two substrates, aminoacyl-tRNA and peptidyl-tRNA, participate:

$$\text{Pept(n)-tRNA}' + \text{Aa-tRNA}'' \rightarrow \text{Pept(n+1)-tRNA}'' + \text{tRNA}'$$

Correspondingly, the binding sites of these two substrates on the ribosome have been designated as A and P sites. Hence, the strict operational definition of A and

P sites is that they are *the sites occupied by the substrates reacting with each other* in the ribosome-catalyzed transpeptidation reaction.

According to the classical two-site model (Watson, 1964; Lipmann, 1969), at stage I the aminoacyl-tRNA in the complex with EF1A (EF-Tu or eEF1) and GTP enters the ribosome and binds to the vacant template codon located therein. At this time the peptidyl-tRNA is in the P site. The binding of the aminoacyl-tRNA ends in GTP hydrolysis on the ribosome and the release of the EF1A:GDP complex and orthophosphate into solution. At stage II the newly entered aminoacyl-tRNA located in the A site reacts with the peptidyl-tRNA in the P site; this results in the peptide C-terminus being transferred to the aminoacyl-tRNA. Now, the elongated peptidyl-tRNA (its tRNA residue) is occupying the A site while the deacylated tRNA formed in the reaction is located in the P site. At stage III the ribosome interacts with EF2 (EF-G or eEF2) and GTP, and this catalyzes the displacement of the peptidyl-tRNA (its tRNA residue) along with the template codon from the A site to the P site, as well as the release of the deacylated tRNA from the P site. During these events GTP undergoes hydrolysis, and then EF2, GDP, and orthophosphate are released from the ribosome. This again leads to the situation whereby the peptidyl-tRNA is located in the P site while the next template codon is located in the A site; thus the A site is ready to accept the next aminoacyl-tRNA molecule. Translation of the whole coding sequence of the template polynucleotide and corresponding polypeptide elongation on the ribosome are achieved by the repetition of the cycles. It should be pointed out that both the initiation and termination of

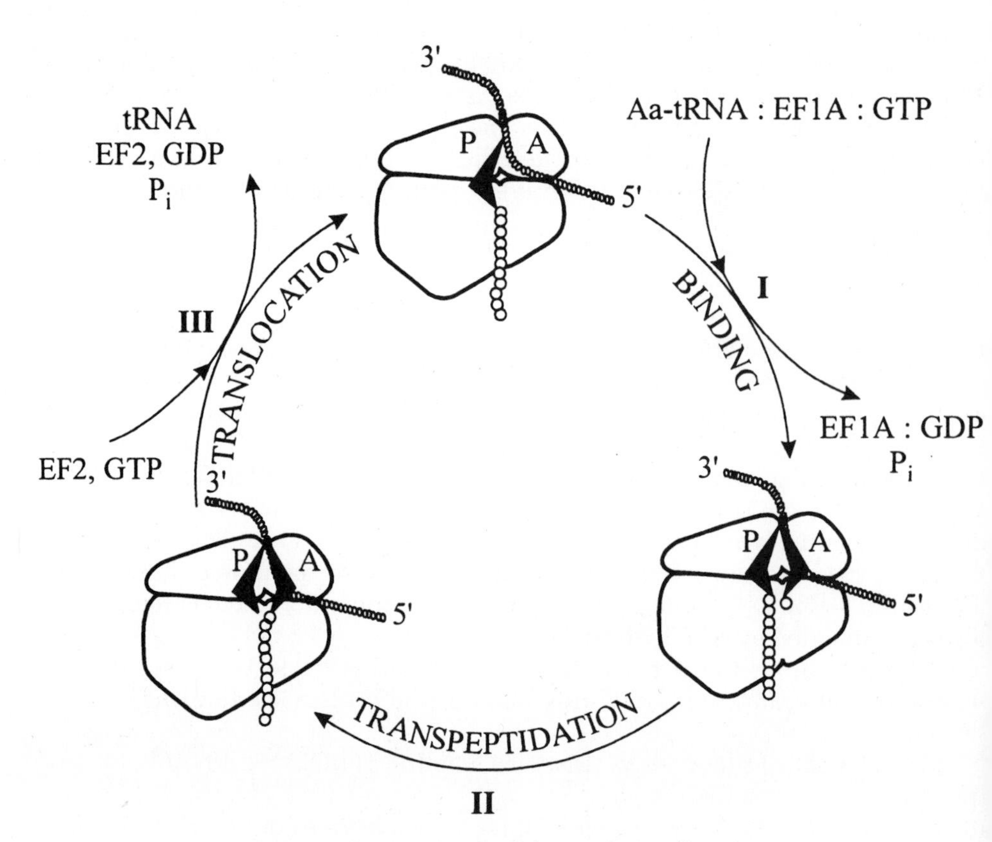

FIG. 9.1. Elongation cycle of the translating ribosome.

translation are simply modifications of the ribosomal working elongation cycle outlined above (see Chapters 14 and 15).

The scheme sketched above describes only the most principal stages of the elongation cycle and omits many intermediate states. In particular, the problem of the "entry site" where aminoacyl-tRNA may be transiently present prior to the ultimate settling in the A site, and that of the "exit site" which may temporarily accommodate deacylated tRNA after its translocation from the P site, as well as intermediate states of translocation, are not outlined here and will be considered in Sections 9.5.3. and 9.5.4, respectively, and in Chapters 10 and 12.

In any case, an analysis of the ribosomal working cycle demonstrates that the ribosome performs a number of functions in the course of translation, including (1) binding and retention of mRNA, (2) retention of peptidyl-tRNA, (3) binding of aminoacyl-tRNA, (4) binding of translation protein factors, (5) participation in the catalytic hydrolysis of GTP, (6) catalysis of transpeptidation, and (7) the complex of intraribosomal displacements referred to as translocation. Different parts of the ribosome are responsible for performing these various functions. On the whole, the ribosome embodies the dualistic nature of translation: it is a decoding machine operating with a genetic message, and at the same time it is an enzyme synthesizing a polypeptide chain. It is remarkable that the dualistic principle is reflected in the two-subunit construction of the ribosome and the partial functions of the subunits: whereas just the small subunit is involved in the genetic message binding and decoding, the large subunit is entirely responsible for the peptide-synthesizing activity.

9.2. Methodological Approaches to Localization of Ribosomal Functional Sites

Numerous approaches have been applied to determine the ribosomal proteins and rRNA regions that take part in the formation of ribosome functional centers. Among them, site-specific chemical modifications and site-directed mutagenesis were thoroughly explored to selectively inactivate a function under investigation. Also localization of natural mutations leading to resistance against specific inhibitors of ribosomal functions was successfully used.

A powerful approach to identifying proteins and rRNA regions forming ribosomal functional sites makes use of *affinity labeling.* With this technique a chemically active or photoactivatable group is introduced into a corresponding ligand (e.g., mRNA, tRNA, translation factor, guanylic nucleotide, antibiotic) that specifically binds with the ribosome. This group attacks the ribosomal components located nearby and becomes cross-linked with them. Proteins and rRNA regions cross-linked with the ligand may then be identified. Furthermore, in some cases it has become possible to achieve photoactivation of unmodified synthetic and natural ligands (e.g., oligo- and polynucleotides) in order to cross-link them with the nearest neighbors in the ribosome. It is clear, however, that this approach does not allow the components directly forming the mRNA-binding site and the components located nearby to be distinguished.

Further development of this approach, specifically for identification of components of RNA-binding centers of the ribosome, is the *"site-directed cross-linking"* technique. It is based on incorporation into mRNA or tRNA molecules of photoreactive nucleoside derivatives that can form the so-called zero-length cross-links (see Briomacombe *et al.*, 1990; Sergiev *et al.*, 1997). 4-Thiouridine (4-thioU) and 6-thioguanosine (6-thioG) derivatives are the most broadly used in these studies. They are very close analogues of "normal" nucleosides, and the occurrence of single 4-thioU or 6-thioG residues in an RNA molecule does not change its spatial structure. To form the RNA–RNA cross-link a photoreactive base has to be partially stacked with another RNA base. Thus, this approach allows one to identify direct contacts between nucleotides of rRNA and RNA ligands (mRNA or tRNA).

Another powerful methodological approach to localization of ligand binding sites on the ribosomal particles, and specifically on ribosomal RNA, is the *footprinting* technique. It is based on the fact that a ligand may protect the nucleotide residues with which it interacts from chemical modifications. Noller *et al.* (1990) (see also reviews by Noller, 1991, 1998) used a set of chemical probes, such as kethoxal, dimethyl sulfate, or carbidiimide, for modification of accessible bases in ribosomal RNA or ribosomal particles. After treatment with the probes, the RNA was used as a template for the extension of synthetic primer deoxyoligonucleotides by reverse transcriptase. The enzyme stops at the sites of modification causing the premature termination. The products of the reverse transcriptase reaction are analyzed by electrophoresis on DNA sequencing gels. Hence, the protection of a base from chemical modification due to interaction with a ligand is visualized as the absence of a corresponding band on the gels.

9.3. Binding, Retention, and Sliding of the Message (mRNA-Binding Site on the Small Subunit)

The ribosome has an intrinsic affinity for template polynucleotides. It has long been known that vacant ribosomes effectively bind polyuridylic acid. It is likely that the absence of a stable secondary and tertiary structure in poly(U) is an important factor contributing to its effective binding with the ribosomes. In the case of mRNA from natural sources, there are definite preferential sites on the polynucleotide for binding vacant ribosomes. In any case, stable double helices of RNA seem to be unable to serve as binding sites for vacant ribosomes.

At the same time, in the course of translation (elongation) the ribosome passes along the entire coding sequence of mRNA and thus can transiently hold the template at any region of the sequence. The ribosome unfolds the translated template polynucleotide in such a way that the template section hold on the ribosome is devoid of its original secondary and tertiary structure. Codon–anticodon interactions with tRNA undoubtedly contribute to a retention of mRNA on the translating ribosome.

A translating ribosome bound to the template polynucleotide protects a rather long nucleotide sequence from external nucleases and chemical modifications. Early experiments with poly(U) have demonstrated that the ribosome covers the 25-residue-long section, making it inaccessible to pancreatic ribonuclease (Takanami and Zubay, 1964). More recently the ribosome with natural mRNA has been shown to protect against nucleases or chemical modifications an mRNA region 40 to 60 nucleotide residues long (Steitz, 1969; Huettenhofer and Noller, 1994) (Fig. 9.2A). Within the 60-nucleotide region the 5′-proximal part of about 10 to 20 nucleotides long may be less strongly protected. These results suggest that the mRNA-binding site of the ribosome is of a considerable size and apparently extends for more than 100 Å.

The first problem regarding the localization of the functional sites of the ribosome has to do with whether they are assigned to one of the two ribosomal subunits, or to both subunits. In the simplest case the experimental solution of this problem is as follows. Ribosomes are dissociated to yield large and small subunits, the subunits are separated, and the tested ligand is added to each of them (in the presence of a sufficient concentration of magnesium ions, which is required to observe any binding to the ribosome). It has been demonstrated in this type of experiment that the isolated bacterial 30S subunit binds the template polynucleotide whereas the 50S subunit does not (Takanami and Okamoto, 1963). On the basis of this result, it is generally accepted that the mRNA-binding site of the ribosome is located only on the small (30S or 40S) subunit.

The isolated small ribosomal subunit protects an mRNA region of principally the same length as does the full ribosome, provided a tRNA is also bound with the subunit (Fig. 9.2B). In the absence of tRNA, however, the 30S subunit protects about 40 nucleotides: the 3′-section of about a dozen nucleotides long becomes less protected or unprotected (Fig. 9.2C).

Several approaches have been used to identify the ribosomal proteins and the

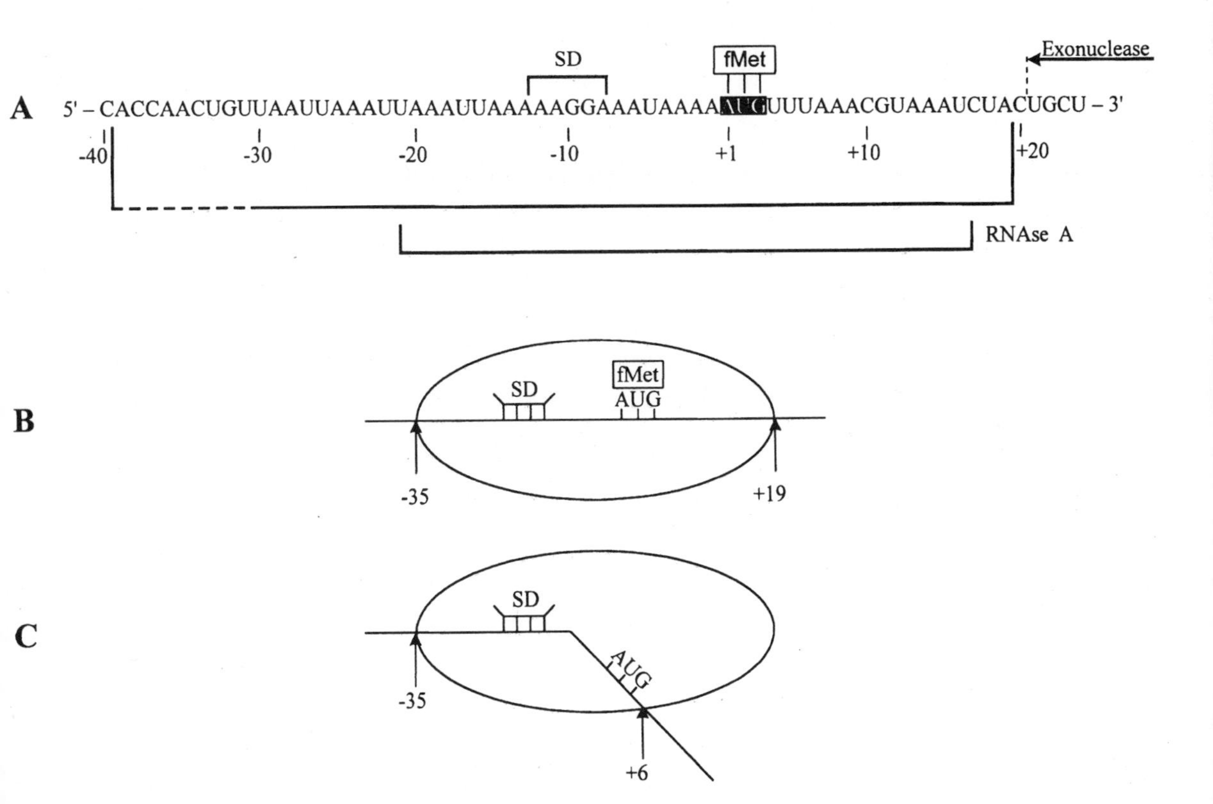

FIG. 9.2. Protection of mRNA by the bacterial ribosome. (**A**) Nucleotide residues protected by the 70S ribosome or the 30S ribosomal subunit with initiator $tRNA_f^{Met}$ against chemical modifications and enzymatic attack (similar results have been obtained for the 30S ribosomal subunit with $tRNA^{Phe}$). A weaker protection of the 5′-portion of the protected region is shown by dotted underlining. (**B**) Schematic representation of the protection of the ribosome-binding site of mRNA by the 30S ribosomal subunit in the presence of codon-interacting tRNA (the initiator $tRNA_f^{Met}$ interacting with initiation AUG codon is shown). (**C**) Schematic representation of the protection of the ribosome-binding site of mRNA by the 30S ribosomal subunit in the absence of tRNA. Reproduced, with modifications, from Hüttenhofer, A., and Noller, H. F. (1994) *EMBO J.* **13**:3892–3901 with permission.

ribosomal RNA regions that take part in the organization of the mRNA-binding site of the 30S ribosomal subunit. The considerable length of the mRNA-binding site suggests a multicenter binding of mRNA to the particle, that is, the participation of several binding points of the ribosome surface. The position of two mRNA codons interacting with anticodons of both substrate tRNAs, that is, the *decoding center* of the small subunit, is of particular interest.

Of the proteins either belonging to the mRNA-binding site of the 30S ribosomal subunit or located nearby, proteins S1, S3, and S5 can be most reliably identified in this way, although at least a dozen other 30S ribosomal proteins have also been reported. On the basis of evidence for the interdomain or interlobe position of proteins S3 and S5 (see Chapters 7 and 8), one may suggest that the mRNA-binding site is located in the region of the grooves dividing the head, the body, and the side bulge of the 30S ribosomal subunit. Protein S1 also appears to occupy an interlobe position being localized on the external surface of the small subunit, in the region of the "neck." Its intrinsic capacity for forming complexes with polynucleotides has been detected. The binding of isolated protein S1 with RNA results in the loosening or unfolding of the RNA secondary structure. Taking into account that protein S1 neighbors mRNA on the ribosome, one may assume that it directly participates in the formation of the mRNA-binding site. Protein S7 located on the 30S subunit head, near the "neck," but on the side opposite to that with protein S3 (see Figs. 8.3, 8.5, and 8.7), is also sometimes mentioned among components adjacent to the mRNA-binding site.

As to the participation of rRNA in the formation of the mRNA-binding site, on the basis of all data available one can define five groups of 16S rRNA regions and individual nucleotide residues that play a key role in the organization of this functional site (Fig. 9.3):

1. The polypyrimidine 3′-terminal region (positions 1535–1541) that has been proved to participate in the formation of complementary complex with the mRNA polypurine sequence (Shine-Dalgarno, or SD sequence) located upstream of the initiator codon at the distance of 5 to 12 nucleotide residues (see Section 15.2.2). This 3′-terminal sequence of the 16S rRNA is sometimes called "anti-SD sequence." (The anti-SD positions of the 16S rRNA are connected with the SD sequence of mRNA by dashed lines in Fig. 9.3).

2. The 16S rRNA regions that interact with the spacer between the initiation codon and the SD sequence of mRNA. They include the 16S rRNA region just adjacent to the anti-SD-region (particularly G1529), the nucleotide A665, and the nucleotide A1360 (they are indicated by long arrows in Fig. 9.3). The nucleotide A665, located at the protein S15 binding site, interacts with an mRNA spacer sequence only at the first stage of initiation of translation, and its contact with mRNA disappears when the initiator tRNA binds to the P site (see Section 15.2.2). On the contrary, the nucleotide A1360 is in contact with an mRNA spacer region only when the initiator tRNA occupies the P site. It is localized at the binding site of protein S7 which may also participate in the organization of the decoding center.

3. The regions interacting with a codon–anticodon duplex at the P site (including positions +1 to +3 of mRNA). This can be considered as the P-site part of the decoding center of the mRNA-binding site. The 16S rRNA nucleotide residues that are crucial for these interactions are highly conservative. They are located in the neighborhood of the nucleotide C1400, and the P-site-bound tRNA protects them from chemical modifications (see filled circles in Fig. 9.3). The nucleotide

C1400 can form a short-range cross-link with the wobble base of the P-site-bound tRNA (Prince *et al.*, 1982). A number of mutations that strongly affect ribosome activity have been produced in the region around nucleotide C1400. As seen in Fig. 8.10, this region is mapped in the groove separating the 30S head from its body, on the side opposite to the bulge (platform). In addition to the footprints around C1400, the helices 938–943/1340–1345 and 923–933/1384–1393, which are known to reside in the 30S subunit head, are also found in the neighborhood of the P-site-bound codon–anticodon duplex, as indicated by the footprints at positions 1338–1339 and by the cross-link of G925 with the P-site-bound initiation codon AUG (Fig. 9.3).

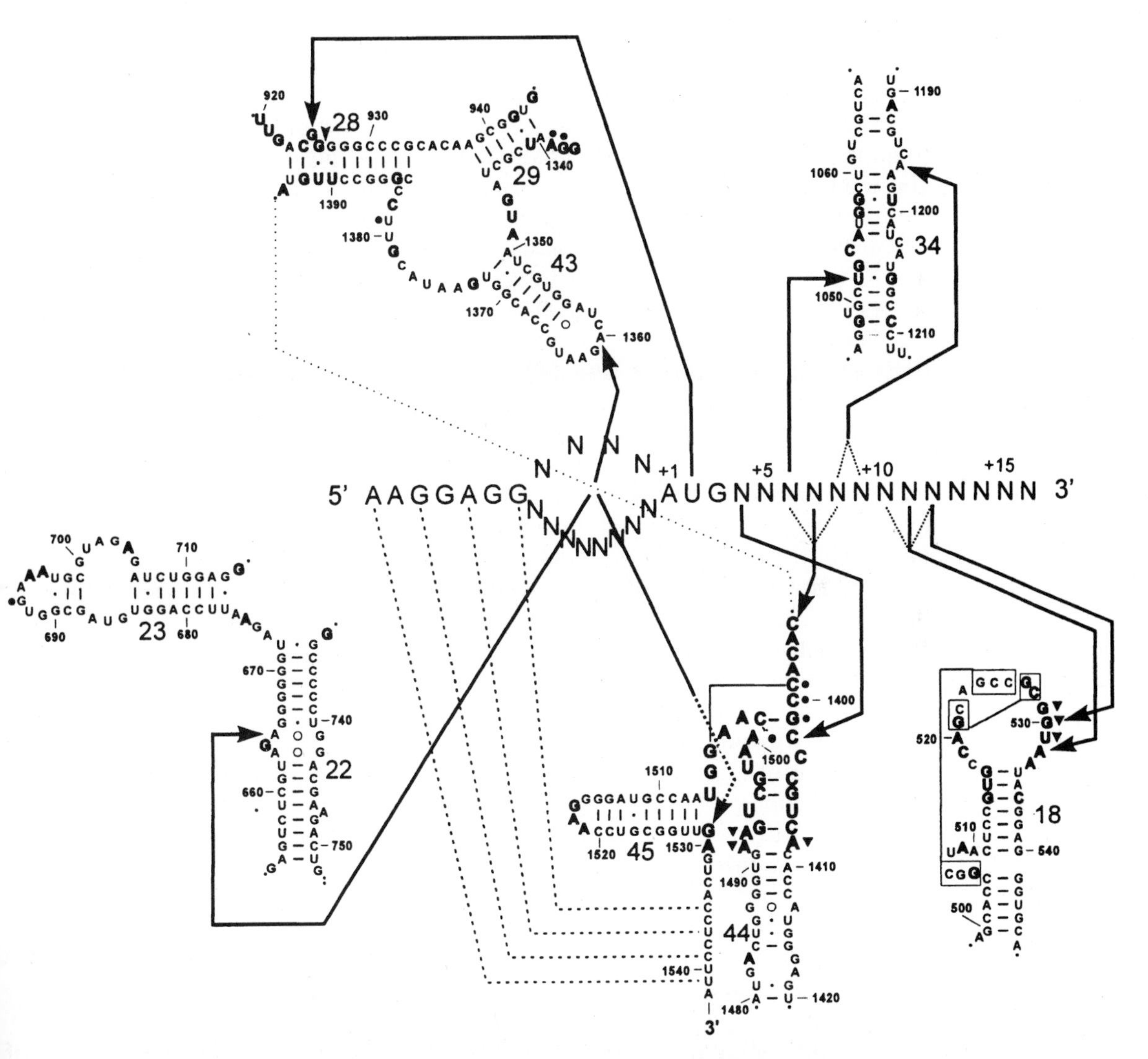

FIG. 9.3. 16S rRNA regions involved in organization of the decoding center of the *E. coli* ribosome. Cross-linking sites are shown by arrows. Nucleotides indicated by bold letters are highly conserved in evolution. Footprint sites from P-site-bound tRNA are indicated by filled circles, and those from A-site-bound tRNA by filled triangles. SD–antiSD complementary interactions are shown by dash lines. Based on the data summarized in Mueller, F., & Brimacombe, R. (1997) *J. Mol. Biol.* **271**:566–587.

4. The regions interacting with the mRNA codon located at the A site. This is the A-site part of the decoding center of the mRNA-binding site. As shown with the use of the site-directed cross-linking technique, the 5′-nucleotide (position +4) of the A-site-bound codon of mRNA is in direct contact with the nucleotide C1402, while the 3′-nucleotide (position +6) of this codon contacts U1052. These nucleotide residues are highly conserved. Their functional importance is confirmed by genetic data: mutations at position 1402 are lethal for bacterial cells, and the nucleotide 1052 is near the positions of well-characterized mutations that affect A-site-related activities of the ribosome. The nucleotide next to the A-site-bound codon (position +7) is in direct contact with the nucleotide C1395 of the 16S rRNA. Deletion of its neighbor, the nucleotide A1394, abolishes the binding of tRNA to the ribosomal A site. It is noteworthy that the A-site-bound section of mRNA is in contact with the nucleotides 1402 and 1052 only when the P site is occupied with tRNA and the A site is free. After binding of tRNA to the A site these contacts disappear. On the whole, the base of the minor 3′-terminal domain of the 16S rRNA, precisely the imperfect double-helical region 1398–1410/1490–1505 (see Figs. 6.1 and 9.3), is often considered the *decoding center of rRNA* proper, including both its A-site and P-site parts, that is, the rRNA region mainly responsible for the retention of the two codon–anticodon duplexes.

5. The 16S rRNA regions interacting with mRNA positions downstream of the codon–anticodon duplexes (positions +8 to +12 of mRNA). These interactions seem to be important to fix mRNA in a correct way at the P and A sites. The first region from this group is near the nucleotide A1196 which can cross-link to positions +8 and +9 in mRNA (Fig. 9.3, long arrow up). The second region from this group is the so-called loop 530. Here, the highly conserved nucleotides G530 and A532 are in direct contact with mRNA positions +11 and +12 (Fig. 9.3, two parallel long arrows down). The loop 530 has a unique spatial structure organized with two pseudoknots (see Fig. 6.1). Any disarrangement of the pseudoknots leads to a decrease in the fidelity of translation. Some mutations in this region are lethal for the cell.

It must be emphasized that all contacts of mRNA and 16S rRNA described in this section are universal and do not depend on mRNA sequence. At the same time, their formation and dissociation depend on binding of tRNA to the decoding center of the ribosome.

Thus, several highly conserved regions of the 16S rRNA universal core (scattered in rRNA primary and secondary structure but apparently clustered in its tertiary structure) form multipoint contacts with mRNA (and tRNAs), providing a specific fixation of the ribosome ligands in the decoding center. A less specific retention of a polynucleotide along an extended path of mRNA on the small ribosomal subunit should not be neglected either.

In order to locate the mRNA-binding site on the morphologically visible surfaces of the ribosome, the immuno-electron microscopy studies of 30S subunits and 70S ribosomes bound with short poly(U) carrying a covalently linked hapten on either the 3′- or 5′-end have been performed (Fig. 9.4). Using this approach the template polynucleotide ends have been detected in the region of the groove ("neck") separating the head and the body of 30S subunit, mainly on its external (facing from the 50S subunit) side and near the side bulge (platform).

Taking all the evidence into consideration, it appears that the mRNA chain binds to and passes along the 30S subunit somewhere on the boundary between its lobes or between the 16S RNA domains. It is likely that the binding site is located in the region of the groove that separates the head from the side bulge and the head from the body (the "neck" region). The extended mRNA-binding region seems to contain many 16S RNA elements (including the 3′-terminal sequence and the minor 3′-terminal domain, helices at the base of the major 3′-domain together with peripheral helices of the central domain, an internal helix of the major 3′-domain, the last hairpin, or loop 530 of the 5′-domain) and several ribosomal proteins (such as S1, S3, S5, and possibly S7). The decoding center retaining the codon section of mRNA seems to be located precisely at the thin "neck" of the small subunit.

In general, it may be assumed that the association of the template polynucleotide with the ribosome permits a slippage of the polynucleotide chain along the mRNA-binding site. This is an obvious requirement for the sequential reading of the mRNA chain in the course of translation. The experiments have demonstrated that the ribosome protects 55 to 60 nucleotide sections of mRNA from attack by hydroxyl radicals (generated by Fe^{2+}-EDTA); since the hydroxyl radicals attack mainly the sugar-phosphate backbone of RNA, the conclusion can be made that the ribosome interacts with mRNA along its sugar-phosphate backbone (Huettenhofer and Noller, 1994). This conclusion is quite consistent with the idea of the slippage of mRNA through the mRNA-binding site. The proposed position of tRNA residues on the ribosome (see below) and the possible trajectories of their displacements during translocation (see Section 12.5.1) suggest that the mRNA slips along the "neck" of the 30S subunit, more or less from outside and the L7/L12 stalk side of the ribosome to the subunit interface and the L1 protuberance side, as shown in Fig. 9.5.

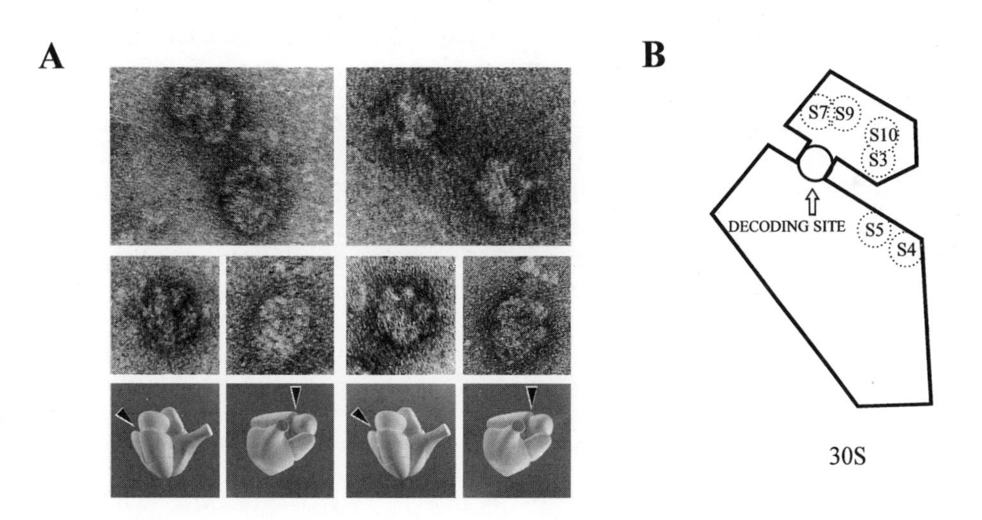

FIG. 9.4. Localization of the mRNA-binding site of the ribosome in the "neck" region of the small ribosomal subunit by immuno-electron microscopy technique. (**A**) Electron microscopy photographs of the complexes of the 70S ribosomes associated with hapten-linked poly(U) with hapten-specific antibodies. (*Upper row*) ribosomes connected by the antibody molecules into dimers. (*Middle row*) single ribosomes interacting with the antibody molecules. (*Lower row*) ribosome model with arrowheads indicating the antibody binding sites on the 70S ribosome in two projections. From Evstafieva, A. G., Shatsky, I. N., Bogdanov, A. A., Semenkov, Y. P., and Vasiliev, V. D. (1983) *EMBO J.* **2**:799–804. Original photos are kindly provided by V. D. Vasiliev. (**B**) Schematic contour drawing of the 30S ribosomal subunit with mRNA-binding (decoding) site in the "neck" region. Positions of several ribosomal proteins are indicated by dotted circles.

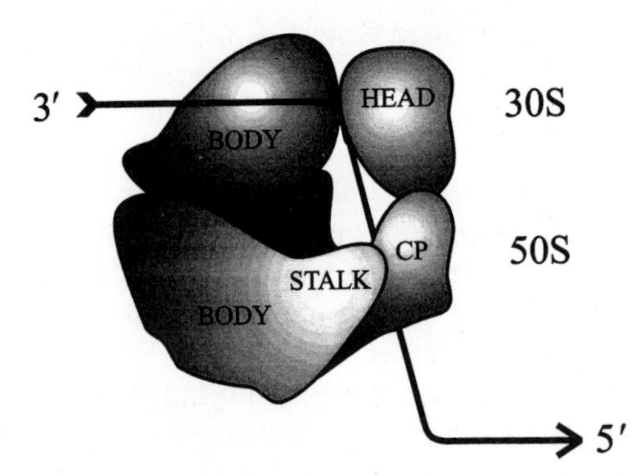

FIG. 9.5. Scheme of the possible trajectory of mRNA chain passage through the ribosome.

9.4. Catalysis of Peptide Bond Formation (Peptidyl Transferase on the Large Subunit)

Peptidyl transferase activity is the main, and seemingly the only catalytic function of the ribosome itself. It is responsible for the formation of peptide bonds during polypeptide elongation. In the translating ribosome, transpeptidation proceeds between the peptidyl-tRNA and the aminoacyl-tRNA. In this reaction the peptidyl-tRNA serves as a donor substrate, and the aminoacyl-tRNA as an acceptor substrate (see Chapter 11, Fig. 11.1):

$$\text{Pept(n)-tRNA}' + \text{Aa-tRNA}'' \rightarrow \text{tRNA}' + \text{Pept(n+1)-tRNA}''$$

However, the ribosome catalyzes transpeptidation not only between these natural substrates, but also others. The antibiotic puromycin is an excellent low-molecular-mass acceptor substrate in the reaction (Nathans, 1964; Traut and Monro, 1964). By its chemical nature, it is an analog of the aminoacylated 3′-terminal adenosine of the aminoacyl-tRNA molecule (Fig. 9.6); a dimethylated amino group at position 6 of the adenine residue, a methylated hydroxyl of the tyrosine residue, and the amide bond between ribose and the aminoacyl residue instead of the ester bond are its characteristic features. The addition of puromycin to the translating ribosomes results in the reaction between the antibiotic as an acceptor sub-

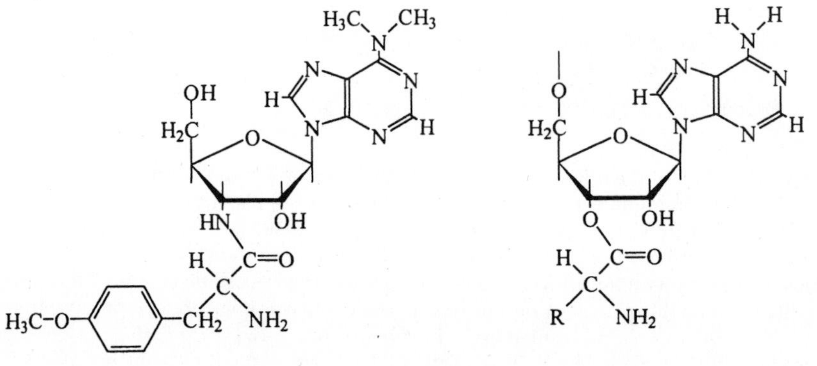

FIG. 9.6. Puromycin (*left*) as an analog of aminoacylated 3′-terminal adenosine of aminoacyl-tRNA (*right*).

strate and peptidyl-tRNA as a donor substrate in the peptidyl transferase center of the ribosome:

$$\text{Pept-tRNA} + \text{PM} \rightarrow \text{tRNA} + \text{Pept-PM}$$

In this way the peptide becomes transferred not to the aminoacyl-tRNA, but to a low-molecular-mass compound that is not retained in the ribosome; as a result, the peptidyl puromycin is released from the ribosome. Thus, puromycin leads to the abortion of the growing peptide.

The use of puromycin has played an important part in studies on the ribosomal peptidyl transferase center (PTC). Its use has made possible the identification of the ribosomal subunit that bears PTC. The isolated large subunit can retain peptidyl-tRNA, as well as show some labile interaction with the 3′-terminal fragments of the N-blocked aminoacyl-tRNA, which serve as donor substrates. The peptidyl transferase reaction occurs when puromycin is added to the large subunits carrying peptidyl-tRNA or its analogs. Hence, it can be concluded that the PTC is located entirely on the 50S subunit or the 60S subunit (Monro, 1967). The small (30S or 40S) ribosomal subunit does not contribute to the catalysis of the reaction at all.

Since two substrates, donor and acceptor, participate in the transpeptidation reaction, two substrate-binding sites should exist within the PTC. These will be referred to as *d* and *a* sites of the PTC, respectively. The simplest substrate for the acceptor-binding site (*a* site) of the PTC is an aminoacylated adenosine, for example, A-Phe, A-Tyr, A-Lys, A-Met, or A-Ala; puromycin is an analog of such a substrate. The effective substrate for the donor-binding site (*d* site) of the PTC is an N-blocked aminoacylated trinucleotide, such as CpCpA-(Ac-Aa) or CpCpA-(F-Aa). If aminoacyl nucleotides (e.g., CpCpA-(F-Met) and A-Phe) are added to the isolated 50S ribosomal subunit, the particle will work as a normal enzyme catalyzing transpeptidation between the low-molecular-mass substrates:

$$\text{CpCpA-(F-Met)} + \text{A-Phe} \rightarrow \text{CpCpA}_{\text{OH}} + \text{A-(F-Met-Phe)}$$

Neither product is retained by the 50S subunit, and both of them are immediately released into solution, again typical of a normal enzymatic reaction. During transpeptidation as a step of the elongation cycle, the PTC binds the 3′-terminal adenosine with the aminoacyl residue of the A-site-bound tRNA at the *a* site, and the 3′-terminal CCA sequence with aminoacyl residue and its peptide group of the P-site-bound tRNA at the *d* site. In this case, however, the products of the reaction cannot be released into solution, but retained by the ribosome.

Naturally, there have been many attempts at isolating the “enzyme” from the 50S or 60S subunit, that is, at finding the ribosomal protein responsible for catalyzing transpeptidation. However, none of these attempts has proved to be successful; the isolated proteins have not showed the presence of such activity. It was concluded that the “enzyme” may consist of several proteins tightly integrated in the 50S ribosomal subunit and that it undergoes disruption in the course of protein isolation.

Various analogs of peptidyl-tRNA and aminoacyl-tRNA which carry a chemically active or photoactivatable group on the aminoacyl residue at the 3′-end of tRNA have been used as affinity labels for identifying proteins located in the region of the PTC (see, e.g., Barta, Kuechler, and Steiner, 1990; Cooperman, Weitzmann, and Fernandez, 1990; Wower, Hixson, and Zimmermann, 1989). Most intense cross-links have been observed in such experiments with protein L27 and, to a lesser extent, L2 and L16; proteins L6, L11, L14, L15, L18, L23, and L33 have also been reported as cross-linkable neighbors of the substrates of the PTC. In experiments on the partial disassembly and reconstitution of 50S ribosomal subunits, proteins L2, L3, L4, L6, L11, L15, and L16 have been found to be essential for peptidyl transferase activity. However, none of the proteins listed has proven to be indispensible for the activity.

At the same time, experiments on the affinity labeling of the PTC by active aminoacyl-tRNA or peptidyl-tRNA analogs repeatedly demonstrated that although proteins were frequently found as cross-linkable neighbors of these analogs, the ribosomal 23S RNA was still the preferred target. Most of the cross-links were concentrated in domain V of the 23S RNA (see Fig. 6.5B), which forms the upper part (the "neck" surroundings, see Figs. 5.8 and 5.9) of the 50S subunit body. Cross-linking of the photoactivated label-carrying acceptor end of tRNA with position 2584 of the bacterial 23S RNA, as well as nucleotide replacements at positions equivalent to 2447–2504 in the 23S RNA accompanying the mutations of the PTC, suggested that the PTC is located in the region of the evolutionarily conservative sequence 2450–2600 of domain V.

The use of the footprinting technique to determine the residues protected by acceptor ends of the two substrate tRNAs (Noller *et al.*, 1990) has demonstrated that the protection sites are mainly within the same sequence, specifically A2439, A2451, U2506, G2553, Ψ2555, U2584, U2585, A2602, U2609, and some others. A classical inhibitor of the bacterial PTC on the ribosome, chloramphenicol, protects positions A2059, A2062, A2451, and G2505, located nearby or in the same region. This region, called the "PTC ring" (Fig. 9.7), is the junction of five helices. It is interesting that this region is especially enriched with modified nucleotide residues.

The tRNA protection sites are found both in the PTC ring itself and in the hairpins connected by the ring. Comparison of the positions protected by A-site-bound and P-site-bound tRNAs, as well as cross-linking sites for these tRNAs, shows that the PTC region of 23S rRNA as a whole is compactly folded. For example, the A-site-bound tRNA protects from chemical modification residues A2451 and U2609 which are quite distant in the rRNA primary (and secondary) structure. Other tRNA protection sites are scattered around the PTC ring. Also, the P-site-bound tRNA cross-links to U2584/U2585 and to A2451 located on the "opposite" sides of the ring. Moreover, the P-site-bound tRNA strongly protects the highly conserved residues G2252 and G2253 outside the PTC ring, in the end loop of the short hairpin 2246–2258 of domain V (see Fig. 6.5). The latter protection is dependent on the presence of the universal 3′-terminal sequence of tRNA. There is experimental evidence that G2252 participates in fixation of the 3′-end sequence CCA of the P-site-bound tRNA in the PTC due to Watson–Crick base pairing with C74 of the tRNA (Samaha, Green, and Noller, 1995).

The formation of a special compact tertiary structure in the PTC region seems to be critical for the activity. It is likely that PTC is organized mainly by self-folding of domain V of the 23S RNA. At the same time, the relevant ribosomal proteins may contribute to stabilization of the proper structure of the peptidyl transferase region and the entire domain V. In any case, up to now nobody has been able to prove unequivocally that protein-free 23S rRNA can catalyze peptide bond formation.

Localization of the PTC on the morphologically visible surfaces of the 50S subunit can be accomplished by knowing the proteins that are complexed with the sequence 2450–2600 of the 23S RNA domain V, and from immuno-electron microscopic detection of specific substrates or inhibitors of the PTC (see Stoeffler and Stoeffler-Meilicke, 1984). The protein L27 has been found to form multiple cross-links with this sequence, and at the same time it has been detected by immuno-electron microscopy under the central protuberance, in the region of the groove ("neck") between it and the rest of the 50S subunit body. Puromycin derivatives as substrates of the peptidyl transferase are also detected under the central protuberance, but more at the side of the L1 ridge. The same place has been indicated by detection of bound inhibitors, such as chloramphenicol and lincomycin.

On the whole, it can be stated that PTC is located at the 50S subunit, on its interface (concave) side, under the head (central protuberance), and more exactly in the region of the groove separating the head from the rest of the body. The likely position of PTC on the 50S ribosomal subunit is shown schematically in Fig. 9.8.

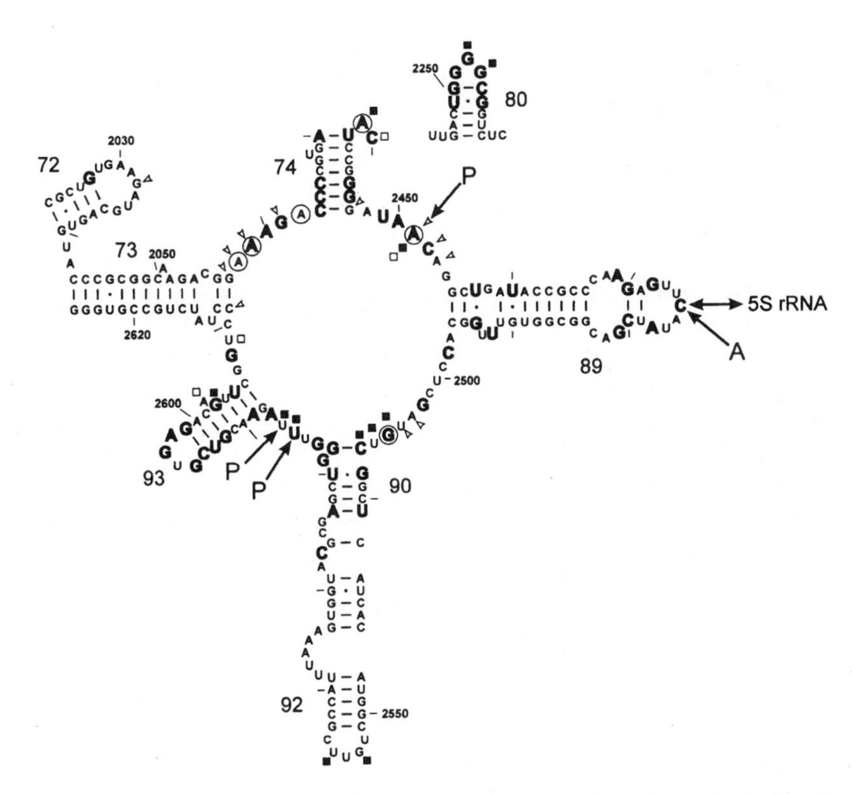

FIG. 9.7. Peptidyl transferase region ("PTC ring" and adjacent hairpins) of domain V of the *E. coli* 23S rRNA (see also Fig. 6.5). Nucleotides indicated by bold letters reflect 100% conservation in evolution. Nucleotides protected by antibiotics against chemical modification are encircled. Open arrowheads point to nucleotides, the methylation or mutation of which confer resistance to antibiotics. See Garrett, R. A., and Rodriguez-Fonesca, C. (1996), in *Ribosomal RNA: Structure, Evolution, Processing and Function in Protein Biosynthesis* (R. A. Zimmermann and A. E. Dahlberg, eds.), pp. 327–355. CRC Press, Boca Raton. Footprint sites from tRNA at the P site (*d* site of PTC) are indicated by filled squares, and those from tRNA at the A site (*a* site of PTC) are indicated by open squares. See Moazed, D., and Noller, H. F. (1989) **57:**585–697. Cross-linking sites from tRNA at the P site (P) and at the A site (A), and from 5S rRNA are shown by arrows. From Osswald, M., Doereing, T., and Brimacombe, R. (1995) *Nucleic Acids Res.* **23:**4635–4641; Dontsova, O. A., Tishlov, V., Dokudovskaya, S., Bogdanov, A., Doering, T., Rinke-Appel, J., Thamm, S., Greuer, B., and Brimacombe, R. (1994) *Proc. Natl. Acad. Sci. USA* **91:**4125–4129.

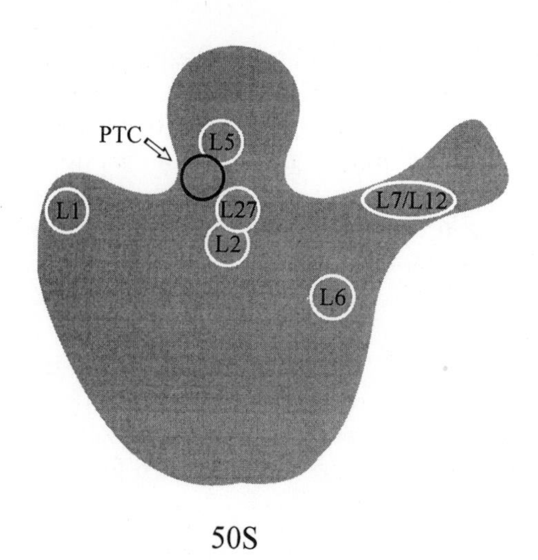

FIG. 9.8. Schematic contour drawing of the 50S ribosomal subunit with a plausible position of the peptidyl transferase center (PTC) in the "neck" region (under the central protuberance) on the contact surface (interface side) of the subunit. Positions of some ribosomal proteins are also indicated.

9.5. GTP-Dependent Binding of Translation Factors (Factor-Binding Site on the Large Subunit)

Elongation involves the periodic binding and release (once per cycle) of proteins EF1A (EF-Tu) and EF2 (EF-G) by the translating ribosome. Each of these proteins is bound in the complex with GTP, and their release is the result of GTP hydrolysis. The binding and release cycle of EF1A takes place during aminoacyl-tRNA binding, whereas the binding and release of EF2 proceeds during the translocation stage. Also, initiation of translation involves the ribosomal binding of protein IF2 (or eIF2 in the case of eukaryotes) with GTP and the release of this protein as a result of GTP hydrolysis. Finally, in the termination of translation, the ribosome binds and releases the RF proteins; GTP takes part in this process as well. All these proteins interacting with ribosomes in the form of their GTP complexes appear to bind to the same region of the ribosomal particle. It is likely that their binding sites on the ribosome are either identical or at least strongly overlapping. In any case, these proteins compete against each other for the binding site on the ribosome and cannot be present on it simultaneously.

Since all of the proteins mentioned are easily released from the ribosome after GTP hydrolysis, the study of their binding *in vitro* may be most conveniently performed if the GTP is replaced by a non-hydrolyzable analog, such as guanylyl methylene diphosphonate (GMP-PCP) or guanylyl imidodiphosphate (GMP-PNP) (Fig. 9.9). The complex of the protein with such an analog interacts with the ribosome and is retained on it.

The study of the binding of protein EF-G to bacterial 70S ribosomes has perhaps been the most thorough. EF-G with GMP-PCP may form a complex with both the translating and the vacant ribosome. EF-G with GTP also interacts with the translating ribosome and the vacant ribosome but it is not retained there because GTP undergoes hydrolysis and EF-G and GDP are released from the particle. In the presence of the antibiotic fusidic acid (see Fig. 12.4), however, EF-G preserves its affinity for the ribosome even after GTP has been cleaved. The isolated 50S ribosomal subunit behaves in a manner similar to the complete ribosome: EF-G with GMP-PCP, as well as EF-G with GTP (or, to be more accurate, with the product of GTP cleavage) in the presence of fusidic acid, forms a rather stable complex with the subunit; the interaction of EF-G plus GTP with the 50S ribosomal subunits results in GTP cleavage and the release of EF-G and GDP.

In order to identify the 50S subunit proteins forming the factor-binding site, antibodies against various ribosomal proteins have been used. It has been found that antibodies against protein L7/L12 inhibit binding of EF-G, whereas antibodies against a wide variety of other ribosomal proteins do not affect this function

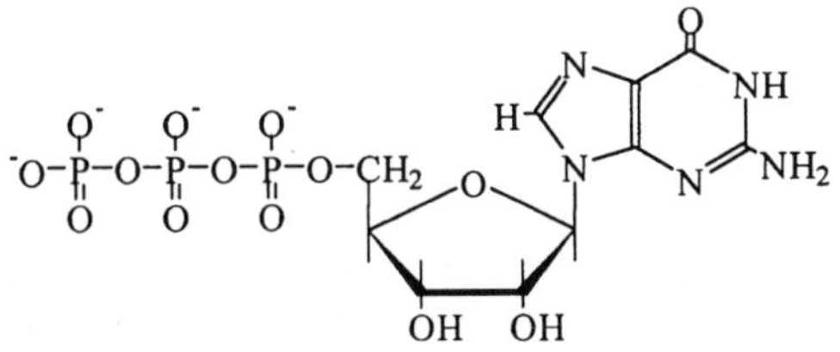

GTP

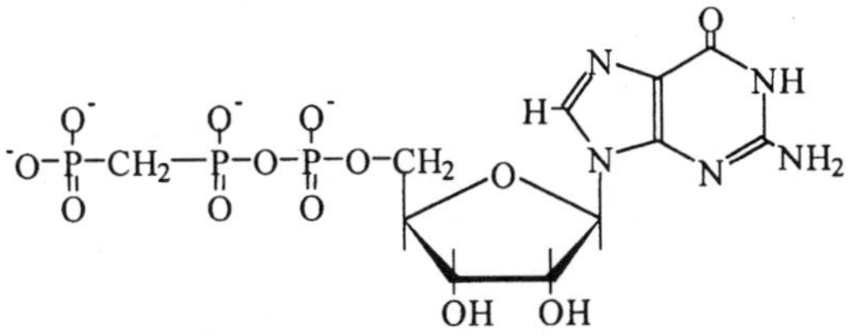

GMP–PCP

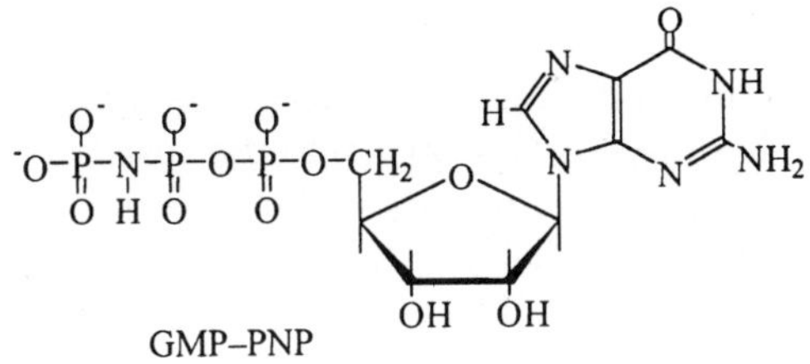

GMP–PNP

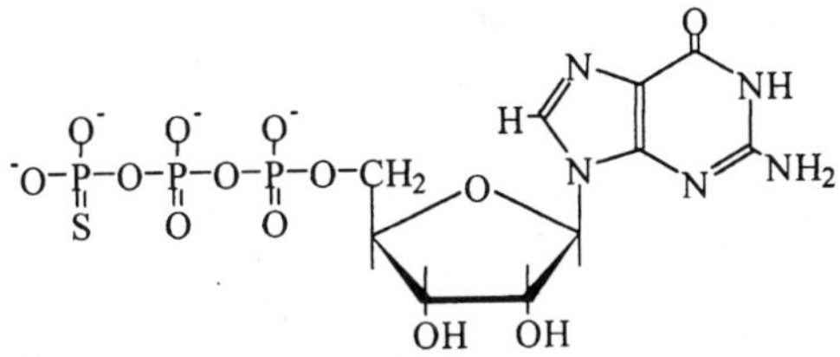

GTP (γS)

FIG. 9.9. GTP and its nonhydrolyzable and slowly hydrolyzable analogs used for studying the functions of the translation factors and the ribosome. GTP, guanosine 5′-triphosphate; GMP-PCP, nonhydrolyzable analog 5′-guanylyl methylene diphosphonate; GMP-PNP, very slowly hydrolyzable analog 5′-guanylyl imidodiphosphate; GTP(γS), slowly hydrolyzable analog guanosine 5′-(γ-thio)-triphosphate.

(Highland *et al.*, 1973). Also, the selective removal of protein L7/L12 from the 50S subunit, achieved by treatment with a 0.5 M NH_4Cl/ethanol mixture, resulted in a markedly decreased binding of EF-G to the ribosome (Hamel, Koka, and Nakamoto, 1972). More recent experiments using the hot tritium bombardment technique (see Section 8.2.4) have demonstrated that the EF-G bound to the 50S subunit shields only the protein L7/L12 and the adjacent L11 (but not L10) from the tritium atom flow. Hence, the L7/L12 stalk and its base seem to be the site of EF-G binding.

The footprinting technique (Noller *et al.*, 1990) has demonstrated that the interaction of the bacterial ribosome with either EF-G or EF-Tu results in protection of the so-called sarcin/ricin loop of domain VI in the 23S RNA. This is the end loop of the first hairpin (positions 2646–2774 in Fig. 6.5B) of domain VI known to be the target of two specific enzymes inactivating the bacterial ribosome: sarcin, which cleaves the internucleotide bond between G2661 and A2662, and ricin, which induces depurinization of A2660 (see Sections 13.5.3 and 13.5.4). This hairpin comprises one of the most highly conserved nucleotide sequences of rRNAs. Until recently, it was thought that the sarcin/ricin loop at positions 2653–2667 had a single-stranded conformation. However, NMR studies have shown that its structure is well ordered and only two nucleotide residues, A2660 and G2661, are not involved in secondary-structure interactions. Both EF-G and EF-Tu have been shown to protect G2655, A2660, and G2661 (Fig. 9.10A). It seems that the sarcin/ricin loop is the main common site of the interaction of the elongation factors with 23S RNA in the 50S ribosomal subunit.

In addition, EF-G protects A1067 in the end loop of a compound hairpin of domain II of the 23S RNA (see Fig. 6.5A). EF-G can also be cross-linked with this loop. This part of the 23S RNA structure (the three-way helical structure at positions 1030–1125) is known to accommodate the protein complex L10:$(L7/L12)_4$ and protein L11, and is the site of the interaction of the antibiotics thiostrepton and micrococcin with the 23S rRNA. These antibiotics (in the presence of protein L11) when bound to the ribosome inhibit the EF-G-dependent GTP hydrolysis. 2'-O-methylation of A1067 confers to the ribosome the resistance against the drugs. The structural element under consideration (Fig. 9.10B) is sometimes called the GTPase region. Thus it is likely that EF-G, and probably other GTP-dependent translation factors, interact with two areas on the large subunit: one is the L7/L12 stalk, including both its protein and rRNA moieties, and the other is the sarcin/ricin loop seemingly located somewhere near the base of the stalk.

This conclusion is confirmed by studies of the site of thiostrepton binding (Cundliffe, 1990). Thiostrepton (see Fig. 10.11) is an antibiotic that prevents the binding of EF-G and EF-Tu to 50S ribosomal subunits. It has been shown that the antibiotics bind to the subunit in the region of protein L11 and the L11-protected 23S RNA sequence 1050–1110. This is the same region where EF-G can be cross-linked and where it protects the nucleotide residue against chemical attack (Fig. 9.10B). The region is in the vicinity of the proteins L7/L12. Thus thiostrepton, a rather large molecule, bound at the base of the L7/L12 stalk may directly block one of the two principal sites of interaction of the ribosome with elongation factors.

The position of an EF-G attachment site on the 50S subunit has been determined using immuno-electron microscopy (Fig. 9.11A). For this purpose, a photoactivatable arylazide derivative of EF-G was prepared and specifically bound to the particle in the presence of GTP and fusidic acid. Then, a covalent cross-link between EF-G and its attachment site was obtained by irradiation. Subunits with such a covalently linked protein (EF-G) were treated with antibodies against EF-G, and the complexes studied by electron microscopy. Fig. 9.11A shows the location of the site of antibody attachment at the base of the L7/L12 stalk, on the interface side of the 50S ribosomal subunit.

It should be pointed out that although the 50S ribosomal subunit is largely responsible for the recognition and binding of EF-G, the EF-G may also come into contact with the 30S subunit. Specifically, protein S12 can be cross-linked by a disulfide bond with EF-G if the complex between EF-G and 70S ribosomes is subjected to oxidation (Girshovich, Bochkareva, and Ovchinnikov, 1981). This result indicates that the ribosome-bound EF-G protrudes from the large subunit to the protein cluster S4-S5-S12 of the small subunit. More recent observations on the interaction of the tRNA-like domain IV of EF-G (see Sections 12.2.1 and 12.2.3) with the decoding center of the 30S subunit, specifically with position 1400 of the 16S rRNA (Wilson and Noller, 1998), confirm this view.

The other elongation factor, EF1A (EF-Tu), is delivered to the ribosome as a complex with the aminoacyl-tRNA and GTP. EF1A interacts with the ribosome when the complex is bound. In experiments with bacterial ribosomes it has been shown that it binds to the 50S ribosomal subunit. The presence of EF-G on the 50S

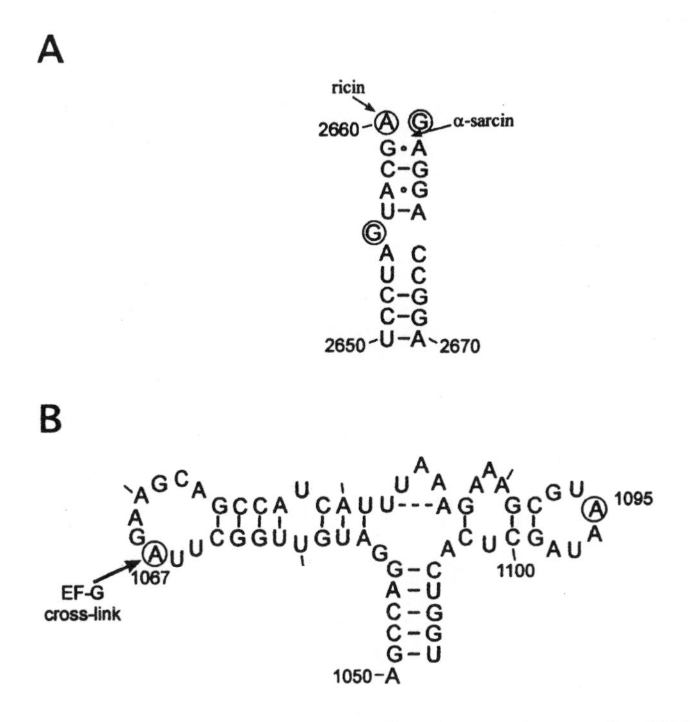

FIG. 9.10. Binding sites of elongation factors Tu and G on the *E. coli* 23S rRNA. (**A**) Secondary structure of the sarcin/ricin region, based on NMR data obtained for the corresponding sequence from eukaryotic 28S rRNA, See Szewczak, A. A., and Moore, P. B. (1995) *J. Mol. Biol.* **247**:81–98. The ricin-catalyzed depurination site and the sarcin cleavage site are indicated by arrows. See Endo, Y., Mitsui, K., Motizuki, M., and Tsurugi, K. (1987) *J. Biol. Chem.* **267**5908–5912; Wool, I. G. (1984) *Trends Biochem. Sci.* **9**:14–17. The sites protected by EF-G and EF-Tu are encircled. See Moazed, D., Robertson, J. M., and Noller, J. F. (1988) *Nature* **334**:362–364. (**B**) Secondary structure of the GTPase region of the 23S rRNA. The major (A1067) and minor (1095) sites protected by EF-G are encircled. See Moazed, D., Robertson, J. M., and Noller, J. F. (1988) *Nature* **334**:362–364.

subunit prevents EF-Tu from interacting with the ribosome, that leads to the conclusion that the EF-Tu-binding site either coincides with or overlaps the EF-G-binding site. As in the case of EF-G, the antibodies directed against protein L7/L12, and only these antibodies, inhibit the interaction between EF-Tu and the ribosome. The removal of protein L7/L12 strongly reduces the interaction between EF-Tu and the ribosome. Immuno-electron microscopy studies have demonstrated that antibodies against EF-Tu can be detected both at the base and at the tip of the L7/L12 stalk (Fig. 9.11B).

The initiation factor IF2, and the termination factor RF3, also compete with EF1 and with EF2 for the binding site. Their interaction with the ribosome again depends on the presence of protein L7/L12. All of this information supports the assumption that the binding of all translation factors, using GTP as an effector, has many common features and that the ribosomes possesses a single factor-binding site at the L7/L12 stalk on the 50S ribosomal subunit.

In all cases GTP must be bound to a translation factor, such as EF1, EF2, IF2, or RF3, *prior* to factor binding to the ribosome. Therefore, it is apparent that the GTP-binding center is located on the factor protein itself. However, hydrolysis of the bound GTP into GDP and orthophosphate takes place *after* the factor is bound to the ribosome. In other words, both the factor and the ribosome in a complex are required to induce the GTPase activity. It is the attachment of the GTP-containing factor to the ribosome that results in GTP hydrolysis.

At the same time, a number of experiments with photoactivatable GTP analogs have demonstrated that if a chemical cross-link between GTP and surrounding groups is induced after the EF-G:GTP ribosome complex has been formed, then EF-G, but not the ribosomal components, is attacked preferentially, regardless of which moiety of GTP carries the photoactivatable group. Moreover, the antibiotic kirromycin (see Section 10.3) has been shown to induce an intrinsic GTPase ac-

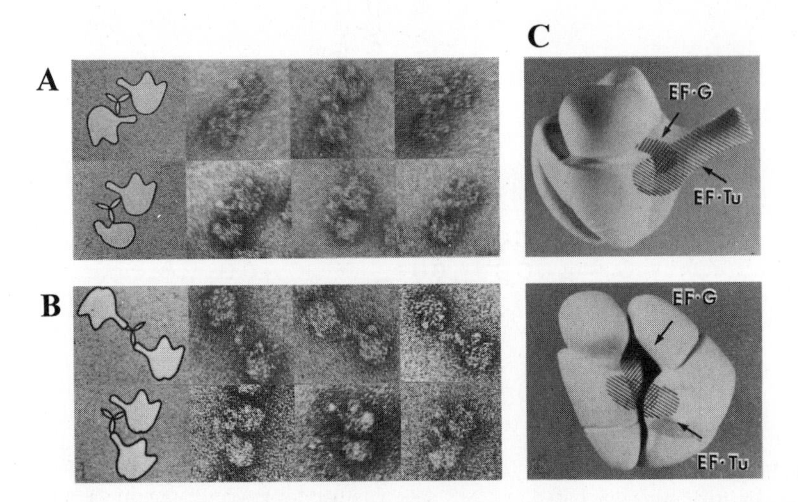

FIG. 9.11. Localization of the elongation factor-binding site on the large ribosomal subunit by immuno-electron microscopy. (**A**) Electron micrographs of 50S ribosomal subunits with EF-G reacted with antibodies. From Girshovich, A. S., Kurtskhalia, T. V., Ovchinnikov, Y. A., and Vasiliev, V. D. (1981) *FEBS Letters* **130**:54–59. (**B**) Electron micrographs of the 50S subunits with EF-Tu reacted with antibodies. From Girshovich, A. S., Bochkareva, E. S., and Vasiliev, V. D. (1986) *FEBS Letters* **197**:192–198. (**C**) The model of the 70S ribosome with the approximate localization of EF-G and EF-Tu at the base of L7/L12 stalk (hatched areas). Original photos were provided by V. D. Vasiliev.

tivity of EF-Tu, in the absence of ribosomes (Chinali, Wolf, and Parmeggiani, 1977). It is therefore now generally held that the attachment of the factor to the ribosome results in an activation of the intrinsic GTPase center of the factor, while the ribosome does not possess either a preexisting GTPase center or any indispensible complement of the GTPase center of the factor.

9.6. Binding of Aminoacyl-tRNA and Retention of Peptidyl-tRNA (tRNA-Binding Sites at the Intersubunit Space)

The ribosome possesses an intrinsic affinity for tRNA. A vacant ribosome can bind any tRNA or its derivative (e.g. aminoacyl-tRNA or peptidyl-tRNA) in the absence of a template polynucleotide. The presence of a template polynucleotide makes this binding specific: only the tRNA corresponding to the template codon—the *cognate tRNA*—will be bound. It is likely that a codon exerts both positive (cognate codon) and negative (non-cognate codon) discrimination effects on the binding of tRNA to the ribosome.

Since two substrates, aminoacyl-tRNA and peptidyl-tRNA, participate in the central chemical reaction of the elongation cycle, the main question is where the corresponding A and P sites are located on the ribosome. The following obvious postulates must be put at the basis of any considerations concerning the tRNA positions in the translating ribosome: (1) The anticodons of the two tRNAs, one in the A site and the other in the P site, must be drawn together, in order to provide their interactions with neighbor codons along the mRNA. (2) The acceptor ends of the two tRNAs also must be in close proximity, in order to provide the transpeptidation reaction. (3) The central cores ("elbows") of the two L-shaped tRNAs may be drawn apart. Thus, the two tRNAs form a tRNA pair (Fig. 9.12) that can be considered a unit in search of its position on the ribosome.

According to all the data available, the codon section of mRNA and, hence, the anticodons of tRNAs, are at the small subunit of the translating ribosome, evidently in the cleft separating the head on one side and the body and the side bulge ("platform") on the other, that is, at the neck of the small subunit; this position is marked by an open cycle at the contour representation of the overlap projection of the ribosome (Fig. 9.13). At the same time, the PTC and, hence, the acceptor ends of the tRNAs are at the large subunit, in the groove under the central protuberance, that is, at the neck of this subunit; a closed cycle marks this site in Fig. 9.13. Therefore, the axis connecting the anticodon regions of the tRNA pair (AC) with the acceptor region (PTC) must be directed more or less perpendicular to the subunit interface. This means that the bodies of both tRNA molecules must be placed in the interface space of the ribosome, between the ribosomal subunits. As already mentioned in Section 5.5, the space seems to be sufficient to accommodate at least two tRNA molecules. Indeed, recent reports on cryo-electron microscopy of ribosomes charged with tRNA molecules directly confirmed that the tRNAs occupy the intersubunit space located in the "inter-neck" pocket of the ribosome (Agrawal et al., 1996; Stark et al., 1997).

In the pair of tRNAs, one being placed in the A site and the other in the P site, two principally different orientations can be considered. The first is the so-called R type orientation (Rich, 1974), when the T loop of the A-site tRNA faces the D loop of the P-site tRNA (Fig. 9.14, P(R)). In this case, after transpeptidation, the translocational movement of the A-site tRNA residue to the P site will proceed

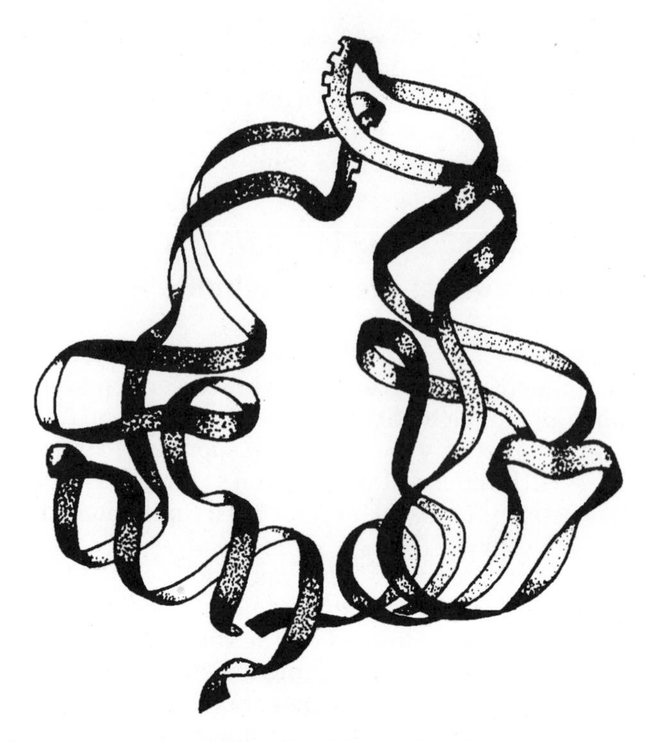

FIG. 9.12. Mutual orientation of two ribosome-bound tRNA molecules represented as ribbon-drawn models: the anticodons are immediate neighbors on the mRNA chain, the acceptor ends are also brought together, while the corners ("elbows") are arranged apart.

clockwise if viewed from anticodons along the axis mentioned (right-hand screw). The alternative is the so-called S type orientation (Sundaralingam et al., 1975): the D loop of the A-site tRNA faces the T loop of the P-site tRNA (Fig. 9.14, P(S)). Here the translocational movement will be counterclockwise (left-hand screw). There are arguments in favor of both possibilities, and the choice between the two alternatives has not been made yet.

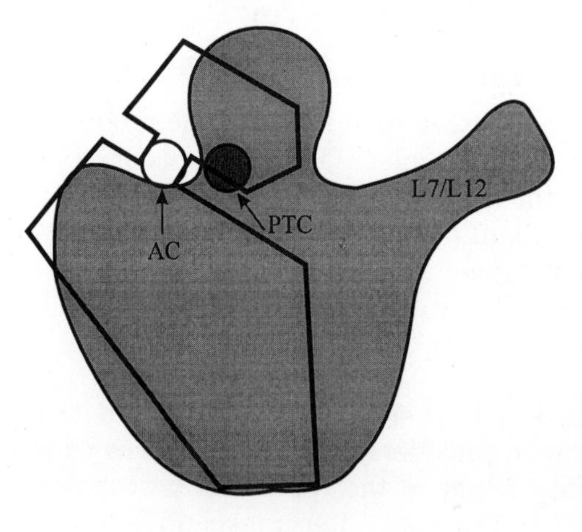

FIG. 9.13. Schematic contour drawing of the 70S ribosome in the overlap projection. The 30S subunit is depicted as an empty figure whereas the 50S subunit is shaded. The decoding site with tRNA anticodons (AC, open circle) on the 30S subunit is positioned over the peptidyl transferase center (PTC, closed circle) with tRNA acceptor ends, so that the axis connecting the anticodons with the acceptor ends is approximately perpendicular to the subunit interface. Redrawn from Lim, V., Venclovas, C., Spirin, A., Brimacombe, R., Mitchell, P., and Müller, F. (1992) *Nucleic Acids Res.* **20**:2627–2637.

Concerning the orientation of the cores ("elbows") of the tRNA pair relative to he ribosomal subunit lobes, the information about the localization of the elonga-ion factors (Section 9.4) appears to be the most relevant to the problem. Thus, :F1A (EF-Tu) is known to be bound with the T-loop side (see below, Section 9.5.3) of the aminoacyl-tRNA which is going into the A site. Since EF1A interacts with he rod-like (L7/L12) stalk and its base on the large subunit, it is likely that the A-ite tRNA is positioned at that sector of the intersubunit space, with the "elbow" t the stalk (Spirin, 1983). Then, in the case of the R type orientation of the tRNAs, he P-site tRNA should be more distal from the heads of the ribosomal subunits nd located between their bodies, as depicted in the top part of Fig. 9.15. If the S ype orientation of the tRNAs is the case, the P-site tRNA will be found at the heads of the subunits, as shown in the lower part of Fig. 9.15.

.6.1. P Site

In cases where tRNA or its derivative is accepted by the vacant ribosome, one of the two tRNA-binding sites is filled first. This seems to be the same site that is occupied by the peptidyl-tRNA prior to transpeptidation in the translating ribo-omes, namely, the P site (see Fig. 9.1).

The retention of tRNA in the P site of the translating ribosome, however, has n important feature. It is vital that the peptidyl-tRNA should not be exchangeable with the medium during translation. Correspondingly, the peptidyl-tRNA bound n the P site of the translating ribosome should not be in equilibrium with exoge-ious tRNA, but rather occluded, that is, its dissociation rate should be very low. n contrast, when the deacylated tRNA or aminoacyl-tRNA occupies the P site, the ite becomes exchangeable and the tRNA may be released. It is likely that the ap-

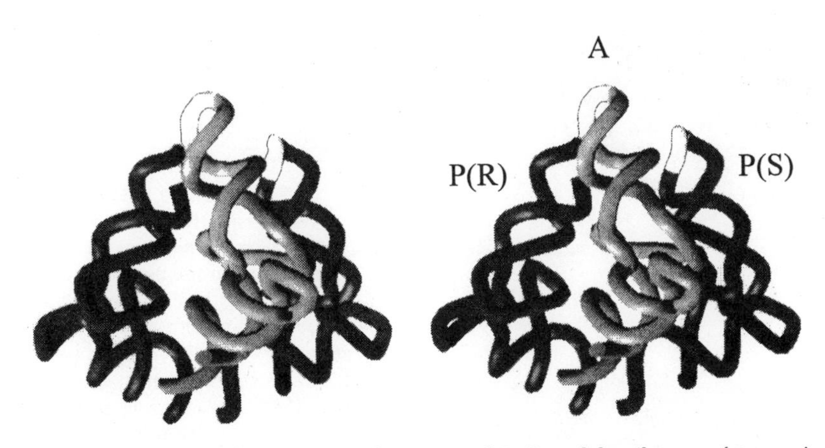

IG. 9.14. Mutual orientation of tRNA molecules in A and P sites of the ribosome (stereo view). The hosphate backbones of the tRNA molecules are depicted (A-site tRNA lighter, P-site tRNA darker, with he anticodon regions in white). Two alternative orientations are shown: "P(R)" indicates the position of a P-site tRNA in the R-orientation according to Rich, A. (1974) in *Ribosomes* (M. Nomura, A. Tissieres, and P. Lengyel, eds.), pp. 871–884, CSHL Press; "P(S)" the corresponding position in the S-orientation according to Sundaralingam, M., Brennan, T., Yathindra, N., and Ichikawa, T. (1975) in *Structure and Conformation of Nucleic Acids and Protein-Nucleic Acid Interactions* (M. Sundar-alingam and S. T. Rao, eds.), pp. 101–115, University Park Press, Baltimore, relative to a common A-site tRNA. Reproduced from Lim, V., Venclovas, C., Spirin, A., Brimacombe, R., Mitchell, P., and Müller, F. (1992) *Nucleic Acids Res.* **20**:2627–2637 with permission.

parent nonequilibrium retention of the peptidyl-tRNA in the P site of the translat ing ribosome is due to the contribution of the peptidyl residue (the C-terminal es ter bond group) that is anchored by the ribosomal particle during elongation. It i the *d* site of the PTC that may be responsible for the anchorage of the C-termina ester group of peptidyl-tRNA.

Experiments with separated ribosomal subunits have demonstrated that botl the small and the large subunit possess a certain affinity for tRNA. The capacity for a codon-specific binding of tRNA, however, is found only for the small (30S o 40S) ribosomal subunit, this being an obvious result of the fact that only this sub unit, but not the large one, can bind and hold the template polynucleotide. At th same time, after dissociation of the translating ribosomes the peptidyl-tRNA ofte remains bound to the large (50S or 60S) subunit. At present there is good reasor to believe that both ribosomal subunits are involved in the formation of the tRNA binding P site. In addition to the retention of the 3′-terminus of tRNA and the C terminal aminoacyl residue with its ester and amide groups in the *d* site of the PTC the rest of the P-site-bound tRNA has been reported to have contacts with domai IV of 23S RNA, as evidenced by protection of positions 1916, 1918, and 1926 fron chemical modifications (Noller *et al.*, 1990).

9.6.2. A Site

When the P site is filled with tRNA the ribosome becomes capable of bindin the second tRNA molecule. This binding takes place at another tRNA-binding site

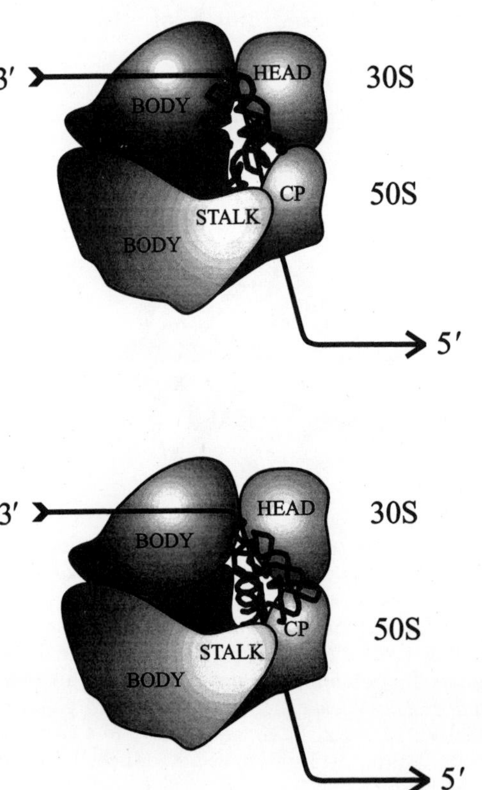

FIG. 9.15. Two possible positions of the tRN pair (A- and P-site tRNAs) in the 70S ribosome The tRNAs occupy the intersubunit space, witl both anticodons in the neck of the 30S subunit the acceptor ends in the groove ("neck") unde the central protuberance of the 50S subunit, an the corners ("elbows") are fixed in the region o the L7/L12 stalk. See Spirin, A. S. (1983) *FEB Letters* **156**:217–221. The *upper* depictio shows the R orientation of the tRNAs, and th *lower* is the S orientation. See Lim, V., Venclo vas, C., Spirin, A., Brimacombe, R., Mitchell, P. and Müller, F. (1992) *Nucleic Acids Res* **20**,:2627–2637; Spirin, A. S., Lim, V. I., an Brimacombe, R. (1993), in *The Translational Ap paratus: Structure, Function, Regulation, Evolu tion* (K. H. Nierhaus, F. Francesci, A. R. Subra manian, V. A. Erdmann, and B. Wittmann- Liebold, eds.), pp. 445–454, Plenum Press, Nev York.

the A site. Binding in the A site is greatly stimulated by the template polynucleotide; in this case binding is codon-specific, that is, only tRNA corresponding to the codon placed in the site becomes bound. The affinity of tRNA toward the A site is approximately one order of magnitude lower than toward the P site. In the course of normal translation, the binding of the aminoacyl-tRNA is specifically stimulated by the EF1A (EF-Tu) protein.

It may be that the A site, like the P site, is formed by both ribosomal subunits. In any case, the tRNA anticodon should be placed in the immediate vicinity of the mRNA codon, that is, on the small (30S or 40S) ribosomal subunit, whereas the acceptor end interacts with the *a* site of PTC, that is, with the large (50S or 60S) subunit (see Section 9.3 and Fig. 9.7).

9.6.3. Entry Site (R or T Site)

When entering the ribosome, aminoacyl-tRNA is complexed with EF1A (EF-Tu). Despite the codon–anticodon interaction at the decoding site of the ribosome, it cannot be a substrate for transpeptidation until GTP is cleaved and EF1A is released. From this it can be postulated that on the way to the A site aminoacyl-tRNA sits first on an intermediate entry site, also called the recognition site (R site), or T site because of the presence of the bound EF-Tu. Indeed, aminoacyl-tRNA can be retained at the entry site if a noncleavable GTP analog, GMP-PCP or GMP-PNP, substitutes for GTP in the Aa-tRNA:EF-Tu complex. According to the foot-printing data available, as well as the results on cross-linking of tRNA with ribosomal components, the contacts of the T-site-bound tRNA with the small (30S) ribosomal subunit are very similar or identical to those of the A-site-bound tRNA. At the same time, no direct contacts with the large (50S) subunit have been detected, except those through EF-Tu with the factor-binding site. It may be concluded, therefore, that the entry site is not a separate tRNA-binding site of the ribosome, but just an intermediate position of aminoacyl-tRNA when it is already bound with the A site on the small subunit and not bound yet with the PTC of the large subunit. This intermediate position has been designated by Noller and co-workers (1990) as the "hybrid A/T site" (the term "A/T position" would be more appropriate, however) (Fig. 9.16A). The subsequent release of EF-Tu allows the acceptor end of the tRNA and its aminoacyl residue to interact directly with the large subunit, specifically with the *a* site of the PTC, thus completing the binding of aminoacyl-tRNA to the A site (acquiring "A/*a* position") (Fig. 9.16B).

9.6.4. Intermediate Positions ("Hybrid Sites")

Following the above terminology, it can be said that the two substrates of the ribosome occupying A and P sites prior to transpeptidation are in positions A/*a* and P/*d* for aminoacyl-tRNA and peptidyl-tRNA, respectively (Fig. 9.16B). Transpeptidation gives two products retained by the ribosome: deacylated tRNA in the P site and elongated peptidyl-tRNA in the A site. However, the acceptor end of the deacylated tRNA is found to change its position as a result of transpeptidation; according to footprinting data, it is now in contact with the sequence 2110–2170 in domain V of 23S RNA, instead of PTC ring forming the *a* and *d* sites. The removal of the reacting group of a product from the reaction site may be the requirement of the reaction course: this provides a direct route to the reaction and prevents its reversibility. In any case, although the main body of the deacylated tRNA after transpeptidation is still in the P site, its 3′-end is shifted to a new site on the large subunit which can be designated as the *e* (*exit*) site. Hence, the deacylated tRNA after transpeptidation occupies the P/*e* position (or "hybrid P/E site" according to Noller's terminology) (Fig. 9.16C).

At the same time, after transpeptidation the newly formed C-proximal peptide group together with the added aminoacyl residue and 3′-terminus of the elongated peptidyl-tRNA is found fixed in the *d* site of the PTC. Thus, the peptidyl-tRNA prior to the translocation occupies A/*d* position (or "hybrid A/P site") (Fig. 9.16C).

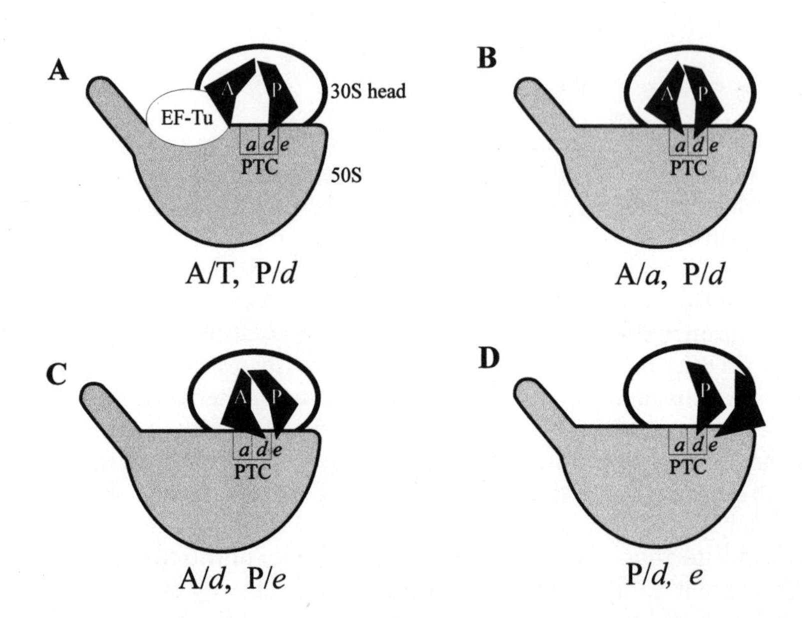

FIG. 9.16. Sites and intermediate positions ("hybrid states") of tRNAs in the ribosome. The ribosome is schematically depicted with the head of the 30S subunit and the central protuberance of the 50S subunit directed to the viewer (thus the L7/L12 stalk being on the left hand). "A" and "P" are the tRNA-binding sites on the 30S subunit, "*a*" and "*d*" are the acceptor- and donor-binding sites, respectively, of the peptidyl transferase center of the 50S subunit, and "*e*" is the site retaining the 3′-terminal adenosine residue of the deacylated tRNA after transpeptidation. (**A**) Positions of tRNAs upon binding of the ternary Aa-tRNA:EF-Tu:GTP complex with the peptidyl-tRNA-occupied ribosome. The tRNA residue of the Aa-tRNA is in the A site on the 30S subunit, but the aminoacylated end is on EF-Tu rather than in the PTC (A/T state). The peptidyl-tRNA occupies the P site on the 30S subunit and the *d* site in the PTC of the 50S subunit (P/*d* state). (**B**) "Non-hybrid" positions of tRNAs after the release of EF-Tu from the ribosome, prior to transpeptidation. The aminoacyl-tRNA sits in the A site of the 30S subunit, with its aminoacylated end in the *a* site of the PTC (A/*a* state). The peptidyl-tRNA resides in the P site of the 30S subunit and the *d* site of PTC (P/*d* state). (**C**) "Hybrid" positions of tRNAs after transpeptidation. The newly formed (elongated) peptidyl-tRNA occupies the A site on the 30S subunit, but its 3′-end with the ester group is caught by the *d* site of the PTC (A/*d* state). The deacylated tRNA sits in the P site of the 30S subunit, but its 3′-terminus with free 3′-hydroxyl is moved to the *e* site of the 50S subunit (P/*e* state). (**D**) Positions of tRNAs after translocation. The peptidyl-tRNA occupies the P site on the 30S subunit and the *d* site of the PTC (P/*d* state). The deacylated tRNA is transiently retained by the *e* site of the 50S subunit, probably without interactions with the 30S subunit (*e* state). Adapted from Moazed, D., and Noller, H. F. (1989) *Nature* **342:**142–148; Noller, H. F., Moazed, D., Stern, S., Powers T., Allen, P. N., Robertson, J. M., Weiser, B., and Triman, K. (1990) in *The Ribosome: Structure, Function, and Evolution* (W. Hill, A. Dahlberg, R. A. Garrett, P. B. Moore, D. Schlessinger, and J. R. Warner eds.), pp. 73–92, ASM Press, Washington, DC.

.6.5. Exit Site (E Site)

As a result of translocation the deacylated tRNA is expelled from the P site Fig. 9.16D). However, this does not necessarily mean that it is immediately re-eased from the ribosome: the deacylated tRNA after translocation may be tran-iently retained on the ribosome in the exit site (E site). The *e* site which binds the '-terminus of deacylated tRNA in the vicinity of the PTC of the large subunit eems to contribute mainly to the retention of the deacylated tRNA after translo-ation. It is not clear yet how other parts of the ribosome, especially the small sub-nit, are involved in the formation of the E site.

.7. Ligand Displacements (Translocation)

Transpeptidation in the course of the ribosome working cycle is followed by imultaneous displacements of the three large ligands: mRNA, peptidyl-tRNA, and leacylated tRNA. This may be defined as the mechanical function of the ribosome. Neither isolated ribosomal subunit is capable of even partially performing this unction. It is likely that the mechanical function requires the ribosome to be con-tructed of two subunits.

In the search for the molecular mechanisms responsible for vectorial dis-lacements of large ligands, we must first consider a possible large-block mobil-ty within the ribosome. Since the ribosome consists of two subunits, which re relatively loosely associated in the absence of ligands, it is possible, in rinciple, that the subunits are capable of moving relative to each other during ibosome functioning (Bretscher, 1968; Spirin, 1969). There are experiments lemonstrating changes in the compactness of the ribosome in the course of ranslocation (Spirin *et al.*, 1987) and electron microscopy observations showing he increase in the intersubunit space in the ribosomes within the cell upon star-ation (Oefverstedt et al., 1994); this may be evidence in favor of the relative novement of the subunits by a swinging ("locking–unlocking") mechanism (Fig. .17).

Another mobile element of the ribosome is the L7/L12 stalk of the 50S ribo-omal subunit (Gudkov et al., 1982). A considerable amount of information sug-ests that the L7/L12 stalk is involved directly in the functions of protein transla-ion factors, and particularly in the EF-G-catalyzed translocation. It would come s no surprise that the mobility of the L7/L12 stalk plays a part in the ligand dis-lacements during translocation, as well as, perhaps, in the course of aminoacyl-RNA delivery into the ribosome.

The possibility of some interdomain (interlobe) mobility within ribosomal ubunits, especially in the small subunit, which seems to be more labile and hangeable, cannot be excluded either.

.8. The Material and Energy Balance of the Elongation Cycle

The consecutive stages of the codon-directed binding of aminoacyl-tRNA, ranspeptidation, and translocation create a cycle, resulting in (1) the determina-ion of the position of one amino acid residue in the polypeptide chain to be synthesized, (2) the formation of one peptide bond, (3) the deacylation of one mole-cule of aminoacyl-tRNA, (4) the hydrolysis of two molecules of GTP to GDP and orthophosphate, and (5) the shift (readout) of one nucleotide triplet of the template polynucleotide relative to the ribosome. Repetitions of this cycle create elongation; the number of cycles during elongation depends on the number of template codons (minus the initiation codon).

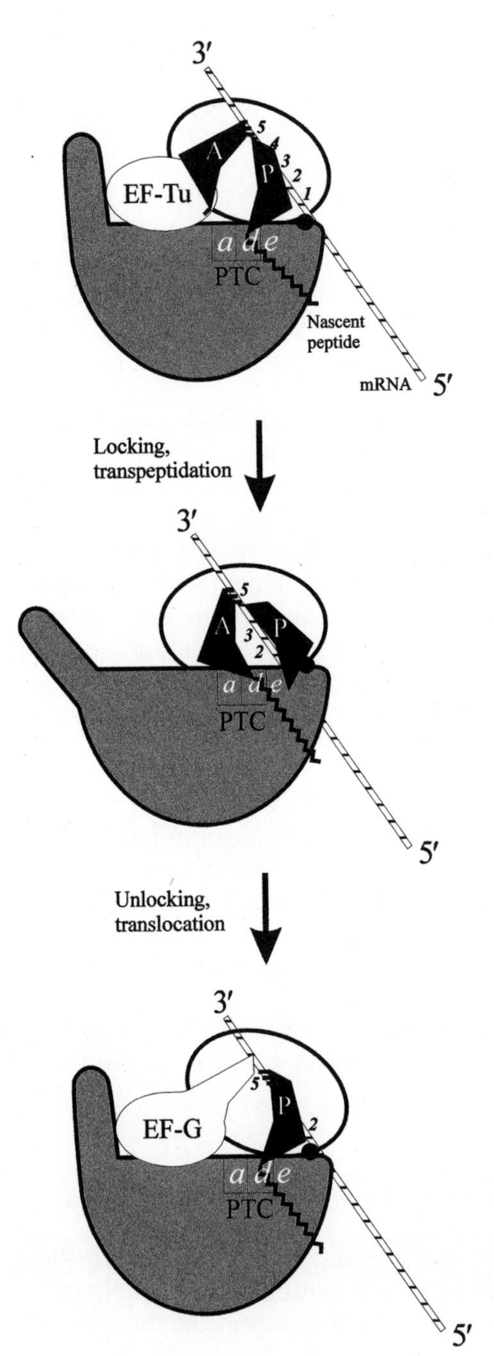

FIG. 9.17. Model of periodical locking-unlocking of the translating ribosome. For the original idea see Spirin, A. S. (1968) *Doklady Akad. Nauk SSSR* **179:**1467–1470; (1969) *Cold Spring Harbor Symp. Quant. Biol.* **34:**197–207. The model pos-tulates that the two ribosomal subunits are mov-ably jointed with each other and capable of draw-ing slightly apart (unlocking) and together (locking). The unlocking opens the functional sites on the subunit interface, such as the A site and provokes ligand displacements including translocation, whereas the locking closes the lig-ands inside the ribosome and brings the sub-strates for transpeptidation together. In other words, the binding of Aa-tRNA requires the un-locked, or open state of the ribosome; it may be that this state is induced by EF1:GTP (the *upper* depiction). The binding is completed by the re-lease of EF1, and the interaction of the aminoacy-lated end of the Aa-tRNA with the *a* site of the PTC may cause the locking of the ribosome; the transpeptidation proceeds in the closed ribosome (the *middle* depiction). The next unlocking can be driven by EF2:GTP, resulting in the translocation-al displacements of tRNAs and mRNA (the *lower* depiction). In particular, the unlocking of the pre-translocation ribosome and drawing the subunits apart will create the situation when the tRNA with the ester group firmly anchored in the *d* site of the PTC on the large subunit drags after itself the mRNA codon with which it interacts and thus displaces the mRNA chain along the small sub-unit (mRNA translocation). After the release of EF2 the posttranslocation ribosome may again close, or rather be in equilibrium between the locked and unlocked states.

With regard to the material balance, one cycle involves the consumption of one molecule of aminoacyl-tRNA and two molecules of GTP (plus two water molecules) from solution. One molecule of deacylated tRNA, two molecules of GDP, and two molecules of orthophosphate are released into solution (Fig. 9.1). Peptide elongation by one residue takes place concomitantly in the ribosome.

From this description the energy balance of the cycle can be summed up. The energy requirements of the cycle appear to be rather modest: they include determination of the position of the amino acid residue in the polypeptide chain ($\Delta G^{0\prime} \leq +2.5$ kcal/mole) and formation of the peptide bond ($\Delta G^{0\prime} \cong +0.5$ kcal/mole). It is clear that these energy requirements of the complete cycle are amply covered by the free energy which is liberated by deacylation of the aminoacyl-tRNA ($\Delta G^{0\prime} \cong -7$ kcal/mole). Nevertheless, the hydrolysis of two molecules of GTP accompanies the cycle (Fig. 9.1), resulting in the liberation of additional large amounts of free energy ($\Delta G^{0\prime} \cong -15$ kcal/mole). Thus, elongation appears to be a wasteful process, noneconomical, and with a low efficiency. The bulk of the free energy liberated during the cycle dissipates into heat.

However, if GTP is excluded even at one stage in the cycle (at the stage of aminoacyl-tRNA binding or at the stage of translocation), the process is greatly slowed and becomes fairly sensitive to unfavorable conditions, drugs, and other impedences. Therefore, a great excess of free energy is needed for the system in order to provide for high rates and high resistance of elongation. It is apparent that economy is not the main advantage providing for the survival of the system and of the corresponding organism in nature.

References

Agrawal, R. K., Penczek, P., Grassucci, R. A., Li, Y., Leith, A., Nierhaus, K. H., and Franck, J. (1996) Direct visualization of A-, P-, and E-site transfer RNAs in the *Escherichia coli* ribosome, *Science* **271**:1000–1002.

Barta, A., Kuechler, E., and Steiner, G. (1990) Photoaffinity labeling of the peptidyltransferase region, in *The Ribosome: Structure, Function, and Evolution* (W. E. Hill, A. Dahlberg, R. A. Garrett, P. B. Moore, D. Schlessinger, and J. R. Warner, eds.), pp. 358–365, ASM Press, Washington, DC.

Bretscher, M. S. (1968) Translocation in protein synthesis: A hybrid structure model, *Nature* **218**:675–677.

Brimacombe, R., Greuer, B., Mitchell, P., Osswald, M., Rinke–Appel, J., Schueler, D., and Stade, K. (1990) Three-dimensional structure and function of *Escherichia coli* 16S and 23S rRNA as studied by cross-linking techniques, in *The Ribosome: Structure, Function, and Evolution* (W. E. Hill, A. Dahlberg, R. A. Garrett, P. B. Moore, D. Schlessinger, and J. R. Warner, eds.), pp. 93–106. ASM Press, Washington, DC.

Chinali, G., Wolf, H., and Parmeggiani, A. (1977) Effect of kirromycin on elongation factor T_u: Location of the catalytic center for ribosome·elongation-factor-Tu GTPase activity on the elongation factor, *Eur. J. Biochem.* **75**:55–65.

Cooperman, B. S., Weitzmann, C. J., and Fernandez, C. L. (1990) Antibiotic probes of *Escherichia coli* ribosomal peptidyltransferase, in *The Ribosome: Structure, Function, and Evolution* (W. E. Hill, A. Dahlberg, R. A. Garrett, P. B. Moore, D. Schlessinger, and J. R. Warner, eds.), pp. 491–501, ASM Press, Washington, DC.

Cundliffe, E. (1990) Recognition sites for antibiotics within rRNA, in *The Ribosome: Structure, Function, and Evolution* (W. E. Hill, A. Dahlberg, R. A. Garrett, P. B. Moore, D. Schlessinger, and J. R. Warner, eds.), pp. 479–490, ASM Press, Washington, DC.

Girshovich, A. S., Bochkareva, E. S., and Ovchinnikov, Y. A. (1981) Elongation factor G and protein S12 are the nearest neighbors in the *Escherichia coli* ribosome, *J. Mol. Biol.* **151**:229–243.

Gudkov, A. T., Gongadze, G. M., Buchuev, V. N., and Okon, M. S. (1982) Proton magnetic resonance study of the ribosomal protein L7/L12 *in situ*, *FEBS Letters* **138**:229–232.

Hamel, E., Koka, M., and Nakamoto, T. (1972) Requirement of an *E. coli* 50S ribosomal protein com ponent for effective interaction of the ribosome with T and G factors and with guanosine triphos phate, *J. Biol. Chem.* **247**:805–814.

Highland, J. J., Bodley, J. W., Gordon, J., Hasenbank, R., and Stoeffler, G. (1973) Identity of the riboso mal proteins involved in the interaction with elongation factor G, *Proc. Natl. Acad. Sci. USA* **70**:142–150.

Huettenhofer, A., and Noller, H. F. (1994) Footprinting mRNA-ribosome complexes with chemica probes, *EMBO J.* **13**:3892–3901.

Lipmann, F. (1969) Polypeptide chain elongation in protein biosynthesis, *Science* **164**:1024–1031.

Monro, R. E. (1967) Catalysis of peptide bond formation by 50S ribosomal subunits from *Escherichia coli, J. Mol. Biol.* **26**:147–151.

Nathans, D. (1964) Puromycin inhibition of protein synthesis: Incorporation of puromycin into pep tide chains, *Proc. Natl. Acad. Sci. USA* **51**:585–592.

Noller, H. F. (1998) Ribosomal RNA, in *RNA Structure and Function,* pp. 253–278, CSHL Press, Cold Spring Harbor, NY.

Noller, H. F., Moazed, D., Stern, S., Powers, T., Allen, P. N., Robertson, J. M., Weiser, B., and Triman, K (1990) Structure of rRNA and its functional interactions in translation, in *The Ribosome: Struc ture, Function and Evolution* (W. E. Hill, A. Dahlberg, R. A. Garrett, P. B. Moore, D. Schlessinger and J. R. Warner, eds.), pp. 73–92, ASM Press, Washington, DC.

Noller, H. F. (1991) Ribosomal RNA and translation. *Annu. Rev. Biochem.* **60**:191–227.

Powers, T., and Noller, H. F. (1995) Hydroxyl radical footprinting of ribosomal proteins on 16S rRNA *RNA* **1**:194–209.

Prince J. B., Taylor, B. H., Thurlow, D. L., Offengand, J., and Zimmermann, R. A. (1982) Covalent cross linking of $tRNA_1^{Val}$ to 16S RNA at the ribosomal P site: Identification of cross-linked residues, *Proc Natl. Acad. Sci. USA* **79**:5450–5454.

Oefverstedt, L.-G., Zhang, K., Tapio, S., Skoglund, U., and Isaksson, L. A. (1994) Starvation in vivo fo aminoacyl-tRNA increases the spatial separation between the two ribosomal subunits, *Cell* **79**:629-638.

Rich, A. (1974) How transfer RNA may move inside the ribosome, in *Ribosomes* (M. Nomura, A. Tis sières, and P. Lengyel, eds.), pp. 871–884, CSHL Press, Cold Spring Harbor, NY.

Samaha, R. R., Green, R., and Noller, H. F. (1995) A base pair between tRNA and 23S rRNA in the pep tidyl transferase centre of the ribosome, *Nature* **377**:309–314.

Sergiev P.V., Lavrik, I.N., Wlasoff, V.A., Dokudovskaya, S.S., Dontsova, O.A., Bogdanov, A.A. and Brimacombe, R. (1997) The path of mRNA through the bacterial ribosome: A site-directe crosslinking study using new photoreactive derivatives of guanosine and uridine, *RNA* **3**:464-475.

Spirin, A. S. (1969) A model of the functioning ribosome: Locking and unlocking of the ribosome sub particles, *Cold Spring Harbor Symp. Quant. Biol.* **34**:197–207.

Spirin, A. S. (1983) Location of tRNA on the ribosome, *FEBS Letter* **156**:217–221.

Spirin, A. S., Baranov, V. I., Polubesov, G. S., and Serdyuk, I. N. (1987) Translocation makes the ribo some less compact, *J. Mol. Biol.* **194**:119–128.

Stark, H., Orlova, E. V., Rinke–Appel, J., Juenke, N., Mueller, F., Rodnina, M., Wintermeyer, W., Brima combe, R., and van Heel, M. (1997) Arrangement of tRNAs in pre- and posttranslocational ribo somes revealed by electron cryomicroscopy, *Cell* **88**:19–28.

Steitz, J. A. (1969) Nucleotide sequences of the ribosomal binding sites of bacteriophage R17 RNA, *Cold Spring Harbor Symp. Quant. Biol.* **34**:621–630.

Stoeffler, G., and Stoeffler–Meilicke, M. (1984) Immunoelectron microscopy of ribosomes, *Annu. Rev Biophys. Bioeng.* **13**:303–330.

Sundaralingam, M., Brennan, T., Yathindra, N., and Ichikawa, T. (1975) Stereochemistry of messenge RNA (codon)-transfer RNA (anticodon) interaction on the ribosome during peptide bond forma tion, in *Structure and Conformation of Nucleic Acids and Protein-Nucleic Acid Interactions* (M Sundaralingam and S. T. Rao, eds.), pp. 101–115, University Park Press, Baltimore.

Takanami, M. and Okamoto, T. (1963) Interaction of ribosomes and synthetic polyribonucleotides, *Mol. Biol.* **7**:323–333.

Takanami, M. and Zubay, G. (1964) An estimate of the size of the ribosomal site for messenger RN binding, *Proc. Natl. Acad. Sci. USA* **51**:834–839.

Traut, R. R., and Monro, R. E. (1964) The puromycin reaction and its relation to protein synthesis, *Mol. Biol.* **10**:63–72.

Watson, J. D. (1964) The synthesis of proteins upon ribosomes, *Bull. Soc. Chim. Biol.* **46**:1399–142

Wilson, K. S., and Noller, H. F. (1998) Mapping the position of translational elongation factor EF-G in the ribosome by directed hydroxyl radical probing, *Cell* **92:**131–139.

Wower, J., Hixson, S. S., and Zimmermann, R. A. (1989) Labeling the peptidyl transferase center of the *Escherichia coli* ribosome with photoreactive $tRNA^{Phe}$ derivatives containing azidoadenosine at the 3′ end of the acceptor arm: a model of the tRNA-ribosome complex, *Proc. Natl. Acad. Sci. USA* **36:**5232–5236.

10

Elongation Cycle, Step I:
Aminoacyl-tRNA Binding

10.1. Codon–Anticodon Interaction

Analysis of the elongation cycle may conveniently begin at the point when the peptidyl-tRNA occupies the P site of the translating ribosome while the A site with the codon of the template polynucleotide positioned there is vacant (Fig. 9.1, top). Such a ribosome is capable of binding the next aminoacyl-tRNA molecule.

Although the binding of the aminoacyl-tRNA to the ribosomal A site appears to involve several binding centers of the site and, correspondingly, several regions of the tRNA molecule, the specificity of the bound aminoacyl-tRNA depends exclusively on the template codon. In other words it is the codon that is responsible for selecting the corresponding aminoacyl-tRNA (cognate aminoacyl-tRNA), that is, the tRNA carrying the aminoacyl residue coded by a given codon.

10.1.1. Adaptor Hypothesis and Its Proof

According to Crick's adaptor hypothesis (see Section 3.1), the structure of the amino acid residue is irrelevant as far as the selection of aminoacyl-tRNA by the codon is concerned. The codon is capable of complementary interaction only with the tRNA residue that plays the part of adaptor. Therefore, the amino acid residue attached to such an adaptor becomes presented to the ribosome without participating in codon recognition.

Direct experimental proof of this postulate of the adaptor hypothesis was obtained in the experiments conducted by Lipmann's and Benzer's groups (Chapeville *et al.*, 1962; von Ehrenstein, Weisblum, and Benzer, 1963). Cys-$tRNA^{Cys}$ was catalytically reduced using Raney nickel; as a result, the cysteine residue was converted into alanine while still bound to the $tRNA^{Cys}$:

$$H_2N{-}CH(CH_2{-}SH){-}C({=}O){-}O{-}tRNA^{Cys} \xrightarrow{H} H_2N{-}CH(CH_3){-}C({=}O){-}O{-}tRNA^{Cys}$$

When the Ala-$tRNA^{Cys}$ was added to a cell-free system containing ribosomes programmed by statistical poly(U,G) copolymer coding for phenylalanine, leucine, valine, glycine, tryptophan, and cysteine, but not coding for alanine, the synthesis of an alanine-containing polypeptide was observed. In another experiment, where the Ala-tRNA^{Cys} was added to the reticulocyte cell-free system of globin synthesis programmed with endogenous mRNA, alanine residues were incorporated into the synthesized polypeptide chain at positions normally occuped by cysteine residues.

10.1.2. The Concept of Anticodon

Codon recognition proceeds by the pairing of the codon with a complementary nucleotide triplet present in the adaptor. Hence, selecting aminoacyl-tRNA should be governed by a complementarity between the codon and this triplet, which is termed the *anticodon*. Experimental proof of the decisive part played by codon–anticodon complementarity in binding aminoacyl-tRNA has been provided by studies of mutationally altered tRNA with nucleotide changes in positions 34–36 (Fig. 3.8) corresponding to the anticodon. For instance, when the GUA anticodon of tRNA^{Tyr} of *E. coli* is changed into CUA, the Tyr-tRNA^{Tyr} no longer recognizes the tyrosine UAC codon but does recognize the termination codon UAG (Goodman *et al.*, 1968).

Thus, selecting the aminoacyl-tRNA for binding to the ribosomal A site is the result of complementary codon–anticodon interaction, implying that the codon and anticodon should form a complex with parameters of the Watson–Crick double helix in the A-like form. Characteristic features of this complex include the antiparallel orientation of chains; the formation of standard A:U, U:A, G:C, and C:G base pairs held by hydrogen bonds; and the stacking interaction of bases. In the case of tyrosine tRNA, the initial anticodon and mutationally altered anticodon should be paired with the tyrosine codon and the termination codon, respectively, as follows:

```
5′      3′        5′      3′
  \    /            \    /
  G : C             C : G
  |   |             |   |
  U : A             U : A
  |   |             |   |
  A : U             A : U
 /     \           /     \
3′      5′        3′      5′
```

10.1.3. Wobble Hypothesis

Strict canonical base pairing, however, does not provide a general rule for the interaction between the first anticodon residue and the third residue of the codon. It has been noted that if an amino acid is coded by two, three, or four codons, the first two nucleotide residues of these codons are always identical; only the third position is different (Figs. 2.1 and 2.2). Thus, a given amino acid is strictly coded by the two first codon positions but less strictly by the third position. On the other hand, it has been found that ribosomes programmed by different codons for the same amino acid may bind the same tRNA species; in other words, a tRNA can recognize more than one codon. For example, the same tRNA^{Phe} recognizes both UUU and UUC.

Upon analyzing this and other facts, Crick (1966) proposed his hypothesis about the pairing of the first nucleotide of the anticodon with the third residue of the codon; he suggested the possibility of *base wobbling* in this position. This proposal implies that, in addition to standard A:U, U:A, G:C, and C:G pairing, as well as I:C pairing (I, a deaminated A derivative, pairs similarly with G), noncanonical pairs may form whose geometric parameters are close to the standard ones. Such pairs include A:G; G:A and I:A; G:U and I:U; U:G; U:U; and U:C and C:U (Fig. 10.1).

The following characteristics of the codon dictionary (Fig. 2.2) should, however, be taken into account in order to fit the proposal to the facts: (1) U and C located in the third position of the codon are always equivalent according to the coding specificity; (2) A and G in the third codon position are often (but not always) equivalent; (3) pyrimidine nucleotides (U or C) and purine nucleotides (A or G) in the third position of the codon sometimes are not equivalent (i.e., can be distinguished).

On the basis of physical considerations, Crick excluded the possibility of A:G and G:A pairs formation since such an interaction would result in guanine NH_2-group dehydration, which is unfavorable from an energy standpoint. Furthermore, taking into account the characteristics of the code dictionary mentioned above, he also excluded the possibility of U:U, U:C, and C:U pairing. (If such pairing were allowed, pyrimidine and purine nucleotides in the third position would always be equivalent, i.e., not distinguished, a contradiction of the experimental data.) All of these considerations were summarized in following rules concerning pairing between the first nucleotide of the anticodon and the nucleotide at the third position of the codon:

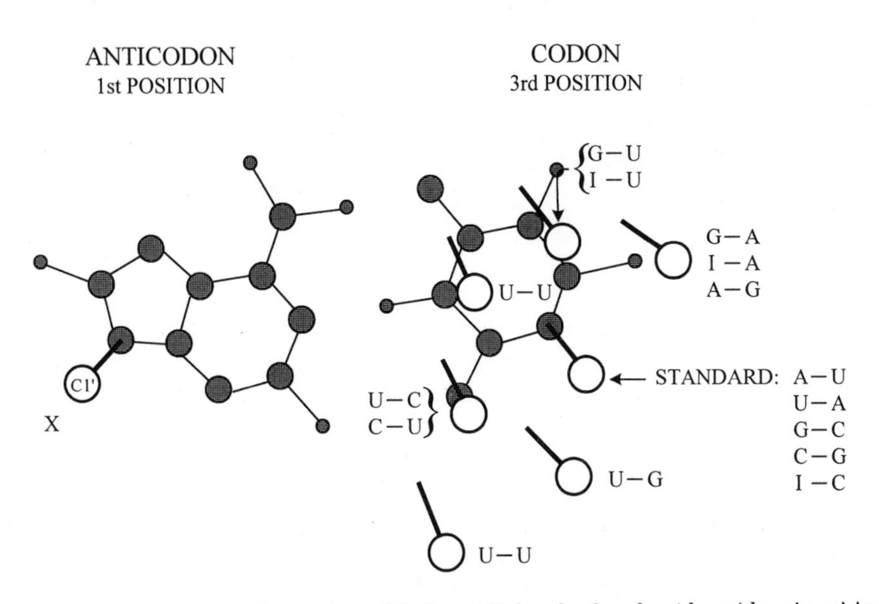

FIG. 10.1. Relative positions and directions of N-glycosidic bonds of nucleoside residues in pairing the first base of the anticodon (*left*) with the third base of the codon (*right*). See Crick, F. H. C. (1996) *J. Mol. Biol.* **19**548–555. The heavy circles indicate the positions of C1′-atoms of the residues; the thick line sections designate N-glycosidic bonds. The standard A:U pair drawn in thinner circles (shaded) and lines is shown for reference. If the C1′-atom of the anticodon residue is fixed in position X, the C1′-atom of the codon residue is found either in the standard position (in the cases of the standard pairs) or in positions deviated from the standard (in the cases of the wobble pairs).

A pairs with U;
G pairs with C,and U;
I pairs with C, U,and A;
U pairs with A,and G;
C pairs with G.

By the time these rules were formulated, it was already known that I is often found in tRNA anticodons, while A in the first anticodon position is not detected and appears always to be converted into I by enzymatic deamination. The types of wobble base pairing proposed by Crick, compared to standard pairing, are given in Fig. 10.2.

Following Crick's hypothesis, one may conclude that no amino acid can be coded by only one codon with A in the third position. Indeed, AUA together with AUU and AUG code for isoleucine; UUA and UUG code for leucine; but UGA does not code for any amino acid. This hypothesis predicts that all three isoleucine codons may be recognized by one tRNA with an IAU anticodon (later this was proved to be the case in eukaryotes), and the two leucine codons mentioned above, by one tRNA which has a UAA anticodon:

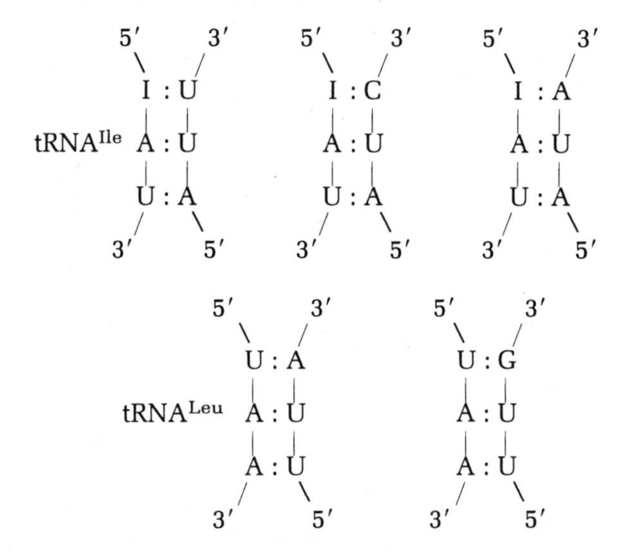

Furthermore, according to the wobble hypothesis, I cannot be present in a tRNA anticodon when the corresponding amino acid is coded by two codons only. Hence, the hypothesis predicts that I could never be present in the first position of the anticodons in phenylalanine, tyrosine, cysteine, histidine, glutamic acid, glutamine, aspartic acid, and asparagine tRNAs.

Finally, when a given amino acid is coded by four codons, no fewer than two tRNAs with different anticodons should be present. One of these tRNA species may recognize U and C, while the other recognizes A and G in the third codon position (if G and U are in the first position of their anticodons, respectively); later it was shown that this situation is typical of prokaryotes (except tRNA^{Arg} with ICG as anticodon). Alternatively, one tRNA recognizes U, C, and A, while the other recognizes only G in the third codon position (if I and C are in the first position of their anticodons); this proved to be more characteristic for codon–anticodon recognition in eukaryotes (except tRNA^{Gly}).

Wobbling at the first position of the anticodon and the third position of the codon, as well as most of the rules and predictions offered by this hypothesis, has been confirmed by subsequent experimental information.

10.1.4. Corrections to Wobbling Rules

Some inconsistencies have been found, however, and so certain rules need to be amended. Most of them are connected with modifications of the wobble position of anticodon (base 34 of tRNA, Fig. 3.8). First of all, in most bacteria and in the eukaryotic cytoplasm, U in the first anticodon position (U34) of tRNAs has been found to be always modified (for review, see Yokoyama and Nishimura, 1995). The modification is typically either a derivative of 5-methyl-2-thiouridine in prokaryotes, or a derivative of 5-hydroxyuridine in eukaryotes, or 5-methyluridine derivatives in

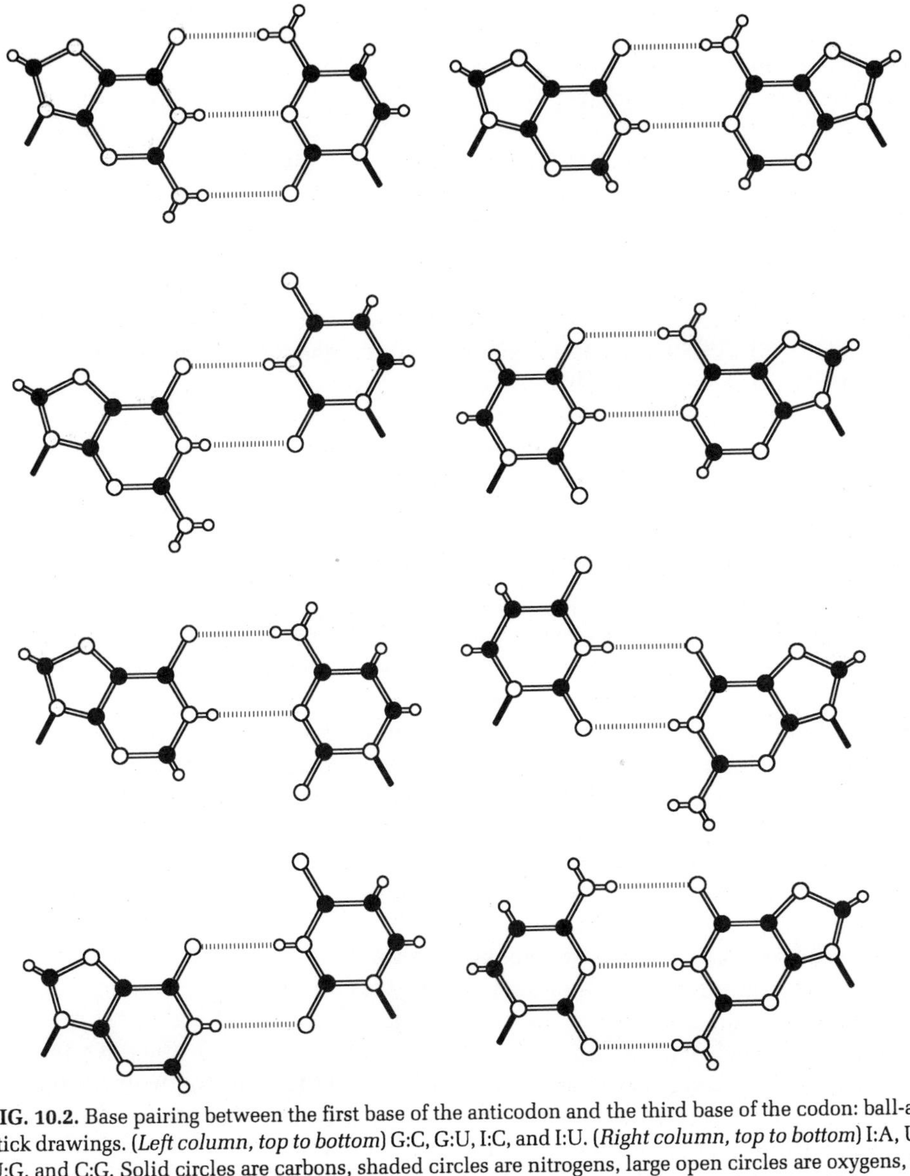

FIG. 10.2. Base pairing between the first base of the anticodon and the third base of the codon: ball-and-stick drawings. (*Left column, top to bottom*) G:C, G:U, I:C, and I:U. (*Right column, top to bottom*) I:A, U:A, U:G, and C:G. Solid circles are carbons, shaded circles are nitrogens, large open circles are oxygens, and small open circles are hydrogens; filled sticks are N-glycosidic bonds between the base and ribose.

both. Methylation of the 2-hydroxyl group of the ribose of the wobble uridine also often occurs. It is the modifications that restrict the recognition in the wobble position in such a way that U pairs with only purines (A and G) rather than with all four nucleotides. U is not modified in a number of tRNAs of mitochondria, chloroplasts, and some primitive bacteria like *Mycoplasma,* and in these cases it recognizes all four bases (A, G, U, and C) in the third position of codon. Correspondingly, only one tRNA species, with anticodon UNN, serves in mitochondria, chloroplasts, and *Mycoplasma* to recognize all four codons of a codon family. For example, $tRNA^{Val}$ with a UAC anticodon recognizes GUU, GUC, CUA, and GUG; $tRNA^{Ala}$ with UGC as an anticodon recognizes GCU, GCC, GCA, and GCG. In cases where the nucleotide in the first position of the anticodon is capable of recognizing only purine nucleotides in codons, U in the mitochondrial tRNA species is also modified.

In bacteria the isoleucine codons are read by two tRNA species, one with anticodon GAU capable of recognizing codons AUC and AUU (according to the classical wobble rules), and the other interacting specifically with codon AUA. Thus, contrary to the original deduction from the wobble rules, there is a situation when a codon ending with A is read alone by a separate tRNA. In this case the wobble position of the anticodon contains a special modification of C where the lysine residue replaces oxygen at position 2 of the pyrimidine ring, the so-called lysidine, that is 4-amino-2-(N^6-lysino)-1-(β-D-ribofuranosyl)-pyrimidinium.

In the original wobble rules it was suggested that A, if it were at the first position of the anticodon, would pair only with U at the third position of the codon. However, A is usually not present at that position of tRNAs but rather modified (deaminated) into I. Nevertheless, in some rare cases, such as in the $tRNA^{Arg}$ of yeast mitochondria and $tRNA^{Thr}$ of *Mycoplasma,* A has been found in the first position of the anticodon. It seems that the unmodified A in the wobble position of the tRNAs can read all four nucleotides (U, C, G, and A) at the third position of codons, rather than just U, as was originally postulated.

10.1.5. Stereochemistry of Codon–Anticodon Pairing

The structure of the anticodon loop described in Section 3.2.2 features the helical arrangement of the chain section including the three anticodon nucleotides and two residues following them toward the 3′-end. The helix parameters are similar to those of a single chain within the standard RNA double helix. Its bases are stacked. The three bases forming the anticodon are oriented such that the groups responsible for pairing through the hydrogen bond formation are exposed (see Fig. 3.6). The anticodon is thus ready to form a double-helical complex with the complementary sequence without its structure being significantly rearranged.

If the L-shaped tRNA molecule is viewed from the outer side of the corner, anticodon up, the groups of the three anticodon bases capable of pairing are turned more or less to the right from the plane containing both limbs of the molecule (see Fig. 3.8). The first anticodon base (wobble position) is located at the very top of the anticodon arm, and the two other anticodon bases and the two subsequent bases of the loop descend from the first base helically, like a spiral staircase, from the left, downward and to the right.

When the tRNA anticodon forms a complex with the mRNA, the resulting segment of the double-stranded helix possesses standard (Watson–Crick) pairing downward from the top of the anticodon arm, and less strict (wobble) pairing at the top. The latter is in agreement with the greater steric freedom of the pair, located at the edge of the base stack, compared to the internal pairs of the helix. (In anticodon–codon pairing the third pair may also be considered to be internal since its conformational freedom is restricted by the adjacent stacked purine base of the anticodon loop, and probably also by the adjacent anticodon of the peptidyl-tRNA.) The mRNA codon interacting with the tRNA anticodon in the A site should also acquire a helical conformation.

The binding of aminoacyl-tRNA to the ribosomal A site during elongation assumes that the P site is occupied by the peptidyl-tRNA. It is likely that the codon–anticodon interaction between the mRNA and the tRNA persists when the tRNA is in the P site. This implies that the triplet next to the codon located at the A site (toward the 5′-end of the mRNA) is also involved in base pairing and has a helical conformation.

At the same time, the orientation of the two tRNA molecules on the ribosome should allow their aminoacyl ends to be brought into close proximity, thus providing for subsequent transpeptidation (see Fig. 9.10). This implies strict requirements regarding the arrangement of the two codon–anticodon duplexes relative to each other. On the basis of steric considerations, it may be assumed that a flexible hinge exists between the two codon–anticodon duplexes to ensure that the two tRNA molecules are aligned so as to bring their acceptor ends together. Such a flexible hinge may be realized by rotations around the bonds of the internucleotide C(3′)–O–P–O–C(5′)–C(4′) bridge connecting the two mRNA codons, which result in a kink being formed; in other words, the two codon–anticodon duplexes may be not co-axial and, therefore, will not be stacked with each other. One of the possible conformations of the two codon–anticodon duplexes with a kink between them is presented in Fig. 10.3.

10.2. Participation of the Elongation Factor 1 (EF-Tu or eEF1A) in Aminoacyl-tRNA Binding

In prokaryotes, the codon-dependent binding of the aminoacyl-tRNA to the ribosomal A site is catalyzed by the protein referred to as elongation factor Tu (EF-Tu); in eukaryotes the corresponding factor is called EF1. It has been proposed that a universal designation EF1A for both prokaryotic and eukaryotic factors be used, just adding the prefix "e" (eEF1A) in the case of eukaryotes. The protein binds GTP

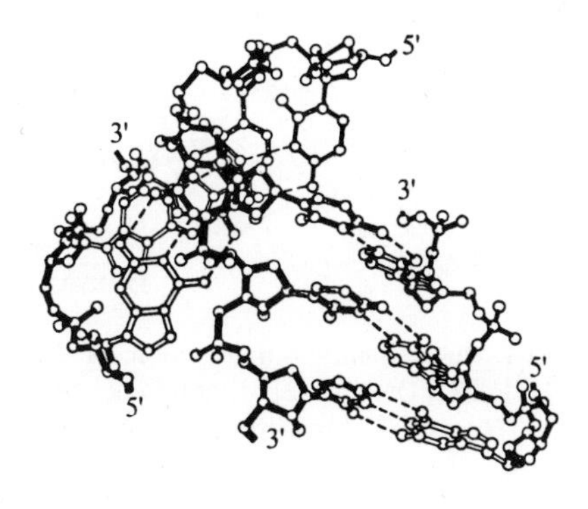

FIG. 10.3. Possible double-helical structures formed by the two codon–anticodon complexes in the ribosome: ball-and-stick drawing (without hydrogens). The mRNA chain (six nucleotide residues) is seen in the middle; the acceptor tRNA anticodon is on the right and the donor tRNA anticodon is on the left. There exists a kink between two mRNA codons such that if the backbone of the 3′-side codon of the mRNA lies approximately on the plane of the figure, the chain direction of the 5′-codon is almost perpendicular to this plane. The broken lines designate hydrogen bonding between the codon and anticodon bases. See Spirin, A. S. and Lim, V. I. (1986) in *Structure, Function, and Genetics of Ribosomes*, (B. Hardesty and G. Kramer, eds.), pp. 556–572, Springer-Verlag, New York.

and aminoacyl-tRNA and comes to the ribosome as a ternary complex EF1A:GTP: Aa-tRNA.

10.2.1. EF1A and Its Interactions

Bacterial EF1A (EF-Tu) is a protein with a molecular mass of about 45,000 daltons (393 and 405 amino acid residues in the cases of *E. coli* and *Thermus thermophilus* or *T.aquaticus*, respectively). It consists of three structural domains (Fig. 10.4).

The N-terminal domain (domain 1, or G domain) is responsible for GTP binding and GTP hydrolysis. It includes about 200 amino acid residues. Structurally it resembles the G domains of all other G proteins (GTPase superfamily) and almost coincides with the *ras* proto-oncogene product, p21. The typical nucleotide-binding motif (Rossmann fold) $\beta\alpha\beta\alpha\beta$, where the α-helices lie on the parallel β-sheet, is an essential feature of this structure. The sequence of the secondary stucture elements of this domain of EF-Tu is as follows:

$$\beta\alpha[\beta\leftrightarrow\alpha]\beta\beta\alpha\beta\alpha\beta\alpha\beta\alpha\alpha$$

All six β-strands are arranged into a twisted sheet (with five parallel strands and one antiparallel strand), whereas three α-helices are on one side of the sheet (facing domain 3) and three on the other (facing outside). GTP has a binding site in the region of the loops connecting strands and helices on the outside surface of domain 1. The "effector loop" (residues 40 to 60) is a region with a changeable conformation ($\beta \leftrightarrow \alpha$); it is located at the edge of the β-sheet and adjacent to the interface between domains 1 and 2. The effector loop with stabilized α-helical conformation is thought to interact with the ribosome.

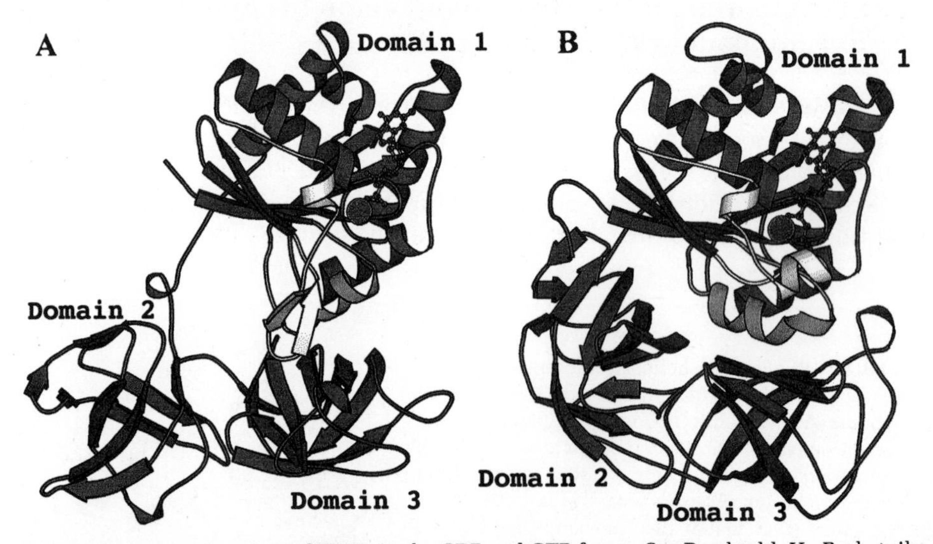

FIG. 10.4. A comparative view of EF-Tu in the GDP and GTP forms. See Berchtold, H., Reshetnikova, L., Reiser, C. O. A., Schirmer, N. K., Sprinzl, M., and Hilgenfeld, R. *Nature* **365**:126–132; Kjeldgaard, M., Nissen, P., Thirup, S., and J. Nyborg (1993) *Structure* **1**:35–50; Polekhina, G., Thirup, S., Kjeldgaard, M., Nissen, P., Lippmann, C., and Nyborg, J. (1996) *Strucutre* **4**:1141–1151; Abel, K., Yoder, M. D., Hilgenfeld, R., and Jurnak, F. *Structure* **4**:1153–1159. (**A**) EF-Tu:GDP complex of *E. coli.* (**B**) EF-Tu:GTP complex of *T. thermophilus.* Reproduced from Abel, K., Yoder, M. D., Hilgenfeld, R., and Jurnak, F. (1996) *Structure* **4**:1153–1159 with permission).

Domains 2 and 3 (about 100 amino acid residues each) consist mainly of antiparallel β-strands forming seven-strand and six-strand barrel structures, respectively (Fig. 10.4). These domains are in tight contact and can be considered as a single structural unit of EF-Tu.

As already mentioned, EF-Tu can bind GTP, as well as GDP, at the surface of domain 1. The conformations of EF-Tu, however, are different depending on whether GTP or GDP is bound. When GDP is bound, or no guanyl nucleotide is present on EF-Tu, the factor is in an inactive form; it cannot bind aminoacyl-tRNA and interact with the ribosome. In the inactive form EF-Tu is characterized by a high degree of interdomain flexibility (relaxed conformation). There is a hole in the middle of the molecule, between domain 1 and domains 2/3, and no tight contact between domains 1 and 2 is observed (Fig. 10.4A). The effector loop is not stabilized and exists mostly in β-strand conformation. When GTP is bound, the two halves (domain 1 and domains 2/3) are rotated relative to each other, the molecule becomes more compact, and the hole disappears (tight conformation). Domains 1 and 2 come into close contact (Fig. 10.4B). The effector loop is stabilized by additional interactions and acquires a partial α-helical conformation (two short α-helices are formed). This is the active state of EF-Tu: the molecule is capable of binding aminoacyl-tRNA and then interacting with the ribosome.

The bacterial EF-Tu, instead of the nucleotides, can interact with another protein which has a molecular mass of about 30,000 daltons and is referred to as the elongation factor Ts (EF-Ts), or EF1B. As a result, a complex between EF-Tu and EF-Ts (EF1A:EF1B) is formed. If EF-Tu is bound to GDP, the complex formation with EF-Ts results in GDP displacement, as follows:

$$\text{EF-Tu:GDP} + \text{EF-Ts} \rightarrow \text{EF-Tu:Ts} + \text{GDP}$$

The conformation of EF-Tu bound with EF-Ts is similar to that in the EF-Tu:GDP complex or in the free EF-Tu, that is, it represents the loose inactive form of EF-Tu. The elongated molecule of EF-Ts interacts with domain 1 and domain 3 of EF-Tu. The interaction induces local conformational changes in domain 1 of EF-Tu resulting in disruption of the GDP-binding site and thus the release of GDP from the complex.

The EF-Tu:Ts complex is capable of interacting with GTP, yielding EF-Tu:GTP and free EF-Ts in the presence of excess GTP:

$$\text{EF-Tu:Ts} + \text{GTP} \rightarrow \text{EF-Tu:GTP} + \text{EF-Ts}$$

Thus, EF-Ts (EF1B) is a factor that displaces GDP from the slowly dissociating EF-Tu:GDP complex and thereby catalyzes the exchange of GDP for GTP according to the following overall equation:

$$\text{EF-Tu:GDP} + \text{GTP} \xrightarrow{\text{EF-Ts}} \text{EF-Tu:GTP} + \text{GDP}$$

Free EF-Tu, as well as EF-Tu:GDP and EF-Tu:Ts complexes, has a low, if any, affinity for aminoacyl-tRNA. Only the addition of GTP induces a strong affinity between EF-Tu and aminoacyl-tRNA, resulting in a ternary complex being formed:

$$\text{EF-Tu:GTP} + \text{Aa-tRNA} \rightarrow \text{Aa-tRNA:EF-Tu:GTP}$$

The association between the aminoacyl-tRNA and EF-Tu is of a multicenter nature and involves all three domains of EF-Tu. The compact active conformation of EF-Tu does not change seriously upon the binding of aminoacyl-tRNA. The EF-Tu bound to the aminoacyl-tRNA covers primarily the left side of the acceptor stem and the T-helix of the L-shaped tRNA molecule when the latter is viewed from the outer side of the corner, anticodon up (Fig. 10.5). Domain 3 is responsible for the binding of the T-stem of aminoacyl-tRNA. It interacts exclusively with the backbone (ribose and phosphate residues) contacting with ribose or phosphate groups of the nucleotide residues 52 to 54 and 63 to 65. Domain 2 forms the pockets to retain the aminoacyl residue with its free amino group and the ester group, as well as the 3′-terminal adenosine with the hydrophobic adenine and the free 2'-OH group. The single-stranded 3′-terminal sequence A73-C74-C75 of the aminoacyl-tRNA molecule binds in the cleft formed by the interface between domains 1 and 2 (Fig. 10.5). The three bases are stacked and continue the helical arrangement of the acceptor double helix; they are turned away from the protein, so that the sequence interacts just by its phosphates with the effector loop (Lys 52) of domain 1. The junction of the three domains of EF-Tu forms a pocket binding the 5′-end of aminoacyl-tRNA.

Two forms of the eukaryotic EF1A (eEF1A) have been described: $eEF1_L$ (light) and $eEF1_H$ (heavy) (for review, see Merrick and Hershey, 1996). $eEF1_L$ is identical to eEF1A and can be regarded as a full analog of bacterial EF-Tu; it is a monomeric protein with a molecular mass of about 50,000 daltons and is capable of binding GTP or GDP and forming the ternary complex Aa-tRNA:EF1A:GTP. The $eEF1_H$ appears to be an aggregated analog of the EF-Tu:Ts complex. It consists of eEF1A and the eukaryotic analog of EF1B (eEF1B) comprising three subunits (α,

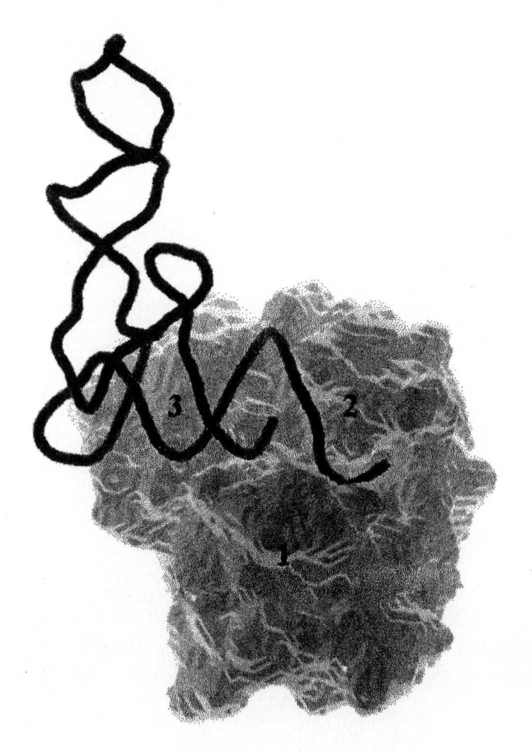

FIG. 10.5. Ternary complex EF-Tu:GTP:Aa-tRNA. The structure of the ternary complex consisting of yeast Phe-$tRNA^{Phe}$, *Thermus aquaticus* EF-Tu, and the GTP analog GMP-PNP was determined by X-ray crystallography at 2.7 Å resolution. See Nissen, P., Kjeldgaard, M., Thirup, S., Polekhina, G., Reshetnikova, L., Clark, B. F. C., and Nyborg, J. (1995) *Science* **270**:1464–1472. Here EF-Tu is shown as a surface representation and Phe-tRNA as a ribbon. It is seen that EF-Tu interacts with the TΨ/acceptor arm of tRNA from the "left" (TΨ loop) side. The anticodon arm of tRNA points away from the protein. Reproduced from Nyborg, J., Nissen, P., Kjeldgaard, M., Thirup, G., Polekhina, G., Clark, B. F. C., and Reshetnikova, L. (1996) *TIBS* **21**:81–82.

β and γ), with molecular masses of about 25,000, 30,000, and 50,000 daltons. The α- and β-subunits of eEF1B seem to play the part of an EF-Ts analog, whereas the role of the larger γ-subunit has yet to be determined. A portion of $eEF1_H$ in mammalian cells has been found to be firmly complexed also with valyl-tRNA synthetase.

10.2.2. Binding the Ternary Complex with the Ribosome

The content of EF-Tu (or eEF1A) in the cell is high, and virtually all cellular aminoacyl-tRNAs exist as ternary complexes with this protein and GTP. Both the tRNA and the protein moiety of the complex show affinity for the ribosome. As a result of codon–anticodon recognition, the complex that carries the tRNA corresponding to the template codon binds productively to the ribosome.

The selective binding of the ternary complex to the ribosome governed by the codon–anticodon interaction induces hydrolysis of the bound GTP into GDP and orthophosphate; orthophosphate is released. It is likely that the settling of the tRNA residue in the A site on the 30S subunit strongly contributes to the induction of the GTP hydrolysis. This indirect effect may be caused by an interdomain shift within EF-Tu as a result of the tRNA residue movement.

After this point EF-Tu (or eEF1A) exists as a complex with GDP (EF-Tu:GDP) and, because this complex has no strong affinity for the ribosome and the aminoacyl-tRNA anymore, EF-Tu:GDP is released from the ribosome. Aminoacyl-tRNA remains bound in the A site and becomes capable of reacting with the peptidyl-tRNA. Thus the following sequence of reactions can be written:

$$\text{Aa-tRNA:EF-Tu:GTP} + \text{RS:mRNA} \rightarrow \text{RS:mRNA:Aa-tRNA:EF-Tu:GTP}$$

$$\text{RS:mRNA:Aa-tRNA:EF-Tu:GTP} + H_2O \rightarrow \text{RS:mRNA:Aa-tRNA:EF-Tu:GDP} + P_i$$

$$\text{RS:mRNA:Aa-tRNA:EF-Tu:GDP} \rightarrow \text{RS:mRNA:Aa-tRNA} + \text{EF-Tu:GDP}$$

It should be mentioned that *in vitro* the codon-dependent binding of the aminoacyl-tRNA to the ribosomal A site may be achieved even in the absence of EF-Tu or eEF1A. Effective *factor-free binding* requires a higher Mg^{2+} concentration in the medium (10 to 15 mM at 100 mM NH_4Cl or KCl in the case of *E. coli* ribosomes) compared to the Mg^{2+} concentration for the EF-Tu- or eEF1A-promoted binding (5 to 10 mM for *E. coli* systems). The factor-free binding is functional: the bound aminoacyl-tRNA can participate in the transpeptidation reaction with the peptidyl-tRNA in the P site. The factor-free (nonenzymatic) binding of the aminoacyl-tRNA differs from the EF-Tu-promoted (enzymatic) binding in that it proceeds at a considerably slower rate. Thus, a slow spontaneous reaction may take place in a cell-free system:

$$\text{RS:mRNA} + \text{Aa-tRNA} \rightarrow \text{RS:mRNA:Aa-tRNA}$$

Nevertheless, the eventual result of this reaction is similar to that observed for the fast binding of the ternary Aa-tRNA:EF-Tu:GTP complex followed by the subsequent cleavage of GTP and the release of EF-Tu.

There is a special case where the binding of the ternary complex with the ribosome strictly depends on a secondary/tertiary structure element of mRNA. This is the case of the UGA-directed binding of selenocysteinyl-tRNA^{Sec} (see Section 3.5.3). In prokaryotes the selenocysteinyl-tRNA^{Sec} is bound by a unique elongation factor SELB, instead of EF-Tu:

$$\text{Sec-tRNA} + \text{SELB:GTP} \rightarrow \text{Sec-tRNA:SELB:GTP}$$

The SELB specifically interacts with a 40-nucleotide structural element of mRNA located immediately downstream of the UGA codon (Fig. 10.6). Thus, the ternary complex is found to be fixed on mRNA at a specific site encoding for selenocysteine. When the translating ribosome encounters this ternary complex on mRNA it accepts the selenocysteinyl-tRNA^{Sec} at the A site with the UGA codon settled there (Fig. 10.7). The subsequent events seem to proceed in the same way as in the case of the standard ternary complex binding: GTP is hydrolyzed, SELB:GDP complex is released from the ribosome, and the selenocysteinyl-tRNA reacts with peptidyl-tRNA to form the next peptide bond. The ribosome continues translation of the mRNA.

In eukaryotes the UGA-directed incorporation of selenocysteine is determined by a special structural element located in the 3′-untranslated region (3′-UTR) of mRNA, far downstream of the UGA codon. This element is much longer than that of prokaryotic mRNAs and has quite different secondary and tertiary structure. The mechanism by which the 3′-UTR-located element affects the binding specificity of the internal UGA is unknown.

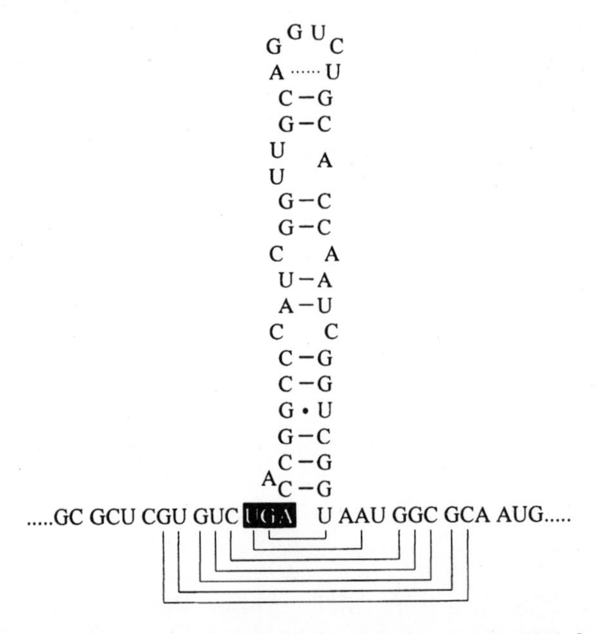

FIG. 10.6. Secondary structure element of *E. coli* formate dehydrogenase mRNA determining the binding of selenocysteinyl-tRNA^{Sec} with the A-site-bound UGA codon. The apical part of the hairpin including the end loop is organized into specific tertiary structure with exposed UU (in the stem) and GU (in the loop) directly responsible for SELB binding. At the same time, the basal part including the UGA codon and the adjacent upstream sequence is also involved in a complex and distorted double-stranded structure with the downstream sequence. See Baron, C., and Böck, A. (1995) in *tRNA Structure, Biosynthesis, and Function* (D. Soll and U. L. RajBhandary, eds.), pp. 529–544, ASM Press, Washington, DC; Hüttenhofer, A., Westhof, E., and Böck, A. (1996) *RNA* **2**:354–366.

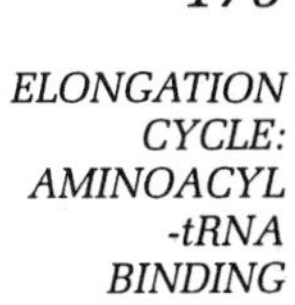

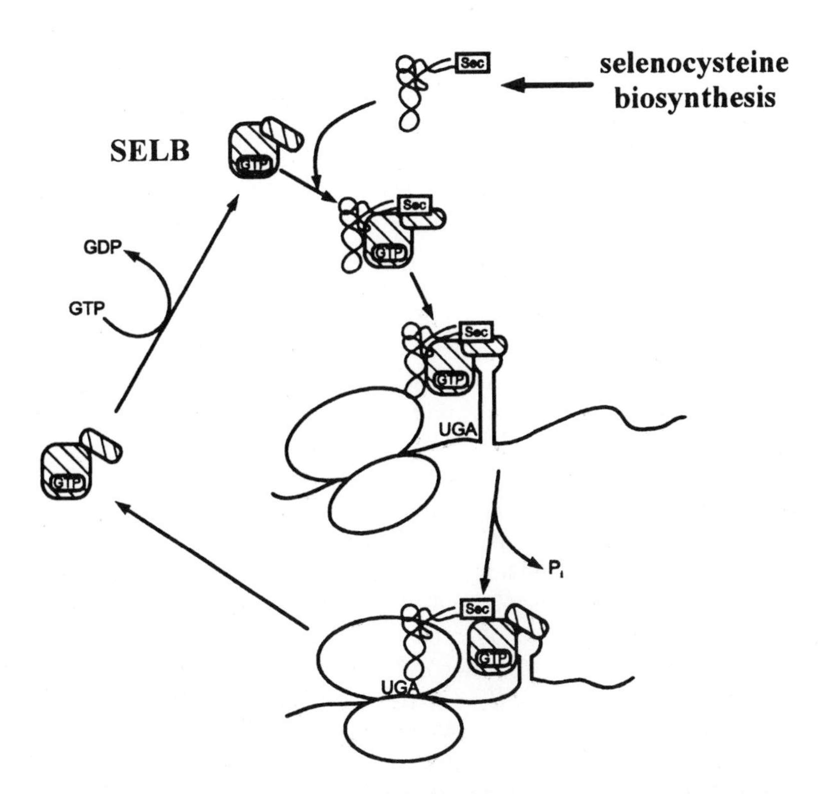

FIG. 10.7. Model for the incorporation of selenocysteine into bacterial proteins. A specialized elongation factor SELB, an analog (and homolog) of EF-Tu, forms the ternary complex with GTP and selenocysteinyl-tRNA^{Sec} (Sec-tRNA), and also binds to a recognition element within mRNA (Fig. 10.6) which is located at the 3' side of the UGA codon. The translating ribosome runs on this quaternary complex and binds it at the A site. Reproduced from Baron, C., and Böck, A. (1995) in *tRNA Structure, Biosynthesis, and Function* (D. Soll and U. L. RajBhandary, eds.), pp. 529–544, ASM Press, Washington, DC, with permission).

10.2.3. Role of GTP and Its Hydrolysis in the Catalysis of Aminoacyl-tRNA Binding

What happens if GTP is replaced by its nonhydrolyzable analogs, such as GMP-PCP or GMP-PNP? EF-Tu (as well as the eukaryotic eEF1A) is capable of interacting with the analog and, as a result of the interaction, the protein acquires an affinity for aminoacyl-tRNA. The protein, the GTP analog, and an aminoacyl-tRNA yield a ternary complex which then binds to the ribosome. The next reaction, however, does not take place. Therefore, EF-Tu continues to retain its affinity for the aminoacyl-tRNA and the ribosome, and the aminoacyl-tRNA remains on the ribosome in the complex with EF-Tu and the nonhydrolyzable GTP analog. In this case the aminoacyl-tRNA cannot serve as the acceptor substrate in the transpeptidation reaction with the peptidyl-tRNA, and the next stage of the elongation cycle becomes blocked.

Indeed, the ternary aminoacyl-tRNA:EF-Tu:CTP or aminoacyl-tRNA:EF-Tu:GMP-PCP complex associates with the ribosome and binds to the template codon via the tRNA anticodon, but the aminoacylated acceptor end of the tRNA moiety still resides on the ribosome-bound EF-Tu rather than interacting with the *a* site of the PTC of the large subunit. Only after GTP is hydrolyzed and EF-Tu:GDP is released, the acceptor end with its aminoacyl residue becomes caught by the PTC. In other words, the aminoacyl-tRNA of the ternary complex cannot fully occupy the A site before GTP is cleaved and EF-Tu is released. Since the location of tRNA on the small subunit, including codon–anticodon interaction, seems to correspond to its A site position, but the tRNA acceptor end still occupies its site on the 50S-subunit-bound EF-Tu rather than on the 50S subunit itself, this state of intermediate codon-depenent aminoacyl-tRNA binding can be referred to as "hybrid A/T" state (see Section 9.5.3 and Fig. 9.14A). When a nonhydrolyzable GTP analog is used, the aminoacyl-tRNA gets stuck in the intermediate A/T position ("hybrid A/T site") and therefore cannot react with the peptidyl-tRNA. From this, GTP hydrolysis could be postulated to provide the energy for transferring the aminoacyl-tRNA from the A/T position to the A/*a* position (or A/A site of Noller) (Fig. 9.14B).

At the same time, the aminoacyl-tRNA bound to the ribosome in the presence of EF-Tu and a nonhydrolyzable GTP analog, such as GMP-PCP, may be rendered active in transpeptidation simply by mechanical washing away of EF-Tu and the GTP analog. Thus, although the aminoacyl-tRNA is also switched from the A/T to A/*a* position, no GTP hydrolysis is involved in this case. On the other hand, it has been shown that GTP hydrolysis by itself fails to provide the switch to the A/*a* position, if EF-Tu remains on the ribosome in the presence of kirromycin, an antibiotic that increases the affinity of EF-Tu:GDP for the ribosome (see Section 10.3.3). Generally, it seems unlikely that a thermodynamically downhill process, such as the accommodation of aminoacyl-tRNA in the A site, would require GTP hydrolysis energy for performing any useful work.

It seems more likely that the GTP hydrolysis during aminoacyl-tRNA binding is used for purely catalytic purposes. If EF-Tu is considered a catalyst for aminoacyl-tRNA binding to the ribosome, the concept of the *transition state,* analogous to the transition state in enzymatic catalysis, may be accepted. Then EF-Tu:GTP has an affinity precisely for the transition state of the aminoacyl-tRNA:ribosome complex. However, in contrast to the case of the enzymatic catalysis of a covalent reaction where the energy for the breakdown of the transition state and the release of a catalyst is provided by the catalyzed reaction itself, EF-Tu uses a destructible effector, GTP: first GTP induces the affinity of EF-Tu for the transition state of the aminoacyl-tRNA:ribosome complex, and then GTP is cleaved, EF-Tu is released, and the transition state converts to the product state. It may be a general rule that a noncovalent catalysis, such as the catalysis of a conformational change or a directed molecular movement, requires a parallel covalent reaction of the ATP or GTP hydrolysis type in order to compensate for the energy of the catalyst affinity and to complete the process catalyzed.

Thus, the codon-dependent binding of the aminoacyl-tRNA with the ribosomal A site is catalyzed by EF-Tu protein plus GTP. The catalysis appears to be the result of (1) the binding of EF-Tu together with the aminoacyl-tRNA to the ribosome (the acquisition of a transition state for the ribosome complex), and (2) the subsequent hydrolysis of the bound GTP and the release of EF-Tu (the attainment of a final, or product state). The significance of GTP hydrolysis lies in destroying the effector (GTP). As a consequence of the GTP hydrolysis, EF-Tu changes its conformation, loses the affinity for the ribosomal complex, and dissociates, thereby allowing the final A-site binding of the aminoacyl-tRNA to be achieved. The free energy of GTP hydrolysis is dissipated into heat.

Numerous specific inhibitors, predominantly antibiotics, are known to block different stages of the ribosomal elongation cycle. Most of these act selectively either upon prokaryotic (bacterial) or eukaryotic ribosomes (for reviews, see Pestka, 1977; Gale *et al.*, 1981; Cundliffe, 1980, 1990).

10.3.1. Tetracyclines

A classic example of specific inhibition of binding of the aminoacyl-tRNA with the A site of a bacterial 70S ribosome is provided by tetracyclines (Fig. 10.8). Tetracyclines also inhibit the codon-dependent binding of the aminoacyl-tRNA with the isolated 30S subunit. In agreement with this, the specific binding site for tetracycline has been detected on the 30S subunit. It is noteworthy that when the ternary Aa-tRNA:EF-Tu:GTP complex interacts with the 70S ribosome in the presence of tetracycline, GTP undergoes hydrolysis and EF-Tu:GDP is released but aminoacyl-tRNA does not remain in a bound state. It is likely that tetracycline bound on the 30S ribosomal subunit in the region of tRNA-binding sites decreases the affinity of the A site for tRNA. On the other hand, a prebound aminoacyl-tRNA impedes the binding of tetracycline to the ribosome. Thus, tetracycline and aminoacyl-tRNA compete for binding to the ribosome.

Since classical tetracyclines such as unsubstituted tetracycline, chlortetracycline, and oxytetracycline (Fig. 10.8) possess relatively low binding constants, their inhibitory effect on ribosomes strongly depends on their concentration in the medium, as well as on the concentration of the competitive aminoacyl-tRNA,

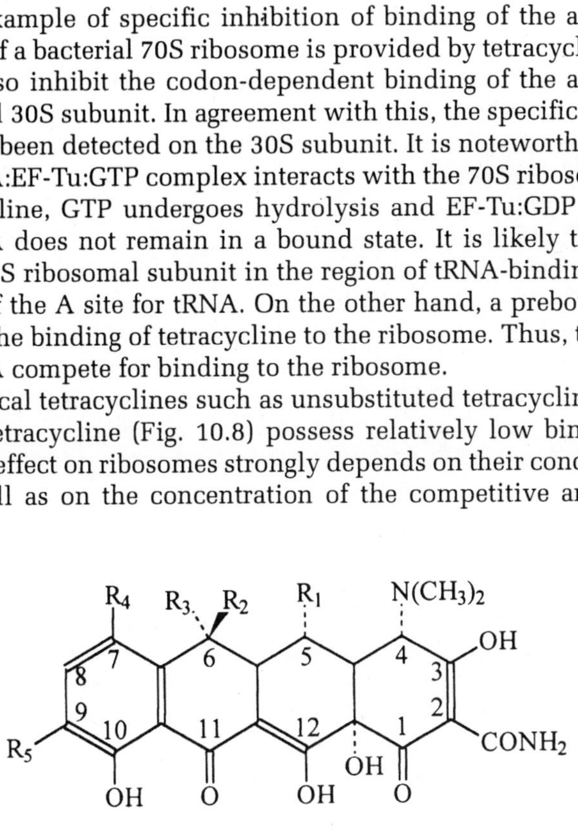

TETRACYCLINES:	R_1	R_2	R_3	R_4	R_5
TETRACYCLINE	—	OH	CH_3	—	—
OXYTETRACYCLINE	OH	OH	CH_3	—	—
CHLORTETRACYCLINE	—	OH	CH_3	Cl	—
MINOCYCLINE	—	—	—	$N(CH_3)_2$	—
DMG-MINOCYCLINE	—	—	—	$N(CH_3)_2$	$NHCOCH_2N(CH_3)_2$
DMG-DEMETHYL-DEOXYTETRACYCLINE	—	—	—	—	$NHCOCH_2N(CH_3)_2$

FIG. 10.8. Tetracyclines.

rather than simply on drug-to-ribosome ratio. Next-generation tetracycline derivatives, such as minocycline, have significantly higher affinity for ribosomes and correspondingly exert stronger inhibitory effects (or are required in lower concentrations). Particularly high affinity for ribosomes is imparted by a dimethylglycyl group at position 9 (the so-called glycylcyclines, or DMG-tetracyclines, Fig. 10.8); such tetracyclines are very efficient against bacterial ribosomes and may overcome tetracycline resistance of many drug-resistant strains.

The resistance of bacteria against tetracyclines is provided by two different mechanisms. One is the acquirement by a bacterium of an efflux system in the bacterial membrane that pumps the drug out from the cell. The high-affinity tetracyclines may overcome this type of resistance due to their efficacy in low concentrations. The other type of resistance against tetracyclines is gained at the ribosomal level: a special protein, called Tet(M), or Tet(O), or Tet(S), depending on the resistant strain, appears in the cell; the protein interacts with the bacterial ribosome in a GTP-dependent manner and removes the bound tetracycline from it. It is interesting that the protein has been found to be a homologue of the bacterial EF-G, ribosomal translocase (see Section 12.2.1). Translocation of tetracycline from the ribosomal A site may be a mechanism of the protective action of Tet(M) and related proteins. The highest-affinity tetracycline derivatives, however, cannot be removed by this protein from the bacterial ribosome.

Since no ribosomal protein has been found to be responsible for tetracycline binding to the ribosome, and no ribosomal protein mutants conferring tetracycline resistance to ribosomes have been detected, it is strongly believed that the antibiotic specifically interacts with a structural element of 16S ribosomal RNA. According to the results of footprinting experiments (Noller *et al.*, 1990), tetracycline protects A892 located in the side loop of the hairpin occupying the interdomain junction of the 16S RNA (see Fig. 6.1). The same base is specifically protected also by the A-site-bound tRNA (but not P-site-bound tRNA). Among ribosomal proteins, protein S7 seems to be the nearest neighbor of the tetracycline-binding site.

There have been several reports on the inhibitory effect of tetracyclines on eukaryotic protein synthesis. These were based on *in vitro* experiments in which high concentrations of the antibiotic were added into eukaryotic cell-free systems. However, tetracycline is a strong Mg^{2+}-chelating agent, so the inhibition could be the result of the depletion of Mg^{2+} in the system, rather than due to a specific action on eukaryotic ribosomes. Indeed, no inhibition of the eukaryotic cell-free systems by tetracyclines was observed at increased Mg^{2+} concentrations.

10.3.2. Aminoglycosides

Another group of antibiotics also affects the binding of the aminoacyl-tRNA to the ribosomal A site, but the action is of a completely different nature. These antibiotics are the so-called aminoglycoside antibiotics, the principal members of which are streptomycin (Fig. 10.9), neomycin, kanamycin, and gentamicin (Fig. 10.10). Antibiotics in this group contribute to the ribosomal retention of noncognate aminoacyl-tRNAs, that is, the aminoacyl-tRNAs that do not correspond to the codon positioned in the ribosomal A site (Davies, Gilbert, and Gorini 1964; Davies, Gorini, and Davis, 1965; Gorini, 1974). Such a *miscoding* results in the synthesis of incorrect polypeptides characterized by a large number of errors, with a consequent cytotoxic or bactericidal effect on cells. Streptomycin specifically affects bacterial 70S ribosomes, and kanamycin and neomycin are known to induce miscoding with eukaryotic 80S ribosomes as well. The main site of the antibiotic binding to the ribosome appears to occur on the small (30S or 40S) ribosomal subunit, although the effect depends on the interaction of both subunits and is manifested only with the complete 70S or 80S ribosome.

Ribosomal protein mutations are known that confer to ribosomes the resistance against streptomycin or other aminoglycosides. Most typically, streptomycin resistance of bacterial ribosomes is acquired as a result of mutations in protein S12. Mutations of the same protein, as well as proteins S4 and S5, are responsible also for streptomycin-dependent phenotypes. Generally, proteins S4, S5, S12, and S17, or their eukaryotic analogs, in the small ribosomal subunit seem to form a protein group that modulates the response of the ribosome to the aminoglycosides.

At the same time the aminoglycosides bind specifically to the ribosomal RNA of the small subunit, rather than to the proteins. When bound, they protect from chemical modification the residues at the regions of the ribosomal RNA that are involved in the formation of the A-site part of the decoding center on the small subunit (see Section 9.3). In the bacterial (*E. coli*) ribosome these are positions 1405, 1408, and 1494 at the base of the long, penultimate hairpin of the 16S rRNA (see Figs. 6.1 and 9.3) in the case of neomycin-type aminoglycosides (Noller *et al.*, 1990; Pirohit and Stern, 1994). Direct contacts of a neomycin-type antibiotic with positions 1492–1494 have been determined also in structural studies of the complex of the antibiotic with the 27-nucleotide-long model of the decoding region of the 16S rRNA (Fourmy *et al.*, 1996). Neomycin protects also C525 in the end loop of the last hairpin of domain I which is known to be in the vicinity of the A site as well (see Fig. 9.3). Streptomycin has been found to protect positions 911–915 at

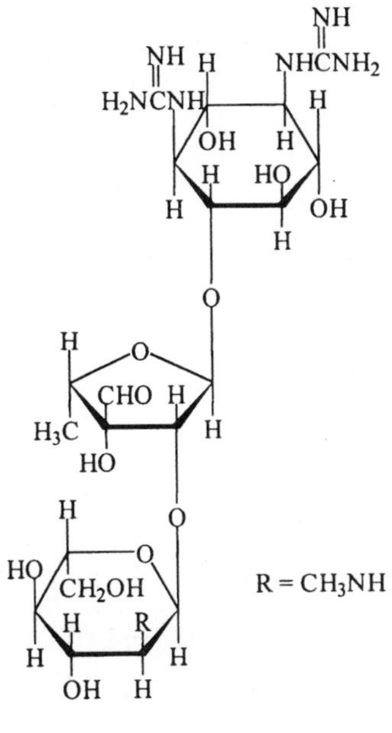

R = CH_3NH

FIG. 10.9. Streptomycin.

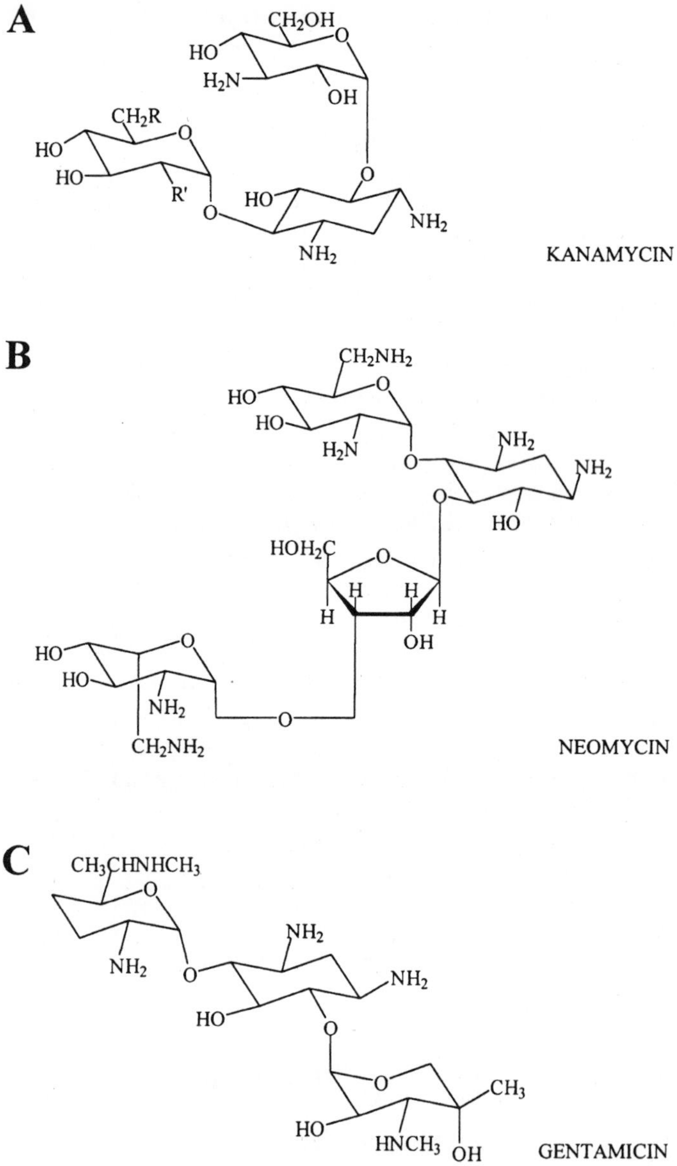

A

CH2OH HO H2N OH CH2R HO HO R' O HO NH2 NH2

KANAMYCIN

B

CH2NH2 HO HO H2N O NH2 NH2 HO HOH2C O H H H H OH HO HO NH2 CH2NH2 O

NEOMYCIN

C

CH3CHNHCH3 NH2 O NH2 NH2 HO O O CH3 HO HNCH3 OH

GENTAMICIN

FIG. 10.10. Aminoglycosides. (**A**) Kanamycin. (**B**) Neomycin. (**C**) Gentamycin.

the interdomain hairpin (see Fig. 6.1), very close to the decoding center mentioned above and to the tetracycline-protected site. Mutations of the ribosomal RNA at the same regions (positions 1409 and 1491 in the case of neomycin-type antibiotics, and 523 and 912 in the case of streptomycin; see Cundliffe, 1990) confer resistance against the aminoglycosides. It is interesting that the mutations of ribosomal proteins that affect the streptomycin sensitivity of the ribosome alter the protection pattern of ribosomal RNA also in the same regions. Thus, the aminoglycosides have an affinity for some structural elements of the ribosomal RNA in the small subunit adjacent to the A site, and their binding affects the selectivity of the A site, ither directly or, more likely, via distortion of the local RNA structure. The ribo- omal proteins of the S4-S5-S12-S17 group are involved in properly maintaining ɿis local structure, and their mutations may be antagonistic or synergistic with ɿe aminoglycoside action.

0.3.3. Some Indirect Inhibitors

There are several known inhibitors that suppress primarily the EF-Tu- or EF1A-promoted binding of the aminoacyl-tRNA to the ribosome by blocking the ıctor-binding site of the 50S or 60S ribosomal subunit.

Antibiotics thiostrepton (Fig. 10.11) and related siomycin provide typical ex- mples of such action. Large molecules of these antibiotics bind tightly to the 50S ubunit in the region of the L7/L12 stalk base and prevent EF-Tu, as well as EF-G, om interacting with the factor-binding site. Protein L11 has been shown to be in- olved in thiostrepton binding. Ribosomes lacking protein L11 are shown to be ınctionally active and insensitive to the antibiotic. It is remarkable that in certain ıutants resistant to the antibiotic protein, L11 is completely absent (rather than eing simply altered). At the same time the site of thiostrepton binding has been dentified as the factor-binding hairpin in domain II of the 23S ribosomal RNA esidues 1050–1100, Fig. 6.6 A; see Section 9.4) where protein S11 is also bound see Section 8.3.2), rather than protein L11 itself (Cundliffe, 1990). It seems that rotein L11 maintains a specific local conformation of the RNA required for hiostrepton binding.

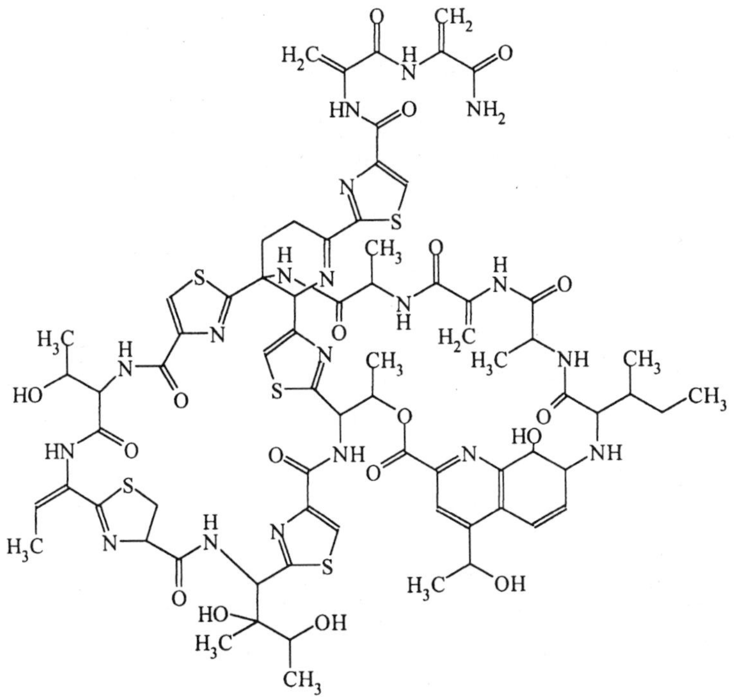

FIG. 10.11. Thiostrepton.

Protein inhibitors (toxins) of plant and fungal origin, such as ricin, abrin, mod eccin, and α-sarcin, inactivate the other factor-binding region of the ribosoma RNA of the large subunit, the so-called ricin/sarcin loop in domain VI (position equivalent to 2653–2667 in *E. coli* 23S RNA, Fig. 6.6A). They act specifically o the eukaryotic ribosomes. The inactivation of the factor-binding loop results fron a very specific enzymatic attack of the protein on this local structural element c RNA. The plant toxins (ricin, abrin, modeccin) are N-glycosidases and catalyz depurinization of the nucleotide residue in the middle of the loop (A4324 of th mammalian 28S RNA, equivalent to A2660 of the *E. coli* 23S RNA), whereas sarci is an endonuclease and splits the polynucleotide chain after the adjacent residu (between the residues equivalent to G2661 and A2662 in *E. coli* RNA; see Section 13.5.3 and 13.5.4, and Fig. 13.6). As a result, the binding of the ternary comple Aa-tRNA:eEF1A:GTP to the ribosome is impaired. The same is true for eEF2:GT binding, and therefore the translocation step is also inhibited.

Fusidic acid (see Fig. 12.4) generally is considered to be a translocation in hibitor but in essence is an inhibitor of aminoacyl-tRNA binding. The antibioti does not affect the ribosome but binds to EF-G. As a result, the protein acquires a increased affinity for the factor-binding site of the 50S ribosomal subunit and re mains complexed with the site after translocation and GTP hydrolysis. The com plex between EF-G:GDP and fusidic acid thus inhibits the interaction of EF-Tu an the aminoacyl end of tRNA with the 50S ribosomal subunit.

Finally, it is worthwhile to consider kirromycin as an antibiotic affecting EF Tu; kirromycin (Fig. 10.12) does not inhibit the binding of aminoacyl-tRNA bu rather blocks the subsequent stage of the cycle. It affects EF-Tu just as fusidic aci affects EF-G. EF-Tu acquires a high affinity for both the ribosome and the aminoa cyl-tRNA, with the result that the factor protein is not released after GTP hydrol ysis and blocks the aminoacyl terminus from participating in transpeptidation. I should be emphasized that the interaction of kirromycin with EF-Tu induces th intrinsic GTPase activity of the protein in the absence of the ribosome. These ob servations provided the basis for concluding that the GTPase center is present o the EF-Tu itself rather than on the ribosome, and thus the ribosome just activate the latent GTPase of EF-Tu.

10.4. Miscoding

Miscoding resulting from the incorrect binding of aminoacyl-tRNA to the ri bosome in the presence of streptomycin and other aminoglycoside antibiotics i not a qualitatively new aspect in the function of translation machinery. A certai low level of erroneous aminoacyl-tRNA binding to the ribosome is always presen in translation, and the aminoglycoside antibiotics only increase the magnitude. A particular template codon located in the ribosomal A site may erroneously bin only a limited set of other (noncognate) aminoacyl-tRNAs of *related* coding speci ficity. Aminoglycoside antibiotics stimulate, as a rule, the binding of the same se of noncognate aminoacyl-tRNAs without contributing new ambiguities.

10.4.1. Misreading of Poly(U)

In the simplest cell-free system of poly(U)-directed polyphenylalanine syn thesis with *E. coli* ribosomes, one can detect a marked misincorporation of leucin

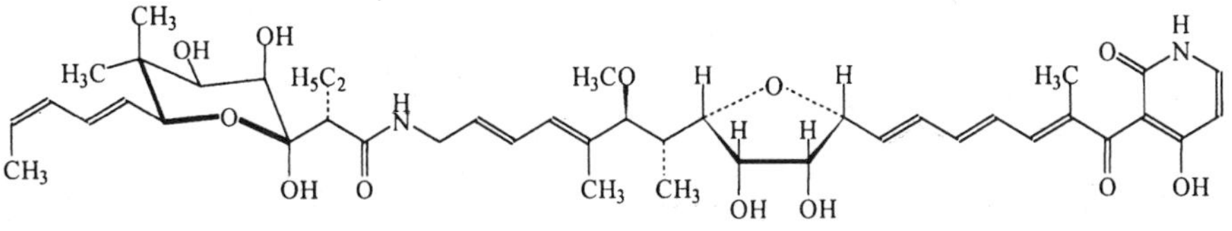

FIG. 10.12. Kirromycin.

nd isoleucine, as well as of serine, tyrosine, and valine in decreasing levels. The ncorporation of these five amino acids, in particular of leucine and isoleucine, is reatly stimulated by streptomycin. Figure 10.13 shows the correct codon–anti-odon pairing of the template with $tRNA^{Phe}$ and the known incorrect types of UUU odon pairing with anticodons of *E. coli* $tRNA^{Ile}$, $tRNA^{Leu}$, $tRNA^{Ser}$, $tRNA^{Tyr}$, and RNA^{Val}. Two of the three nucleotide positions in the codon–anticodon complex hould be paired (including noncanonical pairing at the wobble position) in or-ler to provide perceptible binding of noncognate tRNA in the A site of the ribo-ome.

$tRNA^{Phe}$: 5' G • U 3' / A—U / 3' A—U 5'

$tRNA^{Ile}$: 5' G • U 3' / A—U / 3' A U 5'

$tRNA_2^{Leu}$: 5' U† • U 3' / A—U / 3' G U 5'

$tRNA_3^{Leu}$: 5' G • U 3' / A—U / 3' G U 5'

$tRNA_4^{Leu}$: 5' C U 3' / A—U / 3' A—U 5'

$tRNA_5^{Leu}$: 5' U* U 3' / A—U / 3' A—U 5'

$tRNA_1^{Ser}$: 5' U^V • U 3' / G U / 3' A—U 5'

$tRNA^{Tyr}$: 5' G^Q • U 3' / U U / 3' A—U 5'

$tRNA_1^{Val}$: 5' U^V • U 3' / A—U / 3' C U 5'

$tRNA_2^{Val}$: 5' G • U 3' / A—U / 3' C U 5'

IG. 10.13. Codon–anticodon pairing of poly(U) with cognate and noncognate tRNA species. U^V, 5-arboxymethoxyuridine (cmo^5U or V); U*, 5-carboxymethylaminomethyl-2′-O-methyluridine cmnm^{5}Um); U†, U probably modified. G^Q, queuosine (Q) (see Fig. 3.3).

10.4.2. Leakiness of Stop Codons

An *in vivo* example of miscoding is the reading-through of a stop codon du to misbinding of an aminoacyl-tRNA to the ribosome with the stop codon in the A site. In addition to the well-known cases when such a leakiness of a stop codon re sults from the existence of a special suppressor tRNA species possessing the anti codon complementary to the stop codon, there are observations of misbinding o normal wild-type aminoacyl-tRNAs to the stop codons in the ribosome. In partic ular, it has been demonstrated that the wild-type Trp-tRNA with its anticodon CCA can interact with UGA and be incorporated into the growing polypeptide thus sup pressing this stop signal. The wild-type Gln-tRNA (anticodons CUG and s²UUG was reported to be capable of binding with the stop codon UAG, resulting in th reading-through of normal messages or the suppression of *amber* mutations:

```
           5′    3′                          5′    3′
            \    /                            \    /
            C   A                             C : G
            |   |                             |   |
tRNA^Trp    C : G   Opal codon     tRNA^Gln   U : A   Amber codon
            |   |                             |   |
            A : U                             G   U
           /     \                           /     \
          3′      5′                        3′      5′
```

10.4.3. Principal Types of Mispairing

Analysis of miscoding in the cases of poly(U) and other template polynu- cleotides *in vitro,* as well as *in vivo,* including miscoding induced by various aminoglycoside antibiotics, has demonstrated that errors are largely due to G:U or U:G pairing, as well as to U:U pairing (or juxtaposition) at any position of the an- ticodon–codon duplex. Mispairing or juxtaposition of the U:C or C:U types are less common. Some rare errors are due to the juxtaposition of the C in the anticodon and the A of the codon. In exceptional cases errors involve the formation of A:G or G:A pairs or juxtapositions, as well as of C:C, A:A, and G:G juxtapositions.

It appears that all types of juxtapositions are possible in the wobble position, including I:G, G:A, G:G, U:U, U:C, C:A, C:U, and C:C. This probably explains why all of the isoacceptor tRNAs can recognize in the cell-free system all the four codons of a given codon family, that is, of codons that have the same initial two nucleotides.

10.4.4. Factors Contributing to Miscoding

In addition to aminoglycoside antibiotics, a number of less specific factors in- cluding ionic conditions of the medium, can increase the number of errors in the codon-dependent entry of the aminoacyl-tRNA into the translating ribosomes. Generally, all factors increasing the affinity of the tRNA for the ribosome result in increased miscoding. Increased Mg^{2+} concentration in the medium and the addi- tion of diamines (e.g., putrescine) or polyamines (e.g., spermidine) increase the level of errors during cell-free translation. Ethyl alcohol and other hydrophobic agents added in even low concentrations also increase miscoding. Urea, in con- trast, leads to a decrease in the miscoding level. As for general environmental facɔrs, a lower temperature, decreased pH, and low ionic strength also contribute to higher incidence of miscoding.

Structural features of the tRNA itself also play a part in the accuracy of the odon-dependent binding of tRNA in the A site (for a review, see Kurland and hrenberg, 1984). In particular, the structure of the D hairpin may be important. or example, the mutational alteration of G to A in the D arm of $tRNA^{Trp}$ stimu-ates the pairing of this tRNA (which has the CCA anticodon) with the noncognate odons UGA and UGU. There is reason to believe that the alteration of the D arm ıcreases the affinity of tRNA for the A site.

Furthermore, the ribosome structure plays an important role in the accuracy f aminoacyl-tRNA selection. Gorini (1971, 1974) was the first to demonstrate that ertain mutations leading to alterations in ribosomal components may either de-rease or increase the level of miscoding. It has been found that the mutational al-erations in protein S12 resulting in resistance to streptomycin (*strA* mutations) onfer greater fidelity to the bacterial ribosome in the codon-dependent selection f aminoacyl-tRNA. In contrast, the so-called *ram* mutations (ribosome ambiguity ıutations) involving protein S4 make the ribosome less selective and increase the evel of miscoding. Specific mutational alterations in protein S5 also decrease the electivity of the ribosome. Mutations in eukaryotic (yeast) analogs of these pro-eins have been shown to exert the same effects on the fidelity of the eukaryotic ri-osome. More recently it has been demonstrated that mutational alterations of the 6S RNA of bacteria or 18S RNA of eukaryotes (yeast) in the regions around the A ite, namely positions 517, 912, 1054, 1409, 1495 and others (see Fig. 6.1; cited in Ioller, 1991; Green and Noller, 1997), or their equivalents, also affect translation-l accuracy of the ribosome. It may well be that the 30S or 40S subunit components ɔrming the A site and located nearby—particularly the above-mentioned RNA re-ions and the tightly clustered group of proteins S4, S5, and S12—not only are vi-al to the strength of tRNA retention but also may define the degree of structural igidity/flexibility of the tRNA anticodon or mRNA codon positioned on the ribo-ome.

The environmental factors listed above, as well as intrinsic structural factors, ffect the extent of the selectivity of the mRNA-programmed ribosome with respect ɔ tRNA. At the same time, the miscoding level in the system depends not only on electivity as defined by the intrinsic properties of the components under given onditions, but on the ratio of the components as well. It is apparent that as the ra-io of the concentration of noncognate tRNAs to that of cognate tRNA increases, he probability of misbinding becomes higher. It is for this reason that, for exam-le, when poly(U) is translated in the cell-free system and the cognate pheny-alanyl-tRNA is depleted as a result of the synthesis, the incorporation of leucine, soleucine, and other incorrect amino acids into the polypeptide tends to increase. n an extreme case the system with poly(U)-programmed ribosomes may be sup-lied only with leucyl-tRNA, and then pure polyleucine will be synthesized on oly(U) as a template (although, of course, at a markedly slower rate than with olyphenylalanine synthesis). In natural systems the amount of amino acids mis-ncorporated into the synthesized polypeptide may depend greatly on the con-entrations of different aminoacyl-tRNA species. Thus, cell starvation for an amino cid results in amino acids with near-coding specificities extensively replacing his amino acid in the polypeptide chains (Parker *et al.*, 1978; Gallant and Folley, 979). Strong effects on mistranslation were reported in the cases of overproduction of foreign proteins in bacterial cells, due to aberrations of a normal balanc between different amino acids in the intracellular medium (Bogosian *et al.*, 1990)

10.4.5. Miscoding Level *In Vivo* under Normal Conditions

Since miscoding depends largely on a number of environmental and structura factors, it is clear that its level in the cell-free systems varies greatly. Therefore, i is important to estimate the natural miscoding level in normal living cells that ar not subjected to extreme conditions and do not carry mutations affecting the pro tein-synthesizing machinery.

Several attempts have been made at estimating the level of miscoding *in vivo* Loftfield's classical estimates (Loftfield and Vanderjagt, 1972; see also Coons Smith, and Loftfield, 1979) provided data on the rate of misincorporation of valin instead of threonine in the α-chains of rabbit globin; a value of about 2–6 × 10⁻ was obtained. The frequency of cysteine misincorporation, probably instead o arginine, into the completed (folded and assembled) *E. coli* flagellin which nor mally does not contain cysteine was of the same order of magnitude—10^{-4} pe codon, according to the estimate of Edelman and Gallant (1977). Later more direc estimates of the miscoding level *in vivo* principally confirmed the values fron 10^{-4} to 10^{-3} for the average frequency of translational errors per codon (for review see Kurland *et al.*, 1990).

Some codons, however, can be misread more frequently than the above esti mated rate. In particular, miscoding within one codon group (i.e., among codon differing only in the third nucleotide) is more probable than other replacements It follows from the code dictionary (Fig. 2.2) that the most probable amino acid re placements, resulting from miscoding, would be the following: Phe ↔ Leu, Cys ↔ Trp, His ↔ Gln, Ile ↔ Met, Asn ↔ Lys, Ser ↔ Arg, and Asp ↔ Glu. It is expected tha the replacements from left to right (mispairing with U or C in the third codon po sition) are more probable than replacements from right to left (mispairing with A or G in the third codon position). Indeed, lysine substitutions for asparagine ar far more frequent than replacements of lysine; similarly, histidine is often replace by glutamine, whereas glutamine is rarely replaced. For instance, in the experi ments on the *in vivo* translation of MS2 coat protein mRNA, the frequency of mis reading of the asparagine codon AAU leading to the substitution of the positively charged lysine for the uncharged asparagine was about 5×10^{-3} (though the fre quency of misreading of the asparagine codon AAC was nearly an order of magni tude less, about 2×10^{-4}) (Parker *et al.*, 1983). In exceptional cases the level fo some amino acid replacements was reported to reach 10^{-2} per codon.

The level of miscoding is usually much higher in cell-free systems; the re placement frequency may be as great as 10^{-2} and, within the same codon group (Phe → Leu), even 10^{-1} per codon. However, under controlled ionic conditions (i a low Mg^{2+} concentration and with optimal proportions of other components), a level of miscoding approaching the values of 10^{-4} to 10^{-3} errors per codon may be attained.

It is important to note that under certain conditions the level of miscoding *in vivo* can be greatly increased, both in bacterial and in animal cells. This can be achieved by starving for certain amino acids, as well as by adding ethyl alcoho and some other agents to the medium. As already mentioned miscoding in the cel increases in response to aminoglycoside antibiotics.

A certain miscoding level may be of great biological significance and there ore is maintained in evolution. Bacterial mutants with a low miscoding level streptomycin-resistant mutants) can be obtained, but this level is always higher n wild strains that have been isolated from nature and are more adapted to survival. There is no doubt that, in certain circumstances, miscoding contributes to urvival (e.g., in cases of mutations that would otherwise be lethal). Thus, misinding of an aminoacyl-tRNA to the termination codon produced by the mutation f a sense codon may ensure that a functional protein molecule is completed. The esult of this "deception for the sake of salvation" is that the nonsense mutant survives. Similarly, certain mutants with the point amino acid replacements in imortant proteins, otherwise lethal, may survive through miscoding. In addition, the ell may permanently employ an infrequent misbinding of aminoacyl-tRNA to the ormal termination codon; this results in an mRNA read-through beyond its usul coding region (see Section 14.1) and therefore in a synthesis of small amounts f longer polypeptides yielding functionally differing proteins, which may be necssary to some processes in the cell. It is also possible that a cell occasionally relaces an amino acid in a similar manner in order to synthesize some needed varint of a given protein in small amounts. In any event, too much accuracy in coding uring translation would probably restrict the cellular flexibility necessary for survival.

Another consideration concerns the rate of protein synthesis. The point is that he accuracy requires time and energy. Therefore, the attainment of an exceedingy high accuracy will decelarate translation and exhaust the energy resources of he cell (Kurland *et al.*, 1990). That is why all living organisms in their evolution re forced to keep a balance between a reasonable level of accuracy of the translaion machinery and an adequate rate of protein synthesis and cell growth.

0.4.6. Kinetic Mechanisms of Miscoding and Miscoding Correction

It follows that miscoding depends primarily on the affinity of anticodon to odon. Affinity is strongest when the anticodon is complementary to the codon, nd therefore the cognate tRNA binds to the codon preferentially. Affinity is lowr, but still exists, in the case of partial complementarity and, thus, some noncogate ("near-cognate") tRNA species may bind to the codon as well.

Experiments on different polynucleotide complexes have demonstrated that he difference in affinity constants, in the case of complementary or partially comlementary pairing, is largely determined by the differences in the lifetimes of the omplexes (i.e., in the rates of dissociation), whereas the rates of formation do not iffer greatly, if they differ at all. All this suggests that the probability of a wrong nticodon–codon complex being formed is just as high as the probability of a corect one, but the wrong complex decays faster, that is, has a shorter lifetime.

If reversible codon-dependent tRNA binding in the elongation cycle is folowed by an irreversible stage, then the rate of the irreversible process will affect he level of miscoding: the higher the rate of the subsequent irreversible stage the reater the level of miscoding (Ninio, 1974). The role of such an irreversible stage nay be played by transpeptidation with subsequent translocation. In fact, because omplex formation of the codon with the cognate tRNA and with the noncognate RNA are formed at similar rates, the discrimination between the tRNAs is based nly on the fact that the wrong complex decays faster. However, if the rate of the ext stage of the cycle is very fast and is comparable to the rate at which the wrong omplex decays, the probability is high that the noncognate tRNA will be drawn into the cycle. Clearly, the higher the rate of the following stage, then the more th tRNA binding conditions differ from the equilibrium ones, and the greater th probability of the noncognate tRNAs being captured. Conversely, if the subsequen irreversible stage proceeds at a slower rate, the conditions will approach equilib rium and the discrimination between cognate and noncognate tRNA species wil depend to a larger extent on the difference in their affinity constant toward th codon. To reduce the miscoding level, the ribosome may possess a special mech anism of *kinetic delay* at the stage of reversible codon-dependent aminoacyl-tRN binding. EF-Tu (EF1A) may play this role: this protein bound with GTP blocks th subsequent step of the cycle until GTP is hydrolyzed, while the ternary comple remains reversibly associated with the ribosome.

On the other hand, discrimination between the cognate and noncognate tRN on the ribosome may be even further amplified than follows from the simple dif ference in affinity constants. If in the process of aminoacyl-tRNA binding there ar two successive reversible phases which are separated by a virtually irreversibl step, the aminoacyl-tRNA will have two independent chances to dissociate. In thi case the overall dissociation probability is equal to the product of the dissociatio probabilities at the two stages, that is, discrimination between the cognate an noncognate tRNA will be significantly amplified compared to the difference i affinity constants (Hopfield, 1974). This is a mechanism of *kinetic correction*, o "*proofreading*." Again, EF-Tu could perform the role: the first stage of reversibl binding could include the binding of the ternary Aa-tRNA:EF-Tu:GTP complex pri or to GTP hydrolysis; GTP hydrolysis would represent an irreversible separatin step; and the aminoacyl-tRNA would then have another, independent chance o dissociating from the complex with the codon at the stage of EF-Tu:GDP dissocia tion from the ribosome (see Fig. 10.14).

In the case of the cognate aminoacyl-tRNA, the rate of the side reaction 4′ (righ arrow, decay of a codon–anticodon complex prior to EF-Tu:GDP release) will b low, compared to the main reaction 4 (left arrow, release of EF-Tu:GDP), that is, k $\gg k_{4'}$. In the case of a less stable complex with the noncognate tRNA, the side-re action 4′ proceeds much faster and competes with the main reaction 4 (i.e., $k_{4'} >$ k_4, Fig. 10.14). Thus, discrimination between the cognate and noncognate tRNA i accomplished twice: on the basis of differences in the rates at which the codon– anticodon complex decays in reaction 2 (difference in k_{-2}) and on the basis of th difference in the rates at which the codon–anticodon complex decays in reactio 4′ (difference in $k_{4'}$). GTP hydrolysis, which is a virtually irreversible process serves to separate these two stages and, in this sense, the released free energy dis sipating into heat increases translation fidelity.

10.5. Summary: Sequence of Events and Molecular Mechanisms

10.5.1. Scanning of tRNA Species

In the course of elongation, different aminoacyl-tRNA species are present i the solution surrounding the ribosome. The aminoacyl-tRNA corresponding to th template codon positioned in the A site should be selected from this mixture b the ribosome. To achieve this end, a rapid scanning of different aminoacyl-tRNA (Aa-tRNA:EF-Tu:GTP complexes) should be performed and only the complex rec ognized by the ribosome as a codon cognate should remain bound to the ribosome Such scanning assumes multiple collisions between the codon fixed on the ribo ome and the anticodons of different tRNAs. It may be asked whether this is chieved by random diffusional collisions between nonoriented tRNAs and the ribosome or whether there is a rapid formation and decay (a run-through) of short-lived intermediate complexes between the tRNAs and the ribosome, where collisions between the anticodons and the codon occur in the proper orientation.

Studies on fast kinetics in the course of tRNA binding with the ribosome favor the second alternative (Rodnina, Fricke, and Wintermeyer, 1993). Indeed, the kinetics of tRNA binding with the ribosome have been shown to be multiphasic. The first rapid phase can be interpreted as the formation of an intermediate short-lived complex which does not depend on the template codon (Fig. 10.14, step 1). The intermediate short-lived ribosomal complex results from a collision between the ternary (Aa-tRNA:EF-Tu:GTP) complex and the ribosome followed by a specific orientation of the tRNA. Seemingly, the codon–anticodon interaction is not used in the transient complex formation. It can be assumed that the rapid formation and decay of such short-lived complexes is used for the scanning (run-through) of anticodons. Only when the anticodon fits the codon does this result in

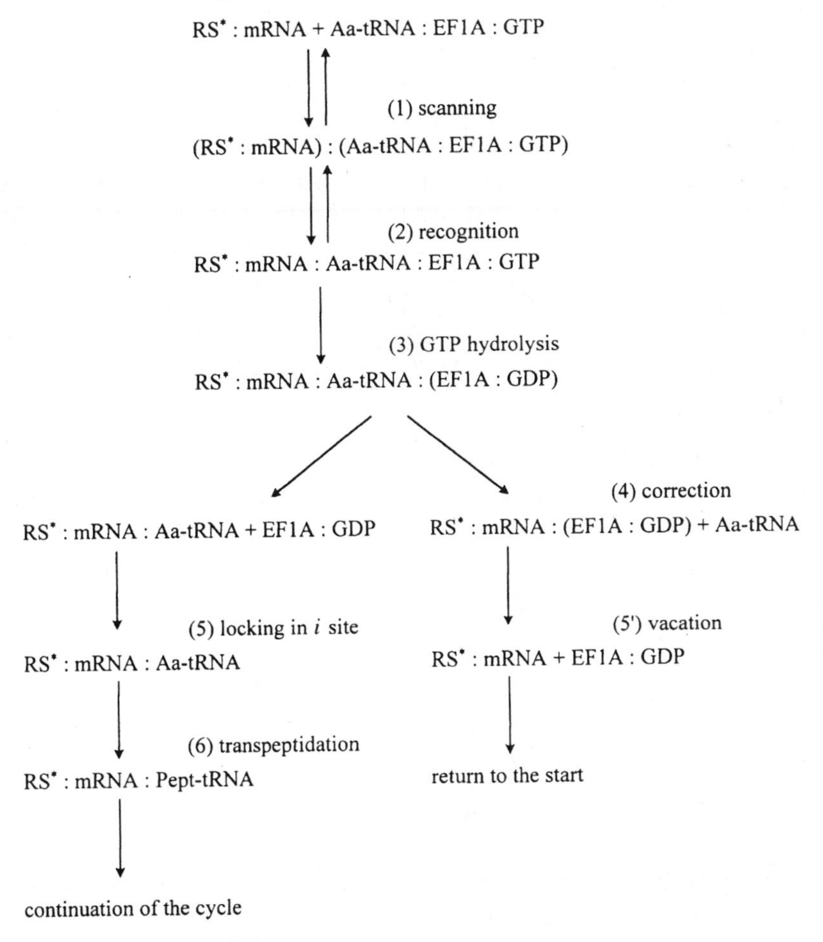

FIG. 10.14. Pathway of aminoacyl-tRNA entry into the elongating ribosome. RS* is an elongating ribosome carrying peptidyl-tRNA in the P site. The bracketed parts of complexes are presumed to be in loose and transient association with the rest of the complex.

recognition which induces the rearrangement of the complex, leading to the next binding phase. It is likely that the EF-Tu moiety of the ternary complex helps in the formation of the intermediate short-lived ribosomal complex: it may transiently interact with the factor-binding site (L7/L12 stalk area) on the 50S ribosomal subunit, thus properly orienting tRNA on the 30S subunit for the codon–anticodon collision. This stage of tRNA binding, prior to codon–anticodon recognition, may be designated as the scanning state, or T state of the ribosomal complex.

10.5.2. Recognition of Anticodon

Studies on the multiphasic kinetics of tRNA binding have demonstrated that when the P site is occupied and only when the cognate codon is present in the A site, the above rapid phase is followed by a slower phase during which the intermediate short-lived complex rearranges into a more stable complex. As a result the anticodon of tRNA becomes specifically bound with the codon in the ribosomal A site. Experiments with nonhydrolyzable or slowly hydrolyzable GTP analogues have demonstrated that, at this stage of codon-dependent binding, the aminoacyl-tRNA or, more accurately, the ternary aminoacyl-tRNA:EF-Tu:GTP complex, is reversibly associated with the ribosome and, therefore, if GTP hydrolysis is not very fast, equilibrium can be approached. Thus, this stage provides the first step of aminoacyl-tRNA selection according to its coding specificity (Fig. 10.14, step 2).

If the interaction underlying the formation of the intermediate short-lived complex continues after codon–anticodon recognition, the aminoacyl-tRNA will then be bound to the 70S ribosome by at least three points: (1) by the anticodon through the mRNA codon with the 30S ribosomal subunit, (2) by the acceptor arm through EF-Tu with the 50S ribosomal subunit, and (3) by an unknown point of the nonspecific interaction with the A site on the 30S ribosomal subunit (possibly through the D arm). This is the so-called A/T state of the ribosomal complex.

10.5.3. GTP Hydrolysis

In this state, hydrolysis of the EF-Tu-bound GTP is induced (Fig. 10.14, step 3). Hydrolysis seems to be induced by the EF-Tu interacting with the factor-binding site of the 50S subunit. The hydrolysis of GTP, an apparently irreversible step (due to a great difference in the thermodynamic potentials of GTP and its hydrolysis products), divides the aminoacyl-tRNA binding process into two phases which are not connected by equilibrium. The shorter the time interval between tRNA recognition and GTP hydrolysis, the lower the selectivity of the recognition phase. If hydrolysis is artificially delayed, for example, by using a slowly hydrolyzable GTP analog, such as guanosine 5′-(γ-thio)-triphosphate (see Fig. 9.8), the discrimination between the cognate tRNA and near-cognate tRNA (e.g., $tRNA^{Phe}$ and $tRNA_2^{Leu}$) increases to a ratio of 10^4 in favor of the cognate species.

When the EF-Tu-bound GTP is hydrolyzed, no intermediate phosphorylated products are detected. The transfer of phosphate proceeds directly to a water molecule. In others words, there is no biochemical coupling of the GTP hydrolysis with an energy-consuming process. All the free energy of GTP hydrolysis seems to dissipate directly into heat. (Meanwhile, it should be remembered that the free energy of the hydrolysis of the factor-bound GTP should not be equal to that of the unbound GTP in solution.)

GTP hydrolysis results in a drastic decrease in EF-Tu affinity for the aminoacyl-tRNA and the ribosome. This is the result of the conformational rearrangement of EF-Tu when it becomes bound to GDP instead of GTP.

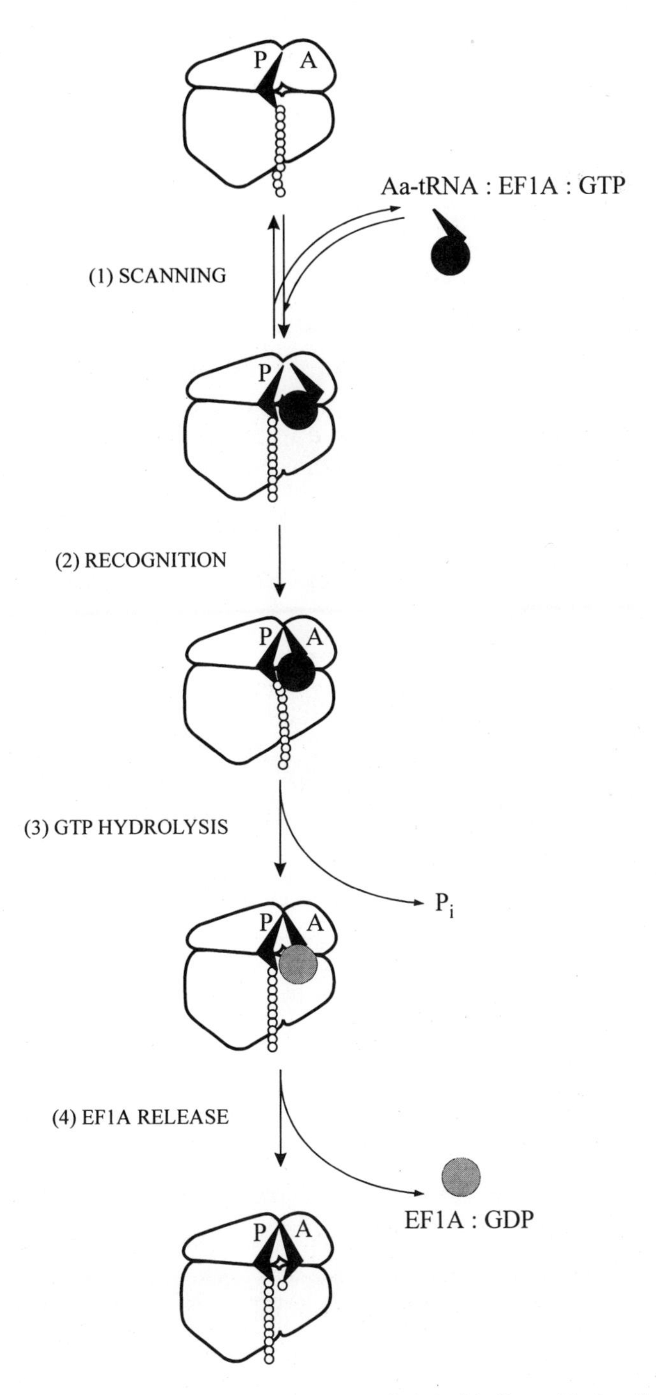

FIG. 10.15. Sequence of events during aminoacyl-tRNA binding with the ribosome.

10.5.4. EF-Tu (EF1A) Release and Correction of Aminoacyl-tRNA Selection

GTP hydrolysis leads to an intermediate ribosomal complex in which EF-Tu no longer has an affinity for the aminoacyl-tRNA. Consequently, the aminoacyl-tRNA has lost its attachment to the ribosome through EF-Tu. It now has another opportunity to dissociate, and if its dissociation rate is particularly high, as is the case for noncognate tRNA, it may be released prior to EF-Tu:GDP leaving the ribosome (Fig. 10.14, 4′, right path). However, if the rate of its dissociation is not that high, because of a correct codon–anticodon interaction, EF-Tu:GDP will be released earlier (Fig. 10.14, 4, left path) and the aminoacyl terminus will be captured at the 50S ribosomal subunit. Thus, after GTP is hydrolyzed, the ribosome is presented with a new opportunity to correct the result of aminoacyl-tRNA selection made prior to GTP hydrolysis.

10.5.5. Locking of Aminoacyl-tRNA in the A Site

After GTP hydrolysis and the subsequent release of EF-Tu:GDP, the acceptor end of aminoacyl-tRNA interacts with the *a* site of the PTC of the 50S subunit. Thus the aminoacyl-tRNA becomes fully locked in the A site. This is the final phase of aminoacyl-tRNA binding to the A site—the so-called A/*a* (or A/A) state of the ribosomal complex (Fig. 10.14, 5). Now the aminoacyl-tRNA is ready to react with the P-site-bound peptidyl-tRNA, and transpeptidation follows.

10.5.6. General Scheme

In the course of enzymatic binding to the mRNA-programmed ribosome, the aminoacyl-tRNA passes through a series of consecutive phases (Fig. 10.14). During transition from one phase to another, certain contacts between the aminoacyl-tRNA and the ribosome may change, but eventually the full binding to the A site (A/*a* state), where this substrate is ready to chemically react with the peptidyl-tRNA in the P site, is attained. The general scheme of the sequence of events described herein is given in Fig. 10.15.

References

Bogosian, G., Violand, B. N., Jung, P. E., and Kane, J. F. (1990) Effect of protein overexpression on mistranslation in *Escherichia coli,RR in The Ribosome: Structure, Function, and Evolution* (W. E. Hill, A. Dahlberg, R. A. Garrett, P. B. Moore, D. Schlessinger, and J. R. Warner, eds.), pp. 546–558, ASM Press, Washington, DC.

Chapeville, F., Lipmann, F., von Ehrenstein, G., Weisblum, B., Ray, W. J., and Benzer, S. (1962) On the role of soluble ribonucleic acid in coding for amino acids, *Proc. Natl. Acad. Sci. USA* **48:**1086–1092.

Coons, S. F., Smith, L. F., and Loftfield, R. B. (1979) The nature of amino acid errors in *in vivo* biosynthesis of rabbit hemoglobin, *Fed. Proc.* **38:**328.

Crick, F. H. C. (1966) Codon-anticodon pairing: The wobble hypothesis, *J. Mol. Biol.* **19:**548–555.

Cundliffe, E. (1980) Antibiotic and prokaryotic ribosomes: Action, interaction, and resistance, in *Ribosomes: Structure, Function, and Genetics* (G. Chambliss, G. R. Craven, J. Davies, K. Davis, L. Kahan, and M. Nomura, eds.), pp. 555–581, University Park Press, Baltimore.

Cundliffe, E. (1990) Recognition sites for antibiotics within rRNA, in *The Ribosome: Structure, Function, and Evolution* (W. E. Hill, A. Dahlberg, R. A. Garrett, P. B. Moore, D. Schlessinger, and J. R Warner, eds.), pp. 479–490, ASM Press, Washington.

Davies, J. E., Gilbert, W., and Gorini, L. (1964) Streptomycin, suppression, and the code, *Proc. Natl. Acad. Sci. USA* **51**:883–890.

Davies, J. E., Gorini, L., and Davis, B. D. (1965) Misreading of RNA codewords induced by aminoglycoside antibiotics, *J. Mol. Pharmacol.* **1**:93–106.

Edelman, P., and Gallant, J. (1977) Mistranslation in *E. coli, Cell* **10**:131–137.

Ehrenstein, von, G., Weisblum, B., and Benzer, S. (1963) The function of sRNA as amino acid adaptor in the synthesis of hemoglobin, *Proc. Natl. Acad. Sci. USA* **49**:669–675.

Fourmy, D., Recht, M. I., Blanchard, S., and Puglisi, J. D. (1996) Structure of the A site of *Escherichia coli* 16S ribosomal RNA complexed with an aminoglycoside antibiotic, *Science* **274**:1367–1371.

Gale, E. F., Cundliffe, E., Reynolds, P. E., Richmond, M. H., and Waring, M. J. (1981) *The Molecular Basis of Antibiotic Action,* John Wiley and Sons, London.

Gallant, J. and Foley, D. (1979) On the causes and prevention of mistranslation, in *Ribosomes: Structure, Function, and Genetics* (G. Chambliss, G. R. Craven, J. Davies, K. Davis, L. Kahan, and M. Nomura, eds.), pp. 615–638, University Park Press, Baltimore.

Goodman, H. M., Abelson, J., Landy, A., Brenner, S., and Smith, J. D. (1968) Amber suppression: A nucleotide change in the anticodon of a tyrosine transfer RNA, *Nature* **217**:1019–1024.

Gorini, L. (1971) Ribosomal discrimination of tRNAs, *Nature New Biol.* **234**:264.

Gorini, L. (1974) Streptomycin and misreading of the genetic code, in *Ribosomes* (M. Nomura, A. Tissières, and P. Lengyel, eds.), pp. 791–803, CSHL Press, Cold Spring Harbor, NY.

Green, R., and Noller, H. F. (1997) Ribosomes and translation, *Annu. Rev. Biochem.* **66**:679–716.

Hopfield, J. J. (1974) Kinetic proofreading: A new mechanism for reducing errors in biosynthetic processes requiring high specificity, *Proc. Natl. Acad. Sci. USA* **71**:4135–4139.

Kurland, C. G., and Ehrenberg, M. (1984) Optimization of translation accuracy, *Progr. Nucleic Acid Res. Mol. Biol.* **31**:191–219.

Kurland, C. G., Joergensen, F., Richter, A., Ehrenberg, M., Bilgin, N., and Rojas, A.-M. (1990) Through the accuracy window, in *The Ribosome: Structure, Function, and Evolution* (W. E. Hill, A. Dahlberg, R. A. Garrett, P. B. Moore, D. Schlessinger, and J. R. Warner, eds.), pp. 513–526, ASM Press, Washington, DC.

Loftfield, R. B., and Vanderjagt, D. (1972) The frequency of errors in protein biosynthesis, *Biochem. J.* **128**:1353–1356.

Merrick, W. C., and Hershey, J. W. B. (1996) The pathway and mechanism of eukaryotic protein synthesis, in *Translational Control* (J. W. B. Hershey, M. B. Mathews, and N. Sonenberg, eds.), pp. 31–69, CSHL Press, Cold Spring Harbor, NY.

Ninio, J. (1974) A semi-quantitative treatment of missense and nonsense suppression in the *str*A and *ram* ribosomal mutants of *Escherichia coli:RR Evaluation of some molecular parameters of translation in vivo, J. Mol. Biol.* **84**:297–313.

Noller, H. F. (1991) Ribosomal RNA and translation, *Annu. Rev. Biochem.* **60**:191–227.

Noller, H. F., Moazed, D., Stern, S., Powers, T., Allen, P. N., Robertson, J. M., Weiser, B., and Triman, K. (1990) Structure of rRNA and its functional interactions in translation, in *The Ribosome: Structure, Function, and Evolution* (W. E. Hill, A. Dahlberg, R. A. Garrett, P. B. Moore, D. Schlessinger, and J. R. Warner, eds.), pp. 73–92 ASM Press, Washington, DC.

Parker, J., Johnston, T. C., Borgia, P. T., Holtz, G., Remaut, E., and Fiers, W. (1983) Codon usage and mistranslation. *In vivo* basal level misreading of the MS2 coat protein message, *J. Biol. Chem.* **258**:10007–10012.

Parker, J., Pollard, J. W., Friesen, J. D., and Stanners, C. P. (1978) Shuttering: High-level mistranslation in animal and bacterial cells, *Proc. Natl. Acad. Sci. USA* **75**:1091–1095.

Pestka, S. (1977) Inhibitors of protein synthesis, in *Molecular Mechanisms of Protein Biosynthesis* (H. Weissbach and S. Pestka, eds.), pp. 467–555, Academic Press, New York.

Pirohit, P., and Stern, S. (1994) Interactions of a small RNA with antibiotics and RNA ligands of the 30S subunit. *Nature* **370**:659–662.

Rodnina, M. V., Fricke, R., and Wintermeyer, W. (1993) Kinetic fluorescence study on EF-Tu-dependent binding of Phe-tRNA^{Phe} to the ribosomal A site, in *The Translational Apparatus* (K. H. Nierhaus, F. Franceschi, A. R. Subramanian, V. A. Erdmann, and B. Wittmann–Liebold, eds.), pp. 317–326, Plenum Press, New York.

Yokoyama, S., and Nishimura, S. (1995) Modified nucleosides and codon recognition, in *tRNA: Structure, Biosynthesis, and Function* (D. Soell and U. L. RajBhandary, eds.), pp. 207–223, ASM Press, Washington, DC.

11

Elongation Cycle, Step II: Transpeptidation (Peptide Bond Formation)

11.1. Chemistry of the Reaction

After the peptidyl-tRNA has occupied the P site and the aminoacyl-tRNA has been fully bound at the A site, the 3′-ends of the two tRNA residues are found close to each other in the region of the peptidyl transferase center (PTC) of the large ribosomal subunit. This is followed by a nucleophilic attack *on the carbonyl group* of the ester bond between the peptide residue and the tRNA moiety of the peptidyl-tRNA molecule *by the amino group* of the aminoacyl-tRNA molecule. As a result, an amide (peptide) bond is formed between the peptide residue and the aminoacyl-tRNA molecule (Fig. 11.1). The peptidyl-tRNA and the aminoacyl-tRNA are the substrates of this reaction. The peptidyl-tRNA is a *donor* substrate, and the aminoacyl-tRNA plays the part of an *acceptor* substrate. The products of the reaction are the deacylated tRNA in the P site and the peptidyl-tRNA with the peptide moiety elongated by one aminoacyl residue attached to the tRNA in the A site, as is shown schematically in Fig. 11.2.

It has already been mentioned (Section 9.3) that the 3′-terminal fragments of the aminoacyl-tRNA, the aminoacyl esters of the adenosine, as well as puromycin, can play the part of acceptor (or nucleophilic) substrate in the reaction with the peptidyl-tRNA, instead of the normal aminoacyl-tRNA molecule.

The reaction proceeds only with the NH_2 group in the α-position of an aminoacyl residue. The amino acid residue should be of L-configuration. The aminoacyl residue accommodated in the *a* site of the PTC must be attached to the 3′-position (not to the 2′-position) of the terminal adenosine of tRNA to serve as an acceptor substrate. The latter requirement warrants a special comments. Depending on the specificities of different ARSases, aminoacyl residues become originally attached either to 2′- or 3′-position of the ribose residue in the tRNA terminal adenosine (see Section 3.4 and Table 3.1). In solution, however, the aminoacyl residue migrates between the 2′- and 3′-positions. In PTC the aminoacyl residue is fixed only in the 3′-position.

The free 2′-hydroxyl of the acceptor substrate ribose is not essential for transpeptidation; if it is substituted (e.g., methylated) or is absent (2′-deoxyderivatives), the acceptor activity of the substrate is retained. The 2′-hydroxyl is, however, indispensible to the activity of the donor substrate.

The catalytic mechanism of ribosomal transpeptidation has been the subject of numerous studies and discussions (for reviews, see Monro *et al.*, 1996; Kra-yevsky and Kukhanova, 1979; Garrett and Rodriguez-Fonseca, 1995). There is evi-dence that the histidine imidazole of some ribosomal protein may be involved in the catalysis. However, all attempts at isolating an acyl-enzyme (acyl-ribosome) in-termediate of the kind found in proteinases catalyzing hydrolysis and transpepti-dation have been unsuccessful. Perhaps such an intermediate should not exist in principle, because peptidyl-tRNA is already an activated macromolecular acyl-de-rivative which may be regarded as a functional analog of the acyl-enzyme group. Therefore, many investigators believe that transpeptidation in the ribosome is cat-alyzed simply by an appropriate spatial orientation and alignment of the amino-acyl-tRNA and peptidyl-tRNA reacting groups without the catalytic involvement of special nucleophilic groups of PTC. This hypothesis is strongly supported by experiments showing a low specificity of the ribosomal PTC with respect to the types of bonds formed in it. Indeed, if a hydroxyacyl residue (HO–CHR–CO–) rather than an aminoacyl residue (H_2N–CHR–CO–), is attached to the tRNA or its 3′-end analog, the hydroxyderivative serves as a good acceptor substrate, and the ester bond is produced by the ribosome (Fahnestock *et al.*, 1970). Similarly, when the acceptor substrate is a thioacyl derivative the ribosomal PTC catalyzes the for-mation of a thioester bond (Gooch and Hawtrey, 1975). Moreover, the donor sub-strate can also be modified and the ribosome is capable of catalyzing the attack on the thioester group of the donor by the amino group of the acceptor, thus forming a thioamide bond (Victorova *et al.*, 1976). Finally, the phosphinoester analog of the donor has been shown to react with aminoacyl-tRNA in the ribosomal PTC, with the formation of an unnatural phosphinoamide bond (Tarsussova *et al.*, 1981). The

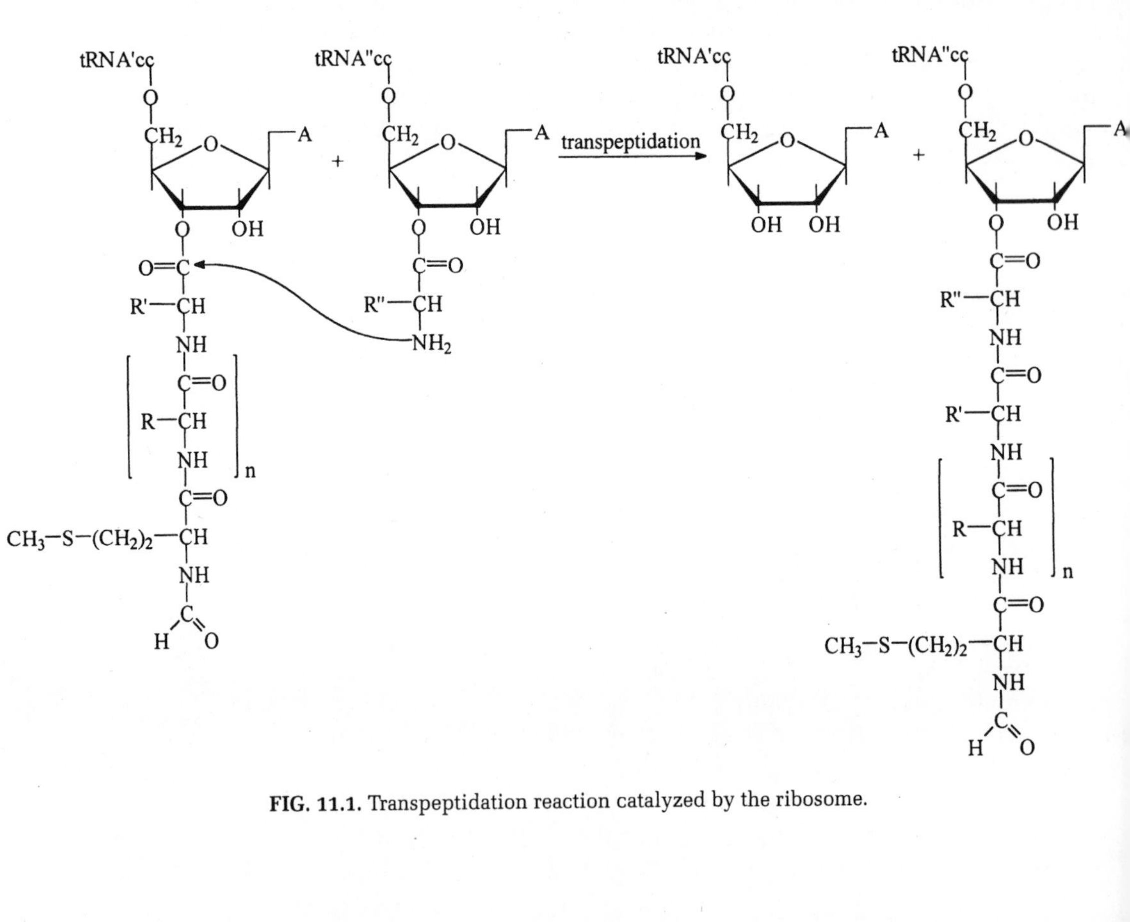

FIG. 11.1. Transpeptidation reaction catalyzed by the ribosome.

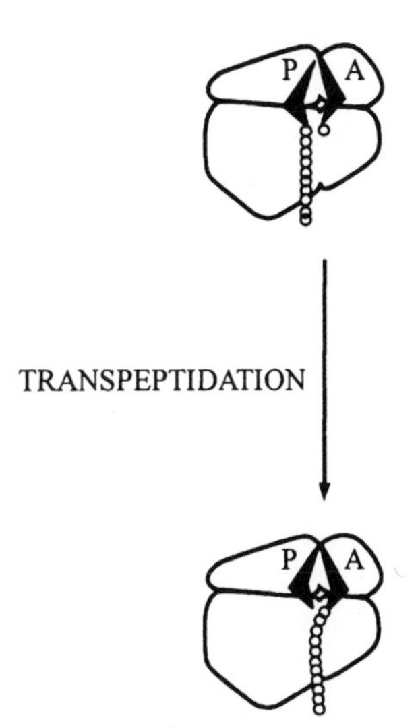

FIG. 11.2. Schematic representation of transpeptidation reaction between aminoacyl-tRNA and peptidyl-tRNA on the ribosome.

latter fact is especially important, since the geometry of the attacked group and the transition state of the reaction (trigonal bipyramid) are significantly different from those in the case of a normal donor (see Section 11.3); the nucleophilic catalysis is hardly compatible with the realization of the reactions of both types by the same enzymatic center.

At the same time, however, the ribosomal PTC under certain conditions is capable of using water and low-molecular-weight alcohols, such as methanol and ethanol, as acceptor substrates, thus performing hydrolysis or alcoholysis of the peptidyl-tRNA. These observations cannot be explained easily within the framework of the purely orientational mechanism of ribosomal PTC action. Most reasonable is the hypothesis that the orientation effect of PTC is indeed instrumental to transpeptidation on the ribosome, but that this effect can be aided by the contribution of specific microenvironments facilitating the reaction. It may well be that some groups located near the substrates properly oriented on the ribosome pull a proton away from the NH_2 group of the acceptor, thereby increasing its nucleophilic nature, on the one hand, and contribute to protonation of the carbonyl oxygen, thereby increasing the electrophilic properties of the attacked carbon of the donor ester group, on the other:

$$\text{tRNA}''\text{—O—}\underset{}{\overset{\text{O}}{\overset{\|}{\text{C}}}}\text{—}\overset{\text{R}''}{\overset{|}{\text{CH}}}\text{—}\overset{\text{H}}{\overset{|}{\text{N}}}\text{—H} \dashrightarrow (\text{X})$$

$$\overset{\delta+}{}$$

$$\text{tRNA}'\text{—O—}\underset{\underset{\underset{\underset{\underset{\text{(Y)}}{|}}{\text{H}}}{\uparrow}}{\text{O}}}{\underset{\|}{\text{C}}}\text{—}\underset{\text{R}'}{\underset{|}{\text{CH}}}\text{—NH-----}$$

Whatever the case may be, it seems certain that the reaction proceeds according to the mechanism of the S_N2 nucleophilic substitution through the so-called *tetrahedral intermediate* (see below, Section 11.3):

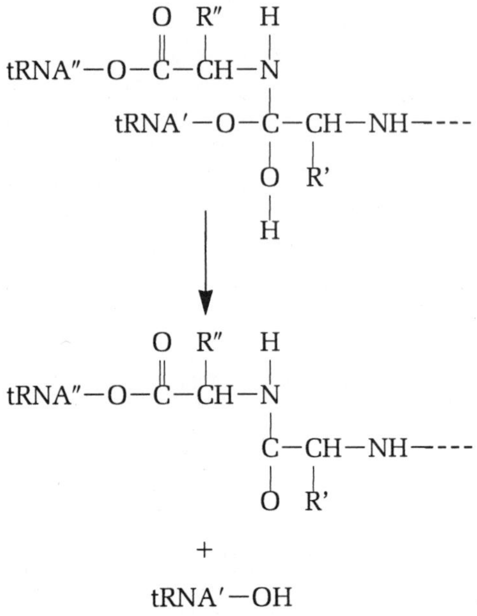

```
                O  R"   H
                ‖  |    |
   tRNA"—O—C—CH—N
                          |
          tRNA'—O—C—CH—NH----
                          |    |
                          O   R'
                          |
                          H
                  |
                  ↓
                O  R"   H
                ‖  |    |
   tRNA"—O—C—CH—N
                          |
                          C—CH—NH----
                          |    |
                          O   R'

                  +

             tRNA'—OH
```

11.2. Energy Balance of the Reaction

The standard free energy of hydrolysis of the ester bond between the tRNA and the carbonyl group of the aminoacyl or peptidyl residue $\Delta G^{0\prime}$ is equal to about −7 to −8 kcal/mole. The standard free energy of hydrolysis of the peptide bond in a polypeptide of infinite length is approximately −0.5 kcal/mole. Thus, if the reaction substrates come directly from solution and the reaction products are released into solution, the net gain of free energy due to transpeptidation under standard conditions should be about −6.5 to −7.5 kcal/mole:

$$\text{Pept(n)-tRNA}' + \text{Aa-tRNA}'' \rightarrow \text{tRNA}' + \text{Pept(n+1)-tRNA}'' \ (-7 \pm 0.5 \text{ kcal})$$

Such a formal calculation is usually put forward as an argument for ribosomal transpeptidation being fully supplied with free energy and therefore thermodynamically spontaneous.

It should be pointed out, however, that such an approach is valid only in the evaluation of ribosome-catalyzed model reactions between low-molecular-mass substrates. For example:

$$\text{F-Met-(3}'\text{)ACC(5}'\text{)} + \text{Phe-A} \xrightarrow{\textit{ribosome}} \text{(5}'\text{)CCA(3}'\text{)} + \text{F-Met-Phe-A}$$

Here the substrates come to PTC directly from solution and the products are immediately and spontaneously released into solution.

In the elongation cycle, one substrate is always associated with the ribosome while the other comes to PTC from a prebound state; one of the reaction products is not released into solution until translation has been completed, while the other product is released not as a result of transpeptidation but only after the next step of the elongation cycle. All this makes it impossible to give even an approximate estimation of the free-energy change in the course of transpeptidation proceeding in the normal ribosomal elongation cycle. Of course, the reaction is fast, suggesting that there is a significant decrease in the free energy of the ribosomal complex in the course of transpeptidation. This decrease, however, must be less than −7 kcal/mole under standard conditions. For one thing, the free energy of hydrolysis of the bound substrate, provided hydrolysis products are released, should be lower than that of the free substrate since some portion of the energy had to be released upon binding if binding was spontaneous. In addition, the free energy of substrate hydrolysis, provided the product remains bound, should be lower than that when the product leaves the complex since the bound state of the product is associated with the accumulation of free energy, if the release of the product is thermodynamically spontaneous in principle. It can be assumed that the free energy of −7 kcal/mole, which would be released in transpeptidation if the substrate and the products were free, is in fact partly distributed at the preceding stage of aminoacyl-tRNA binding and the subsequent stage of translocation, thus driving the entire elongation cycle.

11.3. Stereochemistry

In order to understand both the molecular mechanism of the ribosome-catalyzed transpeptidation and the initial conformation of the peptide to be synthesized, it is vital to have knowledge of the conformation of the reacting substrates in the ribosomal PTC. Unfortunately, however, there is no direct evidence regarding the conformations of the tRNA 3′-termini and the adjacent aminoacyl residues at the reaction site.

Regarding the acceptor substrate of the reaction, some information could be obtained by studying puromycin and its analogs. The conformation of puromycin in crystal has been solved by X-ray analysis (Fig. 11.3) and confirmed by studies on puromycin in solution. Because puromycin is a good acceptor substrate in ribosomal transpeptidation, its structure may provide some information about the stereochemistry of aminoacyl and adenosine residues in the PTC.

At the same time, on the basis of general considerations, one may propose that all types of aminoacyl residues of the acceptor substrate, on one hand, and C-terminal aminoacyl residues of the donor substrate, on the other, are positioned and presented to each other by the ribosomal PTC in a standard equivalent fashion, independent of their type. Using this principle as a guideline, some conclusions about the conformations of reacting substrates on the basis of purely stereochemical analysis may be attempted.

First, it should be remembered that the ribosome catalyzes transpeptidation with the proline residue as a substrate. In contrast to other amino acids, proline has a sterically limited angle of rotation around the C^{α}–N bond, since this bond is involved in the ring structure. In the case of the proline residue in the donor substrate, this limitation will set a fixed angle, equal to about 60°, between the plane (N_i–C^{α}_i–C'_i) and the plane of the adjacent peptide group (N_i–C'_{i-1}–O'_{i-1}) (Fig. 11.4). In peptide chemistry, the angle given by the rotation around the C^{α}–N bond is designated as φ; in this case its value is taken to be −60° since the plane of the

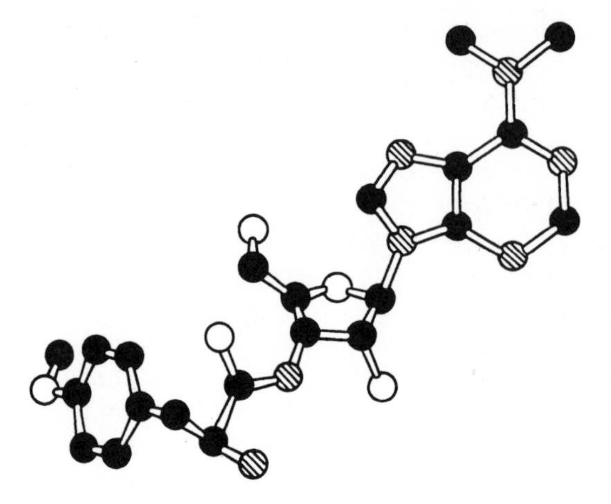

FIG. 11.3. Ball-and-sticks skeletal model (without hydrogens) of puromycin based on X-ray analysis. See Sundaralingam, M., and Arora, S. K. (1972) *J. Mol. Biol.* **71**:49–70. Filled circles are carbons, hatched circles are nitrogens, open circles are oxygen.

peptide group is turned 60° counterclockwise when viewed from the C^{α} atom. Since an amino acid residue should be positioned in the PTC in the standard way, angle φ should be adjusted to the same value by rotating it around the C^{α}–N bond for each of the other 19 types of residues of the donor substrate (the C-terminal residue of the nascent peptide bound to tRNA and participating in transpeptidation is under consideration).

It follows from reaction chemistry that transpeptidation involves a nucleophilic attack on the carbon atom of the carbonyl group realized via the mechanism of S_N2 substitution. Such an attack is known to proceed approximately perpendicular to the plane of the ester group (Fig. 11.5A). The attack should lead to an intermediate wherein the valence bonds of the attacked carbonyl carbon C'_i of the donor substrate are oriented tetrahedrally (tetrahedral intermediate) (Fig. 11.5B).

On the other hand, it should be remembered that the attacking NH_2-group should be deprotonated, that is, the unshared pair of electrons of the nitrogen atom should be free. Hence, the PTC should provide for the deprotonation of the aminoacyl residue of the acceptor substrate. Prior to peptide bond formation the nitro-

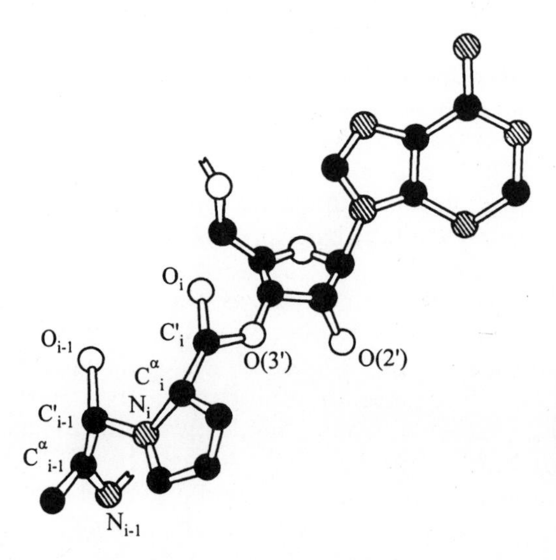

FIG. 11.4. Alanylprolyladenosine residue as a donor substrate in the PTC of the ribosome: ball-and-stick skeletal model (without hydrogens). Atom designations are the same as in Fig. 11.3. Atoms of the C-terminal prolyl residue are marked by the index *i*, those of the preceding alanyl residue by the index *i–1*.

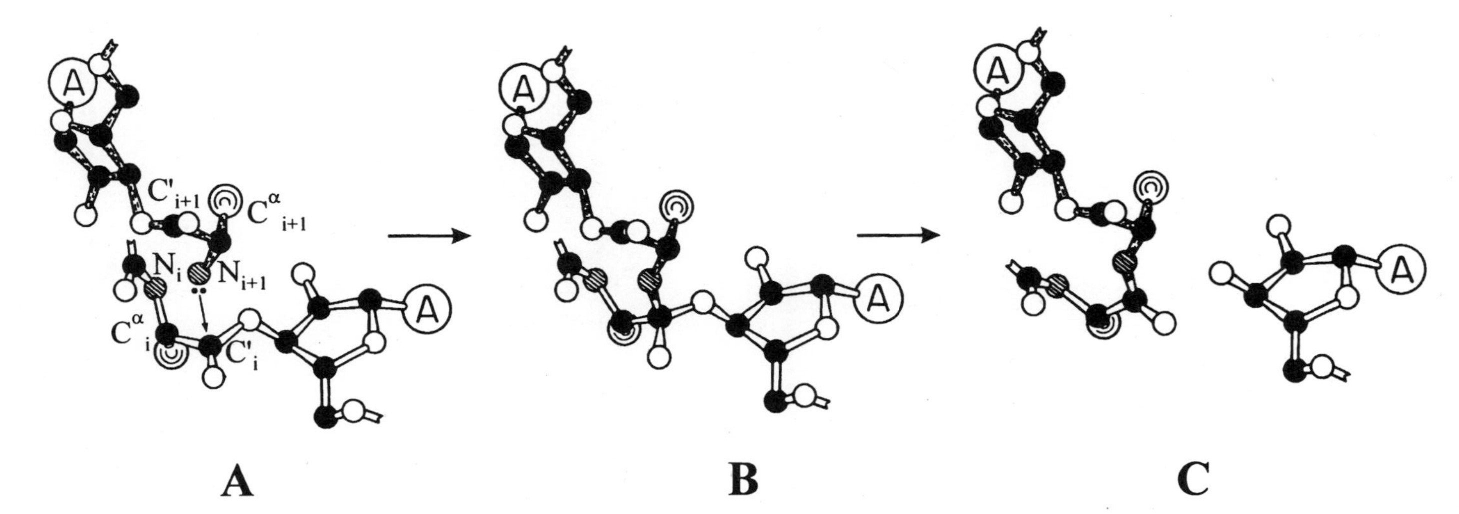

FIG. 11.5. Stereochemistry of the transpeptidation reaction: ball-and-stick drawing (without hydrogens). Atom designations are the same as in Fig. 11.3. Atoms of the acceptor aminoacyl residue are marked by the index *i*+1. A is the adenine residue in both the donor and acceptor substrate. (**A**) Mutual positions of the donor (*lower*) and acceptor (*upper*) substrates in the PTC; the acceptor substrate nitrogen (N_{i+1}) attacks the donor carbonyl carbon (C'_i). (**B**) Tetrahedral intermediate resulting from the attack. (**C**) Products of the reaction: the elongated peptidyl adenosine residue (*left*) and the deacylated adenosine fesidue (*right*). See Lim, V. I., and Spirin, A. S. (1986) *J. Mol. Biol.* **188**:565–574.

gen atom possesses three valence bonds which are directed toward the apexes of a tetrahedron, while the orbital of the unshared pair of electrons is directed toward the fourth apex. It follows from stereochemical analysis that, during the nucleophilic attack, the free valence of the attacking nitrogen atom should have a strictly defined direction: the plane formed by this direction and the N_{i+1}–C^{α} bond should be at an angle of about 120° to the plane (N_{i+1}–C^{α}_{i+1}–C'_{i+1}) of the acceptor aminoacyl residue (Fig. 11.5A); with any other orientations there can be no nucleophilic attack on the carbonyl group because of steric hindrances. After the peptide bond is formed, an angle φ of about −60° is set in the newly added aminoacyl residue of the product (Fig. 11.5C).

Taking all of the above into consideration, stereochemical analysis demonstrates that an effective nucleophilic attack in the ribosomal PTC can take place only if the angle between the plane of the ester group COO and the plane (C'_i–C^{α}_i–N_i) defined by the rotation around the C^{α}_i–C'_i bond in the attacked aminoacyl residue of the donor substrate (peptidyl-tRNA) is about 60° (Fig. 11.5A). After transpeptidation, it becomes an angle ψ (with an approximate value of −60° (60° counterclockwise rotation around the C^{α}–C' bond when viewed from the C^{α} atom) (Fig. 11.5C).

Thus, the conformation of the donor aminoacyl residue in the ribosomal PTC should be similar to that of the aminoacyl residue in the α-helix ($\varphi = -50°$, $\psi = -60°$). Moreover, the acceptor aminoacyl residue attacks the donor residue in such an orientation that the transpeptidation reaction yields an elongated backbone N_i–C^{α}_i–C'_i–N_{i+1}–C^{α}_{i+1} with parameters of the α-helical conformation.

Independent of this result of stereochemical analysis, the rule of the equivalent (universal) positioning of any aminoacyl residue in the ribosomal PTC leads to the conclusion that the residue newly incorporated into the polypeptide chain always acquires a standard conformation of its backbone (i.e., standard values of torsion angles C^{α}–C' and C^{α}–N). This implies that the PTC will generate a helical conformation of the synthesized peptide. It would not come as a surprise, therefore, if the peptide is built in a sterically and energetically favorable helical conformation, such as the α-helix. This initial conformation later rearranges into a unique three-dimensional structure. That the polypeptide chain folding *in vivo* does not, perhaps, begin from a random or extended state but rather is the result of a rearrangement of the helical structure may provide unique opportunities for a quick and accurate search for the final conformations of the protein molecule.

11.4. Movement of Transpeptidation Products

Consideration of the stereochemistry of the transpeptidation reaction (Fig. 11.5) implies that the decay of the tetrahedral intermediate should be accompanied by moving the deacylated ribose group aside. Moreover, in order to prevent reversibility of the decay, and hence reversibility of transpeptidation on the ribosome, the deacylated group should be immediately removed from the reaction site. This is probably the function of the *e* site that fixes the terminus of the deacylated tRNA outside the PTC.

At the same time the ester group with the amino acid residue which had been bound at the *d* site before transpeptidation becomes destroyed as a result of the tetrahedral intermediate decay, and thus this residue loses its affinity for the *d* site.

Instead, the amino acid residue which occupied the *a* site before transpeptidation acquires now the feature (amide group) that provides for its affinity for the *d* site. As a consequence, the newly added amino acid residue with its ester group and amide (peptide) group should displace the previous amino acid residue, that is, move from the *a* site to the *d* site of the PTC.

Indeed, the footprinting analysis by Noller *et al.* (1990) (see also review by Noller, 1991) fully confirms these expectations. It was shown that peptidyl-tRNA bound in the P site of the ribosome (pre-transpeptidation P/*d* state) protects G2252-G2253, U2506, and U2584-U2585 by its CCA terminus on the 50S subunit; thus these positions can be attributed to the *d* site. Aminoacyl-tRNA bound in the A site (pre-transpeptidation A/*a* state) specifically shields G2553, U2555, A2602, and U2609 on the 50S subunit, so that these residues seem to belong to the *a* site. After transpeptidation, prior to translocation, the deacylated tRNA still occupying the P site on the 30S subunit does not protect the *d* site residues anymore and switches its CCA terminus to the residues characteristic of the *e* site: G2112, G2116, A2169, and C2394; this is the pre-translocation P/*e* state of the deacylated tRNA. The newly formed peptidyl-tRNA continues to occupy the A site with its tRNA moiety on the 30S subunit, but now it protects the *d* site residues on the 50S subunit, with the exception of A2602 which is still protected; this is called the pre-translocation A/*d* state of the peptidyl-tRNA. Hence, despite the fixation of the two tRNA moieties at the P and A sites where they have been bound prior to transpeptidation, the CCA termini of the tRNAs move on the 50S subunit during transpeptidation: the deacylated A76 of the P-site-bound tRNA is displaced out of the PTC and caught by the *e* site, while the esterified A76 with the C-terminal amino acid of the A-site-bound tRNA shifts within the PTC from its *a* site to the *d* site (Fig. 11.6).

The potential flexibility of the CCA terminus of the tRNA molecule, and especially the mobility of the 3′-terminal adenosine relative to the rest of the tRNA body, should play the decisive role in the above-mentioned movements of the tRNA end on the large ribosomal subunit. At the same time, conformational changes in the region of the PTC in response to the formation and breakdown of contacts with ligand groups during transpeptidation are very plausible. Also large-block movements, such as slight interdomain and intersubunit shifts, cannot be excluded.

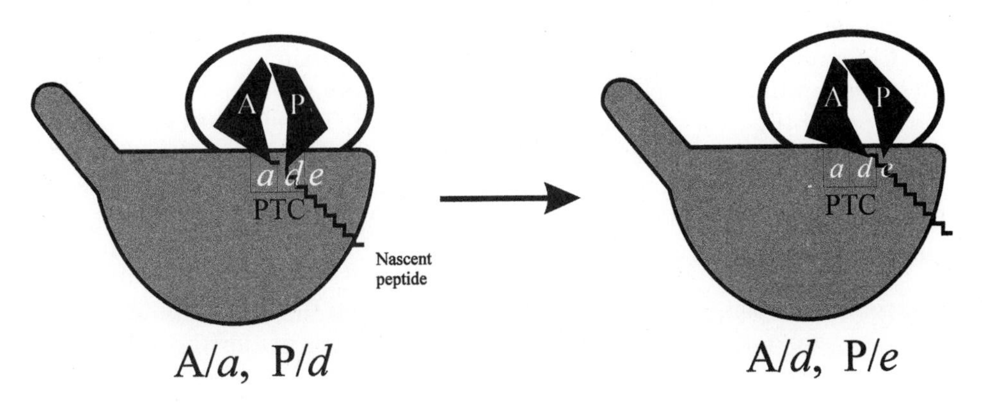

FIG. 11.6. Schematic representation of the *a*, *p*, and *e* sites of the large ribosomal subunit and the product shifts in the ribosomal PTC during transpeptidation.

11.5. Inhibitors

Numerous specific inhibitors of the peptidyl transferase reaction catalyzed by prokaryotic or eukaryotic ribosomes have been described (for reviews, see Gale *et al.*, 1981; Garrett and Rodriguez-Fonseca, 1995). All the inhibitors, as expected, affect the large (50S or 60S, respectively) ribosomal subunit and have an affinity for it. Many antibiotics commonly used for treating bacterial infections are inhibitors of the peptidyl transferase of the prokaryotic 70S ribosome. Kirillov, Garrett and colleagues (1997) have suggested the classification of the peptidyl transferase antibiotics into at least two categories: (1) those inhibitors that are to some degree "co-structural" with the 3′-end of aminoacyl-tRNA, bind to the acceptor (*a*) site of the PTC or nearby, and thus prevent, competitively or noncompetitively, binding or proper settling of the acceptor substrate at the PTC (i.e., at stage A in Fig. 11.5); and (2) those that are bound in the vicinity of the PTC and perturb the movement or positioning of the nascent peptide during transpeptidation (i.e., at stage B → C in Fig. 11.5).

11.5.1. Chloramphenicol

Chloramphenicol (also known as chloromycetin or levomycetin) is the most well-characterized inhibitor of peptidyl transferase in 70S ribosomes (Fig. 11.7). This is a broad-spectrum bacteriostatic antibiotic commonly used in pharmacological practice. It does not affect eukaryotic 80S ribosomes, but may attack mitochondrial ribosomes of eukaryotes. The antibiotic binds loosely to the 70S ribosome or its isolated 50S subunit and can be washed away easily from the particles. Correspondingly, the action of the antibiotic is reversible. The antibiotic binds in the region of the PTC and blocks the site responsible for the interaction with the acceptor substrate (*a* site); puromycin and 3′-terminal fragments of the aminoacyl-tRNAs compete with chloramphenicol for binding to the PTC.

Chemically, chloramphenicol is an analog of N-blocked amino alcohol with an aromatic radical. The dichloromethyl group is not strictly required for activity and can be changed for many moderately massive radicals. The aromatic nitro group may also be replaced by a number of other electronegative groups without resulting in any loss of antibiotic activity. The amide bond and stereochemistry of the –CO–NH moiety with adjacent atoms are noteworthy with respect to the antibiotic action mechanism: this part of the molecule may mimic the peptide group with the adjacent C^{α} atom and side radical.

Chloramphenicol inhibits "natural" transpeptidation between the peptidyl-tRNA and aminoacyl-tRNA in the course of elongation, as well as the reaction of the peptidyl-tRNA or its analogs with puromycin. The simplest explanation for these effects is that chloramphenicol is an inactive analog of the acceptor substrate and after binding to PTC competitively interferes with the interaction of true acceptors. However, there is also evidence for a noncompetitive mode of chloramphenicol inhibition. In accordance with the above, chloramphenicol affects chemical reactivity of residues within the so-called PTC ring of domain V of 23S RNA (see Section 9.3 and Fig. 9.7): it enhances the reactivity of A2058 and protects A2059, A2062, A2451, and G2505. Mutations conferring chloramphenicol resistance are also located in the same PTC ring; their positions have been identified at G2032, G2057, A2058, G2061, A2062, G2447, A2451, C2452, A2503, and U2504.

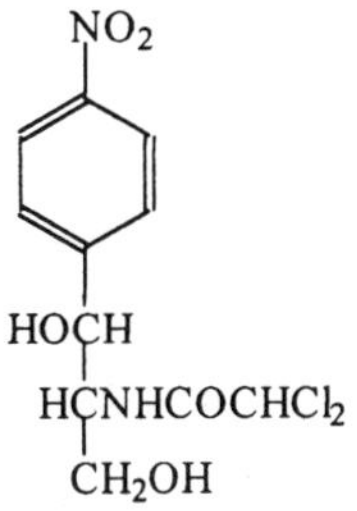

FIG. 11.7. Chloramphenicol.

11.5.2. Lincosamides

This group of drugs includes lincomycin (Fig. 11.8) and its derivatives, such as clindamycin and celesticetin. These are also bacteriostatic antibiotics widely used in therapy. They attack specifically eubacterial 70S ribosomes. The antibiotics interact with the of ribosomes and seemingly block the acceptor-binding site (*a* site). Correspondingly, they compete with chloramphenicol for binding to the ribosome. In any case, the antibiotics inhibit the peptidyl transferase reaction. At the same time the translating ribosomes with long nascent peptides seem to be less sensitive to the drugs, as compared with short-peptide-carrying ribosomes.

The chemical structure of lincomycin, clindamycin, and celesticetin, like that of chloramphenicol, contains an amide bond and a group simulating the peptide group adjacent to the C^{α} atom of the amino acid residue (here, again, the alcohol hydroxyl group is present instead of the acidic one). The mode of action of lincomycin and its derivatives seems to be similar to that of chloramphenicol, though there are indications that both the *a* and *d* sites of the PTC may be affected by the more massive antibiotics. The positions of mutations in 23S RNA which confer the resistance against lincosamides are G2032, G2057, A2058, and A2059, which are near or within the PTC ring, overlapping with the positions of chloramphenicol resistance mutations. When bound, they protect residues A2058, A2059, and G2061 in the same region, as well as A2451 and G2505 in another region of chloramphenicol resistance mutations, also in the PTC ring (Fig. 9.7).

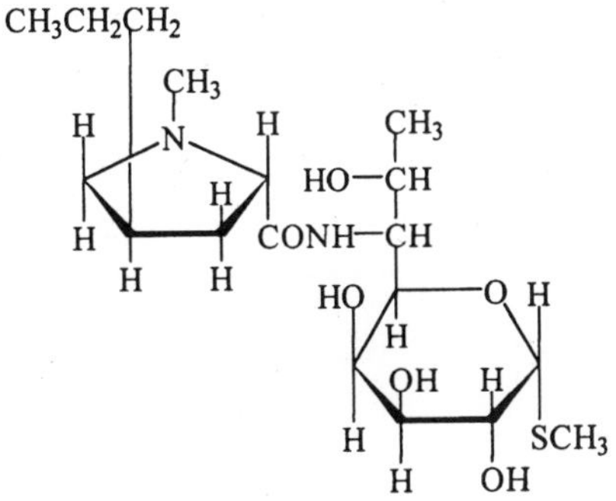

FIG. 11.8. Lincomycin.

11.5.3. Macrolides

Macrolide antibiotics contain a large (12 to 16 atoms) lactone ring with one or more sugar side substitutions. Erythromycin (Fig. 11.9) and its derivatives clarithromycin and azithromycin are most known and widely used in pharmaceutical practice. Other antibiotics in this class include oleandomycin, carbomycin, chalcomycin, niddamycin, tylosin, and spiramycin. Macrolide antibiotics specifically act on prokaryotic ribosomes. They have their binding site on the small (50S) ribosomal subunit. They compete with each other for binding with the ribosome. Their binding site is in the region of the PTC. Indeed, they compete with chloramphenicol, a classical peptidyl transferase inhibitor, and protect the same set of residues in the PTC ring of 23S RNA that chloramphenicol affects: A2058, A2059, A2062, A2451, and G2505. Erythromycin resistance is conferred by mutations of G2057 and its base-paired C2611, as well as A2058, in the PTC ring. It is interesting that the resistance to macrolides, as well as lincosamides and streptogramin B antibiotics, is conferred also by N^6-dimethylation or N^6-monomethylation of A2058, this process being inherent to the producers of the antibiotics.

At the same time macrolide antibiotics do not seem to be patent inhibitors of substrate binding or transpeptidation reaction in the PTC. Although the inhibition of peptidyl transferase activity of prokaryotic ribosomes has been reported for a number of macrolides tested, some results, especially in the case of erythromycin, were discrepant. Under certain *in vitro* conditions erythromycin was shown even to enhance the binding of a donor substrate and stimulate peptide synthesis. The effects of erythromycin on the retention of transpeptidation reaction products (inhibition of deacylated tRNA release and stimulation of peptidyl-tRNA displacement) were mentioned. On the other hand, the interference of erythromycin with nascent peptide was demonstrated. It was reported that erythromycin inhibited the ribosomal PTC preferentially with oligopeptidyl donor substrates, and was less effective with long nascent polypeptides. On the basis of the data available it may be thought that ribosome-bound erythromycin partly overlaps the donor (*d*) site of the PTC and/or the nascent peptide path on the 50S subunit, and thus intervenes in the movement of the newly formed peptide group and the adjacent C-terminal section of the nascent peptide during transpeptidation. Longer nascent polypeptides seem to be capable of overcoming the action of erythromycin.

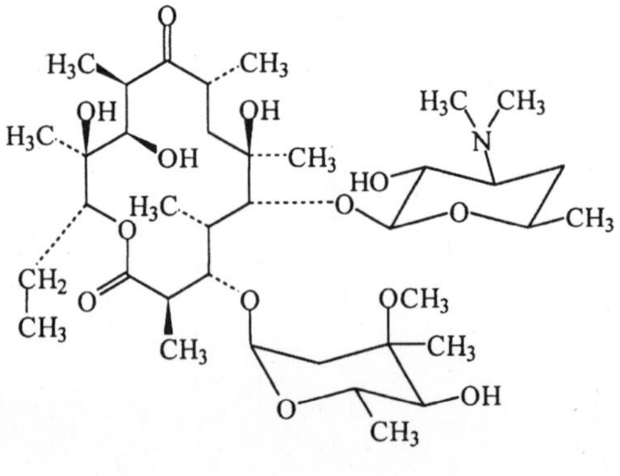

FIG. 11.9. Erythromycin.

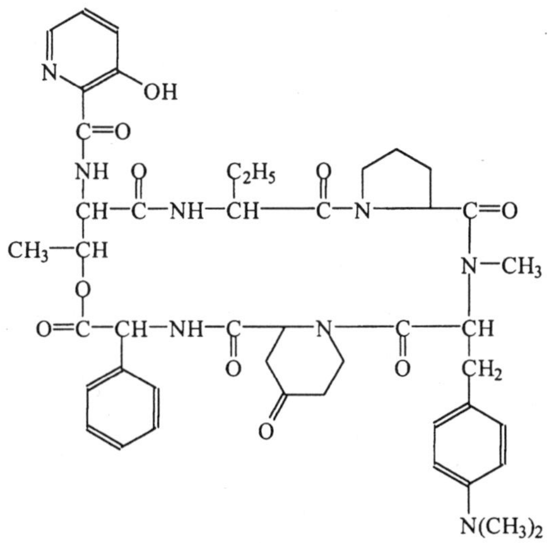

FIG. 11.10. Streptogramin B.

11.5.4. Streptogramin B

Streptogramin B (= mycamycin B = osteogrycin B = pristinamycin I_A = vernamycin B_α = PA114B) and related antibiotics, such as virginiamycin S, doricin, etamycin (viridogrisein), are cyclic hexadepsipeptides containing uncommon amino acids (Fig. 11.10). They bind tightly to eubacterial 70S ribosomes and their 50S subunits, but not to eukaryotic 80S ribosomes. Correspondingly, they inhibit protein synthesis specifically in prokaryotic systems, including mitochondria and chloroplasts, both *in vivo* and *in vitro*. Their target is the PTC. However, as in the case of macrolides, they do not always inhibit transpeptidation reactions under model conditions and sometimes stimulate binding of acceptor and/or donor substrates to the PTC. Also, their binding and inhibitory effect may depend on the presence and the length of the nascent polypeptide. It is likely that this group of antibiotics acts similarly to macrolides, and their main effect can be also the interference with the movement or positioning of the C-terminus-adjacent section of the nascent peptide within the PTC and nearby during transpeptidation.

It is noteworthy that streptogramin B affects chemical reactivity of the same residues of the peptidyl transferase region (PTC ring) of 23S RNA that are known to be protected (or sometimes exposed) by chloramphenicol, lincosamides, and macrolides, namely A2058, A2062, A2451, and G2505. (In addition, streptogramin B protects A2439 near the PTC ring and A752 in domain II of 23S RNA). Also mutations at A2058 confer resistance to streptogramin B and to chloramphenicol, lincosamides, and macrolides. These facts strongly suggest a common mechanism of action on bacterial ribosomes for all the antibiotics discussed, despite their very different chemical structures. It cannot be excluded that such dissimilar structures still have identical contacts within the PTC ring. An alternative explanation could be that the drugs bound within the same area exert a similar "switching" effect on the conformation of PTC; in such a case the protection of A2059–A2062, A2451, and G2505 could be caused by a uniform conformational change ("collapse") of the PTC ring, rather than a result of direct shielding of the residues by the antibiotics. This implies that the three parts of the PTC ring may specifically interact with each other, and they do so in response to the antibiotic binding; mutational changes of the interacting bases prevents the collapse. The inhibition of transpeptidation due to a conformational change of PTC seems to be more consistent with observations on the variability of antibiotic effects and the dependence of the effects on experimental conditions than the direct competitive inhibition mechanism.

11.5.5. Streptogramin A

Streptogramins and related antibiotics are produced by *Streptomyces* species as complexes of two different components designated A (or M) and B (or S). The B group drugs are considered above. The A group antibiotics, such as streptogramin A (Fig. 11.11), virginiamycin M, and griseoviridin, are chemically distinct from the B group drugs; they possess a large nonpeptide ring with several unsaturated C=C bonds and amide, imide, and ester groupings. The A and B components together demonstrate a noticeable synergism of action on bacterial protein synthesis.

Like the members of the B group, streptogramin A and related antibiotics bind tightly to bacterial ribosomes and specifically to their large (50S) subunits. Again long nascent polypeptides counteract their binding. In accordance with this, puromycin reaction of short donor substrates is inhibited by streptogramin A antibiotics, but peptidyl-puromycin formation on polyribosomes is not. Streptogramin A and griseoviridin were reported to impede both donor- and acceptor-binding sites of the PTC of the bacterial ribosome, as well as the binding of aminoacyl-tRNA with the A site (presumably in the A/*a* state). Also the streptogramin A antibiotics prevent the binding of chloramphenicol to ribosomes and 50S ribosomal subunits. At the same time they promote the binding of the B group drugs to bacterial ribosomes. As the synergism, rather than antagonism, between the A and B groups is observed, it is reasonable to assume that their targets are not overlapping and the mechanisms of action are different. It may be that streptogramin A antibiotics block or distort both the *a* and *d* sites of the PTC in translating ribosomes.

11.5.6. 4-Aminohexose Pyrimidine Nucleoside Antibiotics

This group of antibiotics includes such inhibitors of ribosomal transpeptidation as gougerotin, amicetin, blasticidin S, and bamicetin. The chemical structures

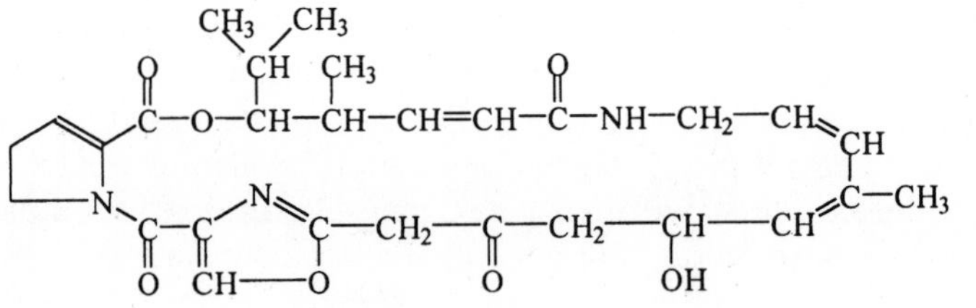

FIG. 11.11. Streptogramin A.

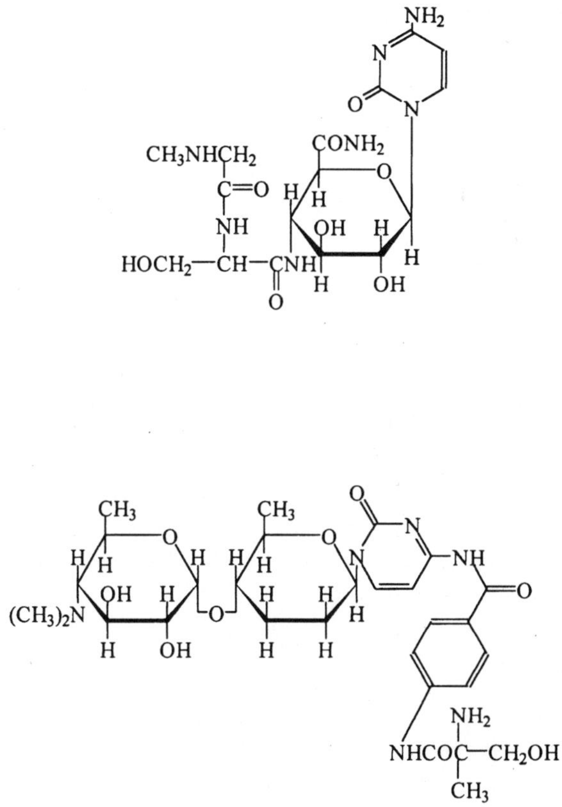

FIG. 11.12. Gougerotin (*upper*) and amicetin (*lower*).

of gougerotin and amicetin are given in Fig. 11.12. The antibiotics of this group possess a nucleoside structure and may be regarded as analogs of the tRNA 3′-terminal adenosine. In addition, gougerotin and blasticidin S show the presence of a structural motif traceable in chloramphenicol and lincomycin, namely, the peptide group with the adjacent C^{α} atom. Antibiotics of this group affect bacterial ribosomes although some of them, such as gougerotin and blasticidin S, can inhibit eukaryotic ribosomes as well. All these antibiotics bind to the 50S ribosomal subunit and seemingly inhibit the interaction between acceptor substrates and the PTC of the ribosome. It is remarkable that the binding of the antibiotics stimulates the interaction of the low-molecular-mass analogs of the donor substrate with the PTC. These antibiotics compete with each other and seem to possess a similar mechanism of action.

Mutations conferring resistance to amicetin were found in position U2438 of 23S RNA that is close to, but not within, the PTC ring. This position does not coincide with the positions of drug resistance mutations in the cases of the other antibiotic groups described above. An interesting fact is that no protection of the 23S RNA residues by the antibiotic of this group, blasticidin S, has been observed in the corresponding footprinting experiments. Thus, the mechanism of action of the antibiotics under consideration seems to be different from that of the above-described groups of lincosamides, macrolides, and streptogramin B, and may be truly competitive.

11.5.7. Sparsomycin

Sparsomycin is a very potent, broad-spectrum antibiotic affecting both prokaryotic (including archaebacterial) and eukaryotic systems. It strongly inhibits peptidyl transferase activity of 70S and 80S ribosomes or their isolated large subunits. The chemical structure of sparsomycin (Fig. 11.13) shows that it contains a pyrimidine residue, amide group, and sulfoxide group that may be relevant to its inhibitory activity. The mode of action of the drug is unusual and different from that of the above-mentioned antibiotics. First, the antibiotic is incapable of binding to ribosomes in the absence of the donor substrate (peptidyl-tRNA or its truncated derivatives); it seems to lack an affinity for ribosomal elements *per se*. Second, the bindings of the antibiotic and the donor substrate are synergistic; the drug stimulates the binding of the donor substrate to the ribosome and fixes it in the *d* site of the PTC. It is possible that the antibiotic forms a ternary complex with the donor substrate and PTC. When bound, the antibiotic blocks or weakens the interaction of the acceptor substrate (aminoacyl-tRNA or puromycin) with the PTC. It is possible that the antibiotic bound with the peptidyl-tRNA-carrying ribosome overlaps with the *a* site of the PTC and thus prevents the binding of the acceptor substrate. Alternatively, it may be hypothesized that sparsomycin fixes the donor substrate, with which it strongly interacts within the PTC, in its original conformation and prevents the formation of the transition complex (tetrahedral intermediate), thus blocking the stage A → B in Fig. 11.5.

11.5.8. Anisomycin

Anisomycin (Fig. 11.14) inhibits transpeptidation specifically in eukaryotic ribosomes, as well as in archaebacterial ribosomes. It binds to the 60S ribosomal subunit in the region of the PTC. It is apparent that anisomycin interferes with the interaction of the acceptor substrate with the PTC. It inhibits puromycin reaction on eukaryotic ribosomes and their isolated 60S subunits *in vitro*. *In vivo*, anisomycin is a powerful inhibitor of the transpeptidation step and may block elongation completely, arresting the movement of ribosomes along the mRNA and thereby "freezing" the polyribosomes. Generally, this antibiotic can be considered as the eukaryotic equivalent of chloramphenicol. Indeed, mutations conferring resistance to anisomycin have been identified in the ribosomal RNA section within

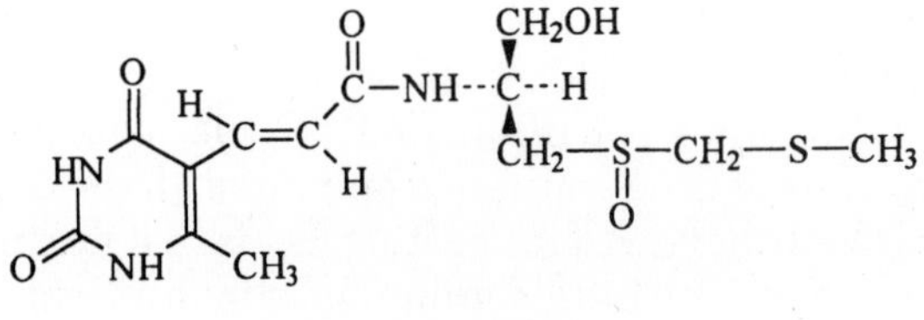

FIG. 11.13. Sparsomycin.

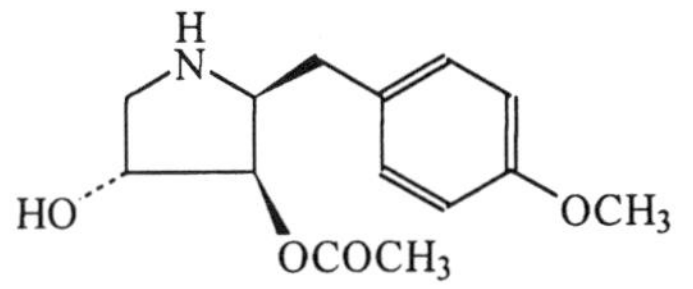

FIG. 11.14. Anisomycin.

the PTC ring coinciding with that in the case of chloramphenicol resistance: the mutation positions are equivalent to G2447, C2452, and A2453 of eubacterial 23S RNA.

References

Fahnestock, S., Neumann, H., Shashoua, V., and Rich, A. (1970) Ribosome-catalysed ester formation, *Biochemistry* **9**:2477–2483.

Gale, E. F., Cundliffe, E., Reynolds, P. E., Richmond, M. H., and Waring, M. J. (1981) *The Molecular Basis of Antibiotic Action,* John Wiley & Sons, London.

Garrett, R. A., and Rodriguez–Fonseca, C. (1995) The peptidyl transferase center, in *Ribosomal RNA: Structure, Evolution, Processing and Function in Protein Synthesis* (R. A. Zimmermann and A. Dahlberg, eds.), pp. 329–357, CRC Press, Boca Raton, FL.

Gooch, J., and Hawtrey, A. O. (1975) Synthesis of thiol-containing analogues of puromycin and a study of their interaction with N-acetylphenylalanyl-transfer ribonucleic acid on ribosomes to form thioesters, *Biochem. J.* **149**:209–220.

Kirillov, S., Porse, B. T., Vester, B., Woolley, P., and Garrett, R. A. (1997) Movement of the 3′-end of tRNA through the peptidyl transferase centre and its inhibition by antibiotics, *FEBS Letters* **406**:223–233.

Krayevsky, A. A., and Kukhanova, M. K. (1979) The peptidyl-transferase center of ribosomes, *Progr. Nucl. Acid Res. Mol. Biol.* **23**:1–51.

Monro, R. E., Staehelin, T., Celma, M. L., and Vazquez, D. (1969) The peptidyl transferase activity of ribosomes, *Cold Spring Harbor Symp. Quant. Biol.* **34**:357–366.

Noller, H. F. (1991) Ribosomal RNA and translation, *Annu. Rev. Biochem.* **60**:191–227.

Noller, H. F., Moazed, D., Stern, S., Rowers, T., Allen, P. V., Robertson, J. M., Weiser, B., and Triman, K. (1990) Structure of rRNA and its functional interactions in translation, in *The Ribosome: Structure, Function, and Evolution* (W. E. Hill, A. Dahlberg, R. A. Garrett, P. B. Moore, D. Schlessinger, and J. R. Warner, eds.) pp. 73–92, ASM Press, Washington, DC.

Pestka, S. (1977) Inhibitors of protein synthesis, in *Molecular Mechanisms of Protein Biosynthesis* (H. Weissbach and S. Pestka, eds.), pp. 468–553, Academic Press, New York.

Tarussova, N. B., Jacovleva, G. M., Victorova, L. S., Kukhanova, M. K., and Khomutov, R. M. (1981) Synthesis of an unnatural P–N bond catalysed with *Escherichia coli* ribosomes, *FEBS Letters* **130**:85–87.

Victorova, L. S., Kotusov, L. S., Azhayev, A. V., Krayevsky, A. A., Kukhanova, M. K., and Gottikh, B. P. (1976) Synthesis of thioamide bond catalysed by *E. coli* ribosomes. *FEBS Letters* **68**:215–218.

12

Elongation Cycle, Step III:
Translocation

12.1 Definition and Experimental Tests

As a result of transpeptidation, the tRNA residue of the newly elongated peptidyl-tRNA occupies the ribosomal A site while deacylated tRNA occupies the P site (Fig. 12.1, *left*). *Translocation* is defined as an intraribosomal movement of the bound tRNA residues accompanied by the shift of the template polynucleotide relative to the ribosome. In sum, the tRNA residue of the peptidyl-tRNA moves from the A site to the P site, the deacylated tRNA is displaced from the P site, the template moves a distance of one codon in the 5′-to-3′ direction, and the A site with the new codon becomes vacant (Fig. 12.1, *right*).

In accordance with this scheme, there are five ways of measuring translocation, based on the following criteria: (1) the transpeptidation reaction with a low-molecular-mass acceptor substrate, such as puromycin; (2) the change of intraribosomal surroundings of the tRNA residue upon its transition from the A site to the P site; (3) the binding of aminoacyl-tRNA to the vacant A site; (4) the release of deacylated tRNA; and (5) the shift of the template polynucleotide along the ribosome.

The puromycin reaction is the simplest and most widely used measurement of translocation. Puromycin has been described as a low-molecular-mass analog of aminoacyl-tRNA (Fig. 9.3) serving as an acceptor substrate for the ribosomal PTC. The amino group of its aminoacyl moiety attacks the ester group of peptidyl-tRNA (or its analog), resulting in transpeptidation. The peptidyl-puromycin formed is released from the ribosome. The peptidyl-tRNA of the post-translocation-state ribosome reacts fast with added puromycin, whereas the peptidyl-tRNA of the pre-translocation state ribosome does not (Traut and Monro, 1964). Therefore, if the peptidyl residue is labeled, the label released from the ribosome in response to puromycin addition can be used as a quantitative measure of the post-translocation state in the population of ribosomes studied. Conversely, lack of reaction with added puromycin implies that the particles are in the pre-translocation state.

If the A-site-bound tRNA carries a fluorescent group, the fluorescence change during translocation can be registered, indicating a change in intraribosomal surroundings of this tRNA residue upon transition from the A site (Wintermeyer *et al.*, 1986).

Binding of the aminoacyl-tRNA may also serve as a quantitative measure of translocation. If aminoacyl-tRNA is labeled, the codon-dependent binding of the label with the translating ribosome is impossible immediately following transpeptidation; the labeled aminoacyl-tRNA will bind only after translocation (Haenni and Lucas–Lenard, 1968). Consequently, an inability to bind aminoacyl-tRNA is evidence of the pre-translocation state whereas aminoacyl-tRNA binding is an indication of the post-translocation state.

Translocation displaces the deacylated tRNA from the P site. The deacylated tRNA after translocation may be still anchored to the ribosome through its 3′-terminus (A76) at the *e* site on the large ribosomal subunit. This binding is transient; under physiological conditions, particularly at lower Mg^{2+} concentrations, the deacylated tRNA is spontaneously released from the translating ribosome after translocation. Therefore, following transpeptidation the deacylated tRNA will remain firmly bound to the ribosome, and only translocation will allow the release of the tRNA into solution (Lucas–Lenard and Haenni, 1969). Thus, for the translating ribosome the firm retention of the deacylated tRNA can be equated with the pre-translocation state, and the release (under proper conditions) or the absence of deacylated tRNA from the ribosome testifies to the post-translocation state.

The movement of the template polynucleotide as a test for translocation is a technically more difficult approach. Measurement of the movement may be indirect: it can be based on the appearance of the capability to bind aminoacyl-tRNA specific to a codon following the codon that was previously positioned in the ribosome. The movement can also be measured directly by analyzing the change in the template region screened (protected) by the ribosome. The direct test has demonstrated that the movement of the polynucleotide template relative to the ribosome by one nucleotide triplet accompanies the appearance of puromycin reactivity and the ability to bind aminoacyl-tRNA (Thach and Thach, 1971; Gupta *et al.*, 1971).

12.2. Participation of the Elongation Factor 2 (EF-G or eEF2) in Translocation

12.2.1. EF2 Structure

Translocation is catalyzed by a large protein, referred to as the elongation factor G (EF-G) in prokaryotes, and as the elongation factor 2 (eEF2) in eukaryotes. The universal designation EF2 has also been proposed and is often used now. The molecular mass of bacterial EF2 (EF-G) is approximately 80,000 daltons. This protein is a single polypeptide chain with a length of 701 or 691 amino acid residues in the cases of *E. coli* and *T. thermophilus*, respectively, folded into several globular domains. The animal eEF2, somewhat larger than the prokaryotic EF-G, has a molecular mass of about 95,000 daltons.

The three-dimensional structure of the bacterial (*T. thermophilus*) EF-G has been determined (Fig. 12.2). The protein consists of five domains. Four of them (I,

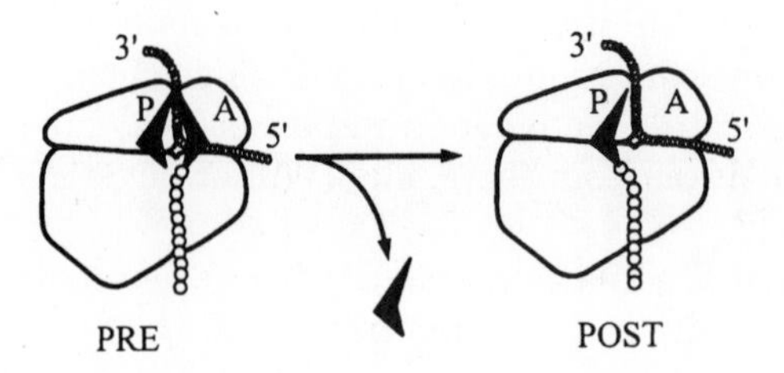

FIG. 12.1. Schematic representation of the pre-translocation (*left*) and post-translocation (*right*) states of the ribosome.

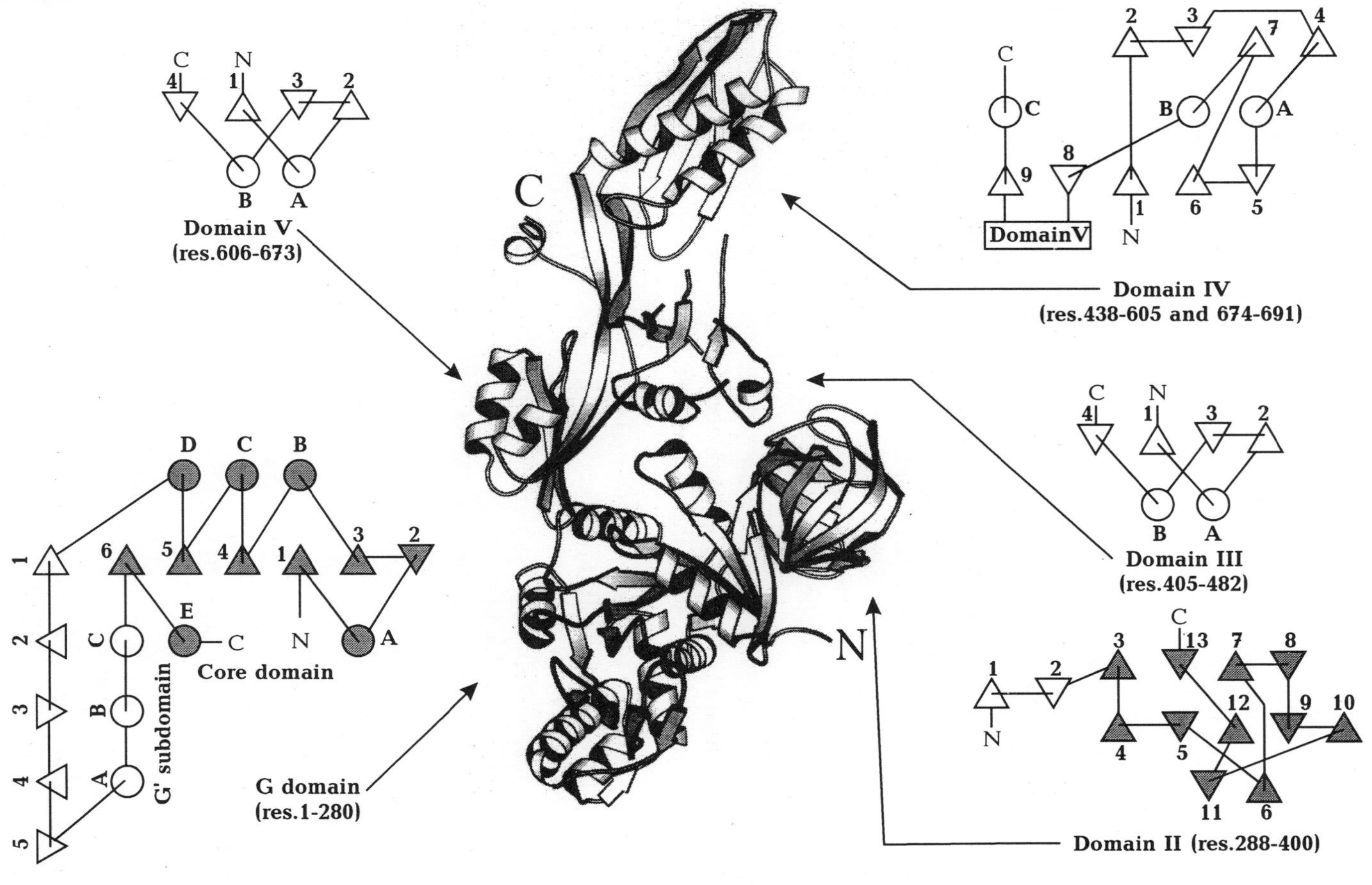

FIG. 12.2. The structure of EF-G as determined by X-ray crystallography. Reproduced from Aevarsson, A., Brazhnikov, E., Garber, M., Zheltonosova, J., Chirgadze, Y. N., Al-Karadaghi, S., Svensson, L. A., and Liljas, A. (1994) *EMBO J.* **13**:3669–3677, with permission. See also Czworkowski, J., Wang, J., Steitz, T. A., and Moore, P. B. (1994) *EMBO J.* **13**:3661–3668.

II, III, and V) are clustered together, whereas the fifth domain (IV) protrudes from the rest making the molecule well extended. The overall dimensions of the protein are 120 × 60 × 50 Å.

The N-terminal domains I (G-domain) and II strongly resemble the homologous domains I (G) and II in EF-Tu (Section 10.2.1), but with a massive insert of subdomain G′ in the C-proximal part (between α-helix D and β-strand 6) of the G domain. Just as in EF-Tu, the "effector loop" between α-helix A and β-strand 2 of domain I (G) is flexible and acquires different conformations depending on functional states of the protein. The sequence of the secondary structure elements in G-domain is as follows (the changeable "effector loop" β↔α and the G′-subdomain are in brackets):

$$\beta\alpha[\beta\leftrightarrow\alpha]\beta\beta\alpha\beta\alpha\beta\alpha[\beta\beta\beta\beta\beta\alpha\alpha\alpha]\beta\alpha$$

Despite the lack of great sequence similarity, domain II of EF-G superimposes very well upon domain II of EF-Tu. It is also an antiparallel β-barrel, but differs from the homologous domain of EF-Tu by the presence of an N-proximal β-hairpin external to the barrel. Domain V of EF-G is adjacent to the G-domain and occupies the position that is similar to that of domain III in EF-Tu relative to domains I and II. No structural homology, however, can be revealed between domain V of EF-G and domain III of EF-Tu; in contrast to all-β-folds of domain III of EF-Tu, domain V of EF-G has a bilayer βαββαβ structure and strongly resembles the ribosomal protein S6. The functions of the group of domains I, II, and V may be partly equivalent to those of the corresponding domains of EF-Tu, such as GTP binding and GTP cleavage (domain I or G), conformational changes affecting the affinity for the ribosome, and interactions with the factor-binding site of the large ribosomal subunit.

Domains III and IV of EF-G have no analogs in EF-Tu and may be considered specific features of the protein necessary to fulfill the function of translocation. Domain III seems to have a bilayer βαββαβ structure similar to that of domain V (and to ribosomal protein S6). Domain IV forms a protruding "tail" on the EF-G molecule and has an unusual topology of the secondary-structure elements. The full sequence of strands and helices is ββββαβββαβ, with an additional C-terminal β-strand and α-helix going from domain V; the peculiarity is that the main sheet is parallel, rather than anti-parallel, and the parallel strands 4 and 7 have an atypical left-handed connection (through helix A and short strands 5 and 6). The most remarkable observation is that the size, the shape, and the orientation of the domain III/domain IV pair in EF-G mimic those of aminoacyl-tRNA in the complex with EF-Tu (Fig. 12.3). Thus, the overall appearance and even some details of the EF-G structure and the structure of the aminoacyl-tRNA:EF-Tu:GTP complex are similar, domain IV and domain III being analogs of the anticodon-dihydrouridylic limb and acceptor-thymidyl-pseudouridylic limb, respectively.

12.2.2. EF2 Interactions

EF2 (bacterial EF-G or eukaryotic eEF2) interacts with GTP and the ribosome. This interaction induces GTPase activity, and GTP is cleaved to GDP and orthophosphate. The interaction (complex formation) of EF2 and GTP with the pretranslocation ribosome results in a quick translocation, upon which EF2, GDP, and orthophosphate are released from the ribosome.

It is domain I, or the G-domain of EF2, that specifically binds GTP (and GDP). The same domain is responsible also for catalyzing the GTP hydrolysis. The organization of the GTP/GDP binding site and the GTPase center of EF2 seems to be very similar to that in the G-domain of EF1 (EF-Tu; see Section 10.2.1). The interaction of EF-G (or eEF2) with GTP, however, is considerably weaker than the analogous interaction of EF-Tu (or eEF1) with GTP. The resultant EF-G:GTP complex is unstable and easily reversible (the eEF2:GTP complex is somewhat more stable). Nevertheless, this complex is being formed and its formation is a prerequisite for the interaction of EF2 with the ribosome. GTP is thought to contribute to a conformational change in the protein (EF-G or eEF2) which results in a strong affinity of the factor for the ribosome. GTP is highly specific in this respect and cannot be substituted for either by other nucleoside triphosphates or any nucleoside mono- or diphosphates. GTP can, however, be replaced by the nonhydrolyzable GTP analogs, such as GMP-PCP or GMP-PNP (see Fig. 9.9).

The EF-G:GTP (or eEF2:GTP, in eukaryotic systems) complex can bind to the functioning (translating) ribosome as well as to the vacant ribosome, or even with the isolated large ribosomal subunit. The binding site of EF2 on the large subunit appears to be located at the base of the L7/L12 stalk (see Section 9.4); in the whole (70S or 80S) ribosome it is found near the interface of the ribosomal subunits, in the region of the tRNA-binding sites. The binding of EF2:GTP to the ribosome is

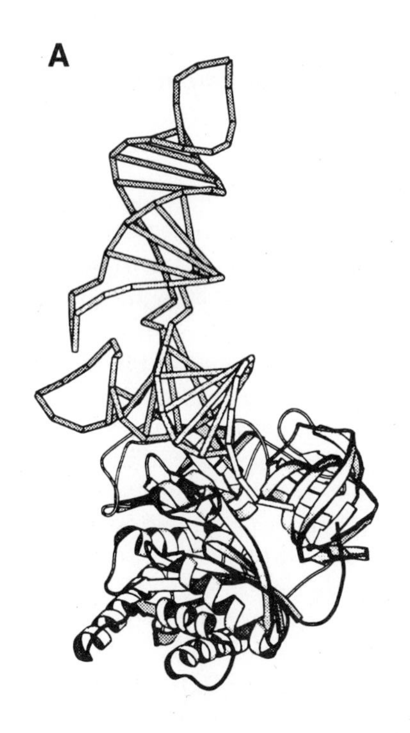

FIG. 12.3. A comparative view of the ternary complex EF-Tu:GTP:Aa-tRNA (*left*) and EF-G (*right*): "molecular mimicry." Reproduced from Nissen, P., Kjeldgaard, M., Thirup, S., Polekhina, G., Reshetnikova, L., Clark, B. F. C., and Nyborg, J. (1995) *Science* **270**:1464–1472. Courtesy of P. Nissen, Molecular Biophysics and Biochemistry, Yale University, New Haven, CT.

prevented by the aminoacyl-tRNA bound in the ribosomal A site, as well as by the presence of the other protein factor, EF1 (EF-Tu or eEF1), on the ribosome.

In all cases, the binding of the EF-G:GTP complex or the eEF2:GTP complex to the ribosome or its isolated large subunit results in cleavage (hydrolysis) of GTP. The ribosome seems to induce GTPase activity of EF2; in other words, it seems that the GTPase center is present on EF2 but is inactive in the absence of the ribosome. If a ribosome is vacant or the large subunit is used instead of the whole ribosome, then the GTP is simply hydrolyzed, without being coupled with any events of elongation.

The EF2-catalyzed hydrolysis of GTP on the ribosome results in EF2 being in a complex with GDP. However, GDP is incapable of supporting the EF2 affinity for the ribosome; therefore, the EF2:GDP complex is released from the ribosome and subsequently dissociates.

Thus, the sequence of reactions is as follows:

$$\text{EF-G} + \text{GTP} \rightleftharpoons \text{EF-G:GTP}$$

$$\text{EF-G:GTP} + \text{RS} \rightleftharpoons \text{RS:EF-G:GTP}$$

$$\text{RS:EF-G:GTP} + H_2O \rightarrow \text{RS:EF-G:GDP} + P_i$$

$$\text{RS:EF-G:GDP} \rightarrow \text{RS} + \text{EF-G:GDP}$$

$$\text{EF-G:GDP} \rightleftharpoons \text{EF-G} + \text{GDP}$$

If the vacant ribosome or its isolated 50S subunit participates in the reactions, the overall process is simply GTP hydrolysis proceeding according to the following overall equation:

$$\text{GTP} + H_2O \xrightarrow{\text{EF-G, RS}} \text{GDP} + P_i$$

In other words, in this case EF-G (eEF2), in combination with the ribosome, serves solely as GTPase. However, if these reactions involve the pre-translocation-state ribosome, then participation of EF-G (eEF2) will result in translocation.

The question then arises as to which of the sequential reactions involving EF2 is directly coupled to the translocation. For a long time translocation was thought to be an energy-consuming process and to be directly coupled with the GTP hydrolysis on the ribosome. Both of these assumptions turned out to be incorrect. First, translocation was shown to be thermodynamically spontaneous accompanied by the release of free energy (see Sections 12.3 and 12.5.2). Moreover, the assumption about the coupling of translocation to GTP hydrolysis has been rejected on the basis of direct experiments wherein a nonhydrolyzable GTP analog was used instead of GTP: the catalysis of translocation was observed when pre-translocation ribosomes interacted with EF-G and GMP-PCP or GMP-PNP, that is without GTP hydrolysis (Inoue–Yokosawa, Ishikawa, and Kaziro, 1974; Belitsina, Glukhova, and Spirin, 1975; Modolell, Girbes, and Vazquez, 1975). It followed that just the attachment of EF-G:GTP or EF-G:GMP-PCP (GMP-PNP) to the ribosome was capable of promoting (catalyzing) translocation. There are two possible explanations here: (1) either the attachment (affinity) of the protein to the ribosome exerts some force, directly shifting the tRNA molecules in the ribosome, or more likely, (2) the kinetic barriers for spontaneous translocational movements decrease when EF-G is attached to the pre-translocation ribosome.

Following the ideology of transition states in both covalent and noncovalent catalysis (see Section 10.2.3), it can be assumed that EF2:GTP as a catalyst has an affinity specifically for the transition intermediate of the translocation reaction. Then, analogously with the EF1-catalyzed binding of aminoacyl-tRNA, the translocation reaction proceeds in two phases: the first is the binding of EF2:GTP with the ribosome that induces the pre-translocation-state ribosome to acquire an intermediate state with bound EF2, and the second is the conversion of the intermediate into the post-translocation-state ribosome as a result of GTP cleavage; the following EF2 release from the ribosome completes the translocation step of the elongation cycle.

Speculating about the molecular aspect of the transition intermediate, the similarity between domain IV of EF-G and the anticodon stem of tRNA of the ternary EF-Tu:Aa-tRNA:GTP complex ("molecular mimicry") should be taken into account (Fig. 12.3). This similarity suggests that domain IV fits the A site on the 30S subunit when EF-G is bound to the ribosome. In such a case, the interaction of EF-G:GTP (more precisely, its central EF-Tu-like body, i.e., domains I, II, and V) with the factor-binding site on the large subunit of the pre-translocation ribosome would displace the tRNA residue of peptidyl-tRNA from the A site and put domain IV there. There are experimental indications that domain IV of the ribosome-bound EF-G does interact with the decoding center on the 30S subunit (see Section 9.4). Now the peptidyl-tRNA occupies an intermediate position, not completely settled into the P site yet. The EF-G-induced conformational change of the ribosomal pre-translocation complex seems to involve changes in the ribosome itself, possibly in the mutual positions of the two ribosomal subunits, as well as in the position of the L7/L12 stalk.

The mechanical removal of the bound EF-G from the ribosomal complex leads immediately to the the post-translocation state. It seems that the translocation intermediate cannot exist in the absence of EF-G and slips down to the post-translocation state concomitantly with the removal of the protein. The natural way of converting the translocation intermediate into the post-translocation state ribosome is the hydrolysis of the EF-G-bound GTP which induces the conformational change of EF-G and the release of EF-G from the ribosome.

12.2.4. Role of GTP and Its Hydrolysis in the Catalysis of Translocation

Thus, three aspects of the contribution of GTP and its hydrolysis to the translocation step can be considered, in accordance with the series of reactions listed in Section 12.2.2. First, GTP serves as an *effector,* inducing the proper conformation of EF-G (or eEF2) that possesses an affinity for the ribosomal translocation intermediate and thus pulls out the ribosomal complex from the pre-translocation state (reactions 1 and 2). There is evidence that the main body of EF-G:GTP binds to the factor-binding site of the large ribosomal subunit, and the protruding domain IV of EF-G falls on the A site of the small subunit. GTP can be replaced by its non-hydrolyzable analog, such as GMP-PCP or GMP-PNP (Fig. 9.5) in performing this function.

Second, GTP is a *destructible* effector. It is hydrolyzed on the ribosome-bound EF-G (eEF2). Thus the effector is abolished, and the resultant relaxed conformation of EF2 allows the translocation intermediate to turn quickly into the post-translocation ribosomal complex, with the peptidyl-tRNA in the P site. In other words, the cleavage of GTP is required for exiting the transition state. The process is analogous to the decay of a transition state intermediate into products in the case of covalent enzymatic catalysis; the cleavage of GTP plays the energy role of the covalent reaction catalyzed by an enzyme.

There is also the third aspect of the problem under consideration: due to reduced affinity of EF2:GDP for the ribosome, the dissociation of EF2 from the post-translocation ribosomal complex is caused by GTP hydrolysis. The use of a nonhydrolyzable GTP analog has demonstrated that translocation can be also catalyzed by EF-G:GMP-PCP or EF-G:GMP-PNP, though slower than with EF-G:GTP (Rodnina *et al.,* 1997). In this case, however, EF-G after translocation remains associated with the ribosome. It may be that the translocation intermediate fixed by the active (tight) conformation of EF-G is capable of spontaneously converting into the post-translocation ribosome, but with EF-G still bound. This relationship is the result of the fact that the effector inducing EF-G affinity for the ribosome is not destroyed and therefore EF-G continues to be retained on the factor-binding site. However, the presence of EF-G on the ribosome blocks the subsequent step of the elongation cycle: the binding of the ternary Aa-tRNA:EF-Tu:GTP complex is precluded due to the occupancy of the factor-binding site on the 50S subunit and, maybe, the A site on the 30S subunit. Therefore, as a result of the translocation effected by EF-G with a nonhydrolyzable GTP analog, peptidyl-tRNA acquires the capacity to react with the acceptor substrate, and deacylated tRNA may be released from the ribosome; however, such post-translocation-state ribosomes cannot bind the next aminoacyl-tRNA and, hence, are unable to continue the elongation. In experiments conducted *in vitro,* EF-G together with the nonhydrolyzable GTP analog have been washed off from such post-translocation ribosomes, resulting in the capacity to bind aminoacyl-tRNA and to continue elongation (Belitsina, Glukhova, and Spirin, 1976; Girbes, Vazquez, and Modolell, 1976). This implies that during the normal process EF-G should become attached to the ribosome in order to induce translocation and then must leave the ribosome to allow the next step to occur.

In this respect, the action of fusidic acid (Fig. 12.4), an antibiotic specifically affecting EF-G, proves to be interesting. EF-G in a complex with fusidic acid nor-

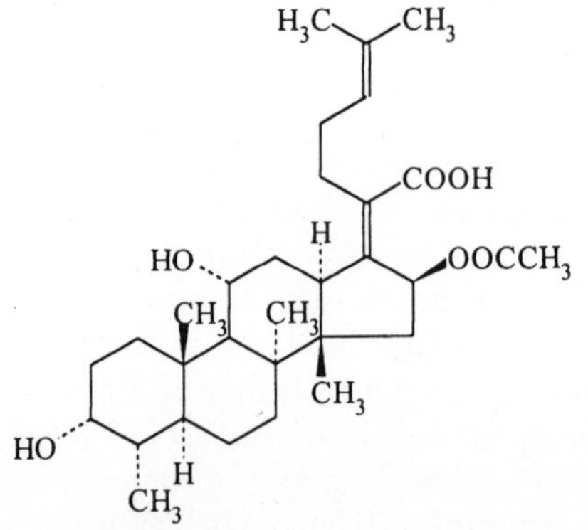

FIG. 12.4. Fusidic acid, an inhibitor affecting EF-G.

mally interacts with GTP and further with the ribosome, the interaction being followed by GTP cleavage to GDP and orthophosphate. Thus, the normal translocation is completed. Fusidic acid, however, acts to increase the affinity of EF-G for the ribosome, and EF-G:GDP is not released after GTP hydrolysis (Bodley, Zieve, and Lin, 1970). As a consequence, despite translocation, the next aminoacyl-tRNA cannot bind with the ribosomal A site, and therefore elongation stops.

12.3. "Nonenzymatic" (Factor-Free) Translocation

It has been established, using bacterial cell-free systems, that translocation can proceed in the absence of elongation factors and GTP (Pestka, 1969a; Gavrilova *et al.*, 1976). This *"nonenzymatic" translocation* takes place far more slowly than the EF-G:GTP-catalyzed process. Nevertheless, it yields a normal post-translocation state of the ribosome, capable of continuing elongation. It can be concluded that translocation is a thermodynamically spontaneous event. The translocation mechanism appears to be an intrinsic property of the ribosome itself, and is not fully provided by the elongation factor.

Again the situation appears to be very similar to that of aminoacyl-tRNA binding: the processes may occur slowly in the factor-free mode, proceeds spontaneously (downhill), and their mechanisms are provided by the ribosome; therefore the elongation factors catalyze only thermodynamically permissible and mechanistically ensured processes.

Factor-free (nonenzymatic) binding of aminoacyl-tRNA, ribosome-catalyzed transpeptidation, and factor-free (nonenzymatic) translocation constitute the factor-free elongation cycle. This cycle is designated by dashed shunting arrows in Fig. 12.5. Repetition of this cycle results in a slow *factor-free elongation*. Bacterial cell-free systems have been used to perform factor-free translation of polyuridylic acid and a number of synthetic heteropolynucleotides. It has thereby been shown that the polypeptide product corresponds completely to the coding sense of the template polynucleotide.

Whereas an increase in the Mg^{2+} concentration stimulates aminoacyl-tRNA binding, a decrease in Mg^{2+} stimulates translocation. At Mg^{2+} concentrations of about 3 mM in a bacterial cell-free system, the rate of factor-free translocation approaches that taking place in the presence of EF-G with GTP, but little aminoacyl-tRNA binding occurs. At Mg^{2+} concentrations of 30 mM, the translocation rate is close to zero, but the factor-free binding of aminoacyl-tRNA is good. Thus, by alternating the Mg^{2+} concentration in the cell-free system, the action of the elongation factors might be simulated and the rate of the factor-free elongation cycle increased. The alternating Mg^{2+} concentration may be considered a way to provide energy for speeding up the factor-free elongation cycle.

When comparing the slow factor-free translocation with the fast EG-G:GTP-catalyzed translocation, evidence has been obtained that the elongation factor does not decrease the heat energy of activation of the process; this suggests that catalysis in this case is primarily of an entropy type (Spirin, 1978). Various inhibitors of translocation (see below) affect the enzymatic as well as the nonenzymatic process, suggesting an identical translocational mechanism in both cases, involving identical targets. Thus, the existence of factor-free (nonenzymatic) translocation indicates that the translocational mechanism is intrinsic to the ribosome and is principally provided with energy without the involvement of GTP.

12.4. Movement of the Template during Translocation

12.4.1. Triplet Translocation

As has been noted, translocation involves a shift of a template polynucleotide by one codon from the 5′-end to the 3′-end. During this shift and after its completion, pairing between the anticodon of peptidyl-tRNA and the template codon is thought to be retained; the codon–anticodon duplex appears to move as a whole, from the A site to the P site of the ribosome (Matzke, Barta, and Kuechler, 1980).

It is natural to wonder what plays the active part in the translocation: movement of the template or movement of the peptidyl-RNA. Several observations suggest that the shift of the template by one codon is driven by the translocational displacement of tRNA: through its anticodon, tRNA pulls the codon of the template. An impressive demonstration of the lack of dependence of translocation on the template polynucleotide was the discovery of ribosomal synthesis of a polypeptide from aminoacyl-tRNA in the absence of any template polynucleotide (Belitsina, Tnalina, and Spirin, 1981); the elongation cycle, including EF-G:GTP-catalyzed translocation, was demonstrated in this case. The most direct evidence of the active (driving) role of tRNA and the passive (driven) role of mRNA in translocation was obtained by Riddle and Carbon (1973); in their experiment, mutant

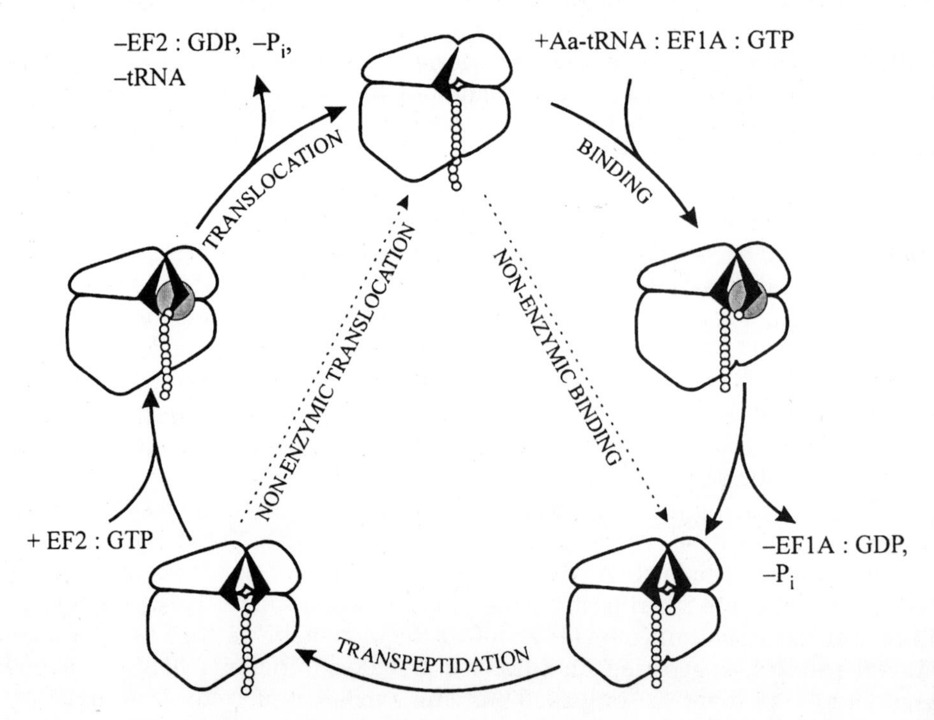

FIG. 12.5. Sequence of events during the factor-catalyzed elongation cycle. Broken lines designate factor-free bypasses: nonenzymatic binding of aminoacyl-tRNA and nonenzymatic translocation.

tRNA with a nucleotide quadruplet instead of a triplet as an anticodon suppressed the (+1) frame shift mutation; in other words, it moved the mRNA in the ribosome, correspondingly, by four (and not by three) nucleotide residues.

Thus, it is likely that the principal event in translocation is the movement of peptidyl-tRNA from the A site to the P site of the ribosome. The anticodon pulls the template codon associated with it, leading to a corresponding shift of the template relative to the ribosome by one triplet (normally). This results in the positioning in the A site of the next (in the direction of the 3′-end) nucleotide triplet of the template, while the preceding (the 5′-side-adjacent) triplet, together with the anticodon of deacylated tRNA, leaves the P site.

At the same time an additional mechanism for actively pulling the mRNA chain through the ribosome in the 5′-to-3′ direction cannot be excluded. A clue comes from the observations of 3′-ward slippage and hopping of translating ribosomes along mRNA under conditions when codon–anticodon interactions weaken or are disrupted (see Section 12.4.3). The idea of an intrinsic capability of the ribosome to count nucleotides or to shift an mRNA chain preferentially by triplets has been proposed (e.g., Trifonov, 1987).

12.4.2. Non-triplet Translocations (Translocation Errors)

The movement of mRNA relative to the ribosome exactly by three nucleotides during translocation is the absolute requirement for keeping the reading frame in the course of translation. Errors, that is nontriplet translocations, are possible, however, and the result is *frameshifting.* After a frameshift within a given mRNA chain, no correct codons are read out anymore, a polypeptide chain having no similarity with the encoded protein is synthesized, and the synthesis terminates soon because of a high probability of stop triplets out of the correct reading frame. Thus, the translocation errors have more severe consequences than miscoding during aminoacyl-tRNA binding.

Generally, an mRNA chain can be occasionally moved by less or more than three nucleotides during translocation, concomitantly with the normal passage of peptidyl-tRNA from the A site to the P site (for reviews, see Weiss *et al.,* 1990; Atkins and Gesteland, 1995). The most frequent translocation errors are those when mRNA is moved either by just two nucleotides (−1 frameshift) or by four nucleotides (+1 frameshift), instead of moving by a regular nucleotide triplet. The average probability of such errors (the level of frameshifting) in bacteria has been estimated to be from 5×10^{-4} to 10^{-5} per codon of mRNA. However, the errors occur mostly at certain codons ("shifty codons"), and certain tRNA species ("shifty tRNAs") may be involved. In addition, the context can play a decisive role in inducing a frameshift and determining the direction of the shift.

12.4.2.1. Frameshifting at the Aminoacyl-tRNA Binding Step

Two different mechanisms leading to the nontriplet movement of mRNA are known. The first one is the consequence of an erroneous mRNA–anticodon pairing at the stage of aminoacyl-tRNA binding, rather than an error at the translocation step of the elongation cycle. One case is the binding of an incoming aminoacyl-tRNA in the A site to the duplet adjacent to an upstream P-site codon, rather than to the triplet. For example, bacterial (*E.coli*) Ser-$tRNA_3$ normally recognizing serine codons AGC and AGU can enter the A site with alanine codon GCA presented and form the duplex between GC of its anticodon and GC of the alanine codon:

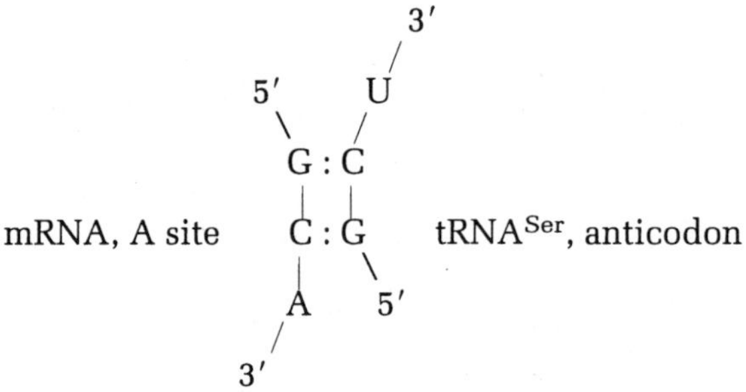

```
                              3'
                             /
                  5'       U
                    \     /
                     G : C
                     |   |
mRNA, A site         C : G       tRNA^Ser, anticodon
                     |    \
                     A     5'
                    /
                  3'
```

After transpeptidation, the subsequent translocation of the tRNA residue from the A site to the P site will drag mRNA, but just by two paired nucleotides. As a result, mRNA will move by two nucleotides, instead of three, and thus a −1 frameshift will occur, for example GCA GCA AAC → GCA *GC* AAA C . . . Another case is the so-called overlapping binding of an incoming aminoacyl-tRNA when its anticodon interacts with the triplet of mRNA extending over the next codon. The same Ser-$tRNA_3$ gives an example of such behavior: when lysine codon AAG, followed by leucine codon CUU, is presented in the A site, Ser-$tRNA_3$ can bind to the triplet AGC overlapping the leucine codon. The result of the subsequent translocation will be a +1 frameshift: AAG CUU → A*AGC* UU... The tRNA species under consideration (*E. coli* $tRNA^{Ser}$ with anticodon GCU) represents a typical "shifty" tRNA capable of producing both −1 and +1 frameshifts due to misbinding at the A site. Some structural peculiarity or flexibility of its anticodon is thought to be responsible for this behavior.

Since aminoacyl-tRNA misbinding underlies the mechanism considered above, the frameshifting produced by this mechanism can be stimulated by all the same factors that are known to stimulate miscoding (see Section 10.4.4), including aminoglycoside antibiotics, increased Mg^{2+} or polyamine concentration, distorted ratio of different tRNAs, starvation for an amino acid or shortage of a tRNA species, and ribosomal *ram* mutations. In particular, the probability of the duplet binding or the overlapping binding increases when the translating ribosome pauses at the empty A site. Such a situation arises with a "hungry" codon (in the case of starvation for a cognate amino acid), or a rare codon (especially when a cognate minor tRNA is exhausted, for example with tandem rare codons) in the A site. In the absence of a cognate aminoacyl-tRNA, a noncognate "shifty" tRNA can enter the A site and produce a frameshift.

Sometimes frameshifting at a specific site of mRNA can be used for regulation of the synthesis of protein. The synthesis of mammalian antizyme of ornithine decarboxylase (ODC) is regulated by frameshifting (Fig. 12.6). Antizyme synthesis is induced by polyamines; the protein tags ODC, a key enzyme in polyamine synthesis, for proteolytic degradation. The antizyme mRNA is organized in such a way that the first 35 codons are in the frame with the initiation codon, but then a stop codon UGA is encountered. The frameshift at the stop codon is strongly induced by polyamines. Thus, when the polyamine level is low, ribosomes read only the beginning of the mRNA sequence and terminate at the UGA; hence, antizyme is not produced, and ODC lives relatively long, supporting the polyamine synthesis. When the polyamine level becomes high, the induced +1 frameshift allows ribosomes to continue translation in the new reading frame and to complete the synthesis of antizyme, provoking ODC degradation. The frameshift sequence of the antizyme mRNA is UCC UGA U. In the case of induction, the anticodon loop of seryl-tRNA in the A site seems to bind quadruplet UCCU, rather than its cognate triplet UCC, and hence the subsequent translocaton by four nucleotides results in setting the triplet GAU, an aspartic acid codon, in the A site. Thus, the initial peptide sequence ended by serine is further elongated by aspartic acid and then by the remaining sequence of the antizyme polypeptide. It is interesting that there is a pseudoknot downstream of the frameshift site that stimulates the +1 frameshifting in some unknown way.

12.4.2.2. Frameshifting at the Translocation Step

The second mechanism of nontriplet movement of mRNA involves translocation events: a strand slippage between an anticodon and mRNA *during* translocation. In typical cases, cognate aminoacyl-tRNA binding takes place in the A site, but after transpeptidation the subsequent translocation of the tRNA residue from the A site to the P site is found to be uncoupled with the codon shift; the tRNA anticodon slides along mRNA or mRNA slides along the anticodon, most often by one nucleotide, thus resulting in −1 or +1 frameshifting, respectively. Such slippage can occur on strings of purine or pyrimidine nucleotides during translocation of tRNA residues with corresponding all-pyrimidine or all-purine anticodons. The simplest examples are the slippage between $tRNA^{Lys}$ with anticodon U*UU and an oligo(A) sequence, or between $tRNA^{Phe}$ with anticodon GAA and an oligo(U) string in mRNA; both +1 and −1 frameshifts are possible. The examples of slippage of $tRNA^{Leu}$ with anticodon GAG on the mRNA sequence CUUU (*CUU*U → C*UUU*: +1 shift), $tRNA^{Leu}$ with anticodon U*AA on the sequence UUUA (U*UUA* → *UUU*A: −1 shift), $tRNA^{Phe}$ with anticodon GAA on the sequence UUUC (U*UUC* → *UUU*C: −1 shift), $tRNA^{Asn}$ with anticodon QUU on the sequence AAAC (A*AAC* → *AAA*C: −1 shift), $tRNA^{Gly}$ on the sequence GGGU (*GGG*U → G*GGU*: +1 shift), and some others, have been also reported.

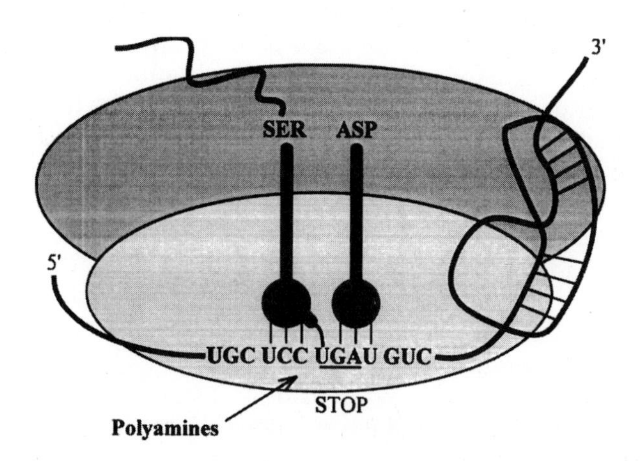

FIG. 12.6. Polyamine-induced frameshifting (+1) during translation of the antizyme mRNA. Reproduced, with minor modifications, from Atkins, J. F., and Gesteland, R. F. (1996) in *Translational Control* (J. W. B. Hershey, M. B. Mathews, and N. Sonenberg, eds.), pp. 653–684, CSHL Press, with permission.

It is interesting that the slippage can be significantly enhanced by putting two slippery codons in tandem. An example is the sequence...U *UUU UUA* in HIV-1 RNA where a −1 shift of $tRNA^{Leu}$ from UUA to UUU is facilitated by the possibility of the preceding $tRNA^{Phe}$ to slide in the same direction (*UUU UUU* A...). Another example is the sequence...A *AAA AAC* in the RNA of mouse mammary tumor virus. Here $tRNA^{Asn}$ slips −1, possibly together with the −1 movement of the preceding $tRNA^{Lys}$ (*AAA AAA* C...). These observations may suggest that the slippage takes place at some intermediate stage of translocation when the A-site tRNA residue and the P-site tRNA are partly pulled out from their sites but both retain their bonds with mRNA. The interactions with mRNA, however, should be weakened at this stage, thus permitting the slippage.

The presence of a stop codon downstream of a slippery codon is also often observed in sequences with a high frequency of frameshifting. For example, the +1 slippage of $tRNA^{Leu}$ from CUU to UUU (see above) becomes especially significant in the sequence CUU UGA C..., resulting in reading the sequence as UUU GAC (LeuAsp) and thus continuing the translation in the new reading frame, instead of termination at the stop codon. It is not clear how a downstream codon can affect the slippage of the A-site tRNA residue during translocation. An increased probability of the slippage of the post-translocation-state ribosome with the empty A site (if the binding of a termination factor RF after translocation is delayed) cannot be excluded.

The strand slippage events can be strongly stimulated by additional elements in the sequence or in the three-dimensional structure of mRNA. In a number of cases such an enhanced slippage at a specific point is used by a living cell or a virus in order to produce variants of polypeptides from the same message, or to regulate the synthesis of a full-size polypeptide. An illustrative case is the translation of the message encoding for one of three termination factors, namely RF2, in *E. coli.* The gene and, correspondingly, the message, consists of two parts: the first, smaller part of the sequence encodes for the N-terminal 25-residue portion of the protein and ends with termination codon UGA; the remainder of the 339-residue protein is encoded by the following sequence that is not in the same frame as the previous part. The reading of the second part of the sequence requires a +1 frameshift at the position of the termination codon: $tRNA^{Leu}$ slides from CUU to UUU (Fig. 12.7). The frequency of the frameshifting in the above sequence can be as high as 30%, resulting in the production of the proper polypeptide chain (protein RF2) with a reasonable yield. The upstream hexanucleotide string AGGGGG has been shown to be critical for providing high-efficiency frameshifting. It has been demonstrated that the complementary interaction between this polypurine sequence and the 3′-terminal polypyrimidine section of the 16S RNA in the ribosome, similar to the Shine–Dalgarno interaction during initiation of translation (see Section 15.2.2), is responsible for enhancing the slippage of the ribosome along the CUUU quadruplet overlapping the stop codon. The frequency of the slippage can be regulated: when the level of RF2 in the cell is sufficient, the protein binds to the stop codon UGA in the ribosomal A site and thus terminates translation; no new RF2 is synthesized. Hence, the frequent slippage and thus the synthesis of RF2 occur only under conditions of RF2 shortage. This is an interesting example of a feedback regulation of the efficiency of translation via a programmed translocation error.

Programmed shifts are used also in eukaryotic systems. The best-studied case is the translation of the *gag-pol* mRNA of retroviruses. The translation of this mRNA can produce either just Pol protein, or Pol-Gag fusion protein which is to be split by a special protease into Pol and Gag. The point is that the mRNA se

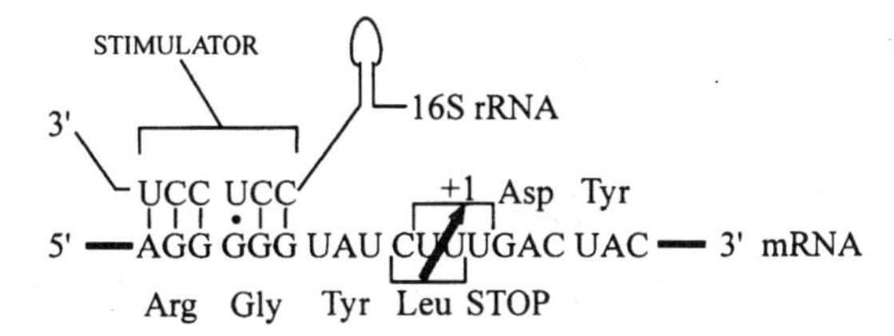

FIG. 12.7. Frameshifting (+1) during translation of the mRNA encoding for termination factor RF2. Reproduced, with minor modifications, from Weiss, R., and Cherry, J. (1993) in *The RNA World* (R. F. Gesteland and J. F. Atkins, eds.), pp. 71–89, CSHL Press, with permission).

quences encoding for Pol and Gag are not in the same frame, and the synthesis of the Gag part requires a −1 frameshift at the border of the two sequences. The scheme of the frameshifting during translation of the *gag-pol* mRNA of HIV-1 is shown in Fig. 12.8. It has been noted that in such cases the sequence immediately downstream from the slippage site can form a stable stem-loop or pseudoknot structure that may be important for increasing the frequency of the frameshift, possibly by creating a structural barrier for mRNA translocation (Fig. 12.9).

12.4.3. Ribosome Hops

Another phenomenon sometimes observed during translation and also violating the principle of triplet translocation of mRNA is the so-called ribosome hopping: the translating ribosome, when it is compelled to pause by a stop codon or a structural barrier, may skip over a stretch of nucleotides in the 3′ direction. Distinct from the slippage mechanism considered above, hopping implies a complete temporary break of anticodon–codon interactions, and then rejoining at another section of mRNA, but always downstream. The rejoining always occurs with a codon recognizing the same tRNA that is translocated from the A site to the P site. One of the first cases reported was the jumping over the amber stop codon during translation of the mutant β-galactosidase mRNA:

Met Lys Ser Leu Gly Tyr Leu Arg Gly Pro
AUG AAA AGC UUA GGG UAU *CUU* **UAG** *CUA* CGG GGC CCU . . .

The frequency of this hopping over six nucleotides (+6 shift) was only 1%. Similar low-frequency hops over two to nine nucleotides can be observed also within

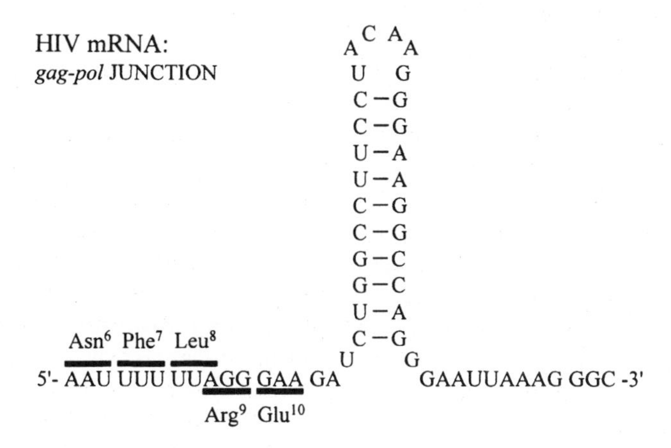

FIG. 12.8. Frameshifting (−1) during translation of HIV mRNA (*gag-pol* junction). Adapted from Weiss, R. B., Dunn, D. M., Shuh, M., Atkins, J. F., and Gesteland, R. F. (1989) *New Biologist* **1**:159–169.

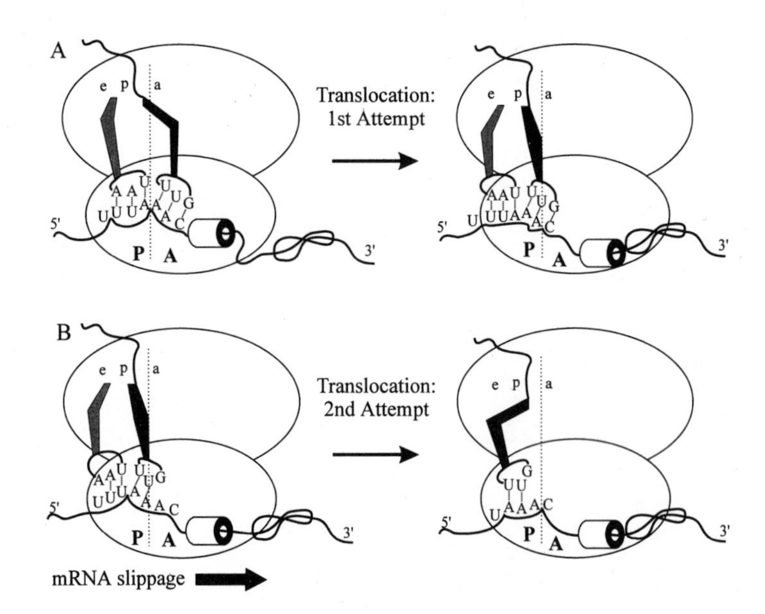

FIG. 12.9. Postulated mechanism for pseudoknot-induced frameshifting. (**A**) First translocation attempt is jammed by pseudoknot structure binding in the 30S "melting site" (a hypothetical mRNA-binding site upstream of the A site responsible for primary unfolding of mRNA): structural barrier. (**B**) Second attempt is successful because both tRNAs have slipped by one nucleotide towards the 5′-end of the mRNA. Reproduced, with minor modifications, from Weiss, R., and Cherry, J. (1993) *The RNA World* (R. F. Gesteland and J. F. Atkins, eds.), pp. 71–89, CSHL Press, with permission.

coding regions of mRNA. Specific mutations of tRNAs or ribosomes may increase the frequency of hopping up to 20% or more.

The discovery of a hop over 50 nucleotides during translation of bacteriophage T4 DNA topoisomerase mRNA (gene 60 transcript) was remarkable (Huang *et al.*, 1988). In this case the translating ribosome with peptidyl-$tRNA^{Gly}$ jumps from the GGA codon at positions 136–138 to an identical codon 50 nucleotides downstream (Fig. 12.10). The efficiency of the jump is close to 100%. A special structure of the 50-nucleotide section is required for efficiently skipping it. Essential elements of this structure are a stop codon 3′-adjacent to the take-off glycine codon, an extremely stable hairpin with a tetraloop at the take-off site, and a properly sized skipped section. In addition, a stretch of the nascent peptide (14 amino acid residues away from the PTC) is found to contribute to the efficiency of the hop. It is likely that the pausing at the stop codon and the subsequent stable hairpin provokes the taking-off, and the ribosome slips over the structured gap as a bulge without its melting. The taking-off, the bypassing of a stretch, and the landing of the ribosome at a new, downstream codon can be considered as reprogrammed genetic decoding, or *recoding* of the message (Gesteland, Weiss, and Atkins, 1992).

Another striking example of ribosome hopping is the case of translation of *E. coli trpR* mRNA (Benhar and Engelberg–Kulka, 1993). The *E. coli trpR* gene codes for the Trp repressor, a 12-kDa protein that regulates transcription of several operons and genes involved in tryptophan metabolism and transport. It has been observed that along with the main product of translation, some amount of a shorter protein, of about 10 kDa, is synthesized during translation of the same mRNA. The N-terminal part of the shorter protein is identical to that of the repressor, but the C-terminal part has been found to be completely different. It has been proven that during translation, at a specific site of the mRNA, ribosomes occasionally (with a frequency of about 5%) jump over an mRNA segment of 55 nucleotides in length, thus resulting in a +1 frameshift and polypeptide shortening. In contrast to the previous case (T4 DNA topoisomerase mRNA), no pair of matched codons at the borders of the gap, no essential secondary structure, and no stop codon at the border in the frame of the gap has been mentioned. Instead, the translation of five specific codons (AUG AGC CAG CGU GAG) preceded by a nonspecific sequence longer than ten codons is required for the jumping. The hypothesis has been put forward that the corresponding sequence of mRNA is specifically looped out in the structure of the mRNA, thus bringing the borders of the bypassed segment into close proximity. It may be assumed that the translating ribosome during translocation can switch over from the ribosome-bound codon to the codon that is spatially adjacent but belongs to a remote sequence along the mRNA chain.

There is a special case when the translating ribosome can hop from one message to another. This phenomenon can be designated "*trans translation*" (Atkins and Gesteland, 1996; Muto *et al.*, 1996). *E. coli* cells and cell-free extracts contain an RNA species called 10Sa RNA (363 nucleotides long) that encodes for a de-

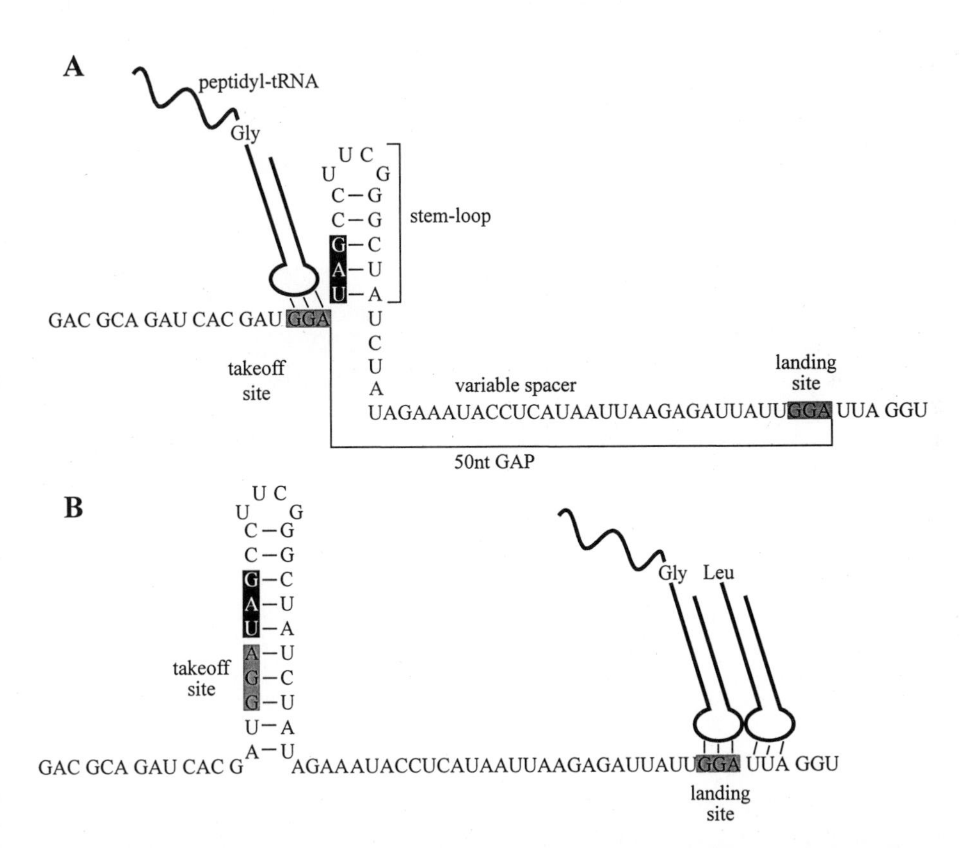

FIG. 12.10. The ribosome hopping over 50 nucleotides during translation of bacteriophage T4 DNA topoisomerase mRNA (gene 60 transcript). See Huang, W. M., Ao, S. Z., Casjens, S., Orlandi, R., Zeikus, R., Weiss, R., Winge, D., and Fang, M. (1988) *Science* **239**:1005–1012. Adapted from Weiss, R., Dunn, D., Atkins, J., and Gesteland, R. (1990) in *The Ribosome: Structure, Function, and Evolution* (W. E. Hill, A. Dahlberg, R. A. Garrett, P. B. Moore, D. Schlessinger, and J. R. Warner, eds.), pp. 534–540, ASM Press, Washington, DC.

capeptide (Ala-Asn-Asp-Glu-Asn-Tyr-Ala-Leu-Ala-Ala) found as a C-terminal extension of various, incomplete (truncated) proteins. This extension serves as a "tag" recognized by a proteolytic degradation system of the bacterium. At the same time the 10Sa RNA has been shown to be aminoacylated at its 3′-end by alanine, like tRNA. When the ribosome reads a truncated mRNA or a synthetic message without a stop codon, it halts at the 3′-end; the A site becomes empty. The 10Sa RNA enters the empty, codon-free A site; as a result of transpeptidation, its alanyl residue is found to be added to the nascent polypeptide encoded by the truncated mRNA, and the elongated peptidyl-10Sa RNA occupies the position of a peptidyl-tRNA in the pre-translocation-state ribosome (Fig. 12.11). Seemingly during translocation of the 10Sa RNA from the A site to the P site the ribosome hops on the first codon of the coding sequence of the 10Sa RNA. Then it normally translates ten codons of this coding sequence and terminates at the stop codon of the 10Sa RNA. Thus, 11 amino acids become added to the C-terminus of a truncated protein, one (non-encoded, "junction" alanine) being brought by the 10Sa RNA as its 3′-bonded residue and the other ten being annexed as the 10Sa RNA-encoded sequence (see the sequence above).

The exclusively downstream movement of mRNA during ribosome hopping may suggest the existence of some mechanism or force for active unidirectional displacement of a message through the ribosome, independent of the translocation of tRNA residues. This cannot be excluded, especially since mRNA is a polar polymer and therefore the forward and backward movements can make difference (a

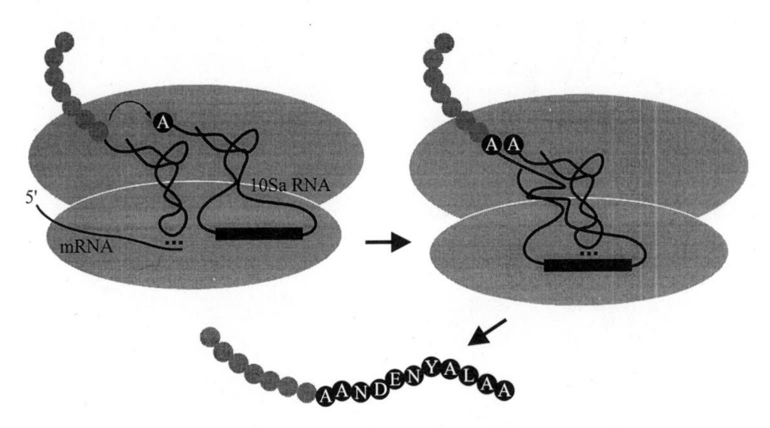

FIG. 12.11. Model for the involvement of 10Sa RNA in tagging truncated proteins by *trans*-translation. Reproduced from Felden, B., Himeno, H., Muto, A., McCutcheon, J. P., Atkins, J. F., and Gesteland, R. F. (1997) *RNA* **3**:89–103, with permission. When the ribosome translates a truncated mRNA or a synthetic polynucleotide without a stop codon, it stalls at the 3′-end of the message without termination. In this case the polypeptide synthesized remains bound to a tRNA in the P site. 10Sa RNA acylated by alanine binds to the vacant A site of the ribosome. Transpeptidation between the peptidyl-tRNA of the P site and the alanyl-10Sa RNA results in the formation of the alanine-elongated peptidyl-10Sa RNA. During translocation the ribosome switches messages and starts to translate the open reading frame of the 10Sa RNA producing the elongation of the polypeptide by an additional amino acid sequence AlaAsnAspGluAsnTyrAlaLeuAlaAla. This sequence serves as a tag for rapid degradation of the released polypeptide by cellular proteases. See Tu, G.-F., Reid, G. E., Zhang, J.-G., Moritz, R. L., and Simpson, R. J. (1995) *J. Biol. Chem.* **270**:9322–9326; Keiler, K. C., Waller, P. R. H., and Sauer, R. T. (1996) *Science* **271**:990–993; Muto, A., Sato, M., Tadaki, T., Fukushima, M., Ushida, C., and Himeno, H. (1996) *Biochimie* **78**:985–991.

ratchet). However, a simple explanation of the downstream "driving force" for mRNA is possible: in the polyribosome, if a ribosome pauses or stops, the following ribosome will approach the preceding one and push it downstream along mRNA (or, pull out mRNA from it) during translocation.

12.5. Mechanics and Energetics of Translocation

12.5.1. Stereochemistry and Mechanics

The evidence available on the structure of tRNA and the ribosome as well as on the properties of pre-translocation and post-translocation complexes can be used to formulate a plausible stereochemical model of the interactions between the ribosome, two tRNAs, and mRNA, and of the changes in these relationships during translocation. As has been noted in stereochemical consideration of the codon–anticodon interaction (Section 10.1.5), the anticodons of two ribosome-bound tRNAs form double-helical structures of the A-form type with two adjacent mRNA codons (e.g., as in Fig. 10.3). The 3′-ends of the two tRNA molecules are brought into close proximity, while their corners are somewhat apart, so that the planes of the two tRNA molecules are at an angle to each other (see Figs. 9.12 and 9.14). This situation continues after transpeptidation, except the 3′-ends of the tRNA residues are now shifted inside a limited region of the PTC and its neighborhood (Section 11.4). Therefore the pre-translocation-state ribosome contains a complex between deacylated tRNA (in the P site, with the 3′-end in the *e* site) and peptidyl-tRNA (in the A site, with the ester group and the 3′-end in the *d* site) joined by a complementary hexaplet of mRNA. In the case of the R-type orientation (Section 9.5), the corner of the peptidyl-tRNA may be positioned close to the heads of the 30S and 50S ribosomal subunits, while the corner of the deacylated tRNA molecule would be located in the region of the base of the L7/L12 stalk of the 50S subunit (Fig. 9.15, *upper*). If the S-type orientation is accepted, the positions of the corners of the two tRNAs should be exchanged (Fig. 9.15, *lower*).

Translocation can be conceived as an operation of the *helical displacement* of the two tRNAs (Rich, 1974; Spirin, 1985). In the case of the R-type orientation, the displacement will include a clockwise turn (if one looks from their anticodons) and translation along the axis connecting the anticodons with the acceptor ends. In the case of the S-type orientation, the turn would be counterclockwise, with a similar axial translation. As a result, the deacylated tRNA is displaced from the P site and dissociates from the complex with its codon; the peptidyl-tRNA is then in the P site, and the A site is now vacant. This is the post-translocation state (Fig. 12.1, *right*).

Which force is responsible for the movement of the peptidyl-tRNA from the A site to the P site during translocation? If the pathway of the displacement of the complex between two tRNAs and mRNA is determined by the construction of the ribosome, then the movement itself may be a consequence of just thermal motion. Since the displacement is followed by dissociation of the deacylated tRNA, it should result in an entropy gain. At any rate, in the case of factor-free (nonenzymatic) translocation, there appear to be no other motive forces except thermal motion. It is likely that thermal motion similarly induces the displacement in the course of EF2:GTP-catalyzed translocation, but the attachment of EF2:GTP to the ribosome creates a specific structural environment wherein steric and energy barriers in the transition pathway are decreased.

12.5.2. Energetics

The energy aspect of translocation was misunderstood for a long time due to various historical factors and due to traditional ways of thinking among biochemists. The participation of GTP in translocation was determined earlier than all other facts concerning this stage of the elongation cycle. Consideration of translocation as the process of mechanical displacement of large molecular masses and the observation of coupled cleavage of GTP into GDP and orthophosphate suggested an analogy to muscle contraction, which proceeds at the expense of energy of the ATP hydrolysis into ADP and orthophosphate. This analogy created a powerful psychological stimulus for inventing special problems of energy supply for translocation, which had to be solved at the expense of GTP cleavage. Most of the models of translocation proposed thus far assume that it is the energy of EF-G-mediated GTP hydrolysis that is used in one way or another for mechanical work involving the active movement of ligands (tRNA, mRNA) along the ribosome or at least the active removal of ribosomal ligands (tRNA) from their binding sites; correspondingly, the function of contractile proteins is often ascribed to EF-G or to protein L7/L12. According to some models, the energy of GTP through EF-G is applied to peptidyl-tRNA occupying the A site, and the developing force moves this tRNA together with its codon toward the P site, displacing deacylated tRNA from the P site. In other models, GTP energy is realized through the EF-G initially for the removal (pushing out) of the deacylated tRNA from the P site; then the peptidyl-tRNA undergoes spontaneous transition from the A site to the vacant P site, for which it has a greater affinity. There is also a class of structurally unrealistic models where EF-G is considered as "an authentic molecular motor," "exerting force," working "like a spring," and directly "ratcheting the mRNA tape."

As already mentioned, there is no apparent coupling between GTP hydrolysis and translocation. In addition, it has been shown that hydrolysis involves the direct transfer of the phosphate residue from GTP to water without the formation of a phosphorylated intermediate that could be responsible for such coupling (Webb and Eccleston, 1981). Thus, another mechanism should be assumed: translocation is coupled with the adsorption of EF2 on the ribosome whereas GTP hydrolysis is required for the desorption of EF2. If work had been required to effect the transition of the ribosome from the pre-translocation to the post-translocation state, it might be thought that the work is performed at the expense of the energy of the complex formation between the ribosome and EF2:GTP, and the adsorption energy is then compensated by the energy of GTP hydrolysis. Such a mechanism would imply that the energy of GTP hydrolysis eventually is responsible for the work done, not through direct coupling but by "lending" with the subsequent return.

In reality, it has been demonstrated that translocation can proceed spontaneously, without EF-G and GTP (nonenzymatic translocation). This implies that the process is thermodynamically permissible (downhill process) or, in other words, that the thermodynamic potential (free energy) of the pre-translocation state is higher than that of the post-translocation state. There is no need to explain that in this situation, energy expenditure for performing work (increasing potential) is not required. Thus, any thermodynamic contribution of EF2 with GTP in translocation should be rejected.

Nevertheless, in EF2-catalyzed translocation, EF2 with GTP binds to the ribosome, the GTP is subsequently hydrolyzed, and additional free energy is expended. But what is the purpose of this energy expenditure? It is apparent that energy generally can be expended either on some useful work against thermodynamic potential (uphill process), or on overcoming barriers in a spontaneous (downhill) process without accumulation of productive work. If the first of the two alternatives is excluded, it has to be recognized that the contribution of GTP is a purely kinetic one: at first, the interaction of GTP with EF2 provides for the attachment of EF2:GTP to the ribosome and thereby decreases barriers in the course of translocation; thereafter, GTP hydrolysis removes the barrier created by EF-G itself for the completion of translocation (the decay of the translocation intermediate). Thus, GTP energy is expended solely to overcome barriers, and eventually it dissipates completely into heat (Chetverin and Spirin, 1982). This process is called the catalysis of translocation. A peculiar feature of catalysis in this case is its energy dependence, which is similar to the catalysis of aminoacyl-tRNA binding with the participation of EF-Tu.

12.6. Inhibitors of Translocation

As expected from the involvement of both ribosomal subunits in translocation, this step of the elongation cycle can be inhibited by agents specifically acting either on the small or the large subunit. Three types of inhibitory mechanisms can be anticipated: (1) fastening the tRNA residues at their A and/or P sites on the small subunit; (2) preventing the EF2:GTP interaction with the factor-binding site on the large subunit; and (3) directly intervening in the events of intraribosomal molecular movements, including possible intersubunit movements. A number of specific inhibitors of translocation are described among antibiotics (for reviews, see Pestka, 1969b; Cundliffe, 1980, 1990; Gale *et al.*, 1981).

12.6.1. Aminoglycosides and Aminocyclitols

Neomycin, kanamycin and gentamicin (Fig. 10.10) were reported as inhibitors of translocation, together with their action on the aminoacyl-tRNA binding step and their miscoding effect (see Section 10.3.2). Prokaryotic ribosomes seem to be markedly more sensitive to the drugs than are eukaryotic ribosomes. The inhibitory effect on translocation is connected with their specific binding to the small ribosomal subunit. Concerning the mechanism of this effect, it may be that these polycationic antibiotics bound on the ribosome increase the affinity of the ribosomal A site for tRNA and thus hamper the exit of the tRNA residue of peptidyl-tRNA from the A site during translocation.

Spectinomycin (Fig. 12.12) is representative of a related group of antibiotics, the aminocyclitols. Unlike the aminoglycosides, spectinomycin inhibits only bacterial ribosomes. Its binding site is also localized on the 30S ribosomal subunit, and again the rRNA is mainly responsible for the binding. Spectinomycin protects positions C1063-G1064 of helix 34 near the A-site part of the decoding center (Fig. 9.3) (Noller *et al.*, 1990). Mutations at position C1192 in the same helix confer re-

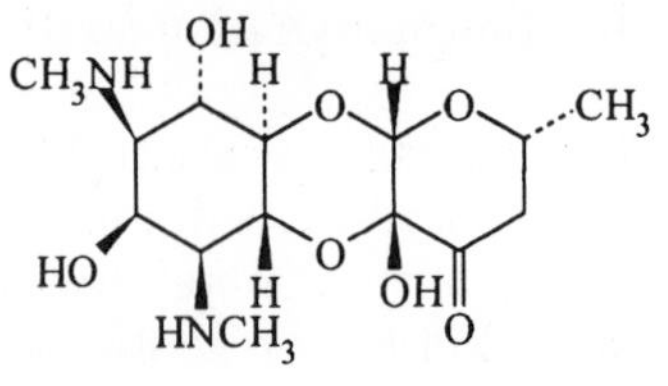

FIG. 12.12. Spectinomycin.

sistance against the drug (Cundliffe, 1990). Mutations of ribosomal protein S5 also confer resistance against the antibiotic. The drug neither inhibits aminoacyl-tRNA binding to the ribosome, nor induces misbinding. It is believed to affect the translocation step of the elongation cycle. The mechanism of action is unknown, but, by analogy with aminoglycosides, it can be thought to impede the exit of the tRNA residue from the A site during translocation.

12.6.2. Viomycin (Tuberactinomycin)

Viomycin is a cyclic basic polypeptide (Fig. 12.13). It is a potent inhibitor of bacterial ribosomes. The drug has no apparent effect on the interaction of EF-G with the ribosome. The target of the antibiotic seems to be the 30S ribosomal subunit, though a firm binding to the 50S subunit has been also reported. The result of the binding of the drug to the ribosome is the inhibition of translocation. It is likely that viomycin confines the tRNA residue of peptidyl-tRNA to the ribosomal A site. Thus, the inhibition of translocation may be the result of either a mechanical block of the exit of tRNA from the A site by the drug, or an enhanced affinity of the A site for tRNA in the presence of the antibiotic. The second alternative

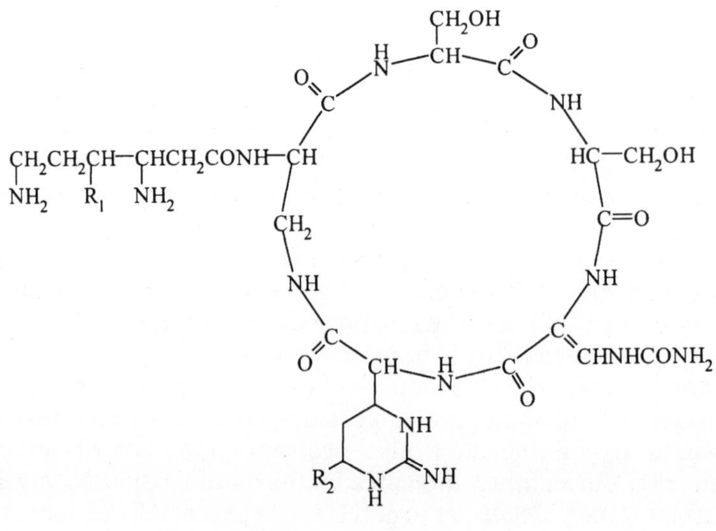

FIG. 12.13. Viomycin (Tuberactinomycin).

seems to be more plausible, and the polycationic nature of the antibiotic may contribute to this effect.

12.6.3. Thiostrepton

Thiostrepton (see Section 10.3.3.) and related antibiotics (siomycin, thiopeptin, sporangiomycin) are big cyclic compounds with thiazole rings and peptide bonds (Fig. 10.11). The antibiotics inhibit the binding of EF-G:GTP, as well as EF-Tu:GTP, to the large subunit of the bacterial ribosome. They interact with the ribosome in the region of the base of the L7/L12 stalk, that is, in approximately the same region where binding of EF-G (and EF-Tu) has been demonstrated (see Section 9.4). Protein L11 and its rRNA site, that is, the three-way helical structure shown in Fig. 9.10B (the so-called GTPase region), are responsible directly for thiostrepton binding. Thiostrepton protects from chemical modifications two hairpins, at positions 1057–1083 and 1087–1102 (Fig. 9.10B). It is remarkable that some thiostrepton-resistant mutants and strains are devoid of protein L11 as a component of the 50S ribosomal subunit. Specific enzymatic methylation of A1067 also confers resistance against the antibiotic and abolishes the affinity of the ribosome for it. The antibiotic directly protects A1067 from chemical modification and covers protein L11 upon binding.

Thiostrepton binding seems to require a proper conformation of the 23S rRNA region under consideration, and protein L11 maintains and stabilizes this structure. The antibiotic binds directly to the rRNA, rather than to the protein. It is likely that the massive molecule of firmly bound thiostrepton mechanically blocks the site of the interaction of the ribosome with EF-G (and EF-Tu). As a result, EF-G cannot interact with the pre-translocation ribosome, and so the catalysis of translocation is not realized. Due to the absence of the interaction, the ribosome-dependent EF-G-catalyzed hydrolysis of GTP, including the uncoupled GTP hydrolysis, is also inhibited by thiostrepton antibiotics.

12.6.4. Fusidic Acid

Fusidic acid is a steroid antibiotic (Fig. 12.4) effective against bacterial protein synthesis. The antibiotic affects the interaction between EF-G and the ribosome, but the target of the drug is EF-G, rather than the ribosomal factor-binding site. Accordingly, fusidic acid-resistance mutations are localized in EF-G. Fusidic acid does not prevent the binding of EF-G:GTP with the ribosome and the subsequent GTP hydrolysis. Instead, the antibiotic inhibits the release of EF-G:GDP from the ribosome after GTP hydrolysis. EF-G-catalyzed translocation, or at least its first stage (formation of the translocation intermediate), seems to proceed normally in the presence of fusidic acid. Hence, the inhibition of translation may result mainly from the delay of EF-G:GDP on the factor-binding site of the post-translocation-state ribosome: the presence of EF-G blocks the next step of the elongation cycle, namely aminoacyl-tRNA binding.

The mechanism of the increased affinity of EF-G:GDP for the ribosome in the presence of fusidic acid is not quite clear. There are two possibilities: either fusidic acid freezes the GTP conformation of EF-G even after GTP hydrolysis, resulting in continued affinity for the ribosome, or EF-G acquires the GDP conformation but fusidic acid imparts to the protein an additional affinity for the ribosome.

12.6.5. Glutarimides

The typical antibiotic of this group is cycloheximide, formerly called actidione (Fig. 12.14). It consists of a β-glutarimide ring and cyclic ketone connected by a hydroxyethyl bridge. The drug specifically inhibits eukaryotic ribosomes. The large (60S) ribosomal subunit has been identified as the target of the antibiotic. Mutations of some proteins of the large subunit of yeast and mammalian ribosomes were reported to be responsible for the resistance against the antibiotic. The drug does not inhibit the peptidyl transferase reaction. Since the drug is known to stabilize eukaryotic polyribosomes and to prevent the release of nascent peptides by puromycin, it is believed to inhibit the translocation step. The mechanism of action is unclear.

12.6.6. Nonspecific Inhibitory Agents

Translocation appears to be the most vulnerable step of the elongation cycle and thus it may be inhibited by a large variety of nonspecific agents and medium conditions. For instance, an elevated Mg^{2+} concentration in the medium strongly inhibits translocation. In the *E. coli* cell-free system the increase in Mg^{2+} concentration to 30 mM (at 100 mM KCl) stops the elongation cycle specifically at the translocation step. In other words, high Mg^{2+} freezes the pre-translocation state of the ribosome. Conversely, low Mg^{2+} stimulates translocation. It is possible that the main factor in these Mg^{2+} effects is the direct dependence of tRNA binding to the ribosomal tRNA-binding sites on Mg^{2+}: both affinities (binding constants) and kinetics (rate constants) are strongly affected by Mg^{2+} concentration. Thus a high Mg^{2+} can simply stick the tRNA residues in the A and P sites and hence prevent their translocation.

A low temperature acts in the same way. It is the translocation step of the elongation cycle that responds first and becomes blocked under lower temperatures in a translation system. At low temperature (e.g., 4°C) the ribosomes that elongate are found stopped mostly at the pre-translocation state.

12.7. Summary: Sequence of Events and Molecular Mechanisms

The plausible sequence of events that take place during EF2-promoted translocation is schematically given in Fig. 12.15. The first event is the collision of the pre-translocation-state ribosome with EF2:GTP. No stable complex between the pre-translocation-state ribosome and EF2:GTP or EF2:GMP-PCP, however, has been observed under experimental conditions.

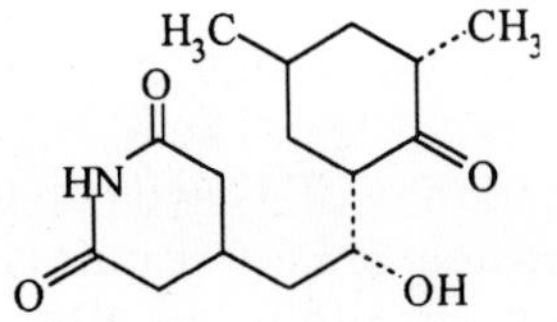

FIG. 12.14. Cycloheximide.

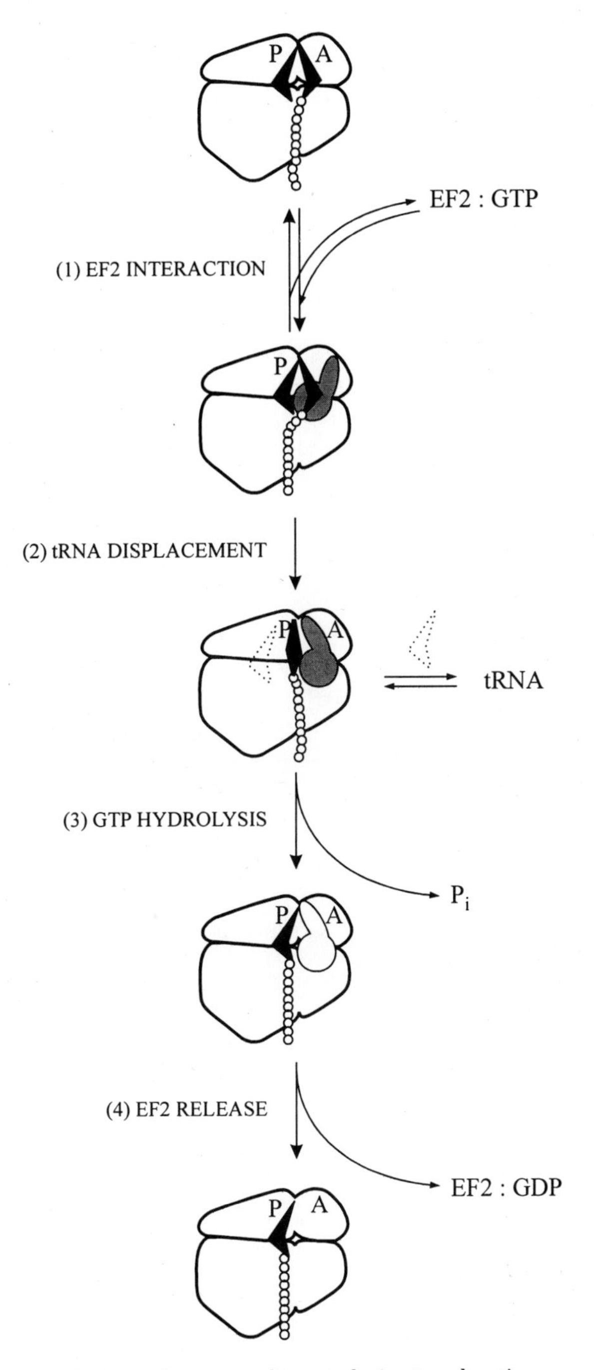

FIG. 12.15. Sequence of events during translocation.

It seems likely that EF2:GTP has a strong affinity specifically for a translocation intermediate (transition state). Therefore, a stable complex is formed between EF2:GTP and the ribosome in this state. In other words, the transition of the pretranslocation ribosome into the translocation intermediate is forced by the interaction with EF2:GTP. This is the second event in the scheme of Fig. 12.15. It may be important to indicate that the A site of ribosome in this state seems to be already freed from the tRNA residue of peptidyl-tRNA and instead occupied by domain IV of EF2.

The third event in the scheme of Fig. 12.15 involves the hydrolysis of GTP. The transition-state ribosome firmly retains EF2 until GTP hydrolysis occurs. If GTP is replaced by a nonhydrolyzable analog, e.g., GMP-PCP, this event does not take place and the intermediate-state ribosome will continue to be in a firm complex with EF2:GMP-PCP. The hydrolysis of GTP and its conversion into GDP as a ligand releases the conformation of EF2, leading to the loss of its strong affinity for the intermediate-state ribosome. As a result, the intermediate state becomes unblocked and allowed to fall downhill to the post-translocation state.

After GTP hydrolysis has taken place, EF2:GDP is found in a weaker complex with the post-translocation ribosome. This complex spontaneously dissociates, that is, EF2 and GDP are released from the ribosome. This is the fourth event in the scheme. Fusidic acid specifically inhibits just this stage of the process by fixing the complex and preventing the release of EF-G:GDP from the bacterial ribosome.

On the whole, the sequence described by the scheme in Fig. 12.15 is strikingly analogous to the sequence of events in EF-Tu-promoted binding of aminoacyl-tRNA (Fig. 10.14): the initial transient interaction is followed by a fast (catalyzed) stage of the main event, which in turn is followed by GTP hydrolysis, and finally by the release of the protein factor together with GDP. Such symmetry of the elongation cycle is seen in Fig. 12.5, where both factor-catalyzed processes are shown as sequences of consecutive events.

References

Atkins, J., and Gesteland, R. (1995) Discontinuous triplet decoding with or without re-pairing by peptidyl tRNA, in *tRNA: Structure, Biosynthesis, and Function* (D. Söll and V. L. RajBhandary, eds.), pp. 471–490, ASM Press, Washington, DC.

Atkins, J. F., and Gesteland, R. F. (1996) A case for *trans* translation, *Nature* **379**:769–771.

Belitsina, N. V., Glukhova, M. A., and Spirin, A. S. (1975) Translocation in ribosomes by attachment-detachment of elongation factor G without GTP cleavage: Evidence from a column-bound ribosome system, *FEBS Letters* **54**:35–38.

Belitsina, N. V., Glukhova, M. A., and Spirin, A. S. (1976) Stepwise elongation factor G-promoted elongation of polypeptides on the ribosome without GTP cleavage, *J. Mol. Biol.* **108**:609–613.

Belitsina, N. V., Tnalina, G. Z., and Spirin, A. S. (1981) Template-free ribosomal synthesis of polylysine from lysyl-tRNA, *FEBS Letters* **131**:289–292.

Benhar, I., and Engelberg–Kulka, H. (1993) Frameshifting in the expression of the *E. coli trpR* gene occurs by the bypassing of a segment of its coding sequence, *Cell* **72**:121–130.

Bodley, J. W., Zieve, F. J., and Lin, L. (1970) Studies on translocation. IV. The hydrolysis of a single round of GTP in the presence of fusidic acid, *J. Biol. Chem.* **245**:5662–5667.

Chetverin, A. B., and Spirin, A. S. (1982) Bioenergetics and protein synthesis, *Biochim. Biophys. Acta* **683**:153–179.

Cundliffe, E. (1980) Antibiotics and prokaryotic ribosomes: Action, interaction, and resistance, in *Ribosomes: Structure, Function, and Genetics* (G. Chambliss, G. R. Craven, J. Davies, K. Davis, L. Kahan, and M. Nomura, eds.), pp. 555–581. University Park Press, Baltimore.

Cundliffe, E. (1990) Recognition sites for antibiotics within rRNA, in *The Ribosome: Structure, Function, and Evolution* (W. E. Hill, A. Dahlberg, R. A. Garrett, P. B. Moore, D. Schlessinger, and J. R. Warner, eds.), pp. 479–490, ASM Press, Washington, DC.

Gale, E. F., Cundliffe, E., Reynolds, P. E., Richmond, M. H., and Waring, M. J. (1981) The molecular basis of antibiotic action, John Wiley and Sons, London.

Gavrilova, L. P., Kostiashkina, O. E., Koteliansky, V. E., Rutkevich, N. M., and Spirin, A. S. (1976) Factor-free ('non-enzymic') and factor-dependent systems of translation of polyuridylic acid by *Escherichia coli* ribosomes, *J. Mol. Biol.* **101**:537–552.

Gesteland, R. F., Weiss, R. B., and Atkins, J. F. (1992) Recoding: reprogrammed genetic decoding by special sequences in mRNAs, *Science* **257**:1640–1641.

Girbes, T., Vazquez, D., and Modolell, J. (1976) Polypeptide-chain elongation promoted by guanyl-5'-yl imidodiphosphate, *Eur. J. Biochem.* **67**:257–265.

Gupta, S. L., Waterson, J., Sopori, M. L., Weissman, S. M., and Lengyel, P. (1971) Movement of the ribosome along the messenger ribonucleic acid during protein synthesis, *Biochemistry* **10**:4410–4421.

Haenni, A.-L., and Lucas-Lenard, J. (1968) Stepwise synthesis of a tripeptide, *Proc. Natl. Acad. Sci. USA* **61**:1363–1369.

Huang, W. M., Ao, S. Z., Casjens, S., Orlandi, R., Zeikus, R., Weiss, R., Winge, D., and Fang, M. (1988) A persistent untranslated sequence within bacteriophage T4 DNA topoisomerase gene 60, *Science* **239**:1005–1012.

Inoue–Yokosawa, N., Ishikawa, C., and Kaziro, Y. (1974) The role of guanosine triphosphate in translocation reaction catalyzed by elongation factor G, *J. Biol. Chem.* **249**:4321–4323.

Lucas–Lenard, J., and Haenni, A.-L. (1969) Release of transfer RNA during peptide chain elongation, *Proc. Natl. Acad. Sci. USA* **63**:93–97.

Matzke, A. J. M., Barta, A., and Kuechler, E. (1980) Mechanism of translocation: Relative arrangement of tRNA and mRNA on the ribosome, *Proc. Natl. Acad. Sci. USA* **77**:5110–5114.

Modolell, J., Girbes, T., and Vazquez, D. (1975) Ribosomal translocation promoted by guanylylimido diphosphate and guanylyl-methylene diphosphate. *FEBS Letters* **60**:109–113.

Muto, A., Sato, M., Tadaki, T., Fukushima, M., Ushida, C., and Himeno, H. (1996) Structure and function of 10Sa RNA: *Trans*-translation system, *Biochimie* **78**:985–991.

Noller, H. F., Moazed, D., Stern, S., Powers, T., Allen, P. N., Robertson, J. M., Weiser, B., and Triman, K. (1990) Structure of rRNA and its functional interactions in translation, in *The Ribosomes: Structure, Function, and Evolution* (W. E. Hill, A. Dahlberg, R. A. Garrett, P. B. Moore, D. Schlessinger, and J. R. Warner, eds.), pp. 73–92, ASM Press, Washington, DC.

Pestka, S. (1969a) Studies on the formation of transfer ribonucleic acid-ribosome complex. VI. Oligopeptide synthesis and translocation on ribosomes in the presence and absence of soluble transfer factors, *J. Biol. Chem.* **244**:1533–1539.

Pestka, S. (1969b) Translocation, aminoacyl-oligonucleotides, and antibiotic action, *Cold Spring Harbor Symp. Quant. Biol.* **34**:395–410.

Rich, A. (1974) How transfer RNA may move inside the ribosome, in *Ribosomes* (M. Nomura, A. Tissières, and P. Lengyel, eds.), pp. 871–884, Cold Spring Harbor Laboratory, Cold Spring Harbor, New York.

Riddle, D. L., and Carbon, J. (1973) Frameshift suppression: A nucleotide addition in the anticodon of a glycine transfer RNA, *Nature New Biol.* **242**:234.

Rodnina, M. V., Savelsbergh, A., Katunin, V. I., and Wintermeyer, W. (1997) Hydrolysis of GTP by elongation factor G drives tRNA movement on the ribosome, *Nature* **385**:37–41.

Spirin, A. S. (1978) Energetics of the ribosome, *Progr. in Nucleic Acid Res. Mol. Biol.* **21**:39–62.

Spirin, A. S. (1985) Ribosomal translocation: Facts and models, *Progr. Nucleic Acid Res. Mol. Biol.* **32**:75–114.

Thach, S. S., and Thach, R. E. (1971) Translocation of messenger RNA and 'accommodation' of fMet-tRNA, *Proc. Natl. Acad. Sci. USA* **68**:1791–1795.

Traut, R. R., and Monro, R. E. (1964) The puromycin reaction and its relation to protein synthesis, *J. Mol. Biol.* **10**:63–72.

Trifonov, E. N. (1987) Translation framing code and frame-monitoring mechanism as suggested by the analysis of mRNA and 16S rRNA nucleotide sequences, *J. Mol. Biol.* **194**:643–652.

Webb, M. R., and Eccleston, J. F. (1981) The stereochemical course of the ribosome-dependent GTPase reaction of elongation factor G from *Escherichia coli*, *J. Biol. Chem.* **256**:7734–7737.

Weiss, R., Dunn, D., Atkins, J., and Gesteland, R. (1990) The ribosome's rubbish, in *The Ribosome: Structure, Function, and Evolution* (W. E. Hill, A. Dahlberg, R. A. Garrett, P. B. Moore, D. Schlessinger, and J. R. Warner, eds.), pp. 534–540, ASM Press, Washington, DC.

Wintermeyer, W., Lill, R., Paulsen, H., and Robertson, J. M. (1986) Mechanism of ribosomal translocation, in *Structure, Function, and Genetics of Ribosomes* (B. Hardesty and G. Kramer, eds.), pp. 523–540, Springer-Verlag, New York.

13

Elongation Rate and its Modulation

13.1. Elongation Rates in Prokaryotes and Eukaryotes

During the elongation stage of translation the elongation cycle, which consists of aminoacyl-tRNA binding, transpeptidation, and translocation (Fig. 9.1), repeats as many times as there are sense codons present in the coding sequence of a message. Thus, the cycle frequency directly determines the polypeptide elongation rate.

13.1.1. Transit Time

The time during which a growing nascent peptide remains attached to the translating ribosome, that is, the time of its elongation plus termination, is called *transit time*. If the termination time is neglected, the number of amino acids in a given protein divided by its transit time is the average rate of elongation on a corresponding mRNA. Hence, the transit time determinations for proteins of known size may give the information about the elongation rate (for review, see Nielsen and McConkey, 1980).

To measure the transit time, usually a radioactive amino acid is provided to a cell. The radioactivity will soon appear in polyribosomes in the form of growing nascent peptides attached to ribosomes. The radioactivity of the polyribosome fraction of the cell will increase as the nascent peptides on ribosomes are elongated by the labeled amino acids (Fig. 13.1). When all the growing peptides become fully labeled, the plateau of radioactivity of the polyribosome fraction will be established. The time to reach the plateau is the transit time.

In practice, it is often more convenient and accurate to determine the transit time from the kinetics of radioactivity incorporation into the ribosome-free fraction of the soluble (completed) proteins as compared with that of the total incorporation into polypeptides. The incorporation into soluble protein will increase first exponentially and then, after reaching the plateau in the polyribosome fraction, linearly at the expense of the release of fully labeled polypeptides. At the same time it is evident that the total incorporation of the radioactive amino acid into polypeptides should be linear almost from the beginning (provided the elongation rate is constant during the experiment). Since after reaching the radioactivity plateau in polyribosomes the increment of the radioactivity both in total peptides and in soluble protein fraction is determined by the same process of releasing labeled polypeptides from ribosomes, the two linear plots should be parallel (Fig. 13.1, *lower panel*). The distance between them along the abscissa corresponds to the time required for the completion of the synthesis of nascent peptides in polyribosomes, that is, to half transit time. (The *average* length of nascent peptides in a polyribosome is always *half* of the full length of completed polypeptides, that is, at each given moment a polyribosome contains half-completed polypeptides on average).

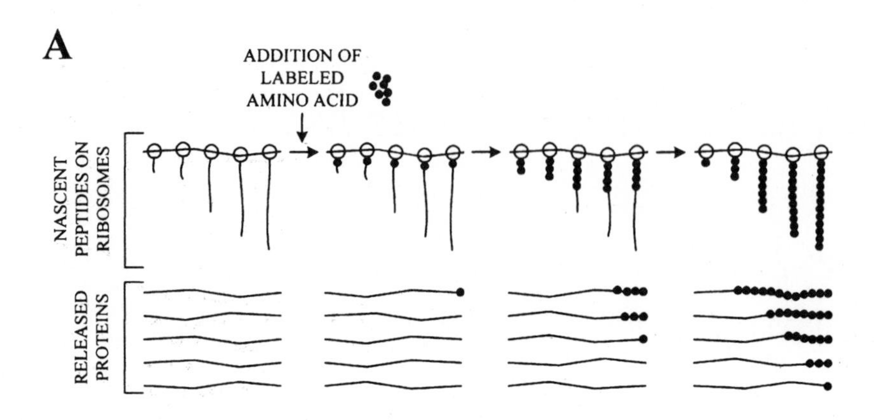

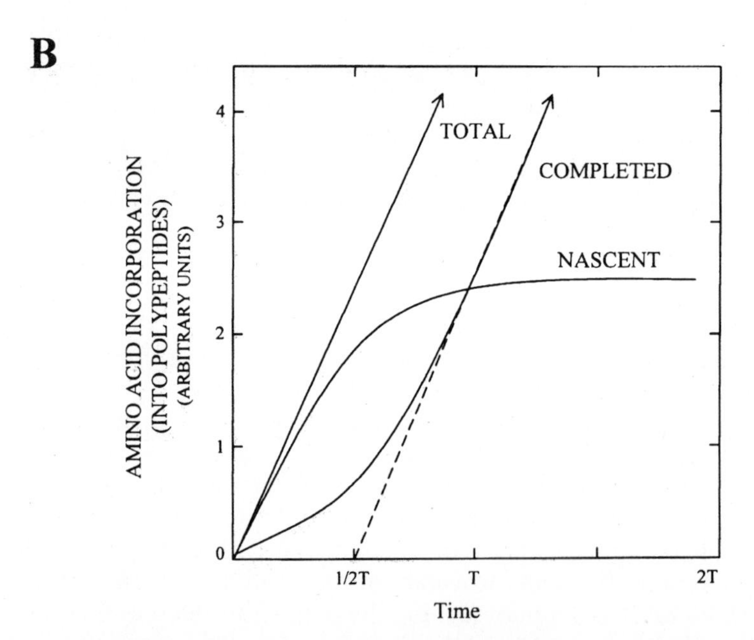

FIG 13.1. Transit time. Reproduced from Spirin, A. S., and Ryazanov, A. G. (1991) in *Translation in Eukaryotes* (H. Trachsel, ed.), pp. 325–350, CRC Press, Boca Raton, FL. (**A**) Schematic representation of the time-course of radioactivity accumulation in the fraction of growing nascent peptides (polyribosome fraction) and in the fraction of completed released proteins (supernatant fraction) after addition of radioactive amino acid. Note that after all the growing peptides are fully labeled the radioactivity in the polyribosome fraction becomes constant. (**B**) Theoretical kinetics curves of incorporation of radioactive amino acid into total polypeptides, completed released protein, and nascent polypeptides attached to ribosomes. T is transit time. Modified from Wu, R. S., and Warner, J. R. (1971) *J. Cell Biol.* **51:** 643–652.

By measuring the transit time, the rates of elongation of different polypeptides (average elongation cycle frequencies on different mRNAs) in both prokaryotic and eukaryotic cells, as well as in cell-free systems, can be estimated. In *E. coli,* up to 15 to 20 codons per second can be read out on some mRNAs at saturating substrate and factor concentrations (rich medium) at 37°C. The rate of ten codons per second is more typical of the bacteria growing in a poorer medium. The same rate of about ten codons per second can be achieved in a bacterial cell-free system under optimized conditions. Thus, the average time of the elongation cycle in bacterial systems varies usually from 0.05 sec to 0.1 sec at 37°C, and it is three times longer at 25°C.

The elongation rate in eukaryotic systems can reach ten codons per second at 37°C, but usually it is lower and varies widely due to the presence of control mechanisms regulating elongation. Typical durations of the eukaryotic elongation cycle range from 0.1 sec to 0.5 sec.

In principle, the speed (rate) of any step of the elongation cycle (aminoacyl-tRNA binding, transpeptidation, or translocation) may be regulated, but realistically aminoacyl-tRNA binding and translocation seem to be the most likely targets for control mechanisms. If the regulation affects a given step of all cycles more or less uniformly the *cycle frequency* will be altered and, hence, the general elongation rate will change. On the other hand, one or another step of specific elongation cycles along mRNA can be intervened; for example, aminoacyl-tRNA binding to a specific codon can be slowed down, or translocation through a specific site of mRNA can be impeded. In this case the elongation becomes uneven (discontinuous), and *translational pauses* may arise, resulting in a change in average elongation rate. Regulation of the cycle frequency and the translational pausing seem to be two main phenomena in the control of elongation rate.

13.1.3. Polyribosome Profile

Absolute rates of elongation are not necessarily required for studies of regulation at the elongation stage of translation. Often it is sufficient to know just relative changes of elongation rates. This can be done by recording a change in the profile of polyribosome distribution upon sucrose gradient centrifugation (Fig. 13.2).

The polyribosome profile depends on the rates of initiation, elongation, and termination. Generally, slowing down the movement of ribosomes along mRNA at a constant initiation rate will result in an increase of heavy polyribosome fraction (i.e., increase in the density of ribosomes on mRNA). It is obvious, however, that the increase in initiation rate at a constant elongation rate will give the same result. Hence, the analysis of polyribosome profiles by itself is not always sufficient to make judgments about changes in elongation rate, and the information about the rate of total translation (the rate of amino acid incorporation) is required.

Three clear cases can be considered: (1) The "light shift" in polyribosome profile (decrease in the number of ribosomes per mRNA) indicates that the elongation rate increases if the amino acid incorporation (translation rate) increases or does not change; (2) The "heavy shift" in polyribosomes indicates that the elongation rate decreases, if the amino acid incorporation goes down or does not change; and (3) the absence of shifts in polyribosome profile with simultaneous increase or de-

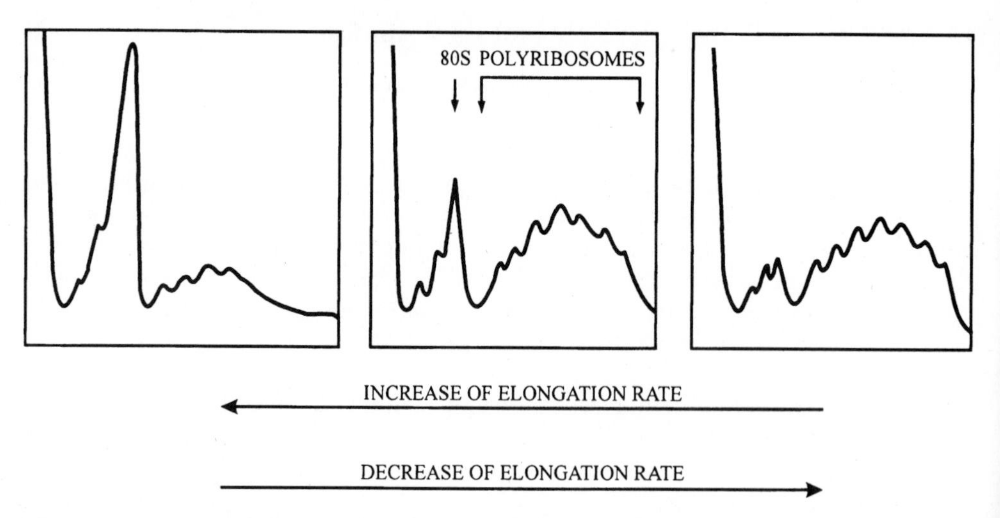

FIG 13.2. Diagrams of mono- and polyribosome distribution profiles upon sucrose gradient centrifugation. Reproduced from Spirin, A. S., and Ryazanov, A. G. (1991) in *Translation in Eukaryotes* (H. Trachsel, ed.), pp. 325–350, CRC Press, Boca Raton, FL.

crease of the amino acid incorporation will give evidence that the elongation rate increases or decreases, respectively.

13.2. Discontinuities in Elongation

As already mentioned, the cycle frequency cited above (Section 13.1.2) can be considered only as an average value for the entire elongation on a given mRNA. There is experimental evidence that the elongation rate may be not constant during translation of an mRNA chain. In other words, the ribosome can move at different speeds at different sites within mRNA.

13.2.1. Translational Pauses

It was reported that during synthesis of the globin chains in rabbit reticulocytes the size distribution of nascent peptides was not continuous but rather displayed discrete size classes, thus suggesting some discontinuities in the process of translation (during elongation) (Protzel and Morris, 1974). Similar results were obtained with the synthesis of bacteriophage MS2 coat protein in MS2 phage-infected *E. coli* cells (Chaney and Morris, 1979). Discontinuous translation was also observed in intact cells and in cell-free systems for a number of other proteins, including silk fibroin (Lizardi *et al.,* 1979), vitellogenin and serum albumin, preprolactin, preproinsulin, procollagen, encephalomyocarditis virus proteins, the large subunit of chloroplast ribulose-1,5-biphosphate carboxylase, chloroplast reaction center protein D1, tobacco mosaic virus proteins, and bacterial colicins (reviewed by Candelas *et al.,* 1987). All these observations have led to the conclusion that the ribosomes move along the mRNA chain during elongation at a non-uniform rate: from time to time, at a few specific sites of mRNA, they can slow down their advance or stop temporarily. In other words, more or less long *pauses* may occur during elongation, at least in some cases.

Pauses at specific sites during polypeptide elongation may play an important role in cotranslational folding of proteins, cotranslational assembly of larger protein or protein-membrane complexes, and cotranslational transmembrane transport (Kim, Klein, and Mullet, 1991; Krasheninnikov, Komar, and Adzhubei, 1991; Komar and Jaenicke, 1995; Young and Andrews, 1996). In the process of cotranslational protein folding, especially in the cases of multidomain proteins, the correct folding of an N-proximal part or domain on the ribosome may require some time in the absence of interactions with a following section of a nascent polypeptide. The pause after the completion of a structurally autonomous or semi-autonomous portion of a nascent polypeptide would provide the delay in producing a next portion of unfolded chain and thus the time necessary for undisturbed self-folding of the completed portion on the ribosome. Pausing at interdomain borders during synthesis of multidomain proteins has been discussed repeatedly. Pausing during elongation of chloroplast reaction center protein D1 is an excellent illustration of the correlation between the pause sites, on one hand, and the stages of cotranslational binding of cofactors such as chlorophyll to D1 and the cotranslational integration of D1 into thylakoid membranes, on the other (Fig. 13.3).

Another role of translational pauses at specific sites may be to provide the targets for regulation of protein synthesis at the level of elongation. First, pausing can be regulated and therefore an overall production of a protein can be modulated by increasing or decreasing the duration of a pause (see Section 13.3). Second, pausing during elongation can stimulate frameshifting at the pause site (Spanjaard and van Duin, 1988), similar to the frameshifting provoked by termination codons (see Section 12.4.2).

Three different mechanisms which may underlie the translational discontinuities (pauses) are discussed in the following sections.

13.2.2. Modulating Codons

The most often considered explanation of ribosome pausing is the shortage of certain aminoacyl-tRNAs; the shortage of these tRNAs can retard ribosome movement at the corresponding codons. In other words, the translating ribosome can be retarded at the codons corresponding to minor (present in small amounts) isoacceptor tRNA species. Analyses of codon usage in different mRNAs demonstrate that the mRNAs coding for abundant cellular proteins selectively use synonymous codons corresponding to isoacceptor tRNA species which are present in the cell in relatively large amounts (Ikemura, 1981). The synonymous codons corresponding to minor tRNA species are used rarely, if at all, in these mRNAs. The very rare codons in *E. coli* are the arginine codons CGA and CGG, as well as AGA and AGG; the isoleucine codon AUA; the leucine codon CUA; and the proline codon CCC. Also the serine codons UCA and UCG, the glycine codons GGA and GGG, the leucine codons UUA and UUG, and CUU and CUC, the proline codon CCU, the threonine codons ACA and ACG, and the serine codon AGU can be qualified as rare codons. It is possible that, when the translating ribosome encounters a rare codon within mRNA, it needs some time to wait for minor tRNA coming from the surrounding medium. In other words, low concentrations of some tRNA species may be a rate-limiting factor, resulting in translational pauses at the corresponding codons.

The rare codons, corresponding to minor tRNAs, can be called *modulating codons*, since they regulate the rate of translation (Grosjean and Fiers, 1982). The

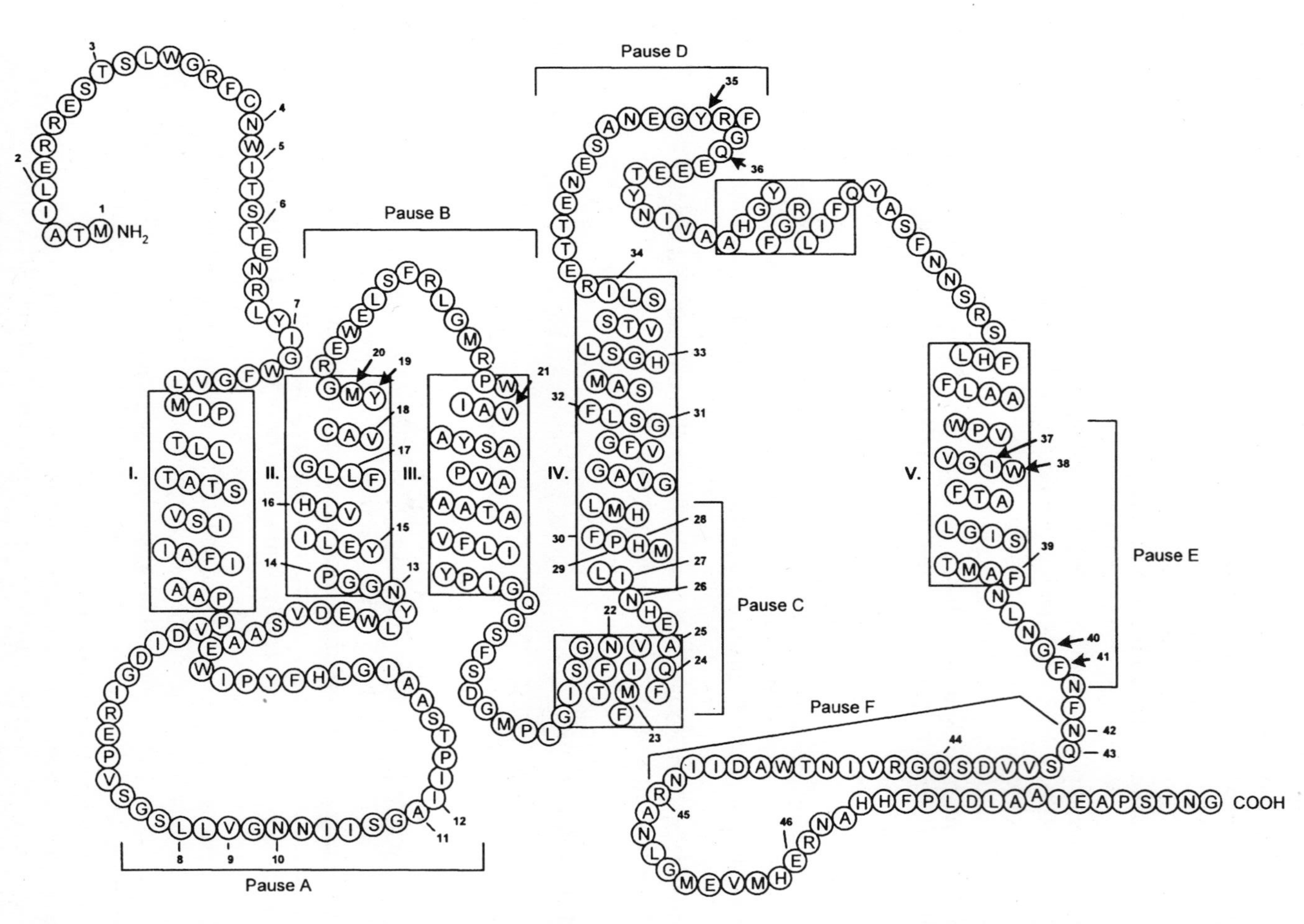

FIG 13.3. Location of ribosome pause sites along the amino acid sequence of membrane-bound chloroplast reaction center protein D1. The pause sites are numbered starting from the N terminus. Strong pause sites are designated by bold arrows, weaker pause sites with lines. Clusters of pause sites are denoted from A to F. α-Helical domains are boxed. The five vertical α-helices (I to V) correspond to membrane-spanning domains (see Chapter 18). The N-terminus of D1 is exposed to the stromal phase, whereas the C terminus is located in the lumen of the thylakoid. Reproduced from Kim, J., Klein, P. G., and Mullet, J. E. (1991) *J. Biol. Chem.* **266:** 14931–14938, with permission.

more modulating codons are contained within an mRNA, the less expression of this mRNA is expected. At the same time, the cell can selectively change the efficacy of expression (translation) of a weakly expressed mRNA through *adaptation* of isoacceptor tRNA concentrations to codon frequencies in the particular mRNA (Elska *et al.*, 1971; Garel, 1976). For example, it is known that during the massive synthesis of fibroin in silk glands of *Bombyx mori* the intracellular pool of tRNA isoacceptors undergoes very significant changes that optimize the availability of tRNA species required for fibroin mRNA translation; in particular, the glycine, alanine, and serine isoacceptor tRNA species become predominant, in accordance with the predominance of the corresponding glycine, alanine, and serine codons in the fibroin mRNA.

The use of silk fibroin mRNA as a message in a heterologous (rabbit reticulocyte) cell-free translation system demonstrated the effect of tRNA set on ribosome pausing: translating ribosomes were found to cease elongation at numerous points, and the full-length product can be observed only when tRNA from silk glands of *Bombyx mori* was added (Chavancy *et al.*, 1981). Since in the silk gland there is a proportionality between the intracellular level of different tRNAs and the amino acid composition of fibroin, as well as a proportionality between the distribution of isoacceptor tRNAs and the frequencies of corresponding synonymous codons in fibroin mRNA, the limitation of certain aminoacyl-tRNAs when the heterologous tRNA was used is the most evident cause of ribosomal pausing.

In a number of *in vivo* and *in vitro* experiments, it was shown, however, that the occurrence of single rare codons along the mRNA chain often does not result in significant ribosomal pausing. A prominent pausing effect is observed when two or more rare codons recognizable by the same minor tRNA species are in tandem (Chen and Inouye, 1990; Rosenberg *et al.*, 1993). It is understandable: the first rare codon binds its cognate minor aminoacyl-tRNA in the A site, and then the codon and the tRNA together are translocated to the P site, so that the ribosome drains the immediate surroundings of the minor aminoacyl-tRNA while the next cognate rare codon is exposed in the A site. As a result, the ribosome is waiting for a rarer chance to meet the minor aminoacyl-tRNA, and hence the corresponding elongation cycle is delayed at the aminoacyl-tRNA binding step.

13.2.3. Structural Barriers along mRNA

The speed of ribosome movement along the mRNA chain may be uneven also because of different stabilities of secondary and tertiary structures at different regions of mRNA. In particular, in order to have a stable structured region of mRNA unwound and open for translation, the ribosome requires more waiting time than during translation of other, less structured regions of mRNA. In such cases discontinuities in elongation will arise. Thus, the sites of the translational pauses will correspond to the entry of ribosomes at stable double helices, pseudoknots, and tertiary interactions. This possibility is consistent with predictions for secondary structures of mRNAs encoding some discontinuously synthesized proteins (see, e.g., Chaney and Morris, 1979).

In this connection, an observation made by Svitkin and Agol (1983) is of special interest. They have demonstrated that during translation of encephalomyocarditis virus RNA, a marked translational barrier causing a significant delay in the time of expression of the subsequent coding sequence exists at a specific site on the RNA. It is remarkable that this barrier can be overcome by addition of eEF2, in excess of the catalytic amounts of eEF2 required for the maximal rates of elongation on the different RNA regions. Hence, the eukaryotic EF2 may possess some regulatory functions in elongation, in addition to the catalysis of translocation. It is possible that this function is related to the RNA-binding capability of eEF2 (absent in the case of prokaryotic EF-G); it is tempting to believe that the RNA-binding capability of eEF2 may serve for unwinding or destabilization of some structural barriers in mRNA, such as stable double-stranded helices or special tertiary interactions.

In the case of globin mRNA translation the distribution of pauses was found to be independent of the tRNA concentrations, as well as of the concentration of the elongation factors. Essentially the same pattern of pauses on globin mRNA was demonstrated in intact rabbit reticulocytes, reticulocyte cell-free translation systems, and wheat germ cell-free translation systems. It is mRNA secondary structure that was suggested to cause the ribosomal pausing in this case (Protzel and Morris, 1974; Chaney and Morris, 1979).

It should be mentioned that, in principle, elongation rate can be affected not only by mRNA secondary and tertiary structure *per se* (e.g., double-helical hairpins or pseudoknots), but also by mRNA-associated proteins which may hinder the elongation at specific sites either directly, or by stabilizing a local secondary or tertiary structure of mRNA.

13.2.4. Inhibitory Nascent Peptides

The possibility of an inhibitory action of a nascent peptide on elongation was demonstrated long ago, when poly(A) was used as a message for the ribosomal synthesis of polylysine in a cell-free system. This proved that after reaching the polypeptide length of about six to seven lysyl residues the ribosome discontinues the elongation, independent of the length of the poly(A) message (Pulkrabek and Rychlik, 1968). Hence, the appearance of an oligolysine stretch of this size at the ribosomal PTC brings the ribosome to a halt. It is likely that the effect is caused by an interaction of the polycationic stretch with rRNA of the PTC.

More recently natural nonbasic peptide sequences also capable of inducing ribosome stalling were found. For example, the synthesis of pentapeptide MetValLysThrAsp or octapeptide MetSerThrSerLysAsnAlaAsp on the *E. coli* ribosome resulted in the cessation of further elongation (Gu *et al.*, 1994). The peptides with these sequences were shown to specifically interact with the rRNA regions of the ribosomal PTC (domains IV and V of the 23S rRNA, see Fig. 6.5B). It is believed that the appearance of such inhibitory peptide stretches during elongation may be the cause of some translational pauses.

A translational pause can be often observed after a leader peptide sequence emerges from the ribosome. For example, such a pause, at the 75th amino acid (the glycine codon GGC), occurs during the synthesis of bovine preprolactin in both wheat germ and rabbit reticulocyte cell-free translation systems (Wolin and Walter, 1988). Limitations in the aminoacyl-tRNA and structural barriers on mRNA are not likely in this case, and a probable cause of the pause is a nascent peptide structure.

In addition to a direct effect of a nascent peptide on the ribosome, some regulatory proteins or small ribonucleoproteins may exist which interact with the nascent peptide of the translating ribosome and selectively stop or impede elongation at defined points. One example of such specific repressors of elongation in ıkaryotes is the 7S RNA-containing ribonucleoprotein particle, the so-called SRP ignal recognition particle) which recognizes a special N-terminal hydrophobic equence of nascent polypeptide on the translating ribosome, attaches itself to the bosomes, and stops further elongation, until the ribosome interacts with the en- oplasmic reticulum membrane (see Section 18.2). It is interesting that the addi- on of SRP to the system where the synthesis of preprolactin was performed in- eased the ribosome pausing specifically at glycine 75 (Wolin and Walter, 1989); is implies that ribosome pausing can be affected by the nascent signal peptide. cannot be excluded that similar mechanisms are used for modulating the elon- ation rate at some other phases of protein synthesis, for example, at some steps f protein folding or protein assembly on the translating ribosome.

3.3. Selective Regulation of Elongation on Different mRNAs

In eukaryotes differential changes in elongation rates on different mRNAs eem to be possible during cell development, as well as in response to hormone eatment or environmental influences. One example is the different elongation ates of two populations of mRNA after heat shock in *Drosophila* cells: elongation n the abundant pre-heat-shock species slows down, while it is fast on the heat- ıock mRNA species (Ballinger and Pardue, 1983). During heat shock of chicken eticulocytes the synthesis of the 70-kDa heat-shock protein (HSP 70), but not glo- in, increased several times due to specific increase in the elongation rate of the SP 70 nascent peptide (Theodorakis, Banerji, and Morimoto, 1988).

Another example of regulated differential changes in the elongation rates was rovided by Ilan and associates (Gerke, Bast, and Ilan, 1981): injection of estrogen ıto chickens resulted in the induction of vitellogenin synthesis in the liver with ıe elongation rate reaching 9 residues per second, while the elongation rate of to- l liver proteins decreased from 7 to 4.5 residues per second and that of serum al- umen was just 2 amino acids per second. A very selective effect was observed in ıltured hepatoma cells: their exposure to dibutyryl cyclic AMP resulted in no ıange in the elongation rate of total protein (about 2 amino acids per second) and the same time a significant stimulation of the elongation rate for tyrosine amino- ansferase (to 10 amino acids per second). In rat liver the elongation rate for or- ithine aminotransferase was shown to be stimulated by the administration of a igh-protein diet, while the elongation rate for total protein was somewhat re- uced.

Thus, differential effects on reading the coding sequences of different mRNAs ın be observed, and specific mRNAs can be regulated selectively. In some case ıe effects can be opposite for different mRNAs of the same cell.

The mechanisms of selective regulation of elongation rates on different RNAs in eukaryotes are unknown. One possibility is based on the fact that the longation rate is non-uniform along mRNA. The number and duration of riboso- ıal pauses are different in various mRNAs, and some of the pauses can be more ensitive to regulation than others. Possible changes in the isoacceptor tRNA pop- lation, as well as changes in concentrations or activities of elongation factors and RSases, could contribute to the selective regulation of elongation, particularly ırough the regulation of pausing time. The participation of hypothetical repres- ɔrs of elongation, such as proteins capable of binding to coding regions of mRNA, annot be excluded either.

Another possibility is that different mRNAs, tRNAs, elongation factors, ARSases, and RNA-binding proteins may be distributed non-uniformly in the cell. A nonrandom distribution of some mRNAs in oocytes (Rebagliati *et al.*, 1985) and the mRNAs encoding for cytoskeleton proteins in fibroblasts (Lawrence and Singer, 1986) can be taken as examples. There is also experimental evidence of non-uniform distribution of elongation factors and ARSases in the cytoplasm of eukaryotic cells (for review, see Ryazanov, Ovchinnikov, and Spirin, 1987). These proteins possess an affinity for polyribosomes and are compartmentalized on them. The dynamic compartmentation of mRNAs, elongation factors, and ARSases suggests the possibility that some polyribosomes, together with associated translational machinery proteins, could be physically separated from other polyribosomes in the same cell, providing a mechanism for differential regulation of elongation rate on various mRNAs.

13.4. Total Regulation of Elongation

13.4.1. Overall Changes in Elongation Rate

The rate of polypeptide chain elongation (i.e., the speed of ribosome movement) *on all translatable cellular mRNAs at the same time* can be slowed down or stimulated at some critical points of cell life in eukaryotes. Among such critical points are egg fertilization, mitosis, transition from quiescence to proliferation and *vice versa,* turning points in cell differentiation, functional activation of differentiated cells, and viral infection.

Many examples where elongation rates for total mRNA were measured illustrate this phenomenon of total regulation of elongation (reviewed by Spirin and Ryazanov, 1991). For example, the fertilization of sea urchin eggs was reported to induce a twofold increase in the total elongation rate. Protein synthesis during mitosis is severely inhibited, at least partly by slowing down the elongation rate at anaphase; this is accompanied by an accumulation of large polyribosomes. A classical example of enhancement of the overall rate of polypeptide elongation is the case of serum stimulation of mammalian cell cultures. During yeast-to-hyphae morphogenesis in the fungus *Mucor racemosus,* a fourfold rise in the elongation rate was observed. The elongation rate on most cellular mRNAs was shown to be reduced in *Drosophila* tissue culture cells upon heat shock. Intraperitoneal glucagon administration produced temporary inhibition of the total elongation rate in rat liver. The injection of estradiol into cockerels was demonstrated to induce a drop in the elongation rate on almost all mRNAs in liver cells.

Virus infection of the eukaryotic cell also can lead to a decline in the rate of elongation of all polypeptides synthesized. This phenomenon of total slowing down of elongation is especially typical for picornavirus infections; it has been reported for poliovirus-infected HeLa cells, mengovirus-infected mouse Ehrlich ascites tumor cells, and encephalomyocarditis virus-infected L cells. Inhibition of elongation has been also reported for vaccinia virus-infected cells, and this effect can be reproduced *in vitro,* in a cell-free system, by addition of vaccinia virus cores. This observation suggests the direct effect of a virus component on the elongation machinery participants.

It should be mentioned that the elongation rate in general can be regulated by nonspecific factors, such as intracellular ionic conditions, pH, and temperature. In particular, the concentration of K^+ was found to be important specifically for elongation: reduced K^+ concentration inhibited predominantly the elongation stage but not the initiation in reticulocyte cells and cell extracts. The increase in intracellular level of Ca^{2+} in the eukaryotic cell also results in inhibition of elongation rate, along with several other effects. The shift of pH from 7.5 to 7.0 leads to a significant deceleration of elongation. In all cases the elongation strongly depends on temperature.

There are grounds to believe that in most of the cases discussed here, the elongation rate is regulated through changes in the activity of elongation factors.

13.4.2. Phosphorylation of Elongation Factor 2

In various mammalian tissue extracts, such as extracts from rabbit reticulocytes or rat liver, eEF2 is the major phosphorylatable protein (Ryazanov, 1987). Phosphorylation of eEF2 is catalyzed by a specific Ca^{2+}/calmodulin-dependent protein kinase (Nairn and Palfrex, 1987) now called eEF2 kinase. The eEF2 kinase, a monomeric 100-kDa protein, was found in virtually all mammalian protein-synthesizing cells and tissues, as well as in a number of invertebrate tissues. The only identified substrate of the kinase is eEF2. The kinase phosphorylates threonine residues located in the N-terminal part of eEF2. The primary phosphorylation site in eEF2 *in vivo* is Thr56, but two other threonine residues, Thr58 and, to a lesser extent, Thr53, can be also phosphorylated, especially during *in vitro* incubation (for reviews, see Spirin and Ryazanov, 1991; Nairn and Palfrey, 1996).

Phosphorylation of eEF2 makes it inactive in protein synthesis (Ryazanov, Shestakova, and Natapov, 1988). In experiments with synthetic and natural messages, the rate of polypeptide synthesis in cell-free translation systems was found to correlate with the fraction of nonphosphorylated eEF2. Protein phosphatase of the 2A type dephosphorylates eEF2 and correspondingly reactivates it. Thus, the fraction of phosphorylated inactive eEF2 depends on the balance of activities of the eEF2 kinase and the 2A-type phosphatase in a cell or cell extract. Okadaic acid inhibits the phosphatase and therefore causes the increase in the fraction of phosphorylated inactive form of eEF2, resulting in inhibition of elongation. The phosphorylation of eEF2 by eEF2 kinase can be suppressed by cAMP, reduced glutathione, an increased pH, and some other factors known as stimulators of protein synthesis in eukaryotes.

Phosphorylation of just a single threonine residue, namely Thr56, has been found to be sufficient to inactivate eEF2. This threonine residue is located in the so-called effector loop of eEF2 (see Section 12.2.1), which is most likely involved in the interaction of eEF2 with the ribosome. At the same time it has been found that phosphorylated eEF2 can form complexes with GTP and ribosomes, but is unable to catalyze the translocation reaction (Ryazanov and Davydova, 1989). Either the binding of phosphorylated eEF2 to the ribosome is not strong enough to induce translocation (it is in fact weaker than that of nonphosphorylated eEF2), or the binding is topographically incorrect.

It is likely that changes of eEF2 phosphorylation *in vivo* under different conditions and effectors provide a direct mechanism for the regulation of the elongation rate (for review, see Spirin and Ryazanov, 1991). As the eEF2 kinase is a Ca^{2+}/calmodulin-dependent enzyme, all hormones and events that increase Ca^{2+} concentration in the cytoplasm must lead to an increase in eEF2 phosphorylation. For example, growth-arrested human fibroblasts respond to mitogenic stimulation with a rapid transient increase in free Ca^{2+} concentration in the cytoplasm as well as with transient increase in the phosphorylation of eEF2. Similar transient increases in eEF2 phosphorylation were demonstrated in the case of thrombin and histamine treatment of human umbilical vein endothelial cells. In experiments with intact rabbit reticulocytes it has been found that the elevation of intracellular Ca^{2+} concentration after treatment of the cells with a Ca^{2+} ionophore results in strong inhibition of the elongation.

Transient decrease in the eEF2 phosphorylation was observed during differentiation of rat pheochromacytoma cells (PC-12 cells) induced by nerve growth factor; the decrease was shown to be caused by inactivation of eEF2 kinase. This example indicates that phosphorylation of eEF2 can be important for regulation of cell differentiation.

When changes in the level of phosphorylated eEF2 were investigated throughout the cell cycle, a dramatic increase in the amount of phosphorylated eEF2 was found in mitosis, and specifically at anaphase. It is preceded by the transient increase in Ca^{2+} concentration during the metaphase–anaphase transition. The result is the inhibition of elongation rate at anaphase.

It should be mentioned that the elevation of Ca^{2+} concentration in the cytoplasm, as a rule, is of a transient nature; in most cases it lasts for seconds and rarely for minutes. Consequently, one can expect that the phosphorylation of eEF2 *in vivo* and the resultant inhibition of protein synthesis should also be transient (about several minutes). Such a transient eEF2 phosphorylation in response to a short Ca^{2+} spike is generally observed in the cases of the transition of quiescent cells into a proliferative state (G_0–G_1 transition), the activation of neurons, and the activation of several types of secretory cells. Hence, mitogenic stimulation or hormonal activation of the cells must be accompanied by transient inhibition of protein synthesis (elongation rate). This inhibition precedes the long-term stimulation of protein synthesis in the stimulated or activated cells. The role of the transient eEF2 phosphorylation and the resultant temporary arrest of protein synthesis may be to switch the cell from one pattern of synthesized proteins to another (reprogramming the cell). The short cessation of translation could be a mechanism for disrupting a process which maintains the cell in the quiescent or inactive state. The most straightforward idea in the case of quiescent cells is that the quiescent state is maintained by the continuous synthesis of short-lived proteins which prevent proliferative events. For example, transcription of immediate-early genes in quiescent cells may be blocked by a short-lived repressor. If this is the case, the protein synthesis inhibition will lead to the disappearance of such a repressor and consequently to the activation of transcription of these genes (Fig. 13.4). Similarly, short-lived proteins may exist which make certain "proliferative" mRNAs unstable in quiescent cells; transient inhibition of protein synthesis will result in stabilization and increased concentration of these mRNAs. Transient inhibition of elongation can also activate translation of some "weak" mRNAs (see Section 17.3.3).

13.4.3. Modifications of Elongation Factor 1

There are numerous observations that eEF1 activity can vary depending on cultivating conditions, hormone action, and aging. The changes in overall protein synthesis rate have been demonstrated to correlate with eEF1 activity. Such a correlation has been reported for serum-stimulated mammalian cells in culture, regenerating tissues, spleen during immune response, toad-fish liver during cold acclimation, sea urchin eggs after fertilization, fungus spores during germination, a well as for mammalian and insect organs and cells during aging (reviewed by Spirin and Ryazanov, 1991). However, despite all these indications of the involvement of eEF1 in regulation of the elongation rate the nature of eEF1 modifications responsible for the activity changes has not been determined.

Concerning the known types of covalent modifications of eEF1, methylation of several lysine residues is typical of this protein. In the case of fungus *Mucor racemosus* the rise of eEF1α methylation correlated with the increase in the overall elongation rate upon yeast-to-hyphae transition, and reduction of methylation accompanied the decrease in eEF1 activity during spore germination. Phosphorylation of eEF1 has been also demonstrated both *in vitro* and *in vivo*. However, no clear effect of the phosphorylation on the factor activity and the elongation rate has been revealed. Finally, eEF1A from various mammalian cells and tissues has been shown to be ethanolaminated at two specific glutamic acid residues. The role of this covalent modification in the eEF1 function is also unknown. Thus, while there is evidence that eEF1 activity can be involved in the regulation of the elongation rate, and several covalent modifications of eEF1 are demonstrated, no convincing information clarifying the interrelationship between these two groups of facts is available.

13.5. Elongation Toxins

Several protein toxins of bacterial, fungal, and plant origin have turned out to be powerful inhibitors of the eukaryotic protein-synthesizing systems. It is the elongation phase of translation that is blocked by these toxins. All of these toxins possess the catalytic (enzymatic) mechanism of action. Their targets are either eEF2 or the factor-binding site of the eukaryotic ribosome.

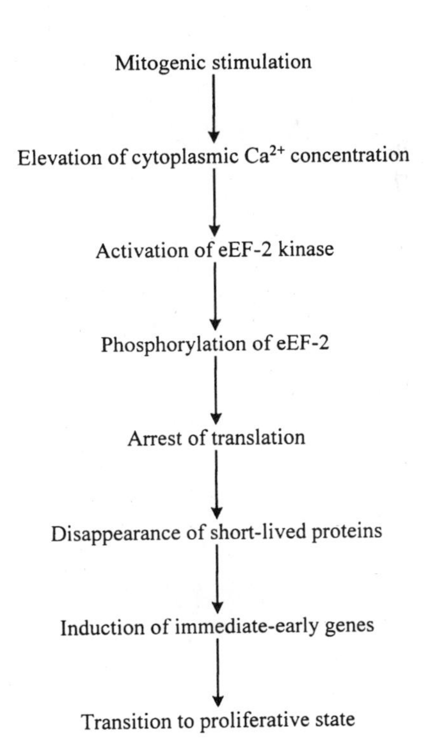

FIG 13.4. A scheme explaining how mitogens can induce the transition of the quiescent cell into the proliferative state through eEF2 phosphorylation. Ryazanov, A. G., and Spirin, A. S. (1993) in *Translational Regulation of Gene Expression 2* (J. Ilan, ed.), pp. 433–455, Plenum Press, New York.

13.5.1. Diphtheria Toxin

Diphtheria toxin (for reviews, see Collier, 1975; Pappenheimer, 1977; van Heyningen, 1980), a protein with a molecular mass of about 60,000 daltons, is secreted by *Corynebacterium diphtheriae* cells carrying a lysogenic bacteriophage β; the protein is encoded by one of the phage genes and not by the bacterial genome. The molecule is a covalently continuous polypeptide chain arranged in at least two relatively independent globular domains (A and B); the domains are additionally connected by a disulfide bridge. The C-terminal domain B, with a molecular mass of about 39 kDa, has a lectin-like action: it is capable of specific binding with a surface receptor of animal cells. The binding of the protein to the cell surface results in the following series of events. The protein enters the cytoplasmic membrane, and then the interdomain peptide bond is proteolytically cleaved, concomitantly with the disulfide bridge reduction. As a consequence, two fragments, A and B, are formed from the original protein. The N-terminal fragment A, with a molecular mass of 21,150 daltons, then enters the cytoplasm. It is this fragment that serves as an inhibitor of protein synthesis in the cell. The fragment is a highly specific enzyme performing ADP-ribosylation of just one amino acid residue in eEF2. Such a modification impairs the normal functioning of eEF2. Since the fragment A possesses catalytic action, one toxin molecule is sufficient to modify all the eEF2 molecules and therefore to kill the cell.

It should be pointed out that the original molecules of diphtheria toxin do not have such an inhibitory effect on protein synthesis; the initial toxin is a zymogen, which is converted into a catalytically active protein (A fragment) only after its cleavage. On the other hand, the fragment A by itself does not possess cytotoxic action since it cannot penetrate into the intact cell.

Nicotinamide adenine dinucleotide (NAD) serves as the donor of the ADP-ribose residue, which is enzymatically transferred to eEF2:

$$\text{eEF2} + \text{NAD} \rightarrow \text{ADP-ribosyl-eEF2} + \text{nicotinamide}$$

The reaction is reversible and, *in vitro,* in conditions of nicotinamide excess, the protein can be de-ADP-ribosylated, the intact eEF2 thereby being recovered.

The aminoacyl residue onto which the ADP-ribose is transferred is a unique histidine derivative, the so-called diphthamide (Fig. 13.5). The ADP-ribosyl residue is transferred to the nitrogen of the imidazole ring (designated by an arrow in Fig 13.5). The diphthamide residue in eEF2 corresponds to His573 in bacterial EF-G which is located at the tip of the tRNA-like domain IV, in the loop connecting β-strand 7 with α-helix B (Fig. 12.2). Hence, upon eEF2 binding to the eukaryotic ribosome, after translocation of tRNA residue from the A site, the diphthamide residue could be found in the A site on the small ribosomal subunit, possibly in the region of codon–anticodon interaction. The appearance of a mas-

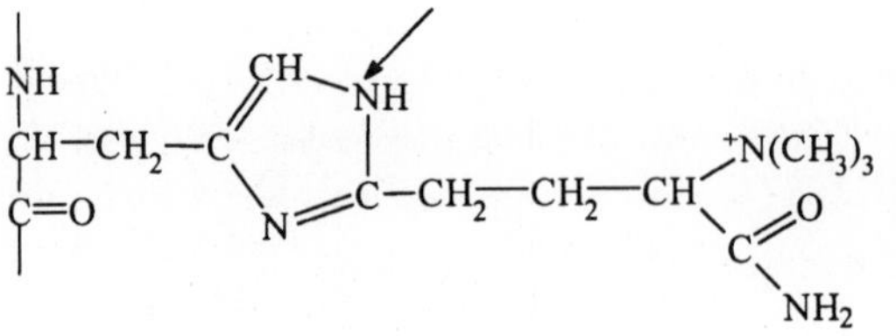

FIG 13.5. Diphthamide residue in eEF2 (in position equivalent to His573 in domain IV of bacterial EF-G). The arrow indicates the site of ADP-ribosylation.

sive charged group as a result of ADP-ribosylation of the diphthamide residue at such a marked site of eEF2 should exert an effect.

However, the mechanism of protein synthesis inactivation resulting from the ADP-ribosylation of diphthamide is still not quite clear. After ADP-ribosylation, the eEF2 is still capable of interacting with GTP as an effector, and the ADPR-eEF2:GTP complex can bind with the ribosome. The binding occurs at the same factor-binding site of the large ribosomal subunit. The interaction of the modified factor plus GTP with the ribosome is accompanied by GTPase activity. At the same time, it has been reported that in the case of ADP-ribosylated eEF2 the affinity of the factor for pre-translocation-state ribosomes is reduced (Nygard and Nilsson, 1985). It may be thought that the modification switches off domain IV of eEF2: this tRNA-like domain becomes incapable of interacting with the A site on the small ribosomal subunit. At least two consequences may be expected: (1) the total affinity of eEF2 to the ribosome may be lowered and (2) the catalysis of translocation may be inefficient, particularly due to the reversibility of the movement of the translocated tRNA residue between the intermediate translocation state and pre-translocation state (return to the vacant A site).

Exotoxin A of *Pseudomonas aeruginosa* has an action mechanism similar to that of diphtheria toxin (van Heyningen, 1980). This protein, with a molecular mass of 71,500 daltons, also interacts with the surface of the eukaryotic cell through its lectin domain. After the protein enters the membrane, it is cleaved to yield fragments A and B, with molecular masses of 27,000 daltons and 45,000 daltons, respectively, and then fragment A passes into the cytoplasm. Fragment A is an enzyme transferring the ADP-ribose residue from NAD to the same diphthamide residue of eEF2 that was discussed previously. This results in inhibition of protein synthesis. There is, however, no immunological cross-reactivity between *Pseudomonas aeruginosa* toxin and diphtheria toxin; in addition, receptors for these two toxins on the cell membrane are different.

13.5.2. Shiga and Shiga-like Toxins

The toxin produced by *Shigella dysenteriae* is also a powerful inhibitor of protein synthesis in the cells of a number of vertebrate animals. This protein toxin consists of one polypeptide A chain, with a molecular mass of 30,500 daltons, and six (or seven) relatively short B chains, with a molecular mass of about 5,000 daltons each (A_1B_6). The B moiety of the protein appears to be responsible for the interaction of the toxin with the cytoplasmic membrane receptor of the animal cell and for the subsequent entry of the toxin into the membrane. Proteolytic cleavage of the toxin A chain in the membrane yields two fragments, A_1 (molecular mass 27,500 daltons) and A_2 (molecular mass 3,000 daltons); the A_1 fragment passes into the cytoplasm and inhibits protein synthesis. The inhibition results from the enzymatic activity of fragment A_1. (It should be pointed out that enzymatic activity is not displayed prior to cleavage.) The enzymatic action in this case is targeted at the 60S subunit of the eukaryotic 80S ribosome. The enzymatic fragment of the toxin proved to be a glycosidase specifically hydrolyzing the N-glycosidic bond of the adenosine residue at position 4324 in mammalian 28S ribosomal RNA (equivalent to A2660 of the *E. coli* 23S RNA) (Endo *et al.*, 1988). Ribosomes with modified 60S subunits are capable of performing transpeptidation; in other words, the PTC is not impaired. The inhibition of protein synthesis by the A_1 fragment of the toxin is due to the damage of the factor-binding site on the 60S ribosomal subunit.

The function of eEF1 in aminoacyl-tRNA binding seems to be impaired to a greater extent than the function of eEF2 in translocation (Obrig, Moran, and Brown, 1987; Furutani *et al.*, 1992).

Shiga-like toxins, called Vero toxins (VT1 and VT2), have been shown to be produced by some pathogenic strains of *E. coli*. The structure of the toxins is similar to that of the Shiga toxin, and the mechanism of their action is the same.

13.5.3. α-Sarcin

α-Sarcin is a toxin produced by the mold *Aspergillus giganteus* (for review, see Kao and Davies, 1995). It is a small basic protein with a molecular mass of about 16,000 daltons. The protein is an exceedingly potent inhibitor of protein synthesis in eukaryotic cells and cell extracts. It is a highly specific ribonuclease: the inhibition is the result of the hydrolysis of a single phosphodiester bond on the 3′-side of G4325 of mammalian 28S ribosomal RNA (equivalent to G2661 in the *E. coli* 23S rRNA, see Fig. 9.10A) (Endo and Wool, 1982). Since the target, a conservative helical structure in domain VI of 28S ribosomal RNA, is involved in the formation of the factor-binding site on the 60S ribosomal subunit (see Section 9.5), the interactions of eEF1 and eEF2 with the ribosome may be impaired. The predominant effect observed in experiments, however, is the inhibition of the binding of aminoacyl-tRNA:eEF1:GTP complex to the ribosome.

13.5.4. Plant Toxins

A number of powerful toxins that inhibit protein synthesis in target cells are found among plant lectins specifically interacting with the D-galactose residues of the glycoproteins present in the cell membrane of animal cells. These toxins include ricin from castor beans, *Ricinus communis;* abrin from *Abrus precatorius;* modeccin from *Modecca digitata;* and viscumin from the mistletoe, *Viscum album* (for review, see Olsnes and Pihl, 1982a). These proteins show a striking similarity to the bacterial toxins in terms of their molecular-functional organization.

Ricin is a two-subunit protein (glycoprotein) with a molecular mass of 62,000 daltons. The B-subunit (molecular mass 31,400 daltons) is a lectin in the strict sense of the word and is capable of binding with galactose residues on the external surface of the animal cell membrane. The A-subunit (molecular mass 30,000 daltons) is responsible for the inhibition of protein synthesis in the cytoplasm. The two subunits are linked by a disulfide bridge. The attachment of the toxin molecule to the membrane is followed by entry of the molecule into the membrane, disulfide bridge reduction, and release of the liberated A-subunit into the cytoplasm. The A-subunit possesses a specific N-glycosidase activity: it depurinates the adenosine residue at position 4324 in domain VI of 28S ribosomal RNA (equivalent to A2660 in the *E. coli* 23S rRNA, see Fig. 9.10A) (Endo *et al.*, 1987). As a result, the function of the factor-binding site on the 60S ribosomal subunit becomes damaged. Correspondingly, the binding of both aminoacyl-tRNA:eEF1:GTP complex and eEF2:GTP has been reported to be affected by ricin. At the same time, the impairment of the eEF1 interactions with the ribosome is more apparent under *in vitro* experimental conditions.

Other plant toxins are organized in a similar way and have similar action, although chemically they are different proteins. The so-called pokeweed antiviral protein (PAP) from *Phytalacca americana* is particularly interesting. Its molecular mass is 27,000 daltons and it is an analog of the A-subunit of ricin. Correspondingly, it does not possess lectin activity and cannot interact with the cell membrane. Therefore it does not affect intact cells but strongly inhibits protein synthesis *in vitro* in eukaryotic cell-free systems due to its specific N-glycosidase activity.

13.5.5. Artificial Chimeric Toxins

Thus, many protein toxins of bacterial and plant origin make use of the same principle of cytotoxic action based on a two-subunit or two-domain structure: one subunit (or fragment) interacts with the membrane and is responsible for the transmembrane transport, while the other is released into the cell and exhibits enzymatic activity there, resulting in the inhibitory modification of some components of the protein-synthesizing system. This principle observed in living nature may be exploited to deliver any enzyme protein inside the cell, if such a protein is artificially conjugated or cross-linked with a suitable membrane-interacting protein (for review, see Olsnes and Pihl, 1982b). For example, using a simple procedure of disulfide exchange, it was possible to conjugate the enzymatic A-fragment of diphtheria toxin or the A-subunit of ricin with a nontoxic plant lectin (e.g., with concanavalin A or lectin of *Wistaria floribunda*) and to obtain a cytotoxic effect; it is clear that the lectin moiety of the chimeric protein was responsible for the delivery of the inhibitory component into the cell. However, just as in the case of the original toxins, the effect was not tissue-specific.

High tissue specificity of such a chimeric toxin can be obtained if the A-fragment of diphtheria toxin or the A-subunit of ricin is conjugated with a peptide hormone interacting with the specific receptor of the cell membrane of a given type (e.g., with the chorionic gonadotropin, or epidermal growth factor, or insulin). In such cases, the enzymatic component is delivered into the cell, inhibits protein synthesis, and kills the cell. Furthermore, the membrane-interacting component may be an antibody (or its Fab fragment) against some surface antigen, which is specific for a membrane of only one cell type. Then, by conjugating the diphtheria toxin A-fragment or ricin A-subunit with such an antibody, an extremely tissue-specific chimeric toxin can be obtained which will selectively kill only specific target cells. When an antibody against the surface antigen of a tumor cell serves as the membrane-binding moiety of the chimeric toxin, such a toxin should selectively kill tumor cells without affecting other cell types.

References

Ballinger, D. G., and Pardue, M. L. (1983) The control of protein synthesis during heat shock in Drosophila cells involves altered polypeptide elongation rates, *Cell* **33**:103–114.

Candelas, G. C., Carrasco, C. E., Dompenciel, R. E., Arroyo, G., and Candelas, T. M. (1987) Strategies of fibroin production, in *Translational Regulation of Gene Expression* (J. Ilan, ed.), pp. 209–228, Plenum Press, New York.

Chaney, W. G., and Morris, A. J. (1979) Nonuniform size distribution of nascent peptides. The effect of messenger RNA structure upon the rate of translation, *Arch. Biochem. Biophys.* **194**:283–291.

Chavancy, G., Marbaix, G, Huez, G., and Cleuter, Y. (1981) Effect of tRNA pool balance on rate and uniformity of elongation during translation of fibroin mRNA in a reticulocyte cell-free system, *Biochimie* **63**:611–618.

Chen, G. T., and Inouye, M. (1990) Suppression of the negative effects of minor arginine codons on gene expression: preferential usage of minor codons within the first 25 codons of the *Escherichia coli* genes, *Nucleic Acids Res.* **18**:1465–1473.

Collier, R. J. (1975) Diphtheria toxin: Mode of action and structure, *Bact. Rev.* **39**:54–85.

Elska, A., Matsuka, G., Matiash, U., Nasarenko, I., and Semenova, N. (1971) tRNA and aminoacyl-tRNA synthetases during differentiation and various functional states of the mammary gland, *Biochim. Biophys. Acta* **247**:430–440.

Endo, Y., Mitsui, K., Motizuki, M., and Tsurugi, K. (1987) The mechanism of action of ricin and related toxic lectins on eukaryotic ribosomes. The site and the characteristics of the modification in 28S ribosomal RNA caused by the toxins, *J. Biol. Chem.* **262**:5908–5912.

Endo, Y., Tsurugi, K., Yutsudo, T., Takeda, Y., Ogasawara, T., and Igarashi, K. (1988) Site of action of a Vero toxin (VT2) from *Escherichia coli* 0157:H7 and of Shiga toxin on eukaryotic ribosomes, *Eur. J. Biochem.* **171**:45–50.

Endo, Y., and Wool, I. (1982) The site of action of alpha-sarcin on eukaryotic ribosomes. The sequence of the alpha-sarcin cleavage site in 28S ribosomal ribonucleic acid, *J. Biol. Chem.* **257**:9054–9060.

Furutani, M., Kashiwagi, K., Ito, K., Endo, Y., and Igarashi, K., (1992) Comparison of the modes of action of a Vero toxin (a Shiga-like toxin) from *Escherichia coli,* of ricin, and of alpha-sarcin, *Arch. Biochem. Biophys.* **293**:140–146.

Garel, J.-P. (1976) Quantitative adaptation of isoacceptor tRNAs to mRNA codons of alanine, glycine, and serine, *Nature* **260**:805–806.

Gehrke, L., Bast, R. E., and Ilan, J. (1981) An analysis of rates of polypeptide chain elongation in avian liver explants following *in vivo* estrogen treatment. II. Determination of the specific rates of elongation of serum albumin and vitellogenin nascent chains, *J. Biol. Chem.* **256**:2522–2530.

Grosjean, H., and Fiers, W. (1982) Preferential codon usage in prokaryotic genes: The optimal codon-anticodon interaction energy and the selective codon usage in efficiently expressed genes, *Gene* **18**:199–209.

Gu, Z., Harrod, R., Rogers, E. J., and Lovett, P. S. (1994) Anti-peptidyl transferase leader peptides of attenuation-regulated chloramphenicol-resistance genes, *Proc. Natl. Acad. Sci. USA* **91**:5612–5616.

van Heyningen, S. (1980) ADP-ribosylation by bacterial toxins, in *The Enzymology of Post-Translational Modification of Proteins* (R. B. Freedman and H. C. Hawkins, eds.), vol. 1, pp. 387–422, Academic Press, London.

Ikemura, T. (1981) Correlation between the abundance of *Escherichia coli* transfer RNAs and the occurrence of the respective codons in its protein genes, *J. Mol. Biol.* **146**:1–21.

Kao, R., and Davies, J. (1995) Fungal ribotoxins: a family of naturally engineered targeted toxins?, in *Frontiers in Translation* (A. T. Matheson, J. E. Davies, P. P. Dennis, and W. E. Hill, eds.), *Biochem. Cell Biol.* **73**:1151–1159.

Kim, J., Klein, P. G., and Mullet, J. E. (1991) Ribosomes pause at specific sites during synthesis of membrane-bound chloroplast reaction center protein D1, *J. Biol. Chem.* **266**:14931–14938.

Komar, A. A., and Jaenicke, R. (1995) Kinetics of translation of gamma-B crystallin and its circularly permutated variant in an *in vitro* cell-free system: Possible relations to codon distribution and protein folding, *FEBS Letters* **376**:195–198.

Krasheninnikov, I. A., Komar, A. A., and Adzhubei, I. A. (1991) Nonuniform size distribution of nascent globin peptides, evidence for pause localization sites, and a cotranslational protein-folding model, *J. Protein Chem.* **10**:445–453.

Lawrence, J. B., and Singer, R. H. (1986) Intracellular localization of messenger RNAs for cytoskeletal proteins, *Cell* **45**:407–415.

Lizardi, P. M., Mahdavi, V., Shields, D., and Candelas, G. (1979) Discontinuous translation of silk fibroin in a reticulocyte cell-free system and in intact silk gland cells, *Proc. Natl. Acad. Sci. USA* **76**:6211–6215.

Nairn, A. C., and Palfrey, H. C. (1987) Identification of the major M_r 100,000 substrate for calmodulin-dependent protein kinase III in mammalian cells as elongation factor 2, *J. Biol. Chem.* **262**:17299–17303.

Nairn, A. C., and Palfrey, H. C. (1996) Regulation of protein synthesis by calcium, in *Translational Control* (J. Hershey, M. B. Mathews, and N. Sonenberg, eds.), pp. 295–318, CSHL Press.

Nielsen, P. J., and McConkey, E. H. (1980) Evidence for control of protein synthesis in HeLa cells via the elongation rate, *J. Cell Physiol.* **104**:269–281.

Nygard, O., and Nilsson, L. (1985) Reduced ribosomal binding of eukaryotic elongation factor 2 following ADP-ribosylation. Difference in binding selectivity between polyribosomes and reconstituted monoribosomes, *Biochim. Biophys. Acta* **824**:152–162.

Obrig, T. G., Moran, T. P., and Brown, J. E. (1987) The mode of action of Shiga toxin on peptide elongation of eukaryotic protein synthesis, *Biochem. J.* **244**:287–294.

Olsnes, S., and Pihl, A. (1982a) Toxic lectins and related proteins, in *Molecular Actions of Toxins and Viruses* (P. Cohen and S. van Heyningen, eds.), pp. 51–105, Elsevier/North-Holland, Amsterdam.

Olsnes, S., and Pihl, A. (1982b) Chimeric toxins, *Pharmacol. Ther.* **15:**355–381.
Pappenheimer, A. M. (1977) Diphtheria toxin, *Ann. Rev. Biochem.* **46:**69–94.
Protzel, A., and Morris, A. J. (1974) Gel chromatographic analysis of nascent globin chains. Evident of nonuniform size distribution, *J. Biol. Chem.* **249:**4594–4600.
Pulkrabek, P., and Rychlik, I. (1968) Effect of univalent cations and role of GTP and supernatant factors during biosynthesis of polylysine chain, *Biochim. Biophys. Acta* **155:**219–227.
Rebagliati, M. R., Weeks, D. L., Harvey, R. P., and Melton, D. A. (1985) Identification and cloning of localized maternal RNAs from *Xenopus eggs, Cell* **42:**769–777.
Rosenberg, A. H., Goldman, E., Dunn, J. J., Studier, F. W., and Zubay, G. (1993) Effects of consecutive AGG codons on translation in *Escherichia coli,RR demonstrated with a versatile codon test system, J. Bact.* **175:**716–722.
Ryazanov, A. G. (1987) Ca^{2+}/calmodulin-dependent phosphorylation of elongation factor 2, *FEBS Letters* **214:**331–334.
Ryazanov, A. G., and Davydova, E. K. (1989) Mechanism of elongation factor 2 (EF-2) inactivation upon phosphorylation. Phosphorylated EF-2 is unable to catalyze translocation, *FEBS Letters* **251:**187–190.
Ryazanov, A. G., Ovchinnikov, L. P., and Spirin, A. S. (1987) Development of structural organization of protein-synthesizing machinery from prokaryotes to eukaryotes, *BioSystems* **20:**275–288.
Ryazanov, A. G., Shestakova, E. A., and Natapov, P. G. (1988) Phosphorylation of elongation factor 2 by EF-2 kinase affects rate of translation, *Nature* **334:**170–173.
Spanjaard, R. A., and van Duin, J. (1988) Translation of the sequence AGG-AGG yields 50% ribosomal frameshift, *Proc. Natl. Acad. Sci. USA* **85:**7967–7971.
Spirin, A. S., and Ryazanov, A. G. (1991) Regulation of elongation rate, in *Translation in Eukaryotes* (H. Trachsel, ed.), pp. 325–350, CRC Press, Boca Raton, FL.
Svitkin, Yu. V., and Agol, V. I. (1983) Translational barrier in central region of encephalomyocarditis virus genome. Modulation by elongation factor 2 (eEF-2), *Eur. J. Biochem.* **133:**145–154.
Theodorakis, N. G., Banerji, S. S., and Morimoto, R. I. (1988) HSP70 mRNA translation in chicken reticulocytes is regulated at the level of elongation, *J. Biol. Chem.* **263:**14579–14585.
Wolin, S. L., and Walter, P. (1988) Ribosome pausing and stacking during translation of a eukaryotic mRNA, *EMBO J.* **7:**3559–3569.
Wolin, S. L., and Walter, P. (1989) Signal recognition particle mediates a transient elongation arrest of preprolactin in reticulocyte lysate, *J. Cell Biol.* **109:**2617–2622.
Wu, R. S., and Warner, J. R. (1971) Cytoplasmic synthesis of nuclear proteins. Kinetics of accumulation of radioactive proteins in various cell fractions after brief pulses, *J. Cell Biol.* **51:**643–652.
Young, J. C., and Andrews, D. W. (1996) The signal recognition particle receptor alpha subunit assembles co-translationally on the endoplasmic reticulum membrane during an mRNA-encoded translation pause *in vitro, EMBO J.* **15:**172–181.

14

Termination of Translation

14.1. Termination Codons

The ribosome reads mRNA triplet by triplet and elongates the polypeptide chain until it comes across one of the following codons: UAA, UAG, or UGA. None of these triplets possess a cognate aminoacyl-tRNA; they all stop translation. Thus, they serve as *termination codons* (Brenner, Stretton, and Kaplan, 1965; Brenner *et al.*, 1967). UAA is the most frequently used of these codons, whereas UAG is rarely used.

The termination codon is always present at the end of the coding region of any naturally occurring mRNA. Termination codons are sometimes present in tandem, for example, at the end of the MS2 bacteriophage coat protein cistron, where the termination codon UAA is followed by UAG (see Section 16.4.1). It is interesting that in eukaryotes a termination codon is often followed by a purine nucleotide, and generally the tetranucleotide stop signals UAA(A/G) and UGA(A/G) are preferred. In bacteria (*E. coli*), the preferred termination sequences are UAAU and UAAG (for review, see Tate and Brown, 1992).

It should be noted that triplets UAA, UAG, and UGA are found far more often in wrong reading frames within mRNA coding regions than in the correct reading frame, where, as a rule, only one such triplet per coding sequence is present. Therefore, an accidental frame shift in the course of elongation does not usually result in the synthesis of a long wrong polypeptide and generally leads to an early termination of this mistranslation. The frequency of termination triplets is also high in the noncoding mRNA regions, including the intercistronic regions of polycistronic mRNAs.

The termination triplet may appear in the reading frame of the mRNA coding region as a result of mutation. For example, a change of G to A in the tryptophan codon UGG results in the appearance of either UAG or UGA; a change of C to U in the glutamine codons CAA or CAG results in either UAA or UAG. Such mutations are referred to as *nonsense mutations;* the appearance of UAG is called the *amber* mutation, UAA the *ochre* mutation, and UGA the *opal* mutation. In contrast to the usual point mutations which result in the replacement of the amino acid in the synthesized polypeptide, these mutations lead to premature termination, which takes place at just the point where the nonsense codon appears. Another mutation changing the anticodon of some tRNA species in such a way that it becomes complementary to the nonsense codon may result in the *suppression* of the nonsense mutation; for example, tyrosine tRNA, in which the anticodon GUA is changed into CUA, recognizes the termination codon UAG and thus suppresses the amber mutations.

Translation may proceed with errors, including ones in which the normal termination codon at the end of the mRNA coding sequence is recognized by the tRNA with an anticodon having partial complementarity to it (for reviews, see Eggertsson and Soell, 1988; Valle and Morsh, 1988). Thus, the normal tryptophan tRNA with its anticodon CCA can occasionally recognize the termination UGA codon, or glutamine tRNA with anticodon CUG or UUG may interact with termination codon UAG or UAA, respectively. A minor tyrosine tRNA with anticodon GΨA isolated from plants and insects has been shown to recognize termination codon UAG. Two leucine tRNAs harboring anticodons CAA and CAG isolated from calf liver have been reported to be capable of interacting with termination codon UAG. A lysyl-tRNA of yeast has been also reported to insert its amino acid residue in response to UAG codon. All this results in a message being occasionally *read through* and a longer polypeptide being synthesized. The formation of such longer products, in addition to the normal translation product, has often been observed in studies of protein synthesis both *in vivo* and *in vitro*. Similar mechanisms underlie the formation of small amounts of the normal product in the nonsense mutants ("leakage", see Section 10.4.2).

The codon UGA is the "weakest" among the three termination codons: it can be read through by a translating ribosome most frequently, seemingly due to its recognition by tryptophanyl-tRNA (Hirsh, 1971). In some cases this termination codon is used in nature to form a small amount of a physiologically important protein from a read-through product, in addition to the main protein product whose translation is terminated by this codon. Such a situation is observed in the case of Qβ phage RNA translation (Fig. 14.1): coat protein cistron is ended by the termination codon UGA which is occasionally read through by ribosomes thus resulting in the synthesis of a small amount of a polypeptide that is much longer than the coat protein (Weiner and Weber, 1973); the read-through polypeptide is a necessary product of the Qβ RNA translation since it is required for the assembly of the normally infectious phage particle. In plant RNA viruses the expression of open reading frames often occurs through read-through of the UAG stop codon which separates two protein coding regions (Pelham, 1978).

The reading beyond a termination codon (read-through) can occur also as a result of frameshifting at the termination codon site. Moreover, it is a termination codon which, in proper structural surroundings, may induce the frameshifting and the resultant reading-through. Thus, in addition to the function of a stop signal, the termination codons can serve as a prerequisite for ribosome slippage or hopping along mRNA (Section 12.4.2).

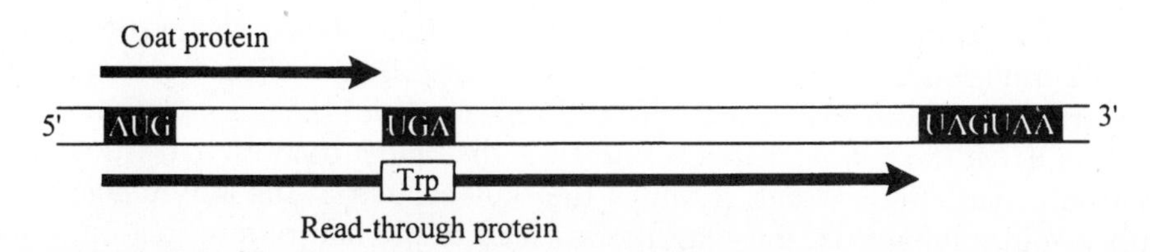

FIG 14.1. Reading-through during translation of bacteriophage Qβ RNA: the termination codon UGA of the coat protein cistron is occasionally recognized by Trp-tRNA, instead of termination factor RF2, resulting in the synthesis of a much longer polypeptide that is required for the assembly of the infectious phage particles.

The termination codon UGA, when followed by a special structural element (in prokaryotic mRNAs), or when mRNA bears a special signal in its 3′-untranslated region (in eukaryotes), encodes for selenocysteine, the 21st amino acid (see Sections 2.3 and 10.2.2). In this case UGA is recognized by $tRNA^{Sec}$ with the complementary anticodon, UCA; the $tRNA^{Sec}$ is aminoacylated by serine, and then the serine residue is enzymatically transformed into a selenocysteine residue on the $tRNA^{Sec}$. In prokaryotes the selenocysteinyl-$tRNA^{Sec}$ is bound by a unique elongation factor SELB (instead of EF-Tu) which interacts with the structural element of mRNA; it is this interaction that determines the use of the upstream UGA triplet as the $tRNA^{Sec}$-binding codon on the translating ribosome.

The mitochondrial genetic code deserves special mention (Table 2.2). The codon UGA does not serve as a termination codon in animal and fungal mitochondria. Rather, it codes for tryptophan, just as UGG does. Instead, AGA and AGG codons in mitochondria of vertebrates do not code for arginine but serve as termination codons. Two genera of primitive eubacteria, *Mycoplasma* and *Spiroplasma,* use UGA also as a tryptophan codon, rather than as a stop codon (Table 2.1).

14.2. Termination Protein Factors

When a termination codon comes to the ribosomal A site, it is recognized by special soluble proteins which bind to the ribosome and induce the hydrolysis of the ester bond between the tRNA and the polypeptide in the peptidyl-tRNA at the P site (Capecchi and Klein, 1969; Caskey *et al.,* 1969). As a result, the polypeptide is released from the ribosome. Proteins recognizing the termination codons and inducing the release of polypeptide are called *termination* or *release factors* (RF).

Prokaryotic organisms have three proteins taking part in termination: RF1, RF2, and RF3 (for reviews, see Caskey, 1977, 1980; Craigen, Lee, and Caskey, 1990). These proteins in *E. coli* have molecular masses of 40,460, 41,235, and 59,460 daltons, respectively (see Grentzmann *et al.,* 1994; Mikuni *et al.*, 1994). The codons UAA and UAG are recognized by RF1; RF2 is specific for the UAA and UGA codons; and RF3 does not participate directly in codon recognition but is necessary for involving GTP/GDP in the termination process. It is interesting that, in accordance with the mitochondrial code, the mammalian mitochondria have no RF2.

Eukaryotes have one termination factor, referred to as eRF or eRF1, to recognize all three termination codons, UAA, UAG, and UGA (Konecki *et al.,* 1977; Frolova *et al.,* 1994). The protein in the monomeric form has a molecular mass of about 55,000 daltons, but in solution it may be present as a dimer. In addition, the second termination factor analogous to prokaryotic RF3 has been shown to exist in eukaryotes; it is designated eRF3 (Zhouravleva *et al.,* 1995). The molecular mass of this protein is about 70,000 daltons (614 amino acid residues). It interacts with GTP and shows GTPase activity in the presence of ribosomes. There is evidence that eRF3 with GTP can form a complex with eRF1. Thus, the complex eRF1: eRF3:GTP may be the functional unit that accomplishes termination on the eukaryotic ribosome in a GTP-dependent manner.

In the process of interacting with the termination codon and the ribosome, the termination factors RF1 and RF2 of prokaryotes, or eRF1 of eukaryotes, appear to simulate the binding of the aminoacyl-tRNA to the A site of the translating ribosome. However, instead of an attack by the amino group of the aminoacyl-tRNA on the peptidyl-tRNA ester bond, this bond is attacked by a water molecule. The transfer of a peptidyl residue to the water molecule, like its transfer to the amino group in transpeptidation, is catalyzed by the ribosomal PTC.

The possibility of protein factor–codon interaction, instead of the codon–anticodon interaction, is intriguing. The protein seems to recognize the nucleotide triplet, and the recognition has a high specificity similar to that for the codon–anticodon recognition. Moreover, in the presence of a suppressor tRNA complementary to the termination codon, the aminoacyl-tRNA and the termination factor compete for binding to the ribosomal A site. Experiments using various modified nucleotides in the termination codons have indicated that the specificity of RF in codon recognition resembles greatly the Watson–Crick base-pairing specificity, including Crick's wobble pairing. Direct contact between RF2 and the stop codon has been confirmed in cross-linking experiments where the stop codon s^4UAA was used as a zero-length cross-linker (Tate, Greuer, and Brimacombe, 1990a). The same stop codon could be cross-linked with A1408 of the 16S rRNA, precisely at the A-site part of the decoding center of the 30S subunit (Fig. 9.3). An interesting idea is that the tertiary structures of the codon-recognizing factors, RF1 and RF2 (as well as eRF1), mimic that of tRNA, including the presence of the anticodon-mimicry element in the proteins. Indeed, an amino acid sequence homology between these termination factors and the C-terminal half (domains III, IV, and V) of EF-G which is known to mimic tRNA (see Section 12.2.1) has been noticed (Ito *et al.,* 1996).

The function of the termination factors, like that of the elongation factors (EF1 and EF2) and one of the initiation factors (IF2 or eIF2), depends on GTP. Among the termination factors, however, only RF3 and eRF3 have an affinity for GTP and GDP; in the presence of the ribosome they show GTPase activity (see Frolova *et al.*, 1996; Freistroffer et al.,RR 1997). At the same time, it has been demonstrated that prokaryotic RF3 with GTP stimulates the recycling of RF1 and RF2 (their release and rebinding to terminating ribosomes), and seemingly eRF3 with GTP does the same with eRF1 and eukaryotic ribosomes. It is likely that the GTP hydrolysis after termination induces the release of the RFs (or eRFs) from the ribosome. In accordance with its GTP/GDP-dependent behavior, RF3 has been found to possess a strong homology with G-proteins, and especially with G-domains of EF-Tu (EF1A) and EF-G (EF2).

The hypothesis may be put forward that the mechanism of the joint action of RFs in the course of termination partly resembles that of the ternary complex Aa-tRNA:EF1:GTP during elongation: RF1 or RF2 mimics the tRNA residue, and RF3 plays the part of EF1A (Ito *et al.,* 1996). Like the ternary complex, the complex RF1:RF3:GTP (or RF2:RF3:GTP) occupies the A site with a proper (termination) codon settled there. Then the hydrolysis of GTP may be induced on RF3, and the dissociation of RF3:GDP from the ribosome and the induction of the hydrolytic activity of the ribosomal PTC follow.

The alternative hypothesis is that RF3 (and eRF3) is translocase, thus acting similarly to EF-G (EF2), rather than to EF-Tu (EF1) (Freistroffer *et al*., 1997; Buckingham, Grentzmann, and Kisselev, 1997). According to this model, RF3 does not promote the binding of RF1 or RF2 to the termination codon at the A site, but displaces them from the A site in a GTP-dependent manner, *after* the peptidyl-tRNA hydrolysis.

The termination step with the participation of RFs is followed by the release of deacylated tRNA from the ribosome and the dissociation of the ribosome into subunits. It has been demonstrated that in bacteria two other proteins are involved in the breakdown of the "termination complex," EF-G and the so-called *ribosome releasing (ribosome recycling) factor* or RRF (Kaji, Igarishi, and Ishitsuka, 1969; Janosi *et al.,* 1996). As in the catalysis of translocation, EF-G requires GTP for performing this function. RRF (or RF4) is a small basic protein with a molecular mass of about 20,000 daltons (185 amino acids). There is evidence that a concerted action of both proteins, EF-G and RRF, is necessary for promoting the dissociation step. The mechanism of the action is unknown. It may be that RRF interacts with the vacant A site, and EF-G:GTP translocates RRF, displacing the deacylated tRNA from the P site. After ejection of the deacylated tRNA the ribosome loses its 50S subunit, and the 30S subunit can slide along mRNA or dissociate from it (Pavlov *et al.,* 1997).

14.3. Ribosomal Site for Binding Termination Factors

As already mentioned, the termination factors RF1 and FR2 (or eRF1) recognize the termination codon positioned in the A site of the ribosome (for reviews, see Caskey, 1977; Tate, Brown, and Kastner, 1990b; Tate and Brown, 1992). Thus, the codon-dependent binding of the RF is possible only after the peptidyl-tRNA has been translocated from the A site to the P site. Under *in vitro* conditions, the AUG triplet and the initiator F-Met-tRNA or Met-tRNA can be bound directly to the P site of the vacant ribosomes, and this may be followed by the addition of the RF and one of the termination triplets, UAA, UAG, or UGA; these in turn bind and induce hydrolysis of the F-Met-tRNA at the P site, yielding free formyl-methionine. It is evident that, in this case, the termination triplet and the RF come into the vacant A site. The fact that the suppressor aminoacyl-tRNA in the complex with EF-Tu and GTP, which possesses an anticodon complementary to the termination codon, competes with the RF in the course of translation also indicates that the RF (RF1, RF2, or eRF1) binds to the A site. Accordingly, the antibiotics affecting the ribosomal A site on the 30S subunit, such as tetracycline and aminoglycosides (Sections 10.3.1. and 10.3.2.), inhibit the codon-dependent binding of the termination factors with the prokaryotic ribosome (Brown, McCaughan, and Tate, 1993).

On the other hand, the EF-G bound to the ribosome in a stable manner (e.g., in the presence of fusidic acid) interferes with the binding of RFs to the ribosome, just as with the binding of the aminoacyl-tRNA:EF-Tu:GTP complex. The protein L7/L12 of the prokaryotic ribosome forming the lateral stalk of the 50S ribosomal subunit contributes to the binding of RF1 and RF2, as well as of the elongation factors. Antibiotic thiostrepton, known to block the factor-binding site on the 50S subunit (Sections 10.3.3 and 12.5.3), is a strong inhibitor of RF binding. The inhibition of the codon-dependent binding of RF1 and RF2 by antibodies against individual ribosomal proteins, such as S3, S4, S5, and S10 of the 30S subunit and L7/L12 of the 50S subunit, confirms that the RF-binding site overlaps considerably with both the A site on the small subunit and the site responsible for binding the elongation factors (the factor-binding site, see Section 9.4) on the large subunit of the ribosome (see Tate *et al.,* 1990b; Tate and Brown, 1992). Together with these observations, immuno-electron microscopy studies strongly indicate that RFs are placed in the space between the two ribosomal subunits (interface cavity between the subunit "necks"), on the side adjacent to the L7/L12 stalk, similar to tRNA residues of the elongating ribosome (Section 9.5). The RF2-binding site is located at the subunit interface, comprising the L7/L12 stalk region of the large subunit and the neck region opposite to the side bulge of the small subunit (Kastner, Trotman, and Tate, 1990).

14.4. Hydrolysis of Peptidyl-tRNA

The codon-dependent binding of the termination factor (RF1, RF2, or eRF1) is necessary for the hydrolysis of the peptidyl-tRNA at the P site and, consequently, for the release of the polypeptide from the ribosome. As has been demonstrated, the hydrolysis of the peptidyl-tRNA in the ribosome is suppressed by the inhibitors interfering with the peptidyl transferase reaction (see Caskey, 1977). Chloramphenicol, amicetin, lincomycin, gougerotin, sparsomycin, and other common inhibitors of the ribosomal peptidyl transferase (Section 11.3) suppress the RF-induced release of the peptide from bacterial ribosomes. These antibiotics, however, do not inhibit the codon-dependent binding of RF to the bacterial ribosome. Hence, the terminating hydrolysis of the peptidyl-tRNA appears to be accomplished by the ribosomal PTC.

There is independent evidence that the PTC is capable of catalyzing the formation of the ester bonds between the C-terminus of the peptide and the hydroxy analogs of the aminoacyl-tRNA or puromycin (Section 11.1). Furthermore, in the presence of methyl or ethyl alcohol and deacylated tRNA, methanolysis or ethanolysis of the peptidyl-tRNA at the ribosomal P site takes place; in other words, the peptide residue is transferred to the alcohol hydroxyl group (Section 11.1). In the presence of other organic solvents, such as acetone, the peptide residue can be transferred to a water molecule; that is, the ester bond of the peptidyl-tRNA is subjected to ribosome-catalyzed hydrolysis. The hydrolysis depends on the presence of deacylated tRNA or its 3′-terminal CCA sequence. In this case as well, the hydrolysis of the peptidyl-tRNA is prevented by inhibitors of the ribosomal PTC.

It is likely that the codon-dependent binding of the termination factor makes the ribosomal PTC more accessible to water, which is a good acceptor substrate; as a result, the peptide is transferred to water, particularly since under these conditions there is no competing amino group of an aminoacyl-tRNA. This accessibility can be achieved, for example, as a consequence of a certain unlocking (drawing apart) of the ribosomal subunits or a slight opening (loosening) of the large subunit that carries the PTC. Other explanations, however, cannot be excluded: either some nucleophilic group of the RF protein may participate directly in an attack on the ester bond and accept transiently the peptide, or the RF-fixed water molecule may serve as a specific acceptor.

14.5. Sequence of Events during Termination

Thus, the termination of translation appears to include several consecutive steps. The first step is the *recognition* of the A-site-bound termination codon by the protein factor of termination, that is RF1 or RF2 in prokaryotes, and eRF1 in eukaryotes. According to one scenario (Ito *et al.*, 1996), the recognition ends in the *binding of the complex of two termination factors,* RF1 and RF3 (or RF2 and RF3) in the case of prokaryotes, or eRF1 and eRF3 in eukaryotes, involving both the codon on the small ribosomal subunit and the factor-binding site on the large ribosomal subunit. The association with the factor-binding site (located at the base of the L7/L12 stalk) requires GTP. Thus, RF1 (or RF2) or eRF1 recognizes the termination codon, while RF3 or eRF3 binds GTP and is responsible for the interaction with the factor-binding site at the base of the L7/L12 stalk. Another scenario, however, is possible (Freistroffer *et al.*, 1997): RF3 (or eRF3) does not come into operation at this stage, but interacts with the terminating ribosome only after the next step, that is after the hydrolysis of peptidyl-tRNA.

The second step involves *hydrolysis* of the ester bond of the peptidyl-tRNA at the P site. As mentioned above, it is still unclear whether the ester bond hydrolysis precedes the RF3 (eRF3) interaction, the hydrolysis of GTP, and the release of RF3 (eRF3), or whether it follows them. In any case, the ester bond hydrolysis is catalyzed by the ribosomal PTC of the large ribosomal subunit, and it is the termination codon-dependent binding of the termination factor to the ribosome that induces the hydrolase activity of the PTC. It seems that the RFs somehow provide for the presence of a water molecule as an acceptor substrate in the PTC. This may be done either by an organized attack of a specifically bound water molecule on the ester bond, or through a conformational change of the ribosome (e.g., unlocking) allowing an access of free water to PTC. As a result, peptide is released from the ribosome. The ribosome remains associated with the mRNA, the deacylated tRNA at the P site, and possibly the termination factor (RF1/RF2 or eRF1) at the A site.

The third step must be the *evacuation* of the ligands from the ribosome. The termination factor is probably the first to leave. By analogy with the EF1 and EF2, it may be assumed that the termination G-protein RF3 and the RF3-catalyzed GTP hydrolysis play the main role in the evacuation of the RFs. It may be that the affinity of the termination factor complex (RF1:RF3 or RF2:RF3) for the ribosome reduces as a result of GTP cleavage, and the RFs are simply released by dissociation. Another possibility is the active translocation of RF1 (or RF2) from the A site by RF3 with a coupled GTP hydrolysis reaction. As to the release of deacylated tRNA from the ribosomal P site, there is evidence that in prokaryotes this event is promoted by a special protein, RRF, together with EF-G and GTP.

Association between the ribosomal subunits in the vacant ribosome is far weaker than in the ribosome carrying the ligands. Therefore, after the ligands are released a reversible dissociation of the ribosome is facilitated. In prokaryotes, after the 50S subunit has gone, the 30S subunit may be retained on mRNA less firmly and can dissociate easily from the template:

$$\text{mRNA:70S} \rightleftharpoons \text{mRNA:30S} + \text{50S} \rightleftharpoons \text{mRNA} + \text{30S} + \text{50S}$$

The 30S subunit, however, should not necessarily leave the template after termination; seemingly it can slip along the mRNA for a while without translation and reinitiate translation at the next cistron of the same polycistronic mRNA.

The sequence of events in the course of termination, with the reservation that the exact stage of RF3 binding and action (either Step 1, as in the figure, or Step 3) is not yet determined, is shown schematically in Fig. 14.2.

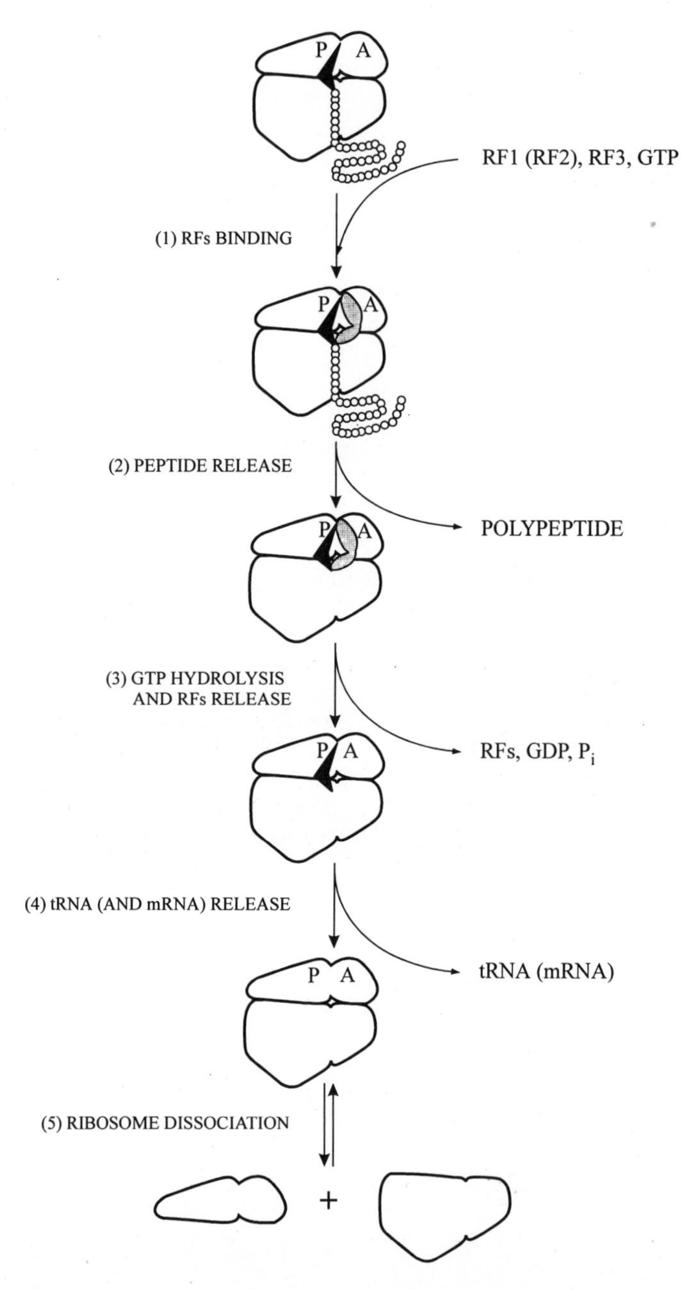

FIG 14.2. Plausible sequence of events during termination of translation. The step where RF3 comes into the terminating ribosome, however, is not clear: it is either Step 1, *before* the peptidyl-tRNA hydrolysis, as indicated in the figure, or Step 3, *after* the peptidyl-tRNA hydrolysis. In the latter case RF3:GTP binds transiently, hydrolyzes GTP and dissociates together with the displacement of RF1 (or RF2) during. Step 3 See Buckingham, R. H., Grentzmann, G., and Kisselev, L. (1997). *Mol. Microbiol.* **24**:449–456.

Brenner, S., Barnett, L., Katz, E. R., and Crick, F. H. C. (1967) UGA: A third nonsense triplet in the genetic code, *Nature* **213**:449–450.

Brenner, S., Stretton, A. O. W., and Kaplan, S. (1965) Genetic code: The 'nonsense' triplets for chain termination and their suppression, *Nature* **206**:994–998.

Brown, C. M., McCaughan, K. K., and Tate, W. P. (1993) Two regions of the *Escherichia coli* 16S ribosomal RNA are important for decoding stop signals in polypeptide chain termination, *Nucleic Acids Res.* **21**:2109–2115.

Buckingham, R. H., Grentzmann. G., and Kisselev, L. (1997) Polypeptide chain release factors, *Mol. Microbiol.* **24**:449–456.

Capecchi, M. R., and Klein, H. A. (1969) Characterization of three proteins involved in polypeptide chain termination, *Cold Spring Harbor Symp. Quant. Biol.* **34**:469–477.

Caskey, C. T. (1977) Peptide chain termination, in *Molecular Mechanisms of Protein Biosynthesis* (H. Weissbach and S. Pestka, eds.), pp. 443–465, Academic Press, New York.

Caskey, C. T. (1980) Peptide chain termination, *Trends Biochem. Sci.* **5**:234–237.

Caskey, T., Scolnick, E., Tomkins, R., Goldstein, J., and Milman, G. (1969) Peptide chain termination, codon, protein factor, and ribosomal requirements, *Cold Spring Harbor Symp. Quant. Biol.* **34**:479–488.

Craigen, W. J., Lee, C. C., and Caskey, C. T. (1990) Recent advances in peptide chain termination, *Mol. Microbiol.* **4**:861–865.

Eggertsson, G., and Soell, D. (1988) Transfer ribonucleic acid-mediated suppression of termination codons in *Escherichia coli, Microbiol. Rev.* **52**:354–374.

Freistroffer, D. V., Pavlov, M. Yu., MacDougall, J., Buckingham, R. H., and Ehrenberg, M. (1997) Release factor RF3 in *E. coli* accelerates the dissociation of release factors RF1 and RF2 from the ribosome in a GTP-dependent manner, *EMBO J.* **16**:4126–4133.

Frolova, L., Le Goff, X., Rasmussen, H. H., Cheperegin, S., Drugeon, G., Kress, M., Arman, I., Haenni, A. L., Celis, J. E., Philippe, M., Justesen, J., and Kisselev, L. L. (1994) A highly conserved eukaryotic protein family possessing properties of polypeptide chain release factor, *Nature* **372**: 701–703.

Frolova, L., Le Goff, X., Zhouravleva, G., Davydova, E., Philippe, M., and Kisselev, L. L. (1996) Eukaryotic polypeptide chain release factor eRF3 is an eRF1- and ribosome-dependent guanosine triphosphatase, *RNA* **2**:334–341.

Grentzmann, G., Brechemier-Baey, D., Heurgue, V., Mora, L., and Buckingham, R. H. (1994) Localization and characterization of the gene encoding release factor RF3 in *Escherichia coli, Proc. Natl. Acad. Sci. USA* **91**:5848–5852.

Hirsh, D. (1971) Tryptophan transfer RNA as the UGA suppressor, *J. Mol. Biol.* **58**:439–458.

Ito, K., Ebihara, K., Uno, M., and Nakamura, Y. (1996) Conserved motifs in prokaryotic and eukaryotic polypeptide release factors: tRNA-protein mimicry hypothesis, *Proc. Natl. Acad. Sci. USA* **93**: 5443–5448.

Janosi, L., Hara, H., Zhang, S., and Kaji, A. (1996) Ribosome recycling by ribosome recycling factor (RRF) - an important but overlooked step of protein biosynthesis, *Adv. Biophys.* **32**:121–201.

Kaji, A., Igarashi, K., and Ishitsuka, H. (1969) Interaction of tRNA with ribosomes: Binding and release of tRNA, *Cold Spring Harbor Symp. Quant. Biol.* **34**:167–177.

Kastner, B., Trotman, C. N. A., and Tate, W. P. (1990) Localization of the release factor-2 binding site on 70S ribosomes by immuno-electron microscopy, *J. Mol. Biol.* **212**:241–245.

Konecki, D. S., Aune, K. C., Tate, W. P., and Caskey, C. T. (1977) Characterization of reticulocyte release factor, *J. Biol. Chem.* **252**:4514–4520.

Mikuni, O., Ito, K., Moffat, J., Matsumura, K., McCaughan, K., Nobukuni, T., Tate, W., and Nakamura, Y. (1994) Identification of the prfC gene, which encodes peptide-chain-release factor 3 of *Escherichia coli, Proc. Natl. Acad. Sci. USA* **91**:5798–5802.

Pavlov, M. Yu., Freistroffer, D. V., MacDougall, J., Buckingham, R. H., and Ehrenberg, M. (1997) Fast recycling of *Escherichia coli* ribosomes requires both ribosome recycling factor (RRF) and release factor RF3, *EMBO J.* **16**:4134–4141.

Pelham, H. R. B. (1978) Leaky UAG termination codon in tobacco mosaic virus RNA, *Nature* **272**:469–471.

Tate, W. P., and Brown, C. M. (1992) Translational termination: 'Stop' for protein synthesis or 'pause' for regulation of gene expression, *Biochemistry* **31**:2443–2450.

Tate, W. P., Brown, C. M., and Kastner, B. (1990b) Codon recognition by the polypeptide release factor, in *The Ribosome: Structure, Function, and Evolution* (W. E. Hill, A. Dahlberg, R. A. Garrett, P. B. Moore, D. Schlessinger, and J. R. Warner, eds.), pp. 393–401, ASM Press, Washington, DC.

Tate, W., Greuer, B., and Brimacombe, R. (1990a) Codon recognition in polypeptide chain termination: site directed crosslinking of termination codon to *Escherichia coili* release factor 2, *Nucleic Acids Res.* **18:**6537–6544.

Valle, R. P. C., and Morsh, M.-D. (1988) Stop making sense, or Regulation at the level of termination in eukaryotic protein synthesis, *FEBS Letters* **235:**1–15.

Weiner, A. M., and Weber, K. (1973) A single UGA codon functions as a natural termination signal in the coliphage Q coat protein cistron, *J. Mol. Biol.* **80:**837–855.

Zhouravleva, G., Frolova, L., Le Goff, X., Le Guellec, R., Inge-Vechtomov, S., Kisselev, L., and Philippe, M. (1995) Termination of translation in eukaryotes is governed by two interacting polypeptide chain release factors, eRF1 and eRF3, *EMBO J.* **14:**4065–4072.

15

Initiation of Translation

15.1. General Principles

The vacant ribosomal particles after termination may be again involved in translation. Several mechanisms of initiation of translation of a new message, or reinitiation of translation of the next coding sequence of the same polycistronic mRNA (in prokaryotes), or reinitiation of translation of the same monocistronic mRNA (in eukaryotes) by vacant post-termination ribosomes have evolved in living matter.

15.1.1. Significance of Initiation Stage

Message-dependent elongation of peptide by the ribosome requires peptidyl-tRNA as a substrate to react with a newly arrived, codon-bound aminoacyl-tRNA. In other words, the donor substrate must be present in the ribosomal P site to be elongated by the next amino acid residue. Hence, in order to initiate translation, a special mechanism must exist that inserts a donor substrate into the P site of an empty (nontranslating) ribosome. This mechanism of the *translation initiation* will be discussed in this section.

The initiation mechanism may be considered a modified elementary elongation cycle (Section 9.1, Fig. 9.1). A special initiator aminoacyl-tRNA in complex with a special initiation factor and GTP binds to an initiator codon of mRNA at the ribosome. This stage looks like the aminoacyl-tRNA:EF1A:GTP binding stage in the elongation cycle. However, in the process of binding, the initiator aminoacyl-tRNA is directed to the P site of the ribosome, rather than to the A site. This correlates with the fact that the corresponding initiation factor (IF2 in the case of prokaryotes) is structurally and functionally similar to EF2 (EF-G in prokaryotes). Thus, elements of translocation can be seen in the process of the initiator aminoacyl-tRNA binding. Then the initiator aminoacyl-tRNA occupying the P site serves as a donor substrate in the subsequent reaction with an A-site-bound elongator aminoacyl-tRNA; that is, it functionally mimics the peptidyl-tRNA in the elongation cycle. In prokaryotes the initiator aminoacyl-tRNA resembles a peptidyl-tRNA chemically: its amino group is formylated, and is thus involved in the amide bond.

Initiation of translation matters not only at the beginning of the peptide elongation process, but also at the start of the message readout at a specific point of a message polynucleotide. Since the start of the coding sequence does not coincide with the 5′-end of the message polynucleotide but is removed from the 5′-end, sometimes by a significant distance, a precise recognition of the first codon is required. A precise start point is particularly important in translation because it sets a proper reading frame for all of the subsequent coding sequence of a given mRNA, for its triplet-by-triplet readout. Thus the finding of the first codon from which both the start and the frame are counted is another fundamental function of the initiation mechanism.

Furthermore, initiation constitutes the principal step at which protein synthesis is controlled at the translational level (Chapters 16 and 17). Regulation at the translational level may be discussed in terms of either permitting or preventing the initiation of the mRNA readout by ribosomes. Selective or preferential translation of certain mRNA species (or mRNA cistrons in prokaryotes) and translational inactivation of other mRNAs are achieved precisely in this way. In addition, differential rates of initiation with different mRNAs (or different cistrons) determine the ratio of production of corresponding proteins. Generally, the initiation mechanism exerts its intrinsic selectivity toward different messages and also serves as a target for the action of positive and negative regulatory signals.

15.1.2. Prokaryotic and Eukaryotic Modes of Initiation

There are two modes of initiation in nature. One requires and starts with the 5′-end of mRNA, though the initiation codon is always positioned at some distance from the end. It is assumed that the ribosome associates with the 5′-end, usually modified ("capped", see Fig. 2.4), and then scans the downstream nucleotide sequence until it encounters the initiation codon (Fig. 15.1A). This mode of *5′-terminal initiation* is predominantly used by eukaryotes. The mechanism includes special mRNA-binding protein factors for the recognition of the capped 5′-end, ATP-dependent unwinding of RNA helices during scanning, and fixation of the ribosome at the initiation codon. Initiation of this type takes place on monocistronic mRNA presynthesized in the nucleus, transported into the cytoplasm, and complexed with protein to form mRNP.

The other initiation mode is *internal initiation,* when the ribosome associates directly with the mRNA structure containing an initiation codon within it or nearby, independently of the mRNA end (Fig. 15.1B). This mode is essential in prokaryotes. Eukaryotes, however, also use the mode of internal initiation in translation of mRNAs of some special classes. In the typical prokaryotic version of the internal initiation, the ribosome can associate with nascent mRNA and start translation during mRNA synthesis (coupled transcription/translation), and the multiple coding sequences (mRNA cistrons) of polycistronic mRNAs can be independently initiated and translated.

15.1.3. Components of Initiation

In all cases the process of initiation of translation involves: (1) the small ribosomal subunit (30S or 40S in prokaryotes and eukaryotes, respectively); (2) mRNA with its ribosome-binding site and initiation codon within; (3) initiator aminoacyl-tRNA; (4) a group of proteins called initiation factors (IFs or eIFs in prokaryotes and eukaryotes, respectively); and (5) the large ribosomal subunit at the final stage of initiation.

Dissociation of ribosomes into subunits is a prerequisite for initiation of translation. It is the small ribosomal subunit that initiates the initiation process. It has specific affinities for mRNA, initiator aminoacyl-tRNA, and protein initiation factors. It organizes all these components on itself prior to association with the large ribosomal subunit.

Though the small ribosomal subunit has an affinity for any polynucleotide sequence, there are some structural elements in mRNAs that bind the ribosomal particle with an increased strength. Thus such elements, called ribosome-binding sites (RBS), are capable of selectively attracting the small ribosomal subunits for initiation. These elements include oligopurinic sequences of Shine–Dalgarno in prokaryotic mRNAs, cap structure with adjacent sequence in eukaryotic mRNAs, special three-dimensional IRES structures in RNAs of picornaviruses, and some other less characterized sequences and structures within prokaryotic and eukaryotic messages. The codon of initiation is usually located on 3′-side of RBS immediately or several nucleotides downstream. The predominant initiation codon in prokaryotes and eukaryotes is AUG. Related triplets, such as GUG, UUG, AUA, AUU, ACG, can also function as initiation codons in some cases, especially in prokaryotes, provided the RBS is strong.

Initiator aminoacyl-tRNA is always a special methionyl-tRNA (Met-$tRNA_f^{Met}$ or Met-$tRNA_i^{Met}$) that cannot participate in elongation and does not interact with EF1A (EF-Tu) or eEF1A. In prokaryotes it is formylated, existing as formylmethionyl-tRNA (F-Met-$tRNA_f^{Met}$). Thus, methionine is the universal starting amino acid residue of all polypeptides and proteins synthesized by ribosomes (later it can be split off by aminopeptidases). The anticodon of the initiator aminoacyl-tRNA is CAU, fully complementary to the main initiation codon AUG, and partly com-

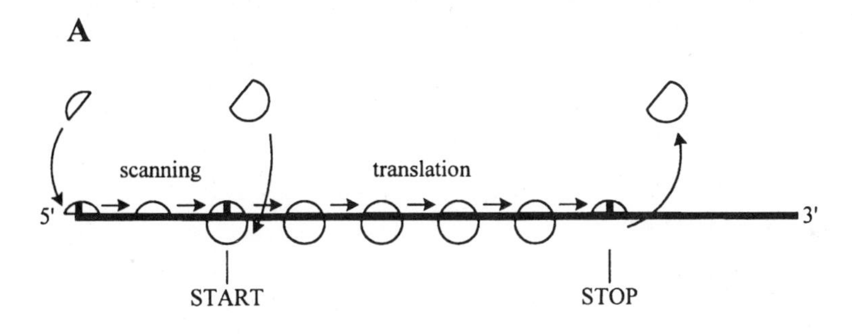

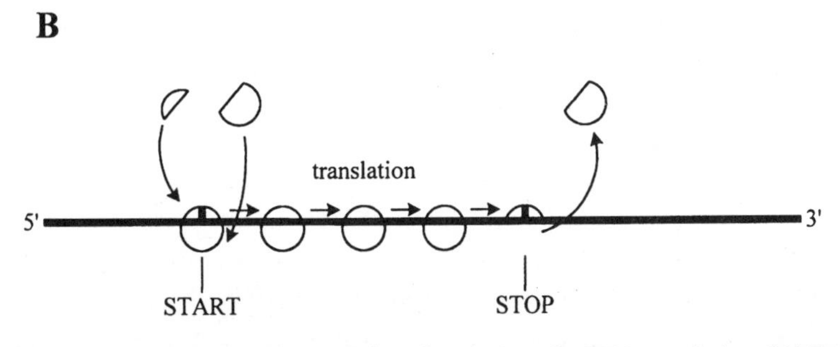

FIG 15.1. Schematic representation of two modes of initiation of mRNA translation. (**A**) Terminal initiation: ribosomal particles (small subunits) bind mRNA at the 5′-end, scan the 5′-untranslated region (5′-UTR) and start translation at the initiation codon (usually AUG); the start is accompanied by association with the large ribosomal subunit. (**B**) Internal initiation: ribosomal particles (small subunits) interact directly with the ribosome-binding site, including initiation codon, at a distance from the 5′-end of mRNA, associate with large ribosomal subunits and start translation from this point.

plementary to the rare initiation codons. The most important properties of the initiator aminoacyl-tRNA are that it is not recognized by elongation factor EF-Tu or eEF1A and has a preferable affinity for the P site of the ribosome, rather than the A site. Prior to the interaction with the ribosome it forms a ternary complex with initiation factor 2 (IF2 or eIF2) and GTP; the complex is the form in which the initiator aminoacyl-tRNA binds with the small ribosomal subunit in the process of initiation.

In addition to IF2 (or eIF2), the small ribosomal subunit cooperates and directly interacts with two more initiation factors: IF1 and IF3, or eIF1 and eIF3, in prokaryotes or eukaryotes, respectively. IF3 (eIF3) prevents the association of the small subunit with the large subunit and thus maintains the dissociated state which is required for initiation. There are grounds to believe that all three initiation factors *in vivo* may be complexed with the small ribosomal subunit prior to its interactions with initiator aminoacyl-tRNA and mRNA; such a complex can be isolated from the cell and is designated as a *native* 30S or 40S subunit, in contrast to the so-called *derived* subunit produced by *in vitro* dissociation of the whole ribosome. The initiation factors sitting on the ribosomal subunit seem to enhance the discrimination of the initiator aminoacyl-tRNA from elongator aminoacyl-tRNAs, and the RBS from other mRNA structures. The increased discrimination may be achieved not only by the involvement of the factors in specific positive interactions with the initiator aminoacyl-tRNA and with RBS, but also by decreasing the affinities of the ribosomal subunit for other aminoacyl-tRNAs and, possibly, for nonspecific mRNA sequences.

In eukaryotes, in addition to the above-mentioned *ribosome-binding* initiation factors, a group of *mRNA-binding* initiation factors exists. This group includes the factors designated as eIF4A, eIF4B, eIF4E, and eIF4F. They have no analogs in prokaryotes and are engaged in preparing the cytoplasmic mRNA (its region upstream from the coding sequence) for initiation.

15.1.4. Steps of Initiation

As mentioned in the previous section, the dissociation of ribosomes into the two subunits is a prerequisite for initiation of translation. The dissociation of the ribosome and the concomitant attachment of the proper initiation factors to the small ribosomal subunit, with the formation of the "native" 30S or 40S particle, can be considered as the first step of initiation (Fig. 15.2). Thus, the "native" small ribosomal subunit with IF3 (in the case of the prokaryotic 30S subunit) or eIF3 (in the case of the eukaryotic 40S subunit) starts the initiation process.

The next step of initiation is either the association of the "native" small subunit with mRNA, or the binding of the initiator aminoacyl-tRNA to the "native" subunit. If the second step is the association with the RBS of mRNA, the ribosomal subunit can seemingly move randomly along the polynucleotide chain for a short distance, as in the case of prokaryotes, or unidirectionally (5′ to 3′) for a longer distance in the case of eukaryotes, searching for the initiation region. It is followed by the binding of the initiator aminoacyl-tRNA which determines the final settlement of the complex (small ribosomal subunit with initiation factors and initiator aminoacyl-tRNA) at the initiation codon; this is the third step of the initiation process. The alternative way is possible when the second step is the binding of initiator aminoacyl-tRNA to the small ribosomal subunit, followed by the association of this complex with mRNA and the search for the initiation codon. In both ways the result is the formation of the so-called *ribosomal initiation complex* (see Fig. 15.2, steps 2 and 3). It is 30S:(IFs):F-Met-tRNA:mRNA in the case of prokaryotes, and 40S:(eIFs):Met-tRNA:mRNA, also called 48S initiation complex, in the case of eukaryotes.

The penultimate step (Fig. 15.2, step 4) is the interaction of the large ribosomal subunit with the above initiation complex. The interaction is accompanied by the hydrolytic cleavage of the IF2 (eIF2)-bound GTP and the release of the initiation factors from the small ribosomal subunit. The step ends with the formation of the whole 70S or 80S ribosome carrying the initiator aminoacyl-tRNA in the P site

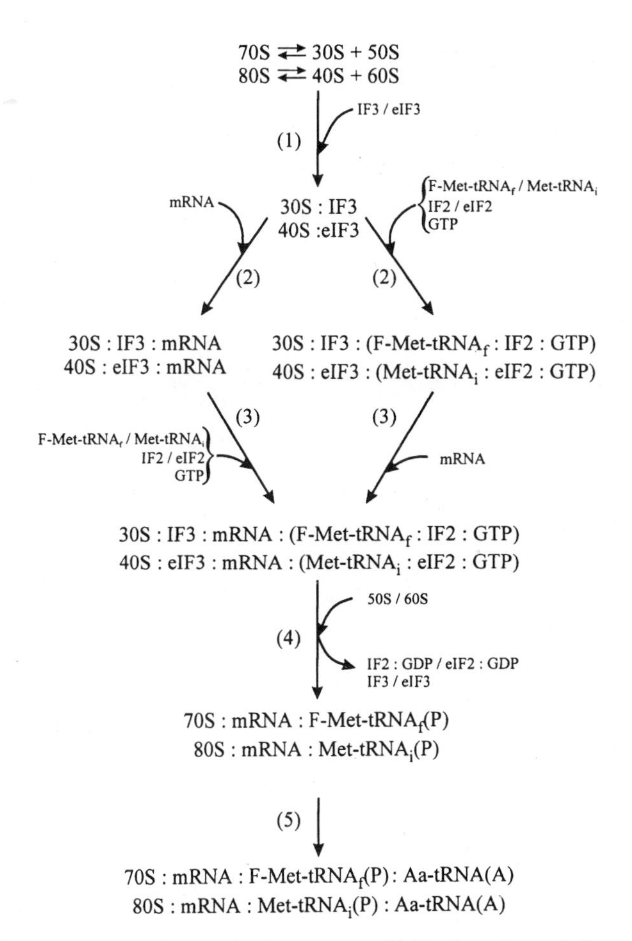

FIG 15.2. Generalized sequence of events in the process of initiation of translation, with two possible pathways of the initiation complex formation. (**1**) Association of the small ribosomal subunit with proper initiation factors (mainly IF3 or eIF3) resulting in the removal of the subunit from the two subunits association/dissociation equilibrium (formation of the "native" subunit). (**2**) Association of the small "native" subunit with either mRNA (*left pathway*) or initiator tRNA (*right pathway*). (**3**) Binding of initiator tRNA with the mRNA-associated subunit (*left pathway*), or association of the initiator tRNA-carrying subunit with mRNA (*right pathway*), both resulting in the formation of the small subunit initiation complex. (**4**) Joining of the large ribosomal subunit to the small subunit initiation complex and the release of initiation factors. Initiator tRNA is positioned in the P site of the ribosome. (**5**) Binding of the first elongator aminoacyl-tRNA to the A site of the ribosome.

and having the A site vacant. In eukaryotes special factors (eIF5 and eIF5A) promote this step.

The final step is the acceptance of the first elongator aminoacyl-tRNA at the A site and the formation of the first peptide bond. At the same time this is the start of elongation.

15.2. Initiation in Prokaryotes

15.2.1. General Characteristics of Prokaryotic Initiation

At least three features of the prokaryotic organization underlie the main peculiarities of the translation initiation process in prokaryotes. (1) Due to the absence of a membrane-isolated nucleus, the processes of transcription and translation are not separated spatially in the cell. Nascent mRNA chains, still attached to the RNA polymerase complex, are accessible for interaction with ribosomal particles and thus ready for initiation of translation. In this way the *coupled transcription/translation* process takes place (Fig. 15.3). (2) Prokaryotic genes are usually organized in polycistronic operons that are transcribed as whole units into corresponding polycistronic polyribonucleotides (*polycistronic mRNAs*). Initiation may occur independently on different coding sequences (mRNA cistrons) of such long polyribonucleotides, requiring the mechanism of *internal initiation* (Fig. 15.1B). Moreover, in order to provide for differential productivity of different coding sequences, the mechanism of discrimination of mRNAs and cistrons within the same polycistronic mRNA by initiating ribosomal particles has been developed in prokaryotes (see Section 16.2). The discrimination is based on differential affinity of the particles for the RBS and results in differential rates of initiation. (3) The prokaryotic mRNA is presented to the translation initiation machinery as a self-folded entity, with its *intrinsic secondary and tertiary structure,* in contrast to the eukaryotic mRNA which is organized in mRNP particles

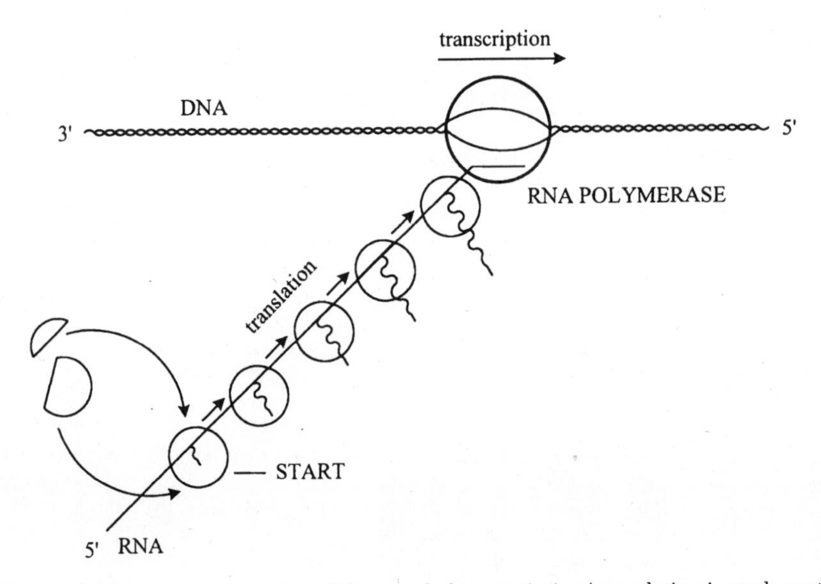

FIG 15.3. Schematic representation of the coupled transcription/ translation in prokaryotes.

where the RNA folding pattern may be significantly modified by RNA-binding proteins.

15.2.2. Ribosome-Binding Site and Initiation Codon

A number of structural requirements should be met for the complex between the initiating 30S ribosomal particle and a section of mRNA to be effectively formed (for reviews, see Steitz, 1980; Stormo, 1986; de Smit and van Duin, 1990; Van Knippenberg, 1990; Gualerzi *et al.*, 1990; Voorma, 1996). First, this section should not be hidden within a *stable* secondary and tertiary structure of mRNA (though *unstable* hairpins often involve the ribosome initiation regions). Ideally, the RBS of mRNA should avoid folds and hairpins, that is, be unstructured.

The most apparent universal structural feature of the ribosome-binding regions of prokaryotic mRNAs is the polypurine nucleotide sequence, the so-called Shine–Dalgarno sequence (Shine and Dalgarno, 1974), or SD sequence, usually located 5 to 12 nucleotides away from the initiation codon toward the 5′-end (upstream). This polypurine sequence preceding the initiation codon is complementary, to a greater or lesser extent, to the pyrimidine-rich 3′-terminal region of the ribosomal 16S RNA (. . . $GAUCACCUCCUUA_{OH}$ in *E. coli*), sometimes called anti-Shine-Dalgarno sequence, or ASD sequence. Between three and nine residues (usually four to six) may be complementary. In most cases it is the CCUCCU sequence of the 3'-terminal region of 16S RNA (in *E. coli*) that is complementary (although often in part) to the polypurine pre-initiation sequence of mRNA. Several examples are given in Fig. 15.4. This complementary pairing of the 3′-terminal region of 16S RNA (ASD) with the SD sequence seems to participate directly in the association between the initiating 30S ribosomal subunit and the RBS of mRNA.

It should be pointed out, however, that the initiation of translation is also possible in the case of mRNAs lacking the SD sequence. For example, the mRNA coding for the cI repressor of the phage λ does not have such a sequence before the initiation codon; even so, initiation is observed, although it is not very effective. Among bacterial (*E. coli*) gene transcripts, about 2% to 3% of presumptive messages seem to have no patent SD sequences. Also bacterial ribosomes may perform initiation with certain eukaryotic mRNAs devoid of the SD sequences. It may be concluded that the SD sequence contributes greatly to the effective association of the ribosomal particles with mRNAs, but probably is not an absolute requirement.

In the case of a polycistronic mRNA, the ribosome after termination at the preceding cistron may pass to the nontranslating state and dissociate into its subunits, but the 30S subunit may remain in association with mRNA for a time. Then, it can either be released from mRNA, or slide without phase along the intercistronic region and reinitiate at the next cistron (see Section 16.3). The chances of reinitiation without dissociation of the 30S subunit from mRNA seem to be greater with shorter and less structured intercistronic regions (de Smit and van Duin, 1990). If the ribosomal particle passes from termination at the preceding cistron to reinitiation at the next cistron without dissociation from mRNA, it may less strictly require the SD sequence for reinitiation. Generally, the nearby termination event may strongly increase the efficiency of the subsequent initiation (reinitiation). In many cases the independent *de novo* initiation (by free ribosomes) of an internal cistron with a poor SD sequence is found to be ineffective, while the reinitiation following termination of the upstream cistron (translational coupling) proceeds at a high rate.

As mentioned, the most frequently used initiation codon is AUG. At the same time, in *E. coli* about one tenth of prokaryotic messages start with GUG as an initiation codon. In about 1% of messages UUG is used. In a representative of Gram-positive bacteria, *Bacillus subtilis,* 30% of the starts are with GUG and UUG, and UUG is even more common than GUG. There are exceptional cases in bacteria where initiation takes place at AUU, AUA, ACG, and CUG. These rare starting triplets are "weak" initiation codons and seemingly may play this role only in combination with "strong" SD sequences upstream and/or other initiation-promoting structural elements.

In addition to the SD sequence and the initiation codon, the RBS of bacterial

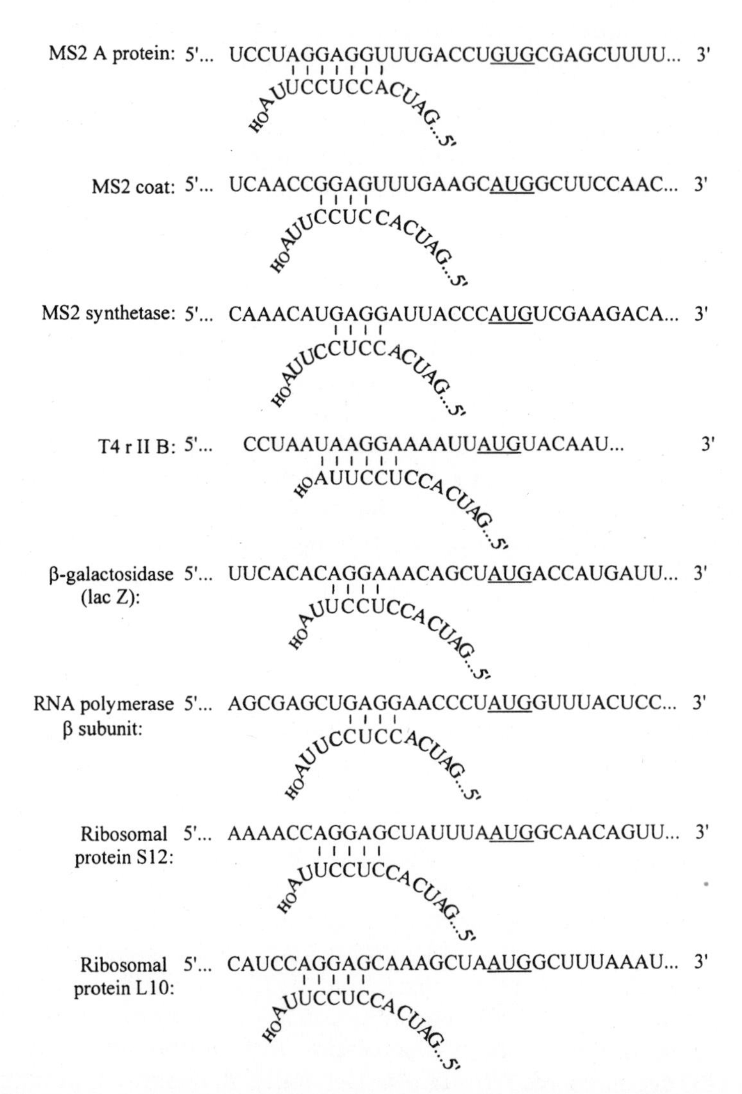

FIG 15.4. Complementary pairing of the 3′-terminal polypyrimidine cluster of ribosomal 16S RNA (*lower sequences*) with polypurine sequences, the so-called Shine-Dalgarno sequences, preceding the initiation codon AUG or GUG in a number of phage and bacterial mRNAs (*upper sequences*). For review, see Steitz, J. A. (1980) in *Ribosomes: Structure, Function, and Genetics,* (G. Chambliss, G. R. Craven, J. Davies, K. Davis, L. Kahan, and M. Nomura, eds.), pp. 479–495, University Park Press, Baltimore.

mRNA comprises both upstream and downstream sequences and covers about 55 to 60 nucleotides, from about position −35 to position +20 (relative to the first nucleotide of the initiation codon). The sequences beyond the SD element and the initiation triplet have been reported to be nonrandom and also contain information that is essential for efficient initiation of translation. In particular, A and U are favored throughout the RBS, and A is especially preferred downstream from the SD element. On the contrary, C is counter-selected in the RBS. Some optimal RBS sequence contexts can be deduced from the data available. For example, the frequent occurrences of A at position −3, GCU and AAA triplets as the codons following the initiation codon, often with A as the next nucleotide, and the sequence UUAA in the fourth and fifth codons have been mentioned.

In any case the region upstream of the SD sequence has been repeatedly reported to be important for efficient initiation. First of all, it must be unstructured in order to facilitate some necessary contacts with the initiating ribosomal particle. Up to 20–30 nucleotides upstream of the first SD residue may be involved in these stimulatory contacts. The existence of stable hairpins in this pre-SD region, even when the SD sequence and the initiation codon are not included, may block the initiation of translation. On the other hand, initiation "enhancers" can exist upstream of the SD element (for review, see McCarthy and Gualerzi, 1990). The well-known enhancing sequence is the epsilon motif UUUAACUUUA in highly expressed late mRNAs of bacteriophage T7 and some other phages, as well as the epsilon-like upstream sequences in some bacterial mRNAs (e.g., UUUUAACU and UAAUUUAC in *atpE* cistron of *atp* polycistronic mRNA of *E. coli,* see Section 16.2).

While the SD sequence of mRNA is complementarily bound with the 3′-terminal sequence of the 16S RNA, other regions of the RBS are believed to interact also with the ribosomal RNA elements of the 30S subunit (McCarthy and Brimacombe, 1994). Thus, according to chemical cross-linking experiments, the region between the SD sequence and the initiation codon is in contact with A665 in the side loop of helix 22 (domain II), A1360 in the end loop of helix 43 (domain III), and G1530 near the base of the ultimate helix 45 of the 16S RNA (see Fig. 9.3). It is interesting that these positions are in the regions of the secondary structure that are universally conserved in all ribosomes, including eukaryotic ones. In addition to the bases of the SD sequence, the bases of mRNA 10 nucleotides and 20–21 nucleotides upstream of the SD sequence have been reported to be protected against chemical modifications due to interaction with the ribosome (probably with the 16S RNA).

The portion of the RBS downstream of the initiation codon, however, does not display such contacts with specific positions of the 16S RNA, until the initiator tRNA is bound. Only after F-Met-RNA_f^{Met} is bound, the downstream regions of RBS become both protected against the hydroxyl attack on the sugar-phosphate backbone and specifically cross-linkable with the 16S RNA. The cross-linkable sites of the 16S RNA seem to be all in the mRNA-binding cleft of the 30S subunit; these are C1395–C1402 in the section between domain III and the long compound hairpin (helix 44) of the 3′-terminal sequence, A532 in the end loop of helix 18 (domain I), and U1052 in helix 34 (domain III) (see Fig. 9.3). The 16S RNA nucleotides cross-linkable with the downstream region of the RBS are universally conserved among ribosomes. The observation that the specific contacts of the downstream region of the RBS with the mRNA-binding cleft are established only after F-Met-RNA_f^{Met} is bound suggests that the mRNA retained by the 30S subunit prior to its fixation by codon–anticodon interaction is rather movable; probably, when it is bound only by ASD, it can slide around on the ribosome by virtue of the flexibility of the 3′-terminal region of the 16S RNA ("stand-by complex").

15.2.3. Prokaryotic Initiator tRNA

The initiation codons of prokaryotic mRNAs are recognized by formylmethionyl-tRNA (F-Met-$tRNA_f^{Met}$), with its anticodon CAU (Marcker and Sanger, 1964; Clark and Marcker, 1966). Thus, the codon–anticodon pairing during initiation may be either fully complementary (AUG), in terms of the classical Watson–Crick complementarity, or partially complementary ("two-of-three"). "Wobbling" at the first position of the initiation codon (GUG and UUG) seems to be the most allowable among the partially complementary codons. (Note that this wobbling principally differs from the Crick wobbling at the third position of codons during elongation).

The structure of the initiator $tRNA_f^{Met}$ is organized similarly to that of the elongator tRNAs: the secondary structure can be presented as a typical clover-leaf fold (Fig. 15.5), and the hairpins are arranged into the L-shaped pattern (like in Fig. 3.7), resulting in the formation of a compact molecule with the anticodon and acceptor protuberances, akin to the paradigmatic yeast $tRNA^{Phe}$ (see Figs. 3.8 and 3.10) (Woo, Roe, and Rich 1980). Some structural differences, however, may be of principal importance for the functions of the initiator tRNA. First, the 5′-terminal nu-

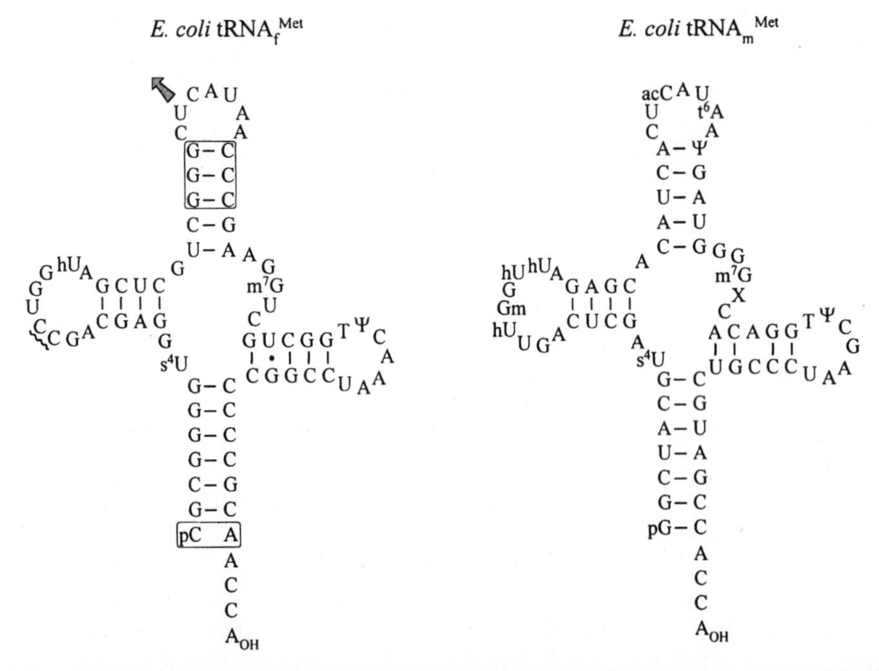

FIG 15.5. Nucleotide sequence and secondary structure of the prokaryotic initiator tRNA (*left*) (See Dube, S. K., Marcker, K. A., Clark, B. C. F., and Cory, S. (1968) *Nature* **218**:232–233), compared with the prokaryotic methionine tRNA participating in elongation (*right*). See Cory, S., Marcker, K. A., Dube, S. K., and Clark, B. F. C. (1968) *Nature* **220**:1039–1040. Note the unpaired 5′-terminal nucleotide residue, a unique CC duplet in the D loop, three consecutive G:C base pairs in the anticodon helix, and the invariant U33 of the anticodon loop turned outside in the initiator tRNA.

cleotide residue of the initiator $tRNA_f^{Met}$ does not form a Watson–Crick base pair with the 72nd nucleotide residue of the 3′-terminal region (Fig. 15.5). It appears to provide greater flexibility for the acceptor end; according to the crystallographic model, the 3′-end of $tRNA_f^{Met}$ curls back toward the 5′-end and does not continue the helical organization of the acceptor stem as in the case of the elongator tRNA species. Thus the formylmethionyl residue may lie on the acceptor stem helix and contribute to the specific recognition of F-Met-$tRNA_f^{Met}$ by IF2.

Second, the structure of the anticodon loop in the initiator $tRNA_f^{Met}$ displays deviations from the classical conformation in yeast $tRNA^{Phe}$: the invariant U in position 33 adjacent to the anticodon from the 5′-side is turned outside (whereas it turns inside the loop and faces the phosphate of the third anticodon residue in the elongator tRNA, see Fig. 3.8). For this reason the stacked conformation of the anticodon in the initiator $tRNA_f^{Met}$ looks distorted compared to the anticodon of the elongator tRNA. The unique presence of the stack of three consecutive G:C base pairs at the end of the anticodon helix (Fig. 15.5) in all initiator tRNAs seems to contribute to a disordered conformation of the anticodon loop.

Third, the nucleotide residues of the dihydrouridylic loop of the initiator $tRNA_f^{Met}$ (positions 16 and 17) are more tightly packed together and with the core of the tRNA, compared to elongator tRNA species where they appear to be loosely accommodated near the corner of the L-shaped molecule.

The differences mentioned may have relevance to the special functional features of the initiator tRNA: (1) it is recognized by methionyl-tRNA transformylase; (2) it is not recognized by EF-Tu in the aminoacylated form, but instead recognized by IF2; and (3) it primarily settles in the P site, rather than in the A site of the ribosome.

The initiator $tRNA_f^{Met}$ has a specific affinity for the normal methionyl-tRNA synthetase and, correspondingly, is capable of accepting methionine. Therefore, two classes of $tRNA^{Met}$ (Fig. 15.5) are acylated with methionine by the same ARSase: the elongator $tRNA_m^{Met}$ recognizing the methionine codon AUG in the course of elongation, and the initiator $tRNA_f^{Met}$ which can recognize the AUG triplet, as well as GUG and UUG (and rarely AUU, AUA, etc.), during initiation. In contrast to Met-$tRNA_m^{Met}$, the initiator Met-$tRNA_f^{Met}$ serves as a substrate for a special formyltransferase which transfers the formyl group from the formyl tetrahydrofolate to the amino group of the methionine residue, yielding formylmethionyl-tRNA:

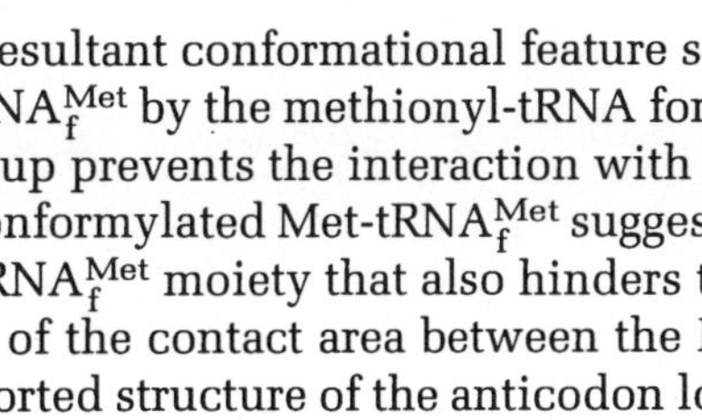

$$\mathrm{H{-}C({=}O){-}NH{-}CH\big((CH_2)_2{-}S{-}CH_3\big){-}C({=}O){-}O{-}tRNA}$$

The unpaired end and/or a resultant conformational feature seems to be required for the recognition of Met-$tRNA_f^{Met}$ by the methionyl-tRNA formylase. In turn, the formylation of the amino group prevents the interaction with EF-Tu. EF-Tu, however, interacts poorly with nonformylated Met-$tRNA_f^{Met}$ suggesting the presence of a structural element in the $tRNA_f^{Met}$ moiety that also hinders the interaction with EF-Tu. The altered structure of the contact area between the D- and T-loops may be such an element. The distorted structure of the anticodon loop may be relevant to the primary positioning of F-Met-tRNA$_f^{Met}$ at the P site and to the unusual wobble interactions of its anticodon with a variety of initiation codons (Varshney, Lee, and RajBhandary, 1993; Dyson, Mandal, and RajBhandary, 1993).

In its aminoacylated and formylated state, F-Met-tRNA$_f^{Met}$ plays a key role in initiation. Correspondingly, formylmethionine is always the first residue of any polypeptide chain to be synthesized by the prokaryotic ribosome. During subsequent elongation the formyl residue is cleaved off by formylase (Adams, 1968). The first methionyl residue is often, although not always, cleaved, also cotranslationally, from the growing polypeptide chain by a special aminopeptidase (Capecchi, 1966).

15.2.4. Prokaryotic Initiation Factors

Three proteins are generally required for initiation of translation in prokaryotic systems; they are referred to as initiation factors—IF1, IF2, and IF3 (for reviews, see Maitra, Stringer, and Chaudhuri, 1982; Gualerzi *et al.*, 1986, 1990; Hartz, McPheeters, and Gold, 1990; McCarthy and Gualerzi, 1990). All of them have an affinity for the 30S ribosomal subunit. It is probable that under *in vivo* conditions they are mostly trapped by free 30S subunits, thus forming the so-called native 30S particles which are competent to enter the initiation process.

IF1 is a small basic protein with a molecular mass of 8 kDa (71 amino acid residues in *E. coli*). This factor seems to be an auxiliary protein taking part in dissociation/association of ribosomes at the early and late initiation steps, respectively, and in stabilization of the interactions of the 30S subunit with the other factors and F-Met-tRNA. IF1 has not been found in some bacterial species.

IF2 is a large acidic protein possessing the GTP-binding domain homologous to those of EF-G and EF-Tu. It has been isolated from *E. coli* in two forms (products of the same gene) differing in molecular mass: IF2α of 97 kDa (889 amino acids) and the truncated form IF2β of 80 kDa (732 amino acids). Both forms appear to function equally in initiation. This protein is the principal initiation factor responsible for the GTP-dependent binding of F-Met-tRNA$_f^{Met}$ to the 30S ribosomal particle and the ribosome-induced GTPase activity. The recognition of F-Met-tRNA$_f^{Met}$ by IF2 seems to involve the T-loop and T-stem of the tRNA$_f^{Met}$.

IF3 is a slightly basic protein with a molecular mass of about 20 kDa (181 amino acids in *E. coli*). When bound with the 30S subunit it prevents the reassociation with the 50S subunit. As a component of the initiating "native" 30S particle, IF3 seems to contribute to IF2 binding, mRNA binding, and, more important, to selective binding of initiator tRNA$_f^{Met}$, probably recognizing its anticodon stem. There is evidence that IF3 destabilizes the complexes of elongator tRNAs with the 30S subunit (i.e., the wrong complexes). Structurally this is an interesting protein: it has a dumb-bell shape and consists of two independent compact α/β domains connected by a long α-helix (Biou, Shu, and Ramakrishnan, 1995; Garcia *et al.*, 1995). The exposed β-sheets of the two domains are separated by 45 Å and thought to interact with two distant regions of the 16S RNA on the 30S ribosomal subunit. It is interesting that one of the binding sites of IF3 on the 16S rRNA (positions 790–793 in the middle domain; see Fig. 6.1) seems to be involved also in the subunit association. This supports the assumption that IF3 shields the 16S rRNA site important for the subunit association, and so has to dissociate from the 30S initiation complex to allow formation of the 70S ribosome.

15.2.5. Sequence of Events

15.2.5.1. Step 1

The dissociation of nontranslating (terminated) 70S ribosomes into 30S and 50S subunits (Section 15.5) precedes the initiation of translation in all cases. Under physiological conditions, the nontranslating ribosomes seem to be in reversible equilibrium with their subunits. IF1 may accelerate the dissociation–association reaction, whereas IF3 binds to 30S subunits and removes them from the equilibrium:

$$
\begin{array}{c}
70S \rightleftharpoons 30S + 50S \\
+ \\
IF3 \\
\downarrow \\
30S{:}IF3
\end{array}
$$

15.2.5.2. Step 2

The 30S ribosomal subunit with IF3, or with all three bound initiation factors ("native" 30S particle), can associate with the initiation region or the RBS of mRNA. This region may be located near the 5′-end or far from it; in the case of polycistronic mRNA there may be several such regions along the mRNA chain. What is important is that this region is accessible for the interaction with the ribosomal particle and contains the polypurine SD sequence, as well as the initiation triplet (AUG is the most preferable) at the proper distance downstream. In the absence of initiation factors, the 30S ribosomal subunit itself is also capable of recognizing the initiation region (RBS) of mRNA, but IF3 probably enhances the interaction:

$$30S{:}(IF3{:}IF1{:}IF2) + mRNA \rightarrow mRNA{:}30S{:}(IF3{:}IF1{:}IF2)$$

The mRNA bound in this complex is retained mainly by virtue of SD:ASD interaction and seems to be not properly established yet in the mRNA-binding cleft of the 30S subunit ("stand-by state").

15.2.5.3. Step 3

According to the above classical scenario, F-Met-tRNA$_f^{Met}$ is bound to the mRNA:30S complex at the next step. The binding is GTP-dependent. It is mediated by IF2. If IF2 is already sitting on the 30S particle, it can be activated by GTP and thus acquire the affinity for F-Met-tRNA$_f^{Met}$, and then the F-Met-tRNA$_f^{Met}$ may join the complex:

$$
\begin{array}{l}
mRNA{:}30S{:}IF3{:}IF1{:}IF2 + GTP + F\text{-}Met\text{-}tRNA \rightarrow \\
\quad \rightarrow mRNA{:}30S{:}IF1{:}IF2{:}GTP{:}F\text{-}Met\text{-}tRNA + IF3
\end{array}
$$

Codon–anticodon interaction of the initiation triplet of mRNA with the bound F-Met-tRNA$_f^{Met}$ is believed to set the mRNA in the mRNA-binding site of the 30S subunit (Section 9.2).

If free IF2 with GTP encounters F-Met-tRNA$_f^{Met}$ in solution, they first form the ternary complex F-Met-tRNA:IF2:GTP, and then this complex binds to the mRNA:30S complex:

$$\text{mRNA:30S:IF3:IF1} + \text{F-Met-tRNA:IF2:GTP} \rightarrow$$
$$\rightarrow \text{30S:IF1:IF2:GTP:F-Met-tRNA} + \text{IF3}$$

There are indications that IF3 dissociates from the ribosomal particle upon F-Met-tRNA$_f^{Met}$ binding.

The alternative pathway has been also discussed when the "native" 30S particle with initiation factors first binds F-Met-tRNA$_f^{Met}$, and then the complex formed interacts with the ribosome-binding region of mRNA:

$$\text{30S:IF3:IF1:IF2} + \text{F-Met-tRNA} \rightarrow \text{30S:IF1:IF2:GTP:F-Met-tRNA} + \text{IF3}$$
$$\text{30S:IF1:IF2:GTP:F-Met-tRNA} + \text{mRNA} \rightarrow$$
$$\text{mRNA:30S:IF1:IF2:GTP:F-Met-tRNA}$$

In any case it is the anticodon of the bound F-Met-tRNA$_f^{Met}$ that searches for and finds the initiation codon downstream from the polypurine sequence and sets the ribosomal particle precisely at the start of the coding sequence.

15.2.5.4. Step 4

Now the initiating 30S complex is ready to join the 50S ribosomal subunit. IF1 seems to be released concurrently with the subunit association. The factor-binding site of the 50S subunit interacts with IF2 and induces the GTPase activity of the factor. GTP is hydrolyzed, resulting in the loss of the affinity of IF2 for F-Met-tRNA$_f^{Met}$ and the ribosome. Thus the 70S particle is formed precisely at the initiation codon of mRNA, with initiator F-Met-tRNA$_f^{Met}$ in the P site:

$$\text{mRNA:30S:IF1:IF2:GTP:F-Met-tRNA} + \text{50S} \rightarrow$$
$$\rightarrow [\text{mRNA:30S:IF2:GTP:F-Met-tRNA:50S}] + \text{IF1} \rightarrow$$
$$\rightarrow [\text{mRNA:30S:IF2:GDP:F-Met-tRNA:50S}] + \text{IF1} + P_i \rightarrow$$
$$\rightarrow \text{mRNA:70S:F-Met-tRNA} + \text{IF1} + \text{IF2} + \text{GDP} + P_i$$

15.2.5.5. Step 5

The initiating 70S complex formed in the above reaction has the vacant A site, with the codon set there that immediately follows the initiation codon. The P site is occupied with an analog of peptidyl-tRNA. Thus the 70S complex is ready to accept the first elongator aminoacyl-tRNA at its A site and form the first peptide bond between the two substrates, initiator F-Met-tRNA$_f^{Met}$ and elongator aminoacyl-tRNA:

$$\text{mRNA:70S:F-Met-tRNA}_f + \text{Aa-tRNA}_e\text{:EF-Tu:GTP} \rightarrow$$
$$\rightarrow \text{mRNA:70S:F-Met-tRNA}_f\text{:Aa-tRNA}_e\text{:EE-Tu:GTP} \rightarrow$$

$$\rightarrow \text{mRNA:70S:F-Met-tRNA}_f\text{:Aa-tRNA}_e + \text{EF-Tu:GDP} + P_i \rightarrow$$
$$\rightarrow \text{mRNA:70S:F-Met-Aa-tRNA}_e + \text{tRNA}_f^{Met} + \text{EF-Tu:GDP} + P_i$$

This is the end of initiation and the beginning of elongation.

Fig. 15.6 presents schematically a tentative sequence of events during prokaryotic initiation of translation.

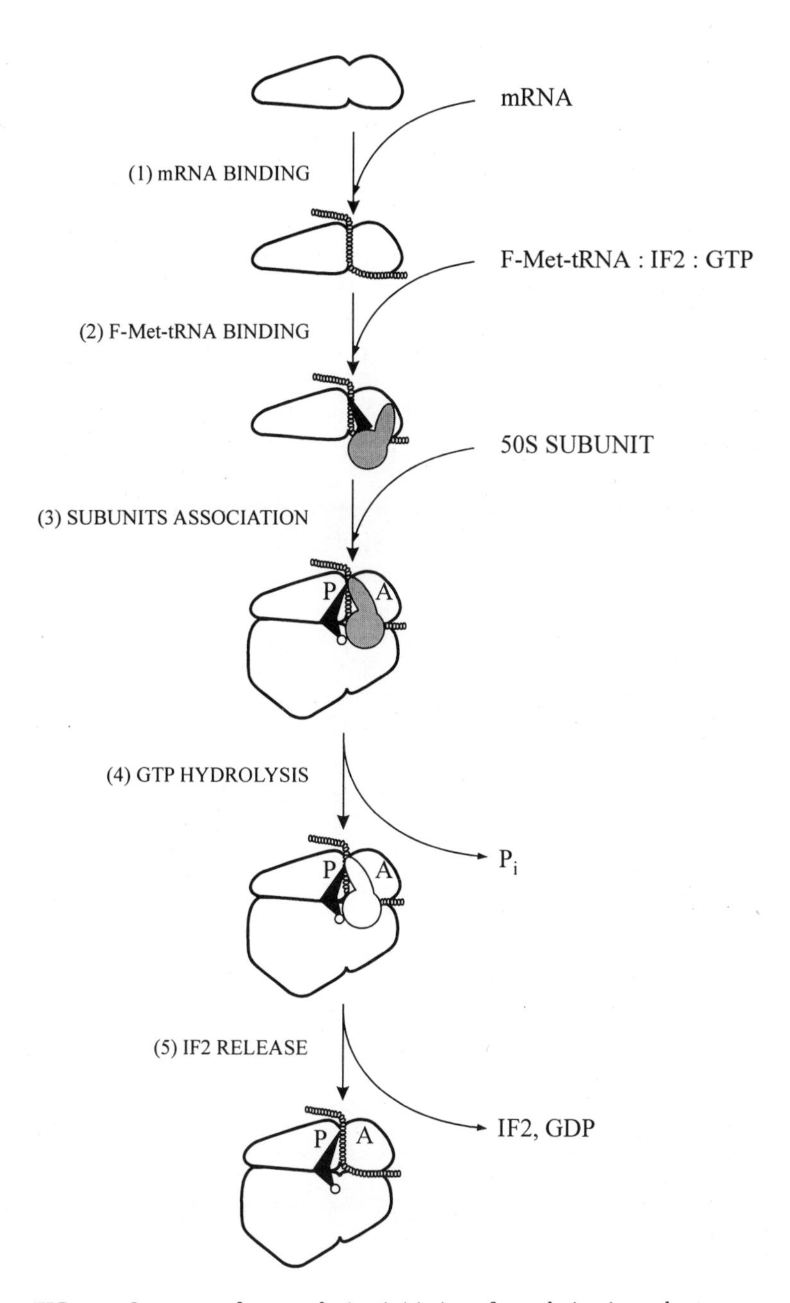

FIG 15.6. Sequence of events during initiation of translation in prokaryotes.

15.3. Initiation in Eukaryotes

15.3.1. Characteristics of Eukaryotic mRNA

The eukaryotic mRNA is synthesized as pre-mRNA in the nucleus. From the beginning it is complexed with a large amount of protein. Nuclear pre-mRNA:protein complexes are processed into mRNA:protein complexes (mRNPs) which go through the nuclear membrane. During the transport from the nucleus to the cytoplasm, the mRNPs strongly change their protein composition. The protein-to-mRNA weight ratio is about 4:1 in cytoplasmic mRNPs that are not yet engaged in translation. After initiation of translation and during elongation the mRNA becomes significantly populated by translating ribosomes and loses part of the proteins. Nevertheless, the mRNA within translating polyribosomes is still found to retain a large proportion of protein, the protein-to-mRNA weight ratio being 2:1 or higher. Thus all eukaryotic mRNA at every stage of its life exists in the form of mRNP (for reviews, see Preobrazhensky and Spirin, 1978; Spirin, 1996). This seems to be the most characteristic feature of the eukaryotic mRNA.

It follows that the initiation machinery in eukaryotes deals with highly protein-loaded mRNPs, not just with mRNA. At least three consequences of this fact may be relevant to the initiation process. First, secondary and tertiary structures of mRNA may be significantly modified by proteins within mRNPs: on one hand some "core" proteins are known to melt the structures of mRNA; on the other hand, local hairpins and folds can be stabilized by interacting with proteins. Second, a direct competition between the proteins of mRNPs and the mRNA-binding initiation factors is possible. Third, the structural organization (quaternary structure) of mRNP particles may modulate the accessibility of mRNA for the participants of the initiation process, such as ribosomes and initiation factors.

Since mRNA can exist in eukaryotic cytoplasm in two forms, untranslated free mRNP particles and translating polyribosomes, the initiation process may be very different depending on whether ribosomal particles and initiation factors interact with the mRNP and open a "virginal" mRNA for translation, or whether they bind to a translated mRNA in line after previously initiated ribosomes. In special cases reinitiation after termination, without dissociation of ribosomes from mRNA, is possible when a short open reading frame with termination codon is present upstream of a principal coding sequence (see Section 17.4). Some observations are compatible with the idea that reinitiation can occur also during routine translation of monocistronic messages in circular polyribosomes: if the termination region of mRNA is in spatial proximity to the initiation region, the ribosome after termination may reinitiate translation of the same mRNA (see Section 15.3.7).

Other characteristic features typical of most, *but not all,* eukaryotic mRNAs should also be mentioned (for reviews, see Bag, 1991; Merrick and Hershey, 1996). First, eukaryotic mRNAs are mostly monocistronic. Second, they are capped at the 5′-end, with few exceptions. The best known exception is genomic RNA of picornaviruses. Third, many eukaryotic mRNAs are characterized by long 3′-untranslated regions (3′-UTRs) often comparable in length with their coding sequences. Finally, the majority of mRNAs in the eukaryotic cytoplasm are polyadenylated at the 3′-end. All these features are also directly relevant to the mechanism of initiation and/or its regulation.

In accordance with the 5′-terminal, cap-dependent initiation as a predominant mode of translation initiation in eukaryotes, the eukaryotic mRNAs do not possess the SD sequence upstream of their initiation codons. As a general rule, no marked complementarity between the 3′-end of the RNA of the small ribosomal subunit and the pre-initiation sequence of mRNA is observed, in contrast to the situation in prokaryotic organisms. It is remarkable that the 3′-terminal 50-nucleotide-long sequences of the ribosomal RNAs of the small ribosomal subunits, including both the 16S RNA of prokaryotes and the 18S RNA of eukaryotes, are very conservative in evolution and show high homology, forming a similar stem-loop structure; however, the polypyrimidine CCUCC block of the prokaryotic 16S RNA (ASD sequence) is absent from the eukaryotic 18S RNA (Fig. 15.7). This leads to the assumption that, for important reasons, eukaryotes had to dispose of the complementary recognition between the ribosomal RNA and mRNA in the initial association of ribosomes with the message and develop another way of recognizing the initiation sequence. The new way appears to include the initial association with the 5′-end of mRNA. (This reasoning may be turned upside down: recognizing the 5′-end could be a more ancient mechanism, and the evolution of prokaryotes, particularly the evolution of operons and polycistronic messages, could induce the appearance of a specialized mechanism for recognizing internal initiation regions; the pairing of the polypurine SD sequence of mRNA with the evolutionarily acquired CCUCC insert near the 3′-end of the ribosomal 16S RNA is part of such a new mechanism).

15.3.2. The Cap Structure and the Initiation Codons

The cap (see Fig. 2.4) is added at the 5′-end of eukaryotic mRNAs during synthesis and processing of pre-mRNA in the nucleus. The cap structure plays a very important part in the initiation of translation. It is the cap structure that attracts initiation factors and ribosomes for the 5′-terminal initiation (see Fig. 15.1A).

The cap is followed by the so-called 5′-untranslated region (5′-UTR) of mRNA. Typically the 5′-UTRs of eukaryotic mRNAs are not very long, up to several dozen nucleotides, although there are many exceptions. The cap and the adjacent section of the 5′-UTR form the RBS where the initiating 40S ribosomal subunit (43S initi-

E. coli 16S RNA:	Rat 18S RNA:
G m_2^6A	G m_2^6A
G m_2^6A	U m_2^6A
G−C	G−C
G−C	G−C
A−U	A−U
U−G	U−G
G−C	G−C
C−G	C−G
C−G	C−G
A−U	U−A
A−U	U−A
U−G	U−G
5'.... GG GAUCA**CCUCC**UUA$_{OH}$	5'.... GG GAUCAUUA$_{OH}$

FIG 15.7. Comparison of nucleotide sequences and secondary structures of the 3′-terminal regions of prokaryotic 16S and eukaryotic 18S ribosomal RNAs. The pyrimidine pentanucleotide insert ("anti-Shine-Dalgarno sequence") of the prokaryotic RNA is in black box. See review by Steitz, J. A. (1980) in *Ribosomes: Structure, Function, and Genetics* (G. Chambliss, G. R. Craven, J. Davies, K. Davis, L. Kahan, and M. Nomura, eds.), pp. 479–495, University Park Press, Baltimore.

ation complex) may bind. There is evidence that eukaryotic mRNAs lacking stable secondary structure in their 5′-UTRs display a higher initiation rate than those possessing such structures, and the inhibition of initiation by stable helical elements is more severe when they are close to the 5′-terminus (to the cap structure).

In contrast to prokaryotes, the initiation codon in eukaryotic mRNAs is usually located some distance away from the RBS along the polynucleotide chain (Fig. 15.8). In order to reach the initiation codon from the RBS, the ribosomal initiation complex must move along the 5′-UTR downstream. According to the classical model of M. Kozak (1978, 1980, 1989b), in the process of the movement the initiation complex scans the sequence of mRNA until it encounters the triplet AUG; thus this triplet becomes the initiation codon. Indeed, in many eukaryotic mRNAs the first AUG triplet from the 5′-end is used as the initiation codon and thus establishes the reading frame of the subsequent sequence.

At the same time there are many cases where the first AUG triplet is not an initiation codon, but the initiation occurs at the second or the third (or the next) AUG, this being not necessarily in frame with the previous ones. The explanation is that the scanning ribosomal particles can skip some AUG triplets if they are not in a proper structural environment. Nearly all of the functional initiation AUG codons of eukaryotic mRNAs are preceded by the triplet beginning with the purine nucleotide, in most cases with A. It is assumed that when the initiating 40S particle scans the template, it preferentially recognizes the sequence PuNNAUG (where N can be any nucleotide residue) as the correct initiation site (Kozak, 1981, 1989a). AUG triplets with the preceding PyNN triplet seem to be "weak" initiators and can be skipped without initiation. In addition, the functional initiation codons have G as a preferential neighbor at the 3′-side and the C-rich pentanucleotide sequence at the 5′-side; for example, the sequence CCACC**AUGG** provides a proper initiating "strength" to the AUG triplet in it.

There are special cases, particularly among some virus-induced mRNAs, where the phenomenon of "two initiation points" is observed: the first AUG triplet is recognized as the initiation codon only by a portion of scanning ribosomes, thus being just a "weak" initiator, while the other ribosomes skip it without initiation and initiate at the next "strong" AUG. In such cases, mRNA behaves functionally

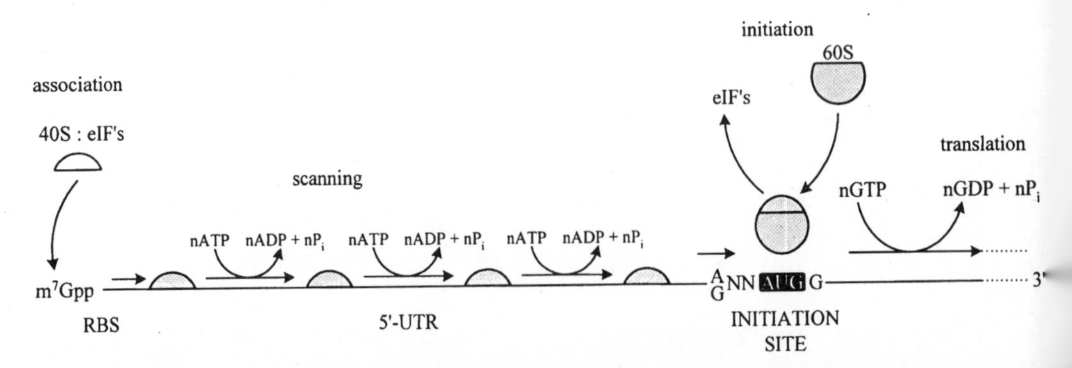

FIG 15.8. Scanning model for initiation of translation in eukaryotes. See Kozak, M. (1978) *Cell* **15:**1109–1123; (1980); **22:**7–8; (1980) 459–467. The ribosomal particle (40S subunit) with initiation factors moves phaselessly along mRNA from the 5′-end downstream and unwinds its secondary/tertiary structure using the energy of ATP hydrolysis until it encounters an AUG triplet in the proper context; this AUG serves as the initiation codon for the following translation.

bicistronic, since the synthesis of two different polypeptides is initiated and proceeds on the overlapping nucleotide sequences.

An alternative to the scanning model has been proposed (Sonenberg, 1991). It presumes that the cap structure and the initiation codon region, being far apart along the sequence, are close to each other in the three-dimensional structure formed by the cap with the 5′-UTR. According to this model, the initiating ribosomal particle recognizes not just the cap structure, but a three-dimensional structural element including both the cap and the initiation codon.

It should be added that the eukaryotic initiation mechanism is much more strict as to the nature of the initiation triplet, in comparison with prokaryotic systems: AUG is almost exclusively used as the initiation codon in eukaryotic mRNAs (reviewed by Kozak, 1983). There are very rare cases of starting at other triplets, such as GUG, UUG, CUG, and ACG.

15.3.3. Internal Ribosome Entry Site

In addition to the cap-dependent 5′-terminal initiation most commonly used in eukaryotes, the internal initiation mechanism is found to be also inherent in eukaryotic systems (for reviews, see Meerovitch, Sonenberg, and Pelletier, 1991; Jackson, 1996; Ehrenfeld, 1996). It has been found that the eukaryotic ribosomes (initiating 40S particles) are capable of recognizing some special three-dimensional structural elements inside mRNA molecules, binding to them and starting either scanning or translation. These elements are designated *internal ribosome entry sites* or IRESes. The best known case of internal initiation in eukaryotic systems is that of translation of picornavirus RNAs. Among cellular mRNAs of the eukaryotic cell the products of genes that control growth and differentiation often have long 5′-UTRs with many AUGs, and they are suspected to possess IRESes and use the internal initiation mechanism; for some of these mRNAs the cap-independent internal initiation has been proven. The best studied cases of cellular mRNAs with internal initiation are the *Antennapedia* mRNA of *Drosophila* and the mammalian BiP mRNA which codes for immunoglobulin heavy-chain binding protein, also called 78-kDa glucose-regulated protein (GRP 78) (Macejak and Sarnow, 1991; Oh, Scott, and Sarnow, 1992).

The structure of IRES is not known, although some primary and secondary structural motifs of the picornavirus IRESes have been identified. The picornaviral IRES comprises a sequence of about 450 nucleotide residues. This segment of RNA appears to form a compact three-dimensional structure having an affinity for the initiating 40S ribosomal particle ("ribosome landing pad"). With respect to the IRES secondary structure, picornavirus RNAs can be divided into two groups with high intergroup conservation, but little similarity between the groups. Thus, enteroviruses (e.g., poliovirus) and rhinoviruses have one type of IRES folding, whereas cardioviruses (e.g., encephalomyocarditis virus) and aphthoviruses (foot-and-mouth disease virus) possess another secondary-structure pattern of their IRESes (Fig. 15.9). Despite very dissimilar secondary and probably tertiary structure, both types of IRESes effectively provide internal initiation, that is the selective association with ribosomal particles and subsequent initiation of translation.

The picornaviral IRES is followed by another important element, the so-called *starting window* (Pilipenko *et al.*, 1994). This is a segment of about 10 nucleotides long located at a fixed distance of 16 or 17 nucleotides from the 3′-boundary of IRES (Fig. 15.10). If an AUG triplet is found within this segment, the ribosomal

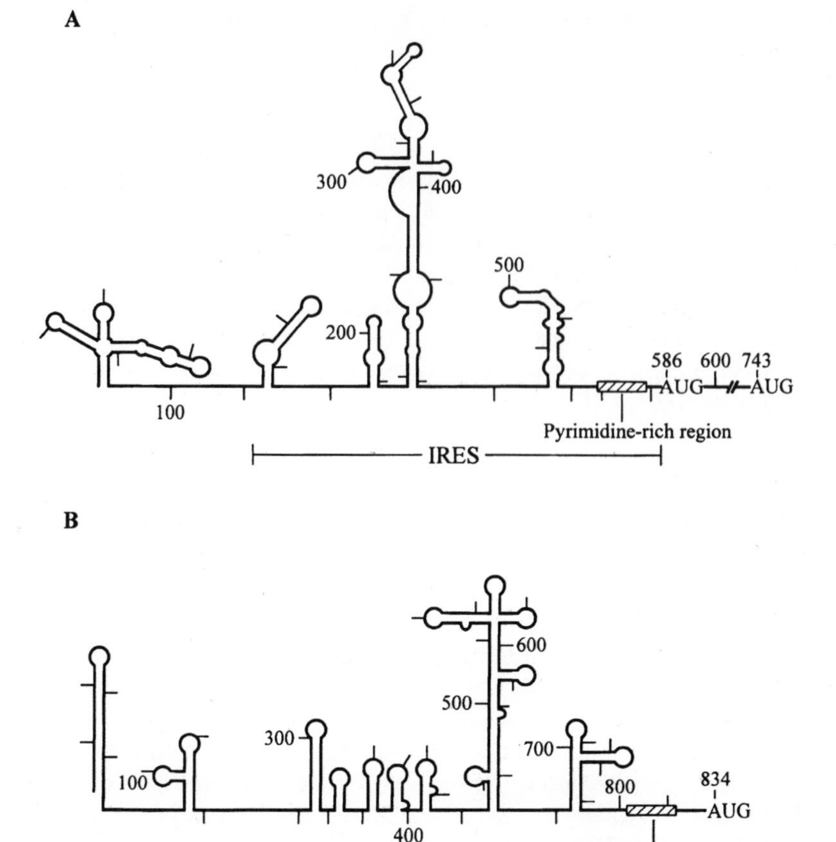

FIG 15.9. Schematic representation of the predicted secondary structures of the 5′-untranslated regions (5′-UTRs) of picornavirus RNAs. (A) 5′-UTR of poliovirus RNA. (B) 5′-UTR of encephalomyocarditis virus RNA. Reproduced, with minor modifications, from Ehrenfeld, E. (1996) in *Translational Control* (J. W. B. Hershey, M. B. Mathews, and N. Sonenberg, eds.), pp. 549–573, CSHL Press, with permission.

particle associated with the IRES starts either translation or movement along RNA to search for the next AUG (scanning). The sequence in this segment is not of great importance for the function of the starting window but it must not be involved in stable secondary or tertiary structure interactions; pairing of the segment in a perfect double-stranded helix closes the window.

In the cases of picornavirus RNAs the spacer between the IRES and the starting window usually includes a conserved polypyrimidine tract, with a sequence (UUUCC) complementary to the 3′-proximal polypurine sequence (GGAAC) of the 18S RNA of the 40S ribosomal subunit (Pilipenko *et al.*, 1992). Thus, the AUG of the starting window is found at a defined distance of about 20 to 25 nucleotides downstream of the polypyrimidine tract (see Figs. 15.9 and 15.10). The situation resembles that with the SD sequence of prokaryotes but in a "reversed" fashion: here *pyrimidines of the pre-initiation sequence* of mRNA may be paired with *purines of the 3′-terminal section* of ribosomal RNA. However, as already mentioned, this AUG does not necessarily serve as the initiation codon. In the case of poliovirus RNA the ribosomal particles scan the sequence further and initiate at the second AUG downstream, whereas in the cardioviruses the entering ribosomal particles initiate translation at this AUG.

Other viral RNAs may also possess IRESes, but of quite different size and structure. The hepatitis C virus (HCV) RNA has an IRES of about 200 to 300 nucleotides long, located just before the initiation codon; it has nothing in common with the picornaviral IRESes. The IRESes recently discovered in some plant viral RNAs are even shorter, about 150 nucleotides long (Ivanov *et al.*, 1997). The IRESes of *Antennapedia* and BiP are also smaller than the picornaviral IRESes. There is no obvious resemblance among these IRESes in their primary and secondary structures. The pyrimidine-rich tract located 25 nucleotides upstream of the effective AUG typical of the picornaviral IRESes is absent from all other known IRESes. It seems that the IRESes of other viral RNAs and cellular mRNAs have their

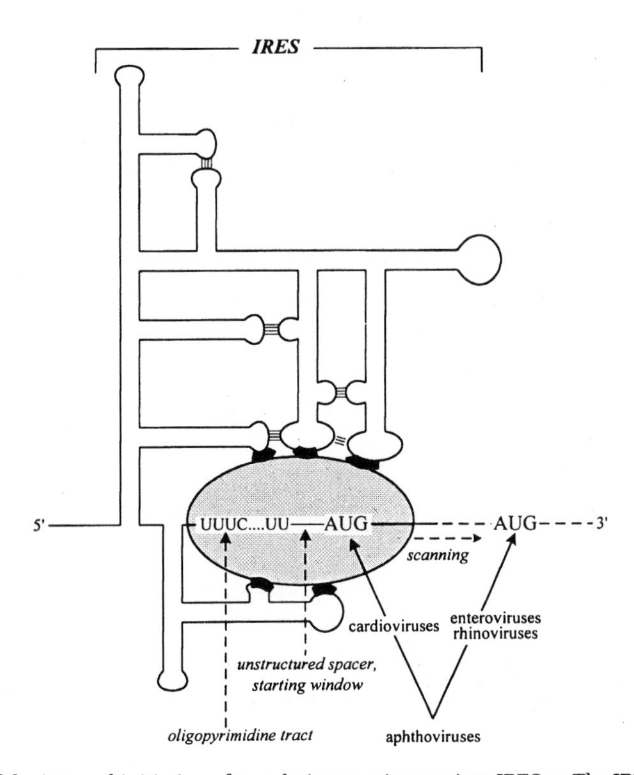

FIG 15.10. Model for internal initiation of translation on picornavirus IRESes. The IRES has a complex tertiary structure with a number of interacting stem loops. This structure presents several sequence motifs, denoted by black bars, for interactions with the initiating ribosomal particle (40S subunit) and the associated initiation factors thus forming an IRES. The 3'-section of IRES includes the oligopyrimidine tract and the unstructured spacer (see the text). The AUG triplet immediately follows the IRES. The AUG at the 3'-end of the IRES serves as the initiation codon in translation of the RNA of cardioviruses (e.g., encephalomyocarditis virus RNA). In the case of the RNAs of enteroviruses (e.g., poliovirus) and rhinoviruses, this AUG accepts the initiating ribosomal particle but is not used as an initiation site; instead, the ribosomal initiating complex seems to start scanning the downstream sequence until it encounters the next AUG that is to be the authentic initiation codon. The aphthoviruses (e.g., foot-and-mouth disease virus) represent an intermediate case, with some of the internally entering ribosomes initiating translation at the AUG at the 3'-end of the IRES and some utilizing the next downstream AUG. Reproduced, with minor modifications, from Jackson, R. J. (1996) in *Translational Control* (J. W. B. Hershey, M. B. Mathews, and N. Sonenberg, eds.), pp. 71–112, CSHL Press, with permission.

own specific three-dimensional structures different from each other and from those of picornavirus RNAs. It is not clear whether any complementary interactions between 18S ribosomal RNA and IRES sequences are involved in the association between the initiating 40S subunit and the IRESes of different types. The structural specificity of IRESes may be connected with the involvement of tissue-specific IRES-binding proteins required for the internal initiation.

15.3.4. Eukaryotic Initiator tRNA

A special tRNA acylated by methionyl-tRNA synthetase serves as the initiator tRNA that recognizes the initiation AUG codon (Smith and Marcker, 1970). The situation with eukaryotes in this respect is similar to that of prokaryotes. However, in contrast to prokaryotic initiator aminoacyl-tRNA, the methionyl-$tRNA_i^{Met}$ of eukaryotes does not undergo formylation. Hence, the only difference between the initiator methionyl-$tRNA_i^{Met}$ and the elongator methionyl-$tRNA^{Met}$ lies in certain structural features of the $tRNA_i$ moiety itself. These features are what make the $tRNA_i$ capable of interacting with initiation factors (eIF2) and the vacant ("native") 40S ribosomal subunit, and incapable of binding to the elongation factor (eEF1A) and entering into the A site during elongation.

The primary and secondary structures, in the cloverleaf form, of the two eukaryotic tRNA species are presented in Fig. 15.11. It can be seen that the differences are not that great. The most salient feature of the cytoplasmic initiator

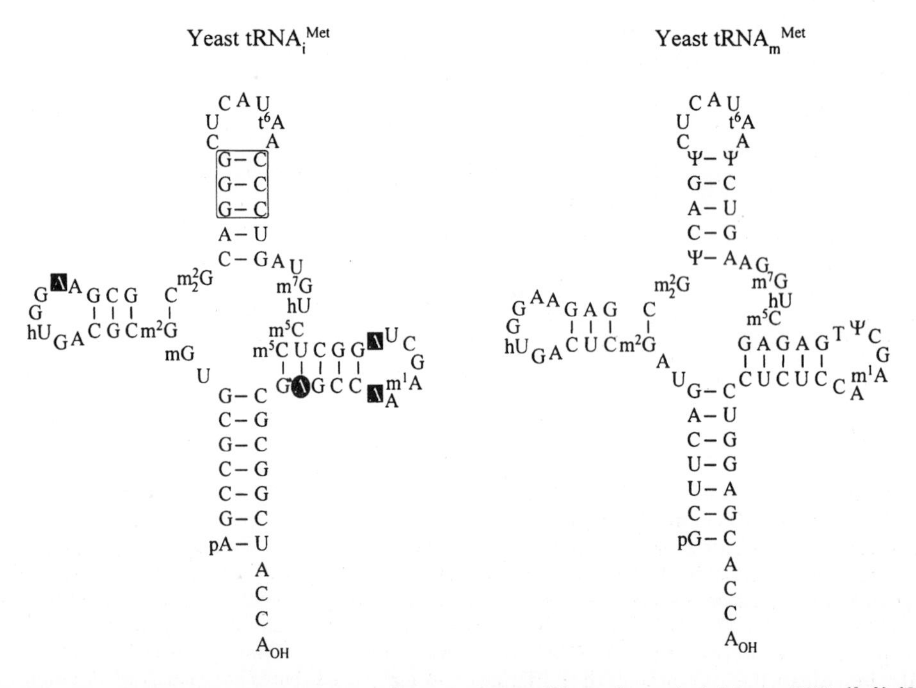

FIG 15.11. Nucleotide sequence and secondary structure of the eukaryotic initiator tRNA (*left*) (See Simsek, M., and RajBhandary, U. L. (1972) *Biochem. Biophys. Res. Commun.* **49**:508–515), compared with the eukaryotic methionine tRNA participating in elongation (*right*) (See Gruhl, H., and Feldman, H. (1975) *FEBS Letters* **57**:145–148). Note a unique 2'-phosphoribosyl adenosine in the T stem (position 64), a "triple A" (positions 20, 54, and 60 clustered together in tertiary structure), and three consecutive G:C pairs in the anticodon stem of the initiator tRNA.

tRNAs of eukaryotes is the unique presence of a "triple A": one A is in the D loop in position 20 (where it is very rare in other tRNAs), and two are in the T loop in positions 54 (instead of the usual T in elongator tRNAs) and 60 (instead of a pyrimidine residue in elongator tRNAs). These three As cluster together in the three-dimensional fold to form a hydrogen-bonded system not seen in elongator tRNAs (Basavappa and Sigler, 1991). The result is the strengthening of the interaction between the D and T loops. It is possible that the tighter contact in the region of the tRNA corner prevents some structural adjustments required for binding at the A site of the ribosome during elongation.

Like in the case of the prokaryotic initiator tRNA, three consecutive G:C pairs are present at the end of the anticodon stem. This structural feature is connected with a distortion of the typical conformation of the anticodon loop and stem, thus probably imparting a preferential interaction of the initiator tRNA with the P site, rather than with the A site of the ribosome.

Cytoplasmic initiator tRNAs of plants and fungi have a hypermodified purine nucleoside residue in the T stem at position 64. This is 2′-phosphoribosyl adenosine in the case of yeast $tRNA_i^{Met}$. It has been demonstrated that the large hydrophilic residue at position 64 prevents the interaction of the $Met\text{-}tRNA_i$ with the elongation factor eEF1A, not affecting the binding with the initiation factor eIF2 (Forster, Chakraburtty, and Sprinzl, 1993). Thus, the hypermodification in the T stem appears to act as a negative discriminant for eEF1A, akin to the formylation of the α-amino group of $Met\text{-}tRNA_i$ in the case of prokaryotes. Exactly which feature of the animal initiator tRNA structure (where position 64 is not modified) takes the part of the negative discriminant for eEF1A remains to be determined.

In contrast to the initiator $tRNA_f^{Met}$ of prokaryotes and mitochondria, the 5′-terminal nucleoside residue of the eukaryotic cytoplasmic initiator $tRNA_i^{Met}$ is paired with the nucleoside residue of the 3′-terminal region, just as in all elongator tRNAs. However, it is always the "weak" A1:U72 pair in the initiator tRNAs of animals, plants, and fungi. Such a pair at the end of a helix can be easily disrupted, like in the case of the interaction of aminoacyl-tRNAs with the class I ARSases (see Section 3.3). An attractive model is that it is disrupted at the interaction with eIF2, and then the 3′-tail of the $Met\text{-}tRNA_i$ bends back and thereby enables the terminal adenosylmethionine residue to lie on the helical acceptor stem (Basavappa and Sigler, 1991). In such a case the methionine side chain could positively contribute to the recognition of $Met\text{-}tRNA_i$ by eIF2.

15.3.5. Ribosomal Initiation Factors

The eukaryotic initiation factors (for reviews, see Merrick, 1992; Merrick and Hershey, 1996) can be conditionally divided into two principal groups: those that bind and operate with ribosomal particles promoting ribosomal subunit dissociation/association, initiator $Met\text{-}tRNA_i$ binding, and mRNA binding; and those that are aimed at mRNA and engaged in preparing its upstream region for initiation. The first group (which is called here *ribosomal initiation factors*) contains the factors analogous to prokaryotic IF1, IF2, and IF3, namely eIF1 (and eIF1A), eIF2, and eIF3, respectively, as well as several additional factors, such as eIF2B and eIF5. The second group which is to be considered in the next section, seems to have no analogs in prokaryotes and includes special mRNA-binding and mRNA-unwinding proteins facilitating initiation of translation; these are the factors of eIF4 group, namely eIF4A, eIF4B, eIF4F, and eIF4E. Mammalian initiation factors are considered below.

eIF1 is a small protein of a molecular mass of 12.6 kDa. It seems to be an auxiliary factor, like the prokaryotic IF1, stimulating formation of and stabilizing the initiation complexes of the small 40S ribosomal subunit. There is another initiation factor, designated now as eIF1A and earlier called eIF-4C, also taking part in the assembly of the initiation complexes of the 40S ribosomal subunit; it is a small acidic protein with a molecular mass of 16.5 kDa.

eIF2, in contrast to its prokaryotic functional analog, is a complex protein consisting of three different subunits: acidic, 36.1 kDa (α); acidic, 38.4 kDa (β); and basic, 51.8 kDa (γ). Like the prokaryotic IF2, it interacts with GTP and initiator Met-$tRNA_i$. The main function of eIF2 is GTP-dependent binding of the initiator Met-$tRNA_i$ to the initiating 40S ribosomal subunit (for reviews, see Voorma, 1991; Trachsel, 1996). eIF2 is also a target of regulatory phosphorylation by specific kinases which results in the inhibition of the rate of initiation complex formation (Section 17.2.1.1); the phosphorylation site is Ser51 on the α-subunit. The β-subunit seems to bear the mRNA-binding domain. The γ-subunit is strongly homologous to EF1A, especially in its G-domain, and may be mainly responsible for both GTP binding and Met-$tRNA_i^{Met}$ binding. At the same time, GTP analogs and Met-$tRNA_i^{Met}$ can cross-link to the β-subunit as well, suggesting its close contact with the γ-subunit and its direct or indirect participation in the binding of the ligands.

Apart from its function to bind GTP and initiator Met-$tRNA_i$, eIF2 has been reported to be capable of recognizing a specific initiation site in some mRNAs. This mRNA-binding activity resides in the α-subunit of eIF2 and is regulated by ATP which interacts also with the β-subunit and switches on its mRNA-binding activity. No ATP hydrolysis takes place in this case. It has been demonstrated that the binding of eIF2 to Met-$tRNA_i$ with GTP and the binding to mRNA with ATP are mutually exclusive, although distinct epitopes of eIF2 are involved in the two binding activities. It can be speculated that, once bound to the 40S ribosomal subunit, eIF2 with ATP may interact directly with mRNA and thus guide the 40S subunit to its binding site in mRNA, and then GTP switches the activity of eIF2 on Met-$tRNA_i$ binding.

eIF2A and eIF2B are additional proteins promoting the functions of eIF2 and having no analogs in prokaryotes. eIF2A is a simple basic protein with a molecular mass of 65 kDa. It may take part in AUG-dependent Met-$tRNA_i$ binding to the initiating 40S ribosomal subunit. eIF2B is a large multisubunit protein consisting of five different subunits with molecular masses of 33.7 kDa (α), 39 kDa (β), 58 kDa (γ), 57.1 kDa (δ), and 80.2 kDa (ε). It is capable of forming a complex with eIF2 outside the ribosome and facilitates GDP/GTP exchange on eIF2 (see Section 15.3.8, Fig. 15.15); the regulation of the availability of the active (GTP-bound) form of eIF2 in the eukaryotic cell may be a possible function of eIF2B.

eIF3 is the largest initiation factor in eukaryotes. It is a complex multisubunit protein with a total molecular mass of about 600 kDa. At least eight different subunits compose the protein; the molecular masses of the subunits are 35 kDa (α), 36.5 kDa (β), 39.9 kDa (γ), 46.4 kDa (δ), 47 kDa (ε), 66 kDa (ξ), 105.3 kDa (η) and 170 kDa. (θ). The protein has a strong affinity for the 40S ribosomal subunit, promotes dissociation of terminated ribosomes into subunits, and renders the "native" 43S particles (complexes with 40S ribosomal subunits) competent to begin the initiation process. Thus, eIF3 is an initiation factor which is necessarily present on the 43S initiation complex. This multisubunit factor is also known as a strong mRNA-binding protein. However, the interaction of eIF3 with mRNAs and other polyribonucleotides has been shown to be nonspecific. Nevertheless, the nonspecific RNA-binding capacity of eIF3 may be exactly what is needed for the interaction of the 43S initiation complex with mRNA, provided the specificity is introduced by other proteins.

A protein of 25 kDa called eIF3A or eIF6 has been reported to form a complex with the dissociated 60S ribosomal subunit ("native" 60S particle).

eIF5 is an initiation factor having no analogs in prokaryotes. Two, seemingly unrelated, forms of eIF5 have been described: p150 and p45. Both forms of eIF5 have been reported to induce GTPase on the 40S-bound eIF2 and the release of eIF2 (and possibly other initiation factors) when the 60S ribosomal subunit joins with the 43S initiation complex. In prokaryotes, these functions are fulfilled by the large ribosomal subunit itself. One more initiation factor, eIF5A, earlier called eIF4D, which is a small acidic protein of molecular mass 15 kDa, may also contribute to the joining of the ribosomal subunit and correct 80S ribosome formation.

15.3.6. mRNA-binding Initiation Factors

The individual mRNA-binding initiation factors of the eIF4 group may form a complex that functions as a whole. Indeed, eIF4E is an essential subunit of eIF4F. eIF4A is another subunit of eIF4F, but bound loosely and also existing in excess as free protein. eIF4B is not considered a part of eIF4F, but the two have significant affinity for each other and function rather in association. Three main successive functions of the eIF4 complex are (1) recognition of the cap structure of mRNA, (2) unwinding of the cap-adjacent sequence of mRNA, and (3) facilitating the proper landing of the initiating 40S ribosomal subunit (the so-called 43S initiation complex) on mRNA (reviewed by Rhoads, 1991). Correspondingly, three activities of the eIF4 complex should be mentioned: ATP-independent cap-binding activity, ATP-dependent RNA helicase activity, and an affinity for the ribosomal 43S initiation complex. It should be added that mRNA-binding capacity is clearly manifested also by eIF2 and eIF3. Being mostly attached to the "native" 40S ribosomal subunits, they may also take part in the proper landing of the ribosomal initiation complex on mRNA.

15.3.6.1. Cap-Binding Complex

Initiation factor eIF4F specifically binds to the cap structure of eukaryotic mRNAs. Three subunits are usually accepted to be components of the mammalian factor: α (p25, or eIF4E) is the subunit responsible for cap recognition; β (p45) is RNA-dependent ATPase and identical to free eIF4A; and γ (p220, or eIF4G; real molecular mass is 153.4 kDa in mammals) is a core protein that also appears to participate in initial mRNA binding. The β-subunit (eIF4A) is loosely bound in the complex and can be lost during isolation and purification procedures. No eIF4A was found in eIF4F preparations isolated from some sources, such as wheat germ. In any case, the eIF4A subunit can be omitted from eIF4F without loss of cap-binding activity, so that the association of eIF4G (p220) with eIF4E can be considered as a minimal cap-binding complex. eIF4E is bound to the N-terminal portion of eIF4G, whereas eIF4A (and eIF3) may associate with the C-terminal part of this large subunit.

It is interesting that in wheat germ cells two different cap-binding complexes have been found. One is similar to the mammalian eIF4F having p26 (eIF4E) and p220 (eIF4G) subunits, whereas the other consists of antigenically unrelated subunits p28 and p80. The "typical" eIF4F is represented in a considerably lower amount in wheat germ cells as compared with the "iso"-eIF4F.

eIF4F is an RNA-binding protein in that it has some affinity for RNA, but it strongly binds specifically to the cap structure of eukaryotic mRNAs due to the presence of the cap-recognizing p25 subunit (eIF4E). This binding is ATP-independent. eIF4E seems to be the most deficient polypeptide among initiation factors and their subunits in eukaryotic cells, so that the cap-binding complex as a whole (eIF4F) is present in a limited amount. In this situation cytoplasmic mRNAs must compete for eIF4F binding.

15.3.6.2. RNA Helicase Complex

Immediately after the cap-binding step, or during the binding, the unwinding of the cap-adjacent region of mRNA starts. This step is ATP-dependent. Thus, the RNA helicase reaction takes place. There are two helicase complexes: the combination of eIF4F(αβγ) with eIF4B, and the combination of free eIF4A with eIF4B. In both cases two polypeptides seem to be directly engaged and critical to the process: eIF4A and eIF4B.

Similarly, the RNA helicase reaction occurs in the process of cap-independent, internal initiation. After the recognition of the IRES on mRNA by the ribosomal initiation complex, the same helicase complexes may go into action and unwind the downstream mRNA region for scanning and initiation. Specifically, eIF4F binds to IRES in a cap-independent manner, probably due to the affinity of the central part of eIF4G for the IRES structure (Pestova, Hellen, and Shatsky, 1996).

Both ATPase and unwinding activities appear to reside in eIF4A. eIF4A, a 45-kDa protein, is the most abundant initiation factor in eukaryotic cells. It exists in a free state and also as a loosely associated subunit of eIF4F. In eIF4F, it is bound with the C-terminal portion of eIF4G (p220, or γ-subunit of eIF4F). eIF4A, being alone or in complex with eIF4G, is capable of binding to single-stranded regions of RNA in an ATP-dependent manner and exhibits an RNA-dependent ATPase activity, but eIF4B strongly stimulates both effects. The continuously repeating cycles consisting of ATP-dependent binding to a single-stranded mRNA site, the mRNA-induced ATP cleavage, and the resultant eIF4A release (Fig. 15.12) may dynamically maintain the single-stranded state of a proper region of mRNA, preventing its refolding and its complexing with competing RNA-binding proteins.

The amino acid sequence of eIF4A demonstrates that it belongs to the so-called DEAD family of proteins that are present in various compartments of the eukaryotic cell and involved in different cellular processes. All are believed to pos-

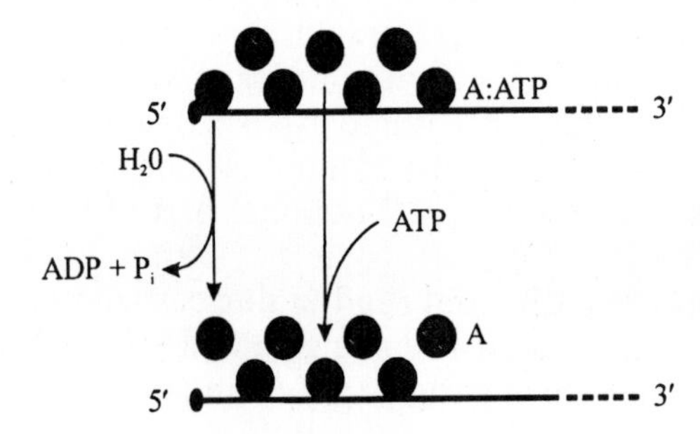

FIG 15.12. A model of the dynamic maintenance of the unwound state of RNA (5′-UTR of mRNA) due to transient ATP-dependent binding of eIF4A (designated as A) to previously unwound regions of the RNA. Solid circles symbolize the eIF4A:ATP complex having an affinity for unwound RNA, and shaded cycles are released eIF4A after ATP hydrolysis.

sess ATPase and RNA-unwinding activities. There are several conserved amino acid sequence regions in common, including two motifs (AXXXXGKT and DEAD) characteristic of ATPases. Thus, eIF4A can be considered the catalytic ATPase/unwinding subunit of the helicase initiation complexes.

eIF4B is a larger protein, of about 80 kDa, rather limited in its amount in the cell. It can be associated with eIF4F but in an even looser way than eIF4A. According to its primary structure, it bears a typical RNA-binding domain (the so-called RNP consensus site, or RNP-CS), with two conserved sequence motifs (RNP-1 and RNP-2), characteristic also of poly(A)-binding protein, hnRNP A1 and C1 proteins (or AU-BP), snRNP U1 70 kDa protein, La protein, Ro protein, and several other RNA-binding proteins. Recently it has been demonstrated that the RNA-binding domain located at the N-terminal part of the protein has a high affinity for a specifically structured RNA element in the 18S ribosomal RNA. On the other hand the C-terminal part of eIF4B has been also shown to possess an RNA-binding activity but it is sequence-nonspecific. Hence, while the C-terminal nonspecific RNA-binding domain may play an important role in the interaction of the helicase complex with mRNA, the N-terminal domain can be responsible for the binding of eIF4B or the entire helicase complex to the initiating 40S ribosomal subunit.

In the experiments studying binding to mRNA, eIF4B is capable of releasing the pre-bound cap-binding complex from mRNA. This may suggest that after the cap recognition step has been performed, eIF4B destabilizes the retention of eIF4F on mRNA; at the same time it combines with eIF4A and initiates the helicase reaction catalyzed by eIF4A. It has been reported also that eIF4B has a preference for AUG, and to a lesser extent for GUG, over other RNA sequences. Hence, recognizing eIF4F and the initiation codon, eIF4B may contribute to specifying the sites on mRNA where the eIF4A-catalyzed helicase reaction is to be performed. At the same time, its specific binding to the 40S ribosomal subunit could target the initiating ribosomal particle to the proper unwound region of the mRNA near the initiation codon.

The general model of the sequence of events during cap binding and the cap-initiated 5′-UTR unwinding is as follows (Fig. 15.13): (1) The eIF4E subunit of the cap-binding complex eIF4F binds to the cap structure in some reversible way. (2) eIF4B forms the helicase complex with eIF4F and thus induces the helicase activity of the eIF4A subunit. The cap-binding complex (its eIF4G subunit) may spread over the cap-adjacent unwound mRNA sequence. (3) In the presence of eIF4B, however, the cap-binding complex is loosely associated with mRNA and tends to be displaced. (4) Free eIF4A interacts now with eIF4B and unwinds the downstream section of mRNA. (5) The repeating acts of interaction of the mRNA-bound eIF4B with eIF4A:ATP progressively unwind the downstream sequence. The cycles, consisting of ATP-dependent eIF4A binding, RNA-induced ATP cleavage, and eIF4A release, maintain the single-stranded state of the 5′-UTR of mRNA. (6) Ribosomal initiation 43S complex binds to the unwound mRNA segment including the initiation AUG codon, thus forming the initiation 48S complex.

15.3.6.3. Ribosomal Complex

It is possible that *in vivo,* where all the components of the protein-synthesizing machinery are present at a high concentration and in proper ionic conditions, the factors mentioned, or at least part of them, do not exist as free proteins but are complexed within mRNPs and with ribosomal particles throughout their life. Affinities of different initiation factors for each other, for RNA, and for ribosomes, as well as their distributions in cell homogenates and during isolation procedures, suggest their preferential compartmentation on larger particles and complexes. More direct experiments demonstrate that the factors may reside in mRNPs and

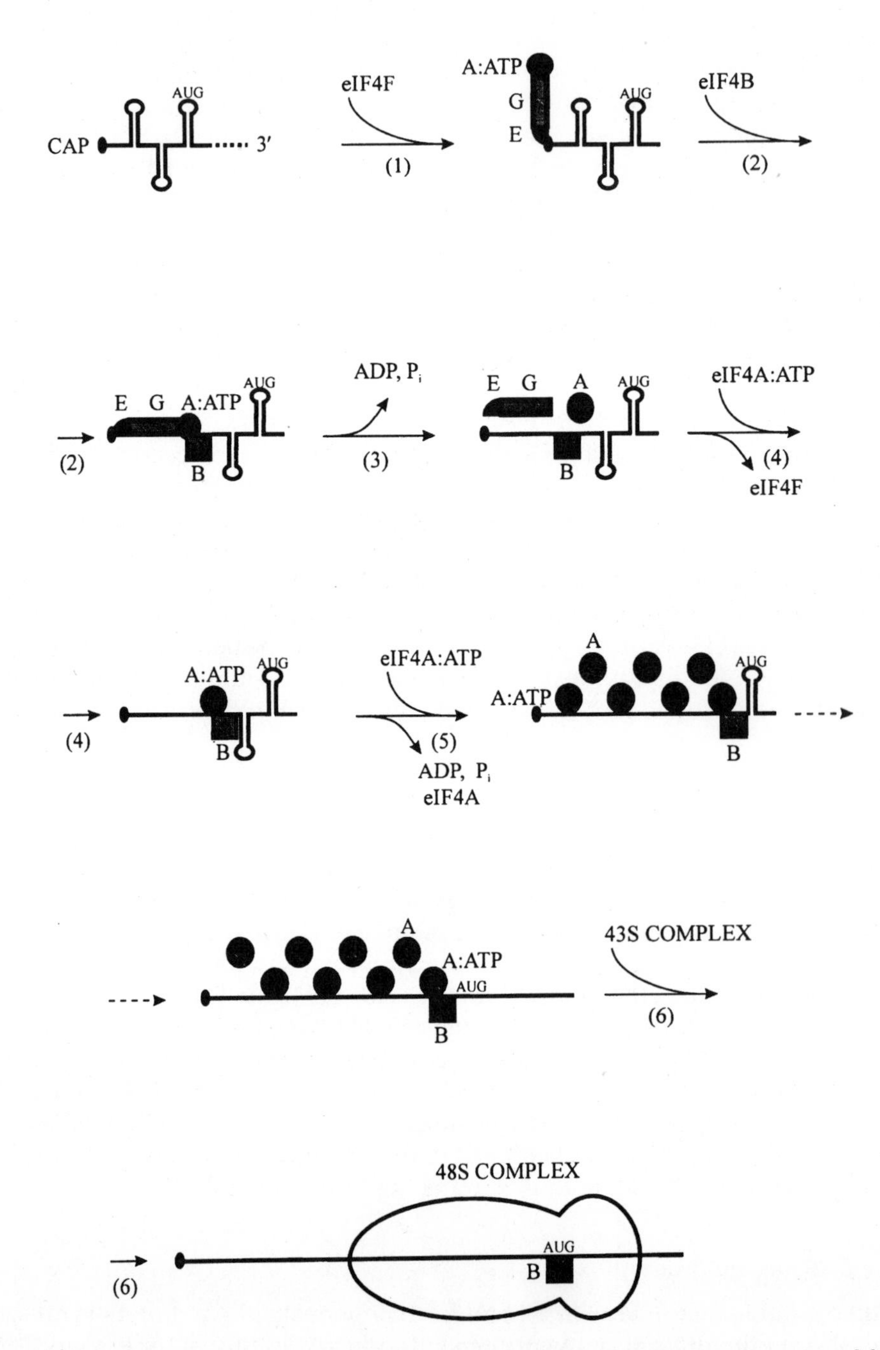

FIG 15.13. A general model for the sequence of events during the cap-dependent formation of the initiation complex with participation of mRNA-binding initiation factors. **A** designates eIF4A; **B,** eIF4B; **E,** eIF4E; and **G,** eIF4G. The consecutive steps from (1) to (6) are explained in the text.

ribosomal initiation complexes (reviewed by Rhoads, 1991). For example, it was found that eIF4E is present in a 1:1 molar ratio in polyribosomal (translated) globin mRNPs, and no eIF4E was detected in free nontranslated mRNPs. The eIF4G (the p220 subunit of eIF4F), as well as eIF4B, was repeatedly reported to be localized in the ribosomal particles fraction. It is noteworthy that eIF3 is capable of forming a complex with the whole eIF4F, and the eIF-G subunit (its C-terminal portion) seems to be responsible for this complex formation.

According to the model depicted in Fig. 15.13, the whole eIF4F(αβγ) first binds to the cap structure, then eIF4B comes and induces eIF4F-catalyzed unwinding of cap-adjacent hairpins on mRNA; free eIF4A or eIF4F then starts to interact with mRNA and eIF4B, thus catalyzing further unwinding of downstream hairpins. The ribosomal 43S initiation complex lands on the unwound region of mRNA in the vicinity of initiation codon. This is the "classical" model generally accepted.

An alternative model seems to be physically more probable under physiological conditions: eIF4E is retained at the cap structure in translatable mRNPs, while eIF4G (p220 subunit of eIF4F), as well as eIF4A and eIF4B are in association with the ribosomal 43S initiation complex; apparently, the eIF4G is complexed with eIF3 on the 40S subunit. In such a case it is the 40S ribosomal subunit carrying also the mRNA-binding initiation factors that lands on the cap structure and uses its associated ATP-dependent helicase activity to move downstream toward the initiation codon (Rhoads, 1991). Thus, the ribosomal initiation complex that is loosely associated with the components of eIF4 (see Fig. 15.16) may be considered an mRNA-binding and mRNA-unwinding machine landing on and moving along mRNA as a whole.

15.3.6.4. IRES-Binding Proteins

In addition to the full set of conventional initiation factors, including eIF2, eIF3, and eIF4F, the cap-independent initiation at the IRES seems to require special IRES-binding proteins. At least two cellular RNA-binding proteins, p52 and p57, are factors involved in IRES-dependent translation of picornavirus RNAs (Jackson, 1996; Ehrenfeld, 1996).

p52 is the protein capable of specifically binding to the downstream (3′-proximal) portion of the poliovirus IRES and providing for the stimulation of the translation initiation at the correct site of the polioviral RNA and the suppression of initiation at incorrect sites. The protein has been found to be identical to the human La antigen earlier detected by antibodies from some patients with autoimmune diseases such as lupus erythematosus. As an RNA-binding protein, it contains a characteristic RNA recognition motif of the so-called RNP-1 type. In normal mammalian cells the La protein is localized predominantly in the nucleus where it may be involved in the maturation of RNA polymerase III transcripts. At the same time, a fraction of the La protein has been shown to reside in the cytoplasm. The virus infection, in particular, poliovirus infection, induces redistribution of a significant portion of the La protein from the nucleus to the cytoplasm, and in this case it comes out as a translation factor for poliovirus RNA. An important question is whether the La protein, or p52, is a translation factor for all picornavirus RNAs, or whether its role is more limited and relates just to enterovirus IRES activity.

p57 is another RNA-binding protein with specific affinity for IRESes of seemingly all picornavirus RNAs. The protein is found to be essential for the initiation of translation of picornavirus RNAs. It seems that the protein binds to multiple sites in the picornaviral IRES, thereby stabilizing a specific active conformation. Again, like in the previous case, p57 has proved to be identical to a nuclear protein, earlier known as "polypyrimidine tract-binding protein" (PTB) that binds preferentially to the polypyrimidine tract near the 3′-end of introns of pre-mRNAs. The protein belongs to the family of RNA-binding proteins with four RNA-binding domains (RBD), but lacks the typical RNP-1 and RNP-2 sequence motifs. It has been localized in the nucleoplasm of interphase cells. At the same time the protein can be detected in the cytoplasm and even in the ribosomal fraction. No information, however, is available about the participation of p57/PTB in internal initiation of cellular mRNAs.

Other RNA-binding proteins, such as p38 and p97, may also participate in the IRES-dependent initiation of translation. Generally, the situation with the interaction between different IRES-binding proteins and the contribution of each of them to the internal initiation process is not yet clear. At the same time, up to now no universal factor essential for internal initiation *on all IRESes* has been found. This suggests that the role of different RNA-binding proteins specifically interacting with different IRESes may be to stabilize the IRES structures, rather than directly participate in the internal initiation process.

15.3.7. 3′-Terminal Enhancers of Initiation

15.3.7.1. Poly(A) Tail

One of the most remarkable features of eukaryotic translation is that it can be stimulated by some elements of mRNA located far away from the initiation site, particularly in the 3′-untranslated region (3′-UTR). The best known and most universal translational enhancer of eukaryotic mRNAs is their poly(A) tail (for reviews, see Sachs, 1990; Jacobson, 1996). The enhancing effect of the poly(A) tail is realized through the stimulation of translational initiation in some unknown way. The enhancement strictly requires the presence and multiple binding of the poly(A)-binding protein (PABP) to the poly(A) tail. There is a hypothesis that the poly(A) tail/PABP complex plays the role of an enhancer of translational initiation by directly contacting the 5′-portion of mRNA. The enhancement has been supposed to be exerted by the tail-bound PABP which facilitates the joining of the 60S ribosomal subunit to the initiating 40S subunit complex (48S complex). Also the cap function was shown to be enhanced over an order of magnitude by the presence of a poly(A) tail, and the poly(A) tail-mediated enhancement of translational efficiency was reported to be strongly dependent on the presence of the cap structure. More recently poly(A) has been shown to attract the cap-binding initiation complex, eIF4F, together with eIF4A and eIF4B, thus suggesting that the initiation factors complex may be compartmentalized on the poly(A) tail and then, during initiation of translation or reinitiation, transferred to the cap structure of mRNA, possibly with the participation of PABP.

All this implies that the poly(A) tail and the tail-bound PABP must be in proximity to the 5′-UTR of mRNA in the translating polyribosome (Fig. 15.14). Indeed, there are several electron-microscopic indications that many polyribosomes in eukaryotic cells have a circular organization (see Fig. 4.5). Along with the possibility of translational enhancement via poly(A)/PABP-mediated tail-to-head interaction in mRNA (Munroe and Jacobson, 1990), it may be that the multimeric poly(A)/PABP complex exerts a general conformational effect on translatable mRNP (Spirin, 1994). The effect can be realized through the maintenance of a more open and admissible conformation of mRNP, rather than through local interactions. Consequently, the longer the poly(A)-tail, the more PABP molecules are cooperatively bound and the better is the stabilization of the global translatable conformation of polyribosomal mRNP.

15.3.7.2. Pseudoknot and tRNA-like Domains

A group of translational enhancers of a more specialized type includes pseudoknot domains and tRNA-like structures at the 3′-UTRs of plant virus RNAs. These RNAs are not polyadenylated, and the 3′-terminal pseudoknot domain with the tRNA-like structure (Fig. 15.15) substitutes for a poly(A) tail as an enhancer (Gallie and Walbot, 1990). Moreover, when the pseudoknot and tRNA-like structure-containing 3′-UTR of plant virus RNA was fused to a coding region of a foreign reporter mRNA, it exerted the enhancing effect both in plant and animal cells (Gallie *et al.*, 1991). Again, like in the case of the poly(A) tail, the synergism was observed between the 3′-UTR structure and the cap structure (Leathers *et al.*, 1993). The fact that the enhancing effect of the plant viral RNA 3′-UTR is displayed in the mammalian cell cytoplasm, as well as in the plant cells, suggests the involvement of a ubiquitous RNA-binding factor, but one probably different from PABP.

The experiments with complex formation of tobacco mosaic virus (TMV) RNA

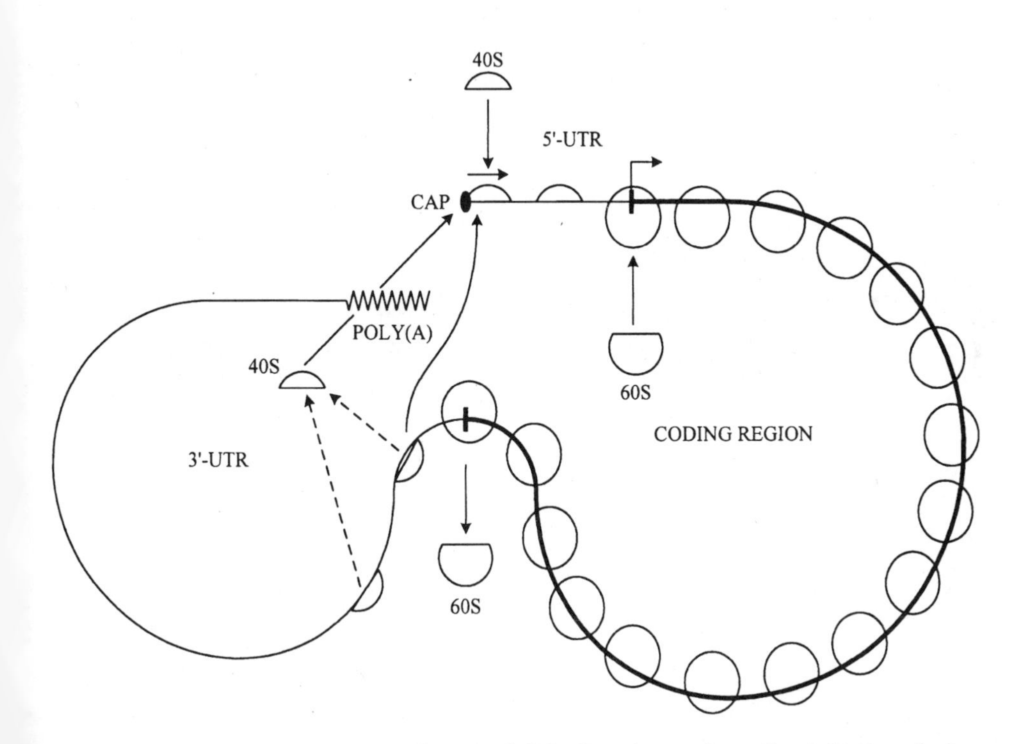

FIG 15.14. A model for the enhancing effect of poly(A) tail on the cap-dependent initiation of eukaryotic mRNA translation. Consisting with the circular organization of eukaryotic polyribosomes, the poly(A) tail complexed with PABP can be in contact with the capped 5′-end of mRNA and serve as a recruiter of free ribosomal subunits and RNA-binding initiation factors (including eIF4E and eIF4F) for the cap structure. (See Munroe, D., and Jacobson, A. (1990) *Mol. Cell. Biol.* **10**:3441–3455; Gallie, D. R., and Tanguay, R. (1994) *J. Biol. Chem.* **269**:17166–17173; Tarun, S. Z., and Sachs, A. B. *Genes & Develop.* **9**:2997–3007).

sequences in plant cell extracts have demonstrated that the 3′-UTR pseudoknot domain of TMV RNA binds a protein from the extract (Leathers *et al.*, 1993). The binding was shown to be specific for the 3′-UTR structure. At the same time the leader sequence of the TMV RNA, that is, the 5′-UTR element, strongly competed for binding with this protein and was found to be even a more efficient competitor than the homologous 3′-UTR sequence. All this suggests that (1) both elements, 3′-UTR pseudoknot domain and 5′-UTR leader domain, are specifically recognized by the same protein factor present in plant cell extract, and (2) the interaction of the binding protein with the 5′-UTR domain is tighter than that with the 3′-UTR domain. One may speculate that the binding protein is one of the ubiquitous RNA-binding initiation factors that normally interact with 5′-UTRs of cellular mRNAs, but also recognize the viral 3′-UTR element.

A potent translational enhancer domain of a different type has been shown in the 3′-UTR of satellite tobacco necrosis virus (STNV) RNA. This is a monocistronic messenger that lacks both the cap structure and the poly(A) tail. A bulged hairpin structure responsible for the enhancing effect is located immediately downstream from the coding sequence (Danthinne *et al.*, 1993; Timmer *et al.*, 1993). The proper structure of the short 5′-UTR is required for the enhancing effect of the 3′-UTR domain. Protein cofactors or initiation factors that could mediate the effect are not known.

Thus, like in the case of the enhancer function of the poly(A) tail, it is tempting to hypothesize that the effect of viral RNA enhancers is achieved through a local, protein-mediated contact of the 3′-UTR structure with the initiation regions of mRNA. The alternative may be the formation of a protein complex at the 3′-por-

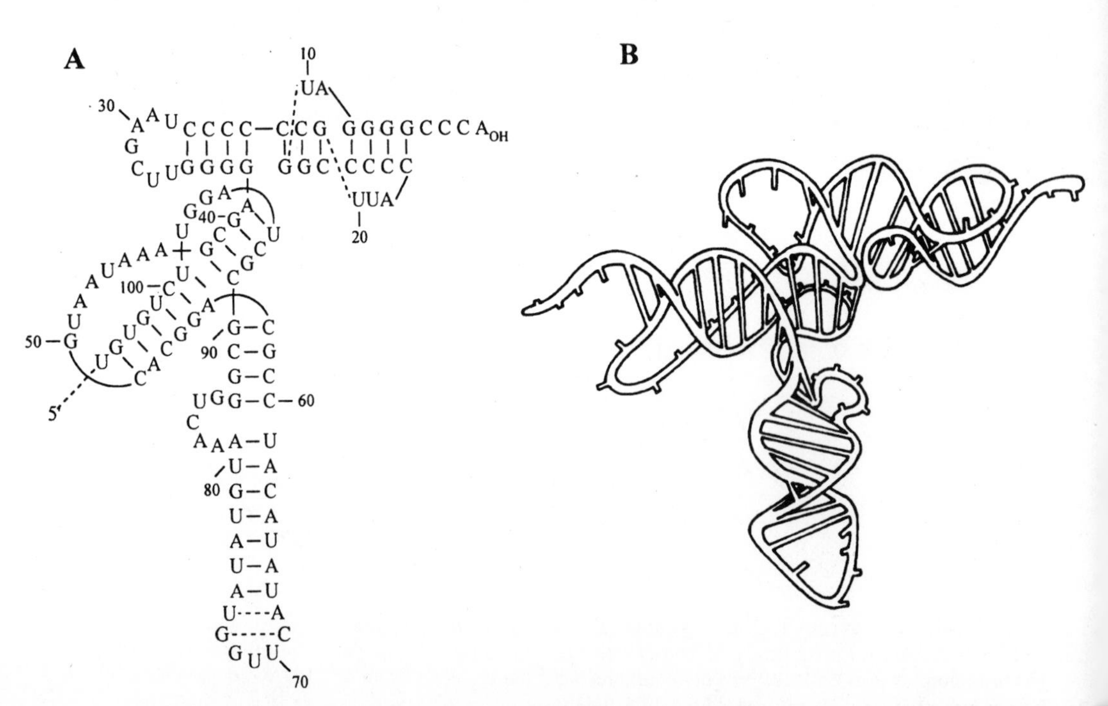

FIG 15.15. tRNA-like structure and pseudoknots at the 3′-end of tobacco mosaic virus (TMV) RNA. (**A**) The predicted secondary structure and its L arrangement. (**B**) The model of three-dimensional folding of the same region. Reproduced from Rietveld, K., Lenschooten, K., Pleij, C. W. A. and Bosch, L. (1984) *EMBO J.* **3**:2613–2619, with permission.

tion of mRNP which fixes the global mRNP structure in a conformation better available for translational initiation (global structural effect).

The enhancing effect of 3′-UTR sequences of cellular mRNAs can also be suspected in some cases. For example, it was reported that the removal of a major portion of the ornithine decarboxylase (ODC) mRNA 3′-UTR inhibits translation (Manzella and Blackshear, 1990), and that the insertion of the ODC 3′-UTR downstream of the termination codon of a reporter mRNA partially relieves the suppression of translation imposed by the ODC 5′-UTR (Grens and Scheffler, 1990). There are no reports of proteins that might mediate these effects.

15.3.8. Sequence of Events

15.3.8.1. Step 1

In order to enter the initiation process the ribosome must be dissociated into the subunits. It is believed that the decisive role in the dissociation and the prevention of reassociation is played by IF1A and eIF3 which interact with the 40S subunit. eIF1 and eIF2 may also join the complex, forming the native 40S ribosomal particle. The 60S subunit interacts with eIF3A (eIF6) and some other proteins, forming the native 60S ribosomal particle.

15.3.8.2. Step 2

The "classical" scenario is the activation of eIF2 with GTP and the subsequent binding of the initiator Met-tRNA$_i$ to the GTP form of eIF2. This event is usually considered to occur as the interaction of the three free components, with the formation of the ternary complex Met-tRNA$_i$:eIF2:GTP in solution:

$$\text{eIF2} + \text{GTP} + \text{Met-tRNA}_i \rightarrow \text{eIF2:GTP} + \text{Met-tRNA}_i \rightarrow$$
$$\rightarrow \text{Met-tRNA}_i\text{:eIF2:GTP}$$

Then the ternary complex attaches to the native 40S particle:

$$\text{40S:eIF3:eIF1A} + \text{Met-tRNA}_i\text{:eIF3:GTP} \rightarrow$$
$$\rightarrow \text{40S:eIF3:eIF1A:eIF2:GTP:Met-tRNA}_i$$

However, since eIF2 is likely to be attached already to the native 40S particle, the formation of its complex with GTP and Met-tRNA may take place rather on the particle than in solution (Fig. 15.16):

$$\text{40S:eIF3:eIF1A:eIF2} + \text{GTP} + \text{Met-tRNA}_i \rightarrow$$
$$\rightarrow \text{40S:eIF3:eIF1A:eIF2:GTP} + \text{Met-tRNA}_i \rightarrow$$
$$\rightarrow \text{40S:eIF3:eIF1A:eIF2:GTP-Met-tRNA}_i$$

The ribosomal complex formed is often designated as *43S initiation complex.*

15.3.8.3. Step 3

The next step is usually considered to be the binding of the 43S initiation complex to mRNA and thus the formation of the so-called *48S initiation complex* (see

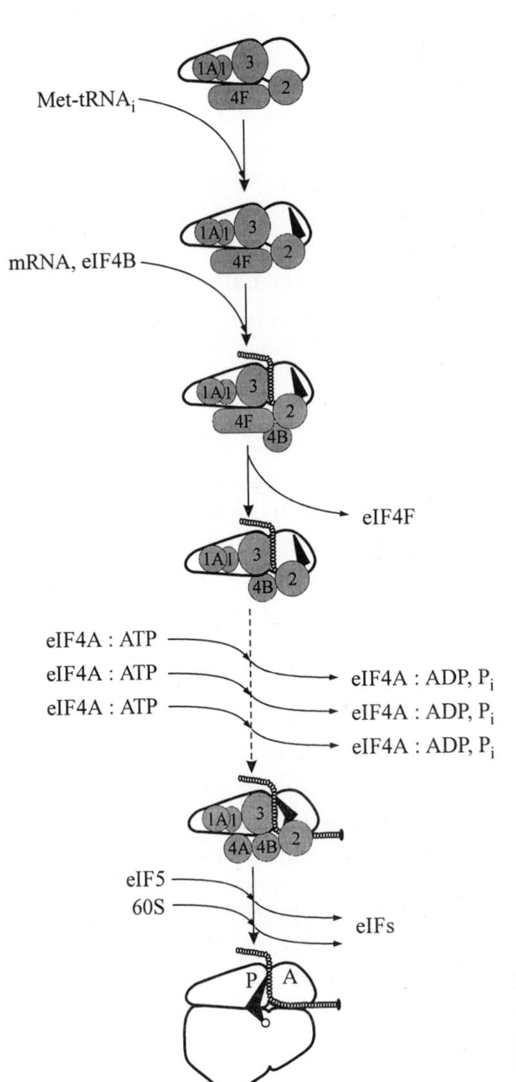

FIG 15.16. Sequence of events during initiation of translation in eukaryotes: the model of the initiating 40S ribosomal subunit with pre-bound initiation factors. The consecutive steps from (1) to (6) are explained in the text.

Fig. 15.13). The mRNA binding proceeds with the participation of eIF4E, eIF4F, eIF4A, and eIF4B. This step includes the primary association of the ribosomal particle with mRNA (with cap structure or IRES), the movement of the ribosomal particle along the 5′-UTR downstream (scanning), and the selection of the initiation codon. The "classical" model presumes that the cap-binding factors eIF4E and eIF4F interact with the cap structure of mRNA, and then eIF4A and eIF4B come and perform ATP-dependent unwinding of the 5′-UTR, this being a prerequisite for landing the 43S initiation complex on mRNA and subsequent scanning.

A more likely model based on information available about the intracellular localization of initiation factors and their complex formation capabilities suggests that the cap-binding complex (eIF4F) and the helicase complex (eIF4A:eIF4B) are *assembled on the ribosomal particle* in the presence of mRNA. Therefore, according to this model, it is the ribosomal particle with the initiation factors that binds to the cap structure and then successively unwinds the downstream sequence thus moving along mRNA and scanning the sequence (Fig. 15.16). The principal role in assembly of the cap-binding and helicase complexes on the native 40S ribosomal particle may belong to eIF3 which is capable of interacting with eIF4F (with eIF4G subunit) and eIF4B. It is possible that eIF4E is pre-bound to the cap structure of mRNA competent for translation, whereas the eIF4G subunit of eIF4F and eIF4B are associated with the native 40S ribosomal particle. In such a case the complete cap-binding complex (eIF4F) is assembled upon binding of the native ribosomal particle with the cap structure, followed by the assembly of the helicase complex (eIF4F:eIF4B or eIF4A:eIF4B) on the particle.

It cannot be excluded that this step (mRNA binding) precedes the step of the initiator Met-$tRNA_i$ binding. In other words, Met-$tRNA_i$ binding may proceed *after* the native ribosomal particle with a full set of initiation factors, including eIF2, associates with mRNA, scans its 5′-UTR, and finds the initiation region. The eIF2:GTP-dependent binding of the Met-$tRNA_i$ may be significantly stimulated by the presence of the initiator AUG codon in the initiation complex.

15.3.8.4. Step 4

The 48S initiation complex formed joins the 60S ribosomal subunit. The process is promoted by eIF5 which seems to react with the 48S complex first and induce the hydrolysis of the eIF2-bound GTP. The GDP-form of eIF2, as well as other initiation factors, are released from the 40S subunit, and the 80S initiation particle with Met-$tRNA_i$ in the P site is assembled at the initiation codon of mRNA (Fig. 15.16).

The eIF2:GDP complex released presents a special problem. It is a stable but inactive form of eIF2. Its reactivation and recycling requires the involvement of eIF2B, a large multisubunit protein. eIF2B interacts with eIF2:GDP resulting in the replacement of GDP by GTP (GDP/GTP exchange reaction). The complex eIF2: GTP:eIF2B may bind Met-$tRNA_i$ to form a heavier complex, Met-$tRNA_i$:eIF2: GTP:eIF2B. The ternary complex Met-$tRNA_i$:eIF2:GTP can be transferred directly from this big complex to the 40S ribosomal subunit. Thus eIF2B appears both as a GDP/GTP exchange factor and a carrier of eIF2. The phosphorylation of eIF2 mentioned above (see Section 14.3.5) makes the complex between eIF2 and eIF2B firmer, thus making the complexed eIF2 and eIF2B unavailable for further func-

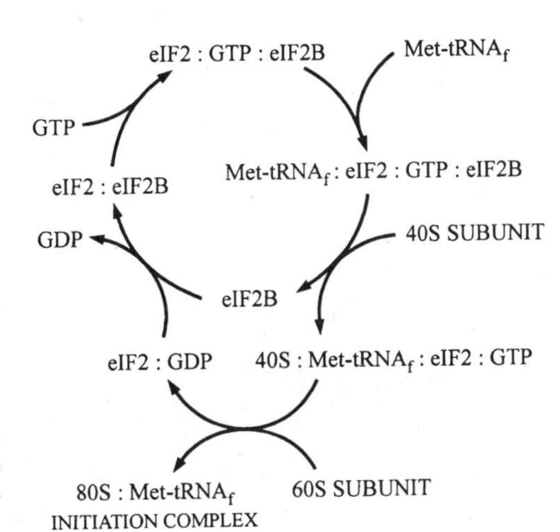

FIG 15.17. A model of the cyclic use of the eukaryotic initiation factors eIF2 and eIF2B. See Salimans, M., Goumans, H., Amesz, H., and Voorma, H. O. (1984) *Eur. J. Biochem.* **145**:91–98.

tional interactions. The schematic representation of the eIF2 cycling through the complex formation with eIF2B is given in Fig. 15.17.

15.3.8.5. Step 5

The vacant A site of the 80S initiation complex accepts the first elongator aminoacyl-tRNA in the form of its ternary complex with eEF1A and GTP. The subsequent GTP cleavage and eEF1A release are followed by the transpeptidation reaction between the initiator Met-$tRNA_i$ and the elongator aminoacyl-tRNA. The formation of the first peptide bond completes the initiation and starts the elongation.

References

Adams, J. M. (1968) On the release of the formyl group from nascent protein, *J. Mol. Biol.* **33**:571–589.

Bag, J. (1991) mRNA and mRNP, in *Translation in Eukaryotes* (H. Trachsel, ed.), pp. 71–96, CRC Press, Boca Raton, Ann Arbor, Boston, London.

Basavappa, R., and Sigler, P. B. (1991) The 3 Å crystal structure of yeast initiator tRNA: functional implications in initiator/elongator discrimination, *EMBO J.* **10**:3105–3111.

Biou, V., Shu, F., and Ramakrishnan, V. (1995) X-ray crystallography shows that translational initiation factor IF3 consists of two compact α/β domains linked by an α-helix, *EMBO J.* **14**:4056–4064.

Capecchi, M. R. (1966) Initiation of *E. coli* proteins, *Proc. Natl. Acad. Sci. USA* **55**:1517–1524.

Clark, B. F. C., and Marcker, K. A. (1966) The role of N-formyl-methionyl-sRNA in protein biosynthesis, *J. Mol. Biol.* **17**:394–406.

Danthinne, X., Seurinck, J., Meulewaeter, F., van Montagu, M., and Cornelissen, M. (1993) The 3' untranslated region of satellite tobacco necrosis virus RNA stimulates translation *in vitro*, *Mol. Cell. Biol.* **13**:3340–3349.

de Smit, M. H., and van Duin, J. (1990) Control of prokaryotic translational initiation by mRNA secondary structure, *Progr. Nucleic Acid Res. Mol. Biol.* **38**:1–35.

Dyson, M. R., Mandal, N., and RajBhandary, U. L. (1993) Relationship between the structure and function of *Escherichia coli* initiator tRNA, *Biochimie* **75**:1051–1060.

Ehrenfeld, E. (1996) Initiation of translation by picornavirus RNAs, in *Translational Control* (J. W. B. Hershey, M. B. Matthews, and N. Sonenberg, eds.) pp. 549–574, Cold Spring Harbor Laboratory Press, New York.

Forster, C., Chakraburtty, K., and Sprinzl, M. (1993) Discrimination between initiation and elongation of protein biosynthesis in yeast: identity assured by a nucleotide modification in the initiator tRNA, *Nucleic Acids Res.* **21**:5679–5683.

Gallie, D. R., Feder, J. N., Schimke, R. T., and Walbot, V. (1991) Functional analysis of the tobacco mosaic virus tRNA-like structure in cytoplasmic gene regulation, *Nucleic Acids Res.* **19**:5031–5036.

Gallie, D. R., and Walbot, V. (1990) RNA pseudoknot domain of tobacco mosaic virus can functionally substitute for a poly(A) tail in plant and animal cells, *Genes and Develop.* **4**:1149–1157.

Garcia, C., Fortier, P.-L., Blanquet, S., Lallemand, J.-Y., and Dardel, F. (1995) Solution structure of the ribosome-binding domain of *E. coli* translation initiation factor IF3. Homology with the U1A protein of the eukaryotic spliceosome, *J. Mol. Biol.* **254**:247–259.

Grens, A., and Scheffler, I. E. (1990) The 5'- and 3'-untranslated regions of ornithine decarboxylase mRNA affect the translational efficiency, *J. Biol. Chem.* **265**:11810–11816.

Gualerzi, C. O., La Teana, A., Spurio, R., Canonaco, M. A., Severini, M., and Pon, C. L. (1990) Initiation of protein synthesis in Procaryotes: Recognition of mRNA by ribosomes and molecular basis for the function of initiation factors, in *The Ribosome: Structure, Function, and Evolution* (W. E. Hill, A. Dahlberg, R. A. Garrett, P. B. Moore, D. Schlessinger, and J. R. Warner, eds.), pp. 281–291, ASM Press, Washington, DC.

Gualerzi, C., Pon, C. L., Pawlik, R. T., Canonaco, M. A., Pace, M., and Wintermeyer, W. (1986) Role of the initiation factors in *Escherichia coli* translational initiation, in *Structure, Function and Genetics of Ribosomes* (B. Hardesty and G. Kramer, eds.), pp. 621–641, Springer Verlag, New York.

Hartz, D., McPheeters, D. S., and Gold, L. (1990) From polynucleotide to natural mRNA translation initiation: Function of *Escherichia coli* initiation factors, in *The Ribosome: Structure, Function, and*

Evolution (W. E. Hill, A. Dahlberg, R. A. Garrett, P. B. Moore, D. Schlessinger, and J. R. Warner, eds.), pp. 275–280, ASM Press, Washington, DC.

Ivanov, P. A., Karpova, O. V., Skulachev, M. V., Tomashevskaya, O. L., Rodionova, N. P., Dorokhov, Yu. L., and Atabekov, J. G. (1997) A tobamovirus genome that contains an internal ribosome entry site functional *in vitro*, *Virology* **232**:32–43.

Jackson, R. J. (1996) A comparative view of initiation site selection mechanisms, in *Translational Control* (J. W. B. Hershey, M. B. Matthews, and N. Sonenberg, eds.), pp. 71–112, Cold Spring Harbor Laboratory Press, New York.

Jacobson, A. (1996) Poly(A) metabolism and translation: The closed-loop model, in *Translational Control* (J. W. B. Hershey, M. B. Matthews, and N. Sonenberg, eds.), pp. 451–480, Cold Spring Harbor Laboratory Press, New York.

Kozak, M. (1978) How do eucaryotic ribosomes select initiation regions in messenger RNA?, *Cell* **15**:1109–1123.

Kozak, M. (1980) Role of ATP in binding and migration of 40S ribosomal subunits, *Cell* **22**:459–467.

Kozak, M. (1981) Possible role of flanking nucleotides in recognition of the AUG initiator codon by eukaryotic ribosomes, *Nucleic Acids Res.* **9**:5233–5252.

Kozak, M. (1983) Comparison of initiation of protein synthesis in prokaryotes, eukaryotes and organelles, *Microbiol. Rev.* **47**:1–45.

Kozak, M. (1989a) Context effects and inefficient initiation at non-AUG codon in eukaryotic cell-free translation systems, *Mol. Cell. Biol.* **9**:5073–5080.

Kozak, M. (1989b) The scanning model for translation: An update, *J. Cell. Biol.* **108**:229–241.

Leathers, V., Tanguay, R., Kobayashi, M., and Gallie, D. R. (1993) A phylogenetically conserved sequence within viral 3' untranslated RNA pseudoknots regulates translation, *Mol. Cell. Biol.* **13**:5331–5347.

Macejak, D. G., and Sarnow, P. (1991) Internal initiation of translation mediated by the 5' leader of a cellular mRNA, *Nature* **353**:90–94.

Maitra, M., Stringer, E. A., and Chaudhuri, A. (1982) Initiation factors in protein biosynthesis. *Annu. Rev. Biochem.* **51**:869–900.

Manzella, J. M., and Blackshear, P. J. (1990) Regulation of rat ornithine decarboxylase mRNA translation by its 5'-untranslated region, *J. Biol. Chem.* **265**:11817–11822.

Marcker, K., and Sanger, F. (1964) N-formyl-methionyl-sRNA, *J. Mol. Biol.* **8**:835–840.

McCarthy, J. E. G., and Brimacombe, R. (1994) Prokaryotic translation: the interactive pathway leading to initiation, *Trends Genet.* **10**:402–407.

McCarthy, J. E. G., and Gualerzi, C. (1990) Translational control of prokaryotic gene expression, *Trends Genet.* **6**:78–85.

Meerovitch, K., Sonenberg, N., and Pelletier, J. (1991) The translation of picornaviruses, in *Translation in Eukaryotes* (H. Trachsel, ed.), pp. 273–292, CRC Press, Boca Raton, Ann Arbor, Boston, London.

Merrick, W. C. (1992) Mechanism and regulation of eukaryotic protein synthesis, *Microbial Reviews* **56**:291–315.

Merrick, W. C., and Hershey, J. W. B. (1996) The pathway and mechanism of eukaryotic protein synthesis, in *Translational Control* (J. W. B. Hershey, M. B. Matthews, and N. Sonenberg, eds.), pp. 31–70, Cold Spring Harbor Laboratory Press, New York.

Munroe, D., and Jacobson, A. (1990) Poly(A) is a 3' enhancer of translational initiation, in *The Ribosome: Structure, Function, and Evolution* (W. E. Hill, A. Dahlberg, R. A. Garrett, P. B. Moore, D. Schlessinger, and J. R. Warner, eds.), pp. 299–305, ASM Press, Washington, DC.

Oh, S.-K., Scott, M. P., and Sarnow, P. (1992) Homeotic gene *Antennapedia* mRNA contains 5'-noncoding sequences that confer translational initiation by internal ribosome binding, *Genes and Develop.* **6**:1643–1653.

Pestova, T. V., Hellen, C. U. T., and Shatsky, I. N. (1996) Canonical eukaryotic initiation factors determine initiation of translation by internal ribosomal entry, *Mol. Cell. Biol.* **16**:6859–6869.

Pilipenko, E. V., Gmyl, A. P., Maslova, S. V., Belov, G. A., Sinyakov, A. N., Huang, M., Brown, T. D. K., and Agol, V. I. (1994) Starting window, a distinct element in the cap-independent internal initiation of translation on picornaviral RNA, *J. Mol. Biol.* **241**:398–414.

Pilipenko, E. V., Gmyl, A. P., Maslova, S. V., Svitkin, Yu. V., Sinyakov, A. N., and Agol, V. I. (1992) Prokaryotic-like cis elements in the cap-independent internal initiation of translation on picornavirus RNA, *Cell* **68**:119–131.

Preobrazhensky, A. A., and Spirin, A. S. (1978) Informosomes and their protein components: The present state of knowledge, *Progr. Nucleic Acid Res. Mol. Biol.* **21**:1–38.

Rhoads, R. E. (1991) Initiation: mRNA and 60S subunit binding, in *Translation in Eukaryotes* (H. Trachsel, ed.), pp. 109–148, CRC Press, Boca Raton, Ann Arbor, Boston, London.

Sachs, A. (1990) The role of poly(A) in the translation and stability of mRNA, *Curr. Opin. Cell Biol.* **2**:1092–1098.

Shine, J., and Dalgarno, L. (1974) The 3′-terminal sequence of *Escherichia coli* 16S ribosomal RNA: Complementarity to nonsense triplets and ribosome binding sites, *Proc. Natl. Acad. Sci. USA* **71**:1342–1346.

Smith, A. E., and Marcker, K. A. (1970) Cytoplasmic methionine transfer RNAs from eukaryotes, *Nature* **226**:607–610.

Sonenberg, N. (1991) Picornavirus RNA translation continues to surprise, *Trends Genet.* **7**:105–106.

Spirin, A. S. (1994) Storage of messenger RNA in eukaryotes: Envelopment with protein, translational barrier at 5' side, or conformational masking by 3' side?, *Mol. Reprod. Develop.* **38**:107–117.

Spirin, A. S. (1996) Masked and translatable messenger ribonucleoproteins in higher eukaryotes, in *Translational Control* (J. W. B. Hershey, M. B. Matthews, and N. Sonenberg, eds.), pp. 319–334, Cold Spring Harbor Laboratory Press, New York.

Steitz, J. A. (1980) RNA-RNA interactions during polypeptide chain initiation, in *Ribosomes: Structure, Function, and Genetics* (G. Chambliss, G. R. Craven, J. Davies, K. Davis, L. Kahan, and M. Nomura, eds.), pp. 479–495, University Park Press, Baltimore.

Stormo, G. D. (1986) Translation initiation, in *Maximizing Gene Expression* (W. Reznikoff and L. Gold, eds.), pp. 195–224, Butterworths, Boston, London.

Timmer, R. T., Benkowski, L. A., Schodin, D., Lax, S. R., Metz, A. M., Ravel, J. M., and Browning, K. S. (1993) The 5' and 3' untranslated regions of satellite tobacco necrosis virus RNA affect translational efficiency and dependence on 5' cap structure, *J. Biol. Chem.* **268**:9504–9510.

Trachsel, H. (1996) Binding of initiator methionyl-tRNA to ribosomes, in *Translational Control* (J. W. B. Hershey, M. B. Matthews, and N. Sonenberg, eds.), pp. 113–138, Cold Spring Harbor Laboratory Press, New York.

Van Knippenberg, P. H. (1990) Aspects of translation initiation in *Escherichia coli,* in *The Ribosome: Structure, Function, and Evolution* (W. E. Hill, A. Dahlberg, R. A. Garrett, P. B. Moore, D. Schlessinger, and J. R. Warner, eds.), pp. 265–274, ASM Press, Washington, DC.

Varshney, U., Lee, C. P., and RajBhandary, U. L. (1993) From elongator tRNA to initiator tRNA, *Proc. Natl. Acad. Sci. USA* **90**:2305–2309.

Voorma, H. O. (1991) Initiation: Met-tRNA binding, in *Translation in Eukaryotes* (H. Trachsel, ed.), pp. 97–108, CRC Press, Boca Raton.

Voorma, H. O. (1996) Control of translation initiation in prokaryotes, in *Translational Control* (J. W. B. Hershey, M. B. Matthews, and N. Sonenberg, eds.), pp. 759–777, Cold Spring Harbor Laboratory Press, New York.

Woo, N. H., Roe, B. A., and Rich, A. (1980) Three-dimensional structure of *Escherichia coli* initiator $tRNA_f^{Met}$, *Nature* **286**:346–351.

16

Translational Control in Prokaryotes

16.1. General Considerations

Protein production in the cell can be controlled principally at three levels: (1) by production of mRNA (transcriptional level); (2) through availability of mRNA for translation and modulation of mRNA translation rate (translational level); and (3) by mRNA elimination (degradation). Although both transcription and mRNA degradation may also depend on ribosomes, only the translational level of protein synthesis regulation will be considered here.

Translational regulation of protein synthesis is accomplished primarily through the control of translation initiation. Under certain circumstances, translation of an individual mRNA or a cistron within a polycistronic mRNA may or may not be started; this case can be classified as an *all-or-none control* of initiation. When initiation is principally permitted, the rate of initiation is different for various mRNAs, that is mRNAs display *differential "strength"* in their entering into initiation process. Furthermore, the rate of initiation, totally or for individual mRNAs, can be *modulated* across a wide range by internal or external signals, thus determining the modulation of protein production of the cell. Both prokaryotes and eukaryotes possess well-developed systems of the translational regulation through the control of initiation. At the same time, the two groups of organisms exhibit such basic differences in the mechanisms of their translation initiation and its control that it is worthwhile to consider them separately.

It is generally accepted that in prokaryotes protein synthesis is controlled mainly at the level of transcription. Indeed, metabolic instability of mRNA in prokaryotic cells, involving its rapid synthesis and rapid degradation, provides for a fast change of templates depending on environmental conditions and cell requirements. At the same time, however, the existence of polycistronic templates in prokaryotes often demands differential control of the individual cistron activities in order to provide for quantitatively different or temporary uncoupled production of proteins encoded by a given polynucleotide. Moreover, in a number of cases the accumulation of excessive amounts of the product of translation may be used to shut down the translation of corresponding mRNA (autoregulation); in this way, a very fine tuning between the level of protein production and the extent of cell requirement in this protein can be achieved. Thus, the translational level of regulation of protein synthesis in prokaryotes may be of great importance in many special cases (for reviews, see, e.g., Stormo, 1987; Gold, 1988; McCarthy and Gualerzi, 1990; Voorma, 1996), though the general pattern of protein production seems to be determined mostly by the activities of genes, that is, at the transcriptional level.

16.2. Discrimination of mRNAs

The discrimination of mRNAs by initiating ribosomal particles is typical of prokaryotes. The prokaryotic 30S ribosomal particle recognizes the RBS containing the SD sequence and initiation codon (Section 15.2.2). The primary structure and the availability of this region for interaction with the initiating ribosomal particle are of primary importance for prokaryotic initiation "strength" (for reviews, see Steitz, 1980; Stormo, 1986; Gold, 1988; Gualerzi *et al.*, 1990; de Smit and van Duin, 1990).

The availability of the RBS depends, first of all, on mRNA secondary and tertiary structure. The intramolecular mRNA folds involving the SD sequence and/or initiation codon, if they are stable enough, can completely block the access of ribosomal particles to the initiation site on mRNA (until some competing interactions melt the fold). More commonly, the stability is not so high, and the availability of the RBS for ribosomal particles will be determined by the competition between the intramolecular secondary/tertiary structure formation and the ribosome–mRNA interaction. In any case, the existence of a secondary or tertiary structure in the RBS region seems to always reduce the initiation rate. The more stable the fold, the more reduction is expected. In the region upstream of the initiation codon and the SD sequence, the absence or low stability of secondary/tertiary structure may also contribute to a higher initiation rate. At the same time, however, some structural elements around or inside the RBS can facilitate the ribosome binding to mRNA, for example, by better exposing the SD sequence and/or the initiation codon, or by better adjusting the distance between them.

The affinity of the 30S ribosomal particle for the available RBS depends on the degree of complementarity between the SD sequence and the 16S RNA 3′-terminal sequence. The distance between the SD sequence and initiation codon, the nature of the initiation codon (AUG or non-AUG), and other structural environments may also contribute to the affinity. Generally, a higher affinity provides a higher initiation rate. Concerning the SD sequence contribution, the most frequent situation among bacterial mRNAs is four to six base pairs (A:U and G:C) between the mRNA sequence and the 16S RNA terminus. Mutations in the SD sequences of mRNAs that decrease the complementarity with the 16S RNA do reduce the initiation rate. Longer SD sequences usually increase the initiation rate. When the spacing between the SD sequence and the initiation codon is more than 12 nucleotides or less than 5 nucleotides, the initiation rate is also decreased. AUG can be qualified as the strongest initiation codon among other initiation codons (GUG, UUG, etc.). Weak initiation was reported for rare initiation codons, such as AUA and AUU. Some special sequences (initiation "enhancers") upstream of the SD sequence may facilitate the initiation rate also kinetically, probably by complementarily interacting with exposed regions of the ribosomal RNA of the 30S subunit; they are believed to fish out the initiating particles from the surroundings and attract them to the proper mRNA sites (McCarthy and Brimacombe, 1994). Thus, different mRNAs have different capacities to bind initiating ribosomal subunits. Hence, when they compete, the strongest win.

Indeed, in bacterial cells some mRNAs are much more expressible than others, due to their higher initiation rates. These are primarily the mRNAs that encode for abundant proteins of the bacterial cell, such as the two major outer membrane proteins, OmpA and lipoprotein, in *E. coli,* the multiple c-subunit of proton-translocating ATPase of plasma membrane, ribosomal proteins, and elongation factors EF-Tu and EF-G. As expected, bacteriophage RNAs encoding for coat proteins are especially expressible. A special emphasis should be made on some highly expressible (very strong) bacteriophage mRNAs, the mRNA transcribed from bacteriophage T7 gene 10 being one of the best studied among them.

The synthesis of eight types of protein subunits of proton-translocating ATPase ($\alpha_3\beta_3\gamma_1\delta_1\epsilon_1a_1b_2c_{10\text{-}15}$) is a good example of how the differential efficiencies of translational initiation at different mRNAs correspond to the required subunit stoichiometry in the final enzymatic complex (McCarthy, 1988, 1990). Five types of subunits (α, β, γ, δ, and ϵ) form the soluble part (F_1) of the enzyme, whereas the membrane-bound part (F_0) comprises three types of subunits (a, b, and c). All the mRNAs for the ATPase subunits are the sections of the same polycistronic transcript (which starts with the message *I* for unknown protein), and therefore the amount of messages is equal for each subunit:

I	*B*	*E*	*F*	*H*	*A*	*G*	*D*	*C*
↓	↓	↓	↓	↓	↓	↓	↓	↓
?	a	c	b	δ	α	γ	β	ϵ

At the same time, the rates of translational initiation for the messages strongly differ, being the highest for the most abundant subunit c encoded for by *atpE,* and the lowest for single-copy subunits, such as a or γ, encoded for by *atpB* and *atpG,* respectively. As demonstrated in direct *in vitro* experiments, the affinity of the 30S ribosomal subunits for *atpE* RBS is high indeed, this being attributed to its open, unstructured and well-balanced nucleotide sequence (including the presence of U-rich and A-rich stretches):

. . . UUUUAACUGAAACAAACU**GGAG**ACUGUC**AUG**
GAAAACCUGAAUAUGGAU . . .

In contrast, the *atpB* RBS has a less favorable SD sequence and the distance between it and the initiation codon is too small:

. . . UAAAA**GG**CAUC**AUG**GCUUCAGAAAAUA . . .

The *atpG,* however, has been predicted to possess a stable secondary structure both upstream of the SD sequence and in the region of the initiation codon and downstream:

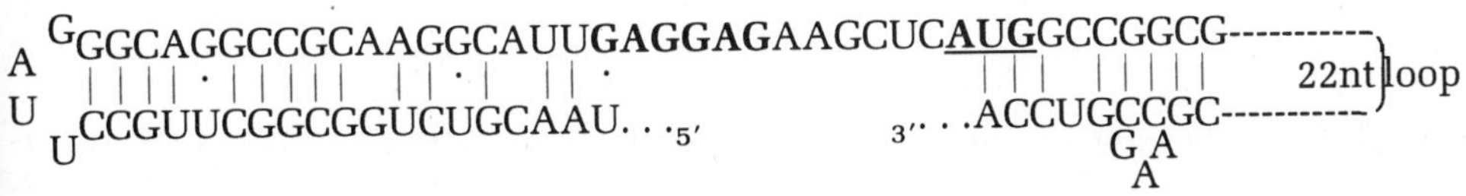

```
 GGGCAGGCCGCAAGGCAUUGAGGAGAAGCUCAUGGCCGGCG----------
A  ||||·|||||| ||·| ||·                ||| |||||     22nt loop
U  CCGUUCGGCGGUCUGCAAU. . .5'       3'. . .ACCUGCCGC----------
 U                                            G A
                                               A
```

Thus, the availability and affinity of initiation regions of different messages signify their "strength" in initiation and hence determine, at least to some degree, the ratio of protein production from them. This can be considered an example of a fixed (constitutive) translational control of the proper proportions in protein synthesis products.

16.3. Translational Coupling

The fully **independent translation initiation** of different messages (cistrons) of a polycistronic mRNA, which determines a fixed proportion of encoded protein products according to the intrinsic initiation "strengths" of the cistrons, is an extreme case among the ways in which polycistronic mRNAs can be read out by ribosomes. In most cases the translation initiation of downstream cistrons *depends on the translation of upstream messages* of a polycistronic mRNA (translational coupling). The cases of such a dependent initiation of translation within a polycistronic mRNA can be classified into two groups: (1) **Initiation induced by upstream translation:** translation of an upstream cistron is required for binding of free ribosomal particles at the RBS of an internal cistron and thus for the initiation of translation of a downstream message. Here the RBS of an internal cistron may be conformationally hidden, and the unfolding of mRNA as a result of translation of a preceding cistron opens it and thus makes it accessible for free ribosomes. One example is the dependence of initiation of replicase synthesis on translation of the preceding coat protein cistron in the systems directed by RNAs of bacteriophages of the MS2 or Qβ type. (2) **Sequential translation via reinitiation:** ribosomes associate with mRNA only at an upstream cistron (at its RBS) and then reinitiate at each subsequent cistron without dissociation after termination. In these cases free ribosomal particles cannot associate with and initiate translation at downstream cistrons. Polycistronic mRNAs encoding for ribosomal proteins are typical examples of this case.

The three situations discussed above (marked by bold face) are sketched in Fig. 16.1. It should be mentioned that their combinations and intermediate situations are also possible. For example, the independent initiation at the cistrons of the proton ATPase mRNA (see Section 16.2) is accompanied by an incomplete translational coupling of various degrees between some cistron pairs. Thus, initiation at the *atpA* cistron has been shown to be significantly enhanced by the translation of the preceding *atpH* cistron, while the initiation of the *atpG* cistron is only slightly increased by the translation of the preceding *atpA* cistron (Hellmuth *et al.*, 1991).

16.3.1. Initiation Induced by Translation of Upstream Cistron

When full-sized polycistronic RNA of bacteriophage MS2 (see Section 16.4.1) encounters free bacterial ribosomes, the initiation occurs only at the RBS of the second cistron *C* encoding for the phage coat protein. Despite the presence of a good SD sequence, a strong initiation codon (AUG), and an optimal distance between them, no initiation takes place at the cistron *S* encoding for the replicase subunit. The RBS of the *S* cistron is inaccessible to ribosomes due to its involvement in the long-range interaction with the upstream parts of the RNA. When ribosomes reading out the upstream *C* cistron reach the interacting section of the *C* cistron message, they melt this structure and thus open the RBS of the *S* cistron for binding with free ribosomes from the surroundings (see Lodish and Robertson, 1969; Weissmann *et al.*, 1973).

Translation of the polycistronic transcript of the *rplJ-rplL* operon encoding for ribosomal proteins L10 and L7/L12, as well as β- and β′-subunits of RNA polymerase (see Section 16.4.2), is another remarkable example of this type of translational coupling (Friesen *et al.*, 1980; Yates *et al.*, 1981; Petersen, 1989). Here the RBS of the *rplL* (L7/L12) cistron is blocked by a long-range base pairing of this region with the region located more than 500 nucleotides upstream, in the beginning of the preceding *rplJ* (L10) cistron; the initiation at the RBS by free ribo-

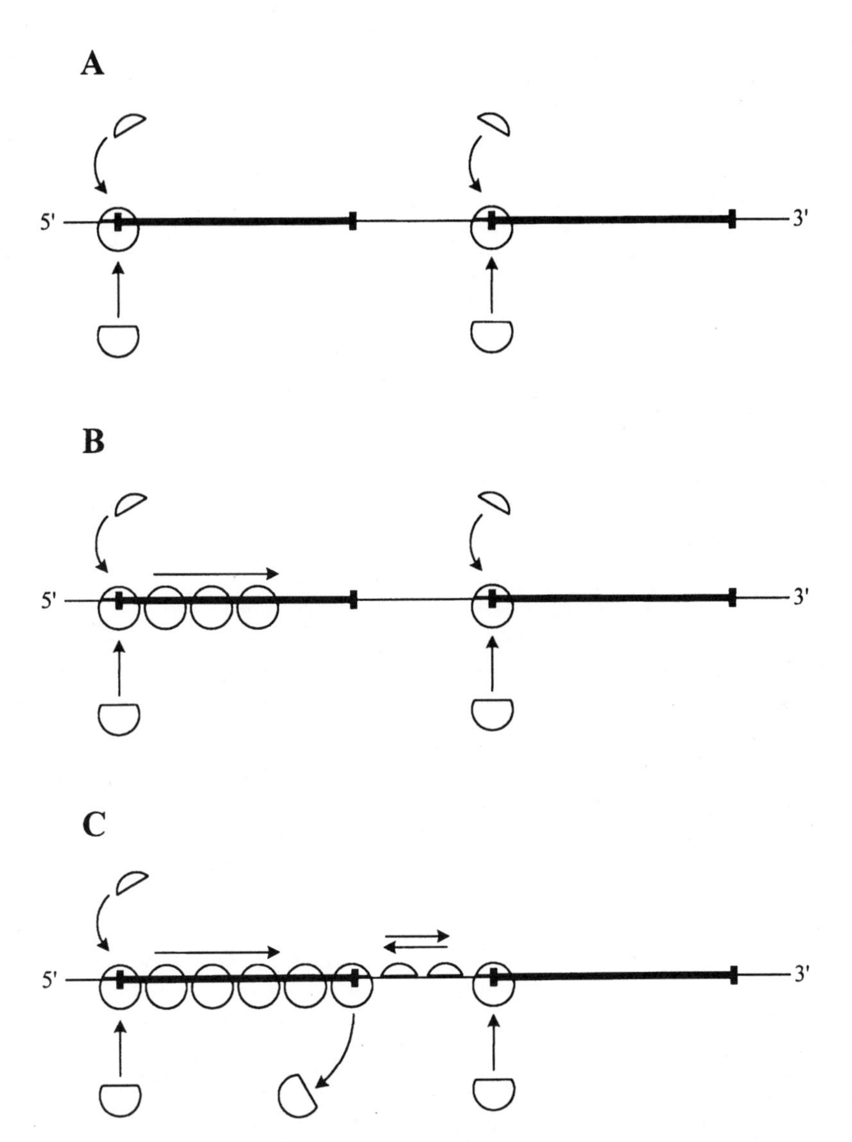

FIG. 16.1. Schematic representation of three different situations with internal initiation on polycistronic mRNAs in prokaryotes. **(A)** Independent initiation on each cistron. **(B)** Initiation on a downstream cistron induced by translation of the preceding cistron. Presumably the translation of the upstream cistron disrupts a long-range interaction within mRNA and thus opens the RBS of the downstream cistron. **(C)** Reinitiation. Downstream cistrons are incapable of initiating with free 30S subunits. Instead, the 30S subunits after termination of translation of the preceding cistron slide phaselessly along mRNA and reinitiate at a nearby RBS, that is, the neighboring downstream cistron.

somes is impossible. When ribosomes initiate at the *rplJ* cistron and then translate the message, they melt this structure and open the *rplL* RBS. The strength of the open RBS is found to be much higher than that of the *rplJ* RBS, providing a very efficient binding of free ribosomes and thus a high level of initiation. As a result, the translation of the *rplL* cistron yields four times more copies of protein L7/L12 than the number of copies of protein L10 produced from the preceding *rplJ* message:

rplJ	*rplL*
↓	↓
L10	4×L7/L12

The proportion 4:1 corresponds to the molar ratio of protein L7/L12 to protein L10 in the ribosome.

The rearrangement of a secondary structure resulting in an appearance of an open RBS can be also induced by ribosomes stalled at a specific position of a preceding coding sequence rather than by actively translating ribosomes (reviewed by de Smit and van Duin, 1990). This mechanism is used in several cases for regulation of the synthesis of some antibiotic-resistance factors. For example, translation of *cat* mRNA encoding for chloramphenicol acetyltransferase (the enzyme that inactivates chloramphenicol) in Gram-positive bacteria is induced by chloramphenicol in the following way. The *cat* mRNA is the second cistron in a bicistronic message and has a hidden RBS involved in the formation of a stable hairpin together with the end of the preceding cistron (Fig. 16.2). The preceding cistron encodes for a short peptide and has an open RBS. In the absence of chloramphenicol, ribosomes translate the first cistron but cannot translate the second (*cat*) cistron: independent initiation is not allowed, and the translation of the preceding cistron until its termination codon opens the *cat* RBS but interferes with the binding of free ribosomes to the opened *cat* RBS due to the very short distance between the termination site and the *cat* RBS. In the presence of chloramphenicol the ribosome translating the first cistron will be stalled by the drug; the ribosome stalled in the middle of the short coding sequence supports the unwound state of the intercistronic hairpin and thus opens the *cat* RBS for initiation.

Another example is the regulation of the synthesis of a specific methylase that modifies A2058 in the ribosomal 23S RNA and thus confers resistance against erythromycin (and other macrolides, lincosamides, and streptogramin B) to the bacterial ribosome (Section 11.3.3). The model of the regulation is as follows (Fig. 16.3). Again a short open reading frame precedes the *erm* cistron which encodes for the methylase. The *erm* RBS is closed in the hairpin (designated as 3–4 hairpin). In the absence of erythromycin, ribosomes translate only the preceding cistron, which ends within another hairpin (1–2 hairpin) located some distance upstream of the *erm* cistron, and do not open the *erm* RBS. In the presence of the antibiotic, the ribosome translating the preceding cistron is stalled in the position around the middle of the coding sequence; as a result the hairpin 1–2 is unwound and the left strand 1 of it is covered by the stalling ribosome. Because of the complementarity between strand 2 and the left strand 3 of the *erm* RBS hairpin 3–4, the rearrangement with the formation of the hairpin 2–3 and the release of the *erm*

RBS takes place. Now free ribosomes can initiate translation of the *erm* cistron and start to synthesize the methylase.

16.3.2. Sequential Translation of Polycistronic Messages via Reinitiation

The longest polycistronic mRNA among those encoding for ribosomal proteins in *E. coli* is that starting with a protein S10 message (see Fig. 16.7). It contains messages for 11 ribosomal proteins. However, they cannot bind the initiating ribosomal particles independently. They are translated sequentially: the association with initiating ribosomal particles takes place at the first cistron (S10), the ribosomes initiate translation and move downstream, and then the ribosomes that have terminated the translation of the preceding cistron do not dissociate from the template but pass to reinitiation at the next cistron. It is believed that such sequential translation of the messages of the same polycistronic mRNA provides for the equimolar production of the ribosomal proteins coded by the polycistronic mRNA (see, e.g., Dean and Nomura, 1980; Nomura, Gourse, and Baughman, 1984). A similar situation is observed with the mRNA starting with protein L14 message

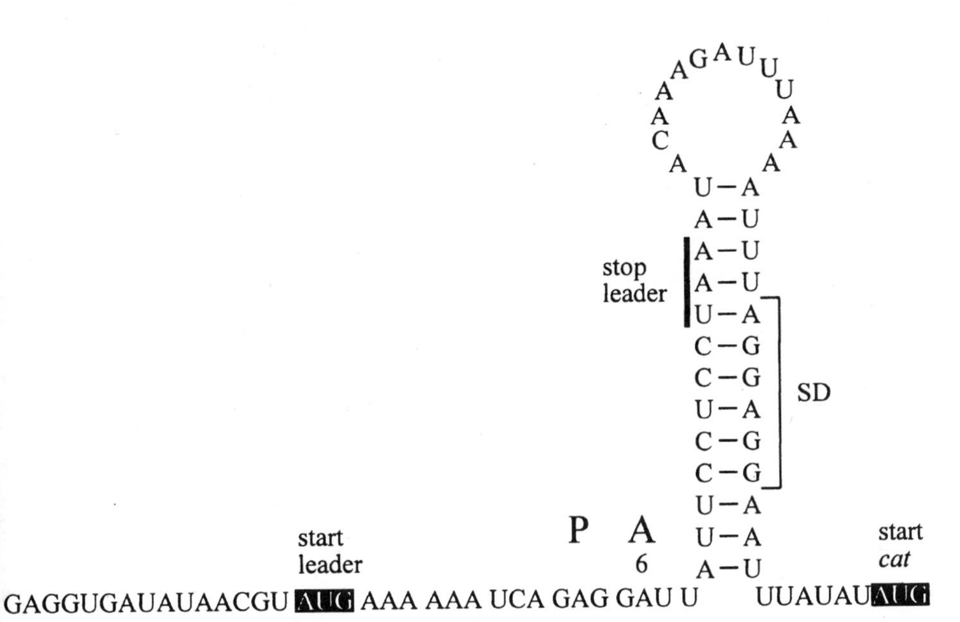

FIG. 16.2. Secondary structure involved in chloramphenicol induction of *cat* mRNA translation. The Shine-Dalgarno (SD) sequence of the *cat* message is firmly closed by base pairing with the end of the preceding short ORF, preventing the independent initiation of *cat* translation. In the absence of chloramphenicol, ribosomes read the upstream ORF till UAA and unwind the helix but shield the *cat* RBS by themselves due to the short distance between the UAA and the RBS. In the presence of chloramphenicol, ribosomes reading the upstream ORF are stalled before reaching the termination codon (for some reason, the preferential stalling point is the sixth codon of the ORF); in this situation the helix is already unwound by the stalled ribosome and, therefore, the *cat* RBS (its SD) is open for initiation. Reproduced, with modifications, from de Smit, M. H., and van Duin, J. (1990) *Progr. Nucleic Acid Res. Mol. Biol.* **38:**1–35, with permission.

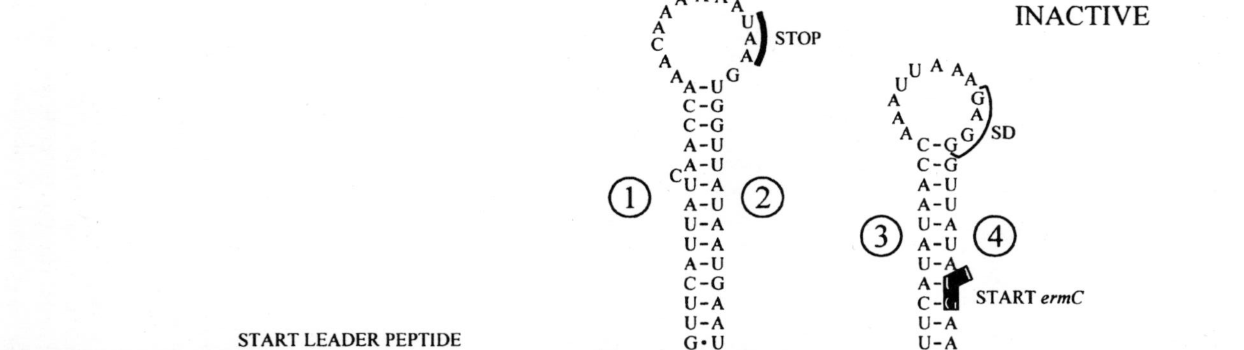

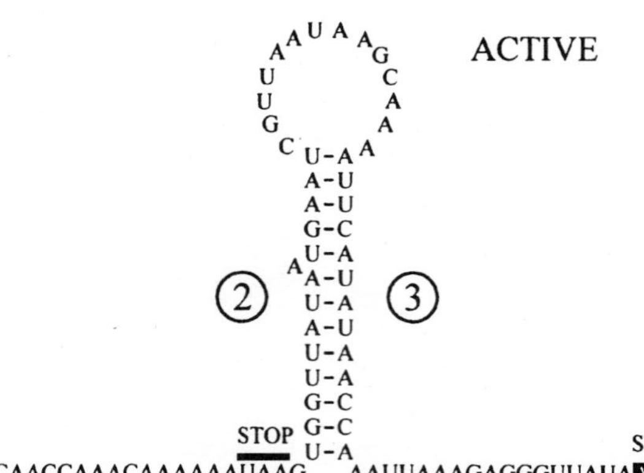

FIG. 16.3. Secondary structures involved in translational regulation of the synthesis of 23S ribosomal RNA methylase on *erm* mRNA. The synthesis of the methylase is induced by erythromycin resulting in the methylation of a specific position of the RNA conferring the erythromycin resistance to ribosomes. **(A)** Inactive structure. In the absence of erythromycin, ribosomes translate the leader cistron, and cannot initiate translation of the *erm* cistron because of the stable pairing between its RBS (strand 4) and the preceding spacer sequence (strand 3). **(B)** Active structure induced by erythromycin-stalled ribosomes. In the presence of erythromycin, a stalled ribosome in the leader message covers strand 1 and thus releases strand 2 for pairing with strand 3. As a result, strand 4 becomes unpaired and the *erm* RBS opens. Reproduced, with modifications, from de Smit, M. H., and van Duin, J. (1990) *Prog. Nucl. Acid Res. Mol. Biol.* **38**:1–35 with permission.

(see Fig. 16.7), but in this case the sequential translation begins from protein L5 cistron whereas the two preceding messages are read out independently.

The tight and efficient translational coupling via termination–reinitiation requires at least three conditions: (1) the distance between the site of termination of the preceding cistron and the restart site must be short, (2) the stable secondary/tertiary structure in the intercistronic region should be absent or melted by ribosomes translating the preceding cistron, and (3) the SD sequence should be present prior to the reinitiation site. Termination and reinitiation sites can be in different reading frames (phases) along the mRNA. It is interesting that the termination and initiation codons can overlap, or even the RBS of the next cistron can be upstream of the termination codon of the preceding cistron.

For example, in the polycistronic transcript of *trp* operon of *E. coli,*

E *D* *C* *B* *A*

there are two examples of adjacent messages with overlapping termination and initiation codons; these are *trpE-trpD* and *trpB-trpA* cistron pairs, both with the intercistronic region UGA**AUG** (Oppenheim and Yanofsky, 1980; Platt and Yanofsky, 1975). Correspondingly, the translational coupling is tight and ensures equimolar production of polypeptides that are constituents of one enzyme complex (Das and Yanofsky, 1984).

In the translationally coupled cistrons VII and IX of the bacteriophage f1 RNA the termination codon UGA of VII overlaps with the initiation codon AUG of IX over two nucleotides, and the latter is shifted upstream relative to the termination codon (Ivey-Hoyle and Steege, 1989):

```
          Met  Ser...
VII  ...CAAAGAUGAGU...   IX
       ...Gln  Arg  Stop
```

In the polycistronic message transcribed from the galactose operon, *galE-galT-galK,* the termination and initiation codons in the *galT-galK* pair are separated by three nucleotides, with the SD sequence upstream of the termination codon:

. . . GAA UCC **GGA G**UG UAA GAA **AUG** AGU . . .

Again, the translational coupling has been shown to be tight and efficient (Schumperli *et al.,* 1982), consistent with the fact that the products of *galT* and *galK* form an enzymatic complex with equimolar proportion of the two proteins (galactose-1-phosphate uridyltransferase and galactokinase, respectively). The mutational analysis demonstrated that reinitiation at *galK* is reduced when termination of the preceding cistron occurs outside the initiation region (upstream of the SD sequence or downstream of the initiation codon), and the more the distance between them, the less the efficiency of reinitiation.

The translation of the message encoding for the lysis peptide (*L*) within the bacteriophage MS2 RNA (Fig. 16.4) is also tightly coupled with the translation of the preceding coat protein cistron (*C*). Due to structural reasons (involvement of the RBS in relatively stable secondary structure), no independent initiation at the

L cistron is possible. The termination at the end of the *C* cistron has been shown to be an absolute prerequisite for the initiation (reinitiation) at the *L* cistron (Adhin and van Duin, 1989). At the same time the efficiency of reinitiation in this case is low, resulting in much lower production of the lysis peptide as compared with that of the coat protein. This correlates with a relatively long distance between the termination codon of the *C* cistron and the reinitiation site of the *L* cistron, the latter being more than 40 nucleotides *upstream* of the *C* cistron terminator UAA (see Fig. 16.5).

Thus, the sequential coupled translation of polycistronic mRNAs implies that the ribosomal particles (seemingly the 30S subunits) after termination are capable of "phaselessly wandering" (Sarabhai and Brenner, 1967) along mRNA around the termination site both downstream and upstream of it. Physically this phenomenon seems to be a lateral (two-dimensional) diffusion of nontranslating, mRNA-associated ribosomal particles along mRNA. If they encounter a structurally available initiation site that includes the SD sequence and the initiation codon with a proper distance between them, they reinitiate translation. When the initiation site overlaps the termination codon of the preceding cistron the reinitiation can be very efficient and approach 100%; this is the case of equimolar production of the encoded polypeptides. At the same time, a high probability of the dissociation of the nontranslating particles from mRNA exists during the "phaseless wandering," so that the greater the distance between the termination codon of the preceding cistron and the RBS of the following cistron, the less the efficiency of the reinitiation (Adhin and van Duin, 1990). A weak RBS can be another factor reducing the efficiency of reinitiation. Therefore, even in the case of tight coupling (strong dependence of the translation of an internal cistron on the translation of the preceding cistron), the efficiency of reinitiation after termination may vary depending on the intercistronic region: in many cases a significant portion of terminated ribosomes dissociate from mRNA, and only a part of the particles reinitiates. Correspondingly, the protein production of the downstream cistron will be decreased compared with that of the preceding cistron.

Generally, the reinitiation by mRNA-associated ribosomal particles is more efficient than the initiation by free ribosomes. That is why relatively weak or poorly exposed initiation sites can be used for reinitiation whereas the initiation by free ribosomes at them is absent or poor. Even in the case when the RBS is buried in a stable secondary structure, the termination of a preceding cistron within the RBS

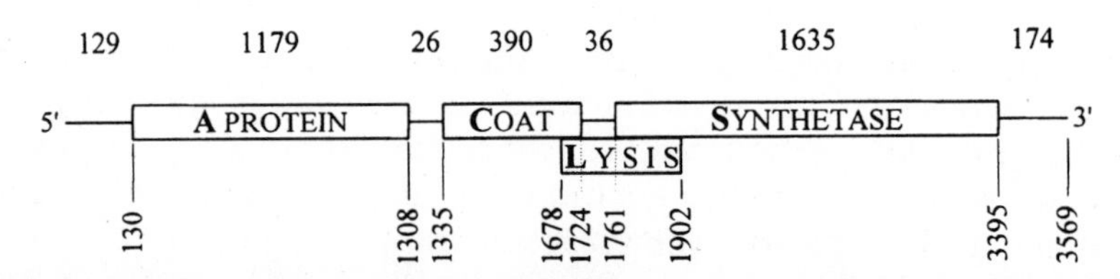

FIG. 16.4. Coding regions (RNA cistrons) in bacteriophage MS2 RNA. See Fiers, W., Contreras, R., Duerinck, F., Haegeman, G., Iserentant, D., Merregaert, J., Min Jou, W., Molemans, F., Raeymaekers, A., Van den Berghe, A., Volckaert, G., and Ysebaert, M. (1976) *Nature* **260**: 500–507; Kastelein, R.A., Remaut, E., Fiers, W., and van Duin, J. (1982) *Nature* **295**: 35–41.

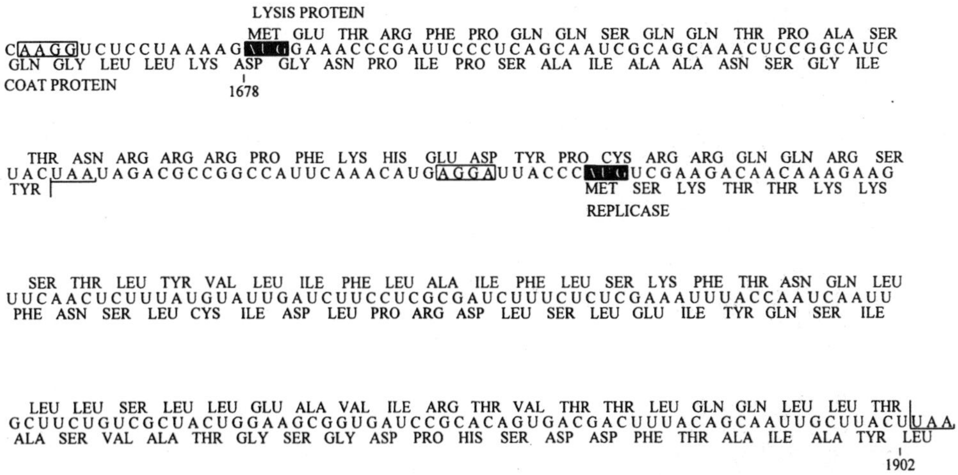

LYSIS PROTEIN
MET GLU THR ARG PHE PRO GLN GLN SER GLN GLN THR PRO ALA SER
C A A G G U C U C C U A A A A G A U G G A A A C C C G A U U C C C U C A G C A A U C G C A G C A A A C U C C G G C A U C
GLN GLY LEU LEU LYS ASP GLY ASN PRO ILE PRO SER ALA ILE ALA ALA ASN SER GLY ILE
COAT PROTEIN 1678

THR ASN ARG ARG ARG PRO PHE LYS HIS GLU ASP TYR PRO CYS ARG ARG GLN GLN ARG SER
U A C U A A U A G A C G C C G G C C A U U C A A A C A U G A G G A U U A C C C A U G U C G A A G A C A A C A A A G A A G
TYR
MET SER LYS THR THR LYS LYS
REPLICASE

SER THR LEU TYR VAL LEU ILE PHE LEU ALA ILE PHE LEU SER LYS PHE THR ASN GLN LEU
U U C A A C U C U U U A U G U A U U G A U C U U C C U C G C G A U C U U U C U C U C G A A A U U U A C C A A U C A A U U
PHE ASN SER LEU CYS ILE ASP LEU PRO ARG ASP LEU SER LEU GLU ILE TYR GLN SER ILE

LEU LEU SER LEU LEU GLU ALA VAL ILE ARG THR VAL THR THR LEU GLN GLN LEU LEU THR
G C U U C U G U C G C U A C U G G A A G C G G U G A U C C G C A C A G U G A C G A C U U U A C A G C A A U U G C U U A C U U A A
ALA SER VAL ALA THR GLY SER GLY ASP PRO HIS SER ASP ASP PHE THR ALA ILE ALA TYR LEU
1902

FIG. 16.5. The nucleotide sequence of the *L* cistron of the bacteriophage MS2 RNA (beginning with nucleotide residue 1678). See Atkins, J.F., Steitz, J.A., Anderson, C.W., and Model, P. (1979) *Cell* **18**: 247–256; Beremand, M.N., and Blumenthal, T. (1979) *Cell* **18**: 257–266. The amino acid sequence of the lysis protein is written *above* the nucleotide sequence, whereas the amino acid sequences of the end of the coat protein and the start of the replicase subunit polypeptide are given *below* the nucleotide sequence.

allows the terminating ribosomes to be immediately captured by the initiation sequences (SD and AUG). This can be considered as a local rearrangement of a termination complex into an initiation complex.

16.4. Translational Repression

Except for the regulation of the synthesis of some antibiotic-resistance factors (chloramphenicol acetyltransferase and 23S RNA methylase, see Section 16.3.1), all the mechanisms discussed in the previous sections control translation by determining fixed (constitutive) rates of initiation at various mRNAs and their functional sections (cistrons). In contrast, repression mechanisms with the participation of special RNA-binding proteins called translational *repressors* provide ways to modulate the rates of initiation in wide ranges depending on external signals (*effectors*), as well as for feedback regulations.

The predominant mechanism of translational repression is the direct binding of a repressory protein to the RBS. The bound protein competes with the binding of ribosomes at the RBS. When the RBS is located within an unstable secondary-structure element, a repressor protein can stabilize this element and thus prevent the interaction of the RBS with the initiating ribosome. Sometimes a repressory protein binds to a region outside the RBS and induces such a rearrangement of a secondary/tertiary structure that RBS becomes closed and inaccessible to initiating ribosomes. The region of mRNA that binds a repressor can be called the *translational operator.*

16.4.1. Regulation of Translation of Bacteriophage MS2 RNA

The bacteriophage MS2 has a spherical shape; its diameter is 250 Å, its molecular mass 3.6×10^6 daltons. The phage particle contains 180 coat protein subunits, each with a molecular mass of 14,700 daltons, one molecule of the so-called A protein having a molecular mass of 38,000 daltons, and one molecule of RNA with a molecular mass of about 10^6 daltons. After infecting the *E.coli* cells or in a cell-free translation system, the phage RNA serves as a template for the synthesis of coat protein, A protein, lysis peptide, and a subunit of RNA replicase with a molecular mass of 62,000 daltons (this subunit and lysis peptide are not components of the phage particle). The location of corresponding cistrons *C, A, L,* and *S* along the MS2 RNA chain is shown schematically in Fig. 16.4.

The chain starts from G, bearing triphosphate at its 5′-position. This is followed by a noncoding sequence with a length of 129 nucleotide residues; this sequence contains AUG and GUG triplets which, however, do not serve as initiation codons. The first initiation codon, GUG, starts the coding sequence of the *A* cistron corresponding to the A protein. The *A* cistron is 1,179 nucleotide residues in length and ends with the UAG termination triplet. It is followed by a noncoding region 26 residues long. The next coding sequence starts from AUG and is 390 nucleotides in length; this is the *C* cistron coding for coat protein. This cistron is terminated by UAA and followed by the second termination codon UAG. The *C* cistron is separated from the *S* cistron, which codes for the RNA replicase subunit, by 36 nucleotides. The *S* cistron begins with AUG, is 1,635 nucleotides in length, and ends with UAG. At a distance of one nucleotide from its termination signal (i.e., out of the reading frame) is found yet another termination triplet, UGA. The 3′-terminal noncoding sequence has a total length of 174 nucleotide residues and ends with an adenosine.

In addition to the three sequentially arranged cistrons *A, C,* and *S* separated by intercistronic spacers, there is the fourth coding sequence *L* overlapping the *C* and *S* cistrons in another reading frame (Fig. 16.4); it codes for the lysis peptide, or L protein. This protein is involved in host cell lysis at the late stage of infection. The *L* cistron begins within the end section of the *C* cistron, contains the entire 36-nucleotide spacer between *C* and *S,* and terminates within the *S* cistron; the reading frame of the *L* cistron is shifted to the right by one residue (+1 shift), so that this cistron is not translated during S protein or C protein synthesis. The *L* cistron has its own initiation codon AUG, which is out of frame with *C* cistron codons, and its own termination codon UAA, which is out of frame with the codons belonging to the *S* cistron (Fig. 16.5).

The three sequentially arranged cistrons *A, C,* and *S* have strong RBSes and their translation is initiated by free ribosomes, independent of termination of a preceding cistron. In contrast, the *L* cistron is translated only as a result of a low-level reinitiation after termination of translation of the *C* cistron and the subsequent "phaseless wandering" (see Section 16.3.2). Despite a high potential of their own initiation, the three nonoverlapping cistrons, *A, C,* and *S,* are strongly dependent on each other in their translation. As already mentioned, the *A* cistron and the *S* cistron of untranslated MS2 RNA are incapable of binding ribosomes because their RBSes are masked in secondary/tertiary structures of the RNA. Only the RBS of the *C* cistron is exposed for immediate interaction with free ribosomal particles and thus can be involved in the initiation of translation, independent of the translation of other cistrons.

In accordance with the above, the translation of MS2 RNA begins with the initiation of the coat protein synthesis. After the translation of the *C* cistron has begun, ribosomes move along its coding sequence in the direction of the *S* cistron and unfold the secondary and tertiary structure in the course of their progression. This results in the opening of the *S* cistron initiation region (Fig. 16.6) (see also Section 16.3.1). Hence, even before the first ribosome has completed translation of the *C* cistron and, thus, before the first coat protein molecule has been synthesized, the initiation region of the *S* cistron becomes accessible for initiation and, correspondingly, synthesis of the RNA replicase subunit is initiated.

In order to form the active RNA replicase molecule, the product of the *S* cistron translation must associate with three host cell proteins. Two of these proteins are the elongation factors EF-Tu and EF-Ts, and the third is the ribosomal pro-

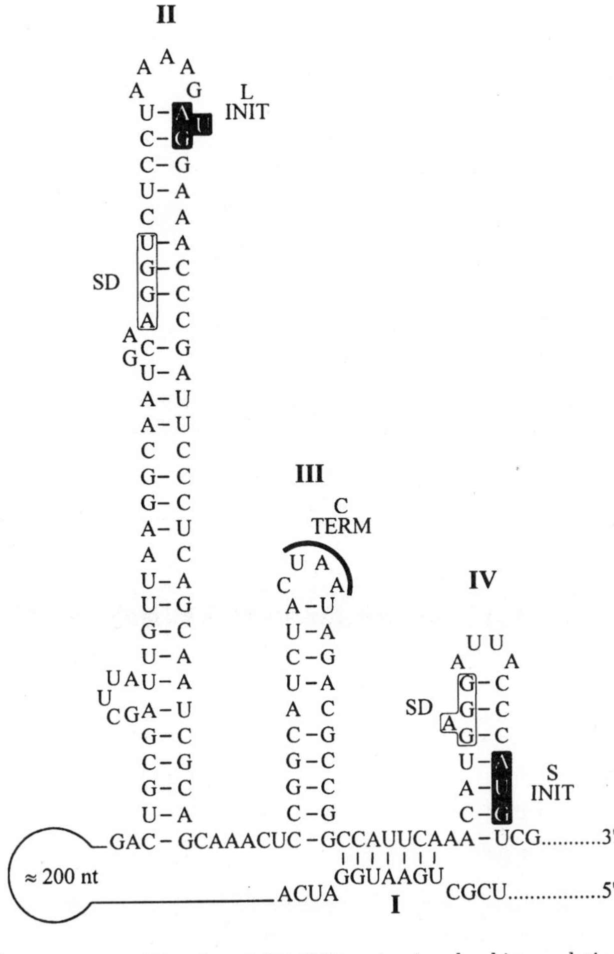

FIG. 16.6. Secondary structure of the phage MS2 RNA region involved in regulation of translation of *L* and *S* cistrons. From de Smit, M.H., and van Duin, J.(1990) *Progr. Nucleic Acid Res. Mol. Biol.* **38**: 1–35.

tein S1. In other words, the complete active RNA replicase is a protein with a quaternary structure consisting of four different subunits (R, S1, EF-Tu, and EF-Ts), and only one of them (R) is coded by the phage RNA. The RNA replicase is a template-specific RNA-dependent RNA polymerase using the original chain ("+" chain) of the MS2 RNA to form the complementary chain ("−" chain) and then, using it as a template to produce numerous copies of the original "+" chain.

The RNA replicase has been also found to be a repressory protein in the translation of MS2 RNA: it specifically recognizes the RBS of the *C* cistron, binds to it, and blocks the initiation (Weber *et al.*, 1972; Meyer, Weber, and Weissmann, 1981). This happens during the early stage of infection when the concentration of the replicase increases and eventually reaches a certain level. The repression of the *C* cistron translation at this stage is aimed to avoid the situation where the RNA serves at same time as template for translation by ribosomes and for replication by the enzyme, that is, to "clear" the RNA of ribosomes. As the replication proceeds, new RNA molecules appear and exceed the number of the RNA replicase molecules, so that the coat protein molecules start to be produced again from nonrepressed *C* cistrons.

The completion of the *C* cistron translation by the ribosomes results in the appearance of free coat protein molecules. As translation proceeds, and as new MS2 RNA molecules become available for the translation, this protein accumulates; eventually it will be used in the self-assembly of mature phage particles. The coat protein, however, in addition to its role in phage particle assembly, possesses a strong specific affinity for the region of MS2 RNA between the *C* and *S* cistrons, including the RBS of the *S* cistron (see Fig. 16.4). The protein binds to this region (operator) and represses the initiation of the *S* cistron translation (Lodish and Zinder, 1966; Bernardi and Spahr, 1972). The repression seems to result from the labile secondary structure (Fig. 16.6, helix IV) being stabilized by the phage coat protein; thus the SD sequence and the initiation codon of the *S* cistron become inaccessible to ribosomes. Hence, after the translation of the *S* cistron has been allowed by the translation of the preceding cistron, the *S* cistron translation is repressed due to accumulation of the protein coded by the preceding cistron. The repression of further synthesis of this protein prevents an unnecessary overproduction of the enzyme. In this way the phage coat protein, which plays the part of the *S* cistron repressor, performs the regulatory function in translation.

The *A* cistron cannot be translated until MS2 replication is started. Its initiation region is masked by the structure of the intact MS2 RNA. It may be exposed for *in vitro* translation and thus can be initiated as a result of some artificial treatments, such as partial nuclease or heat-induced degradation of the intact polynucleotide chain, or mild treatment with formaldehyde, which disrupt base pairings. In the course of RNA replication, however, at the early period of + chain formation when the chain is still growing, the three-dimensional structure of the 5′-terminal section containing the RBS of the *A* cistron is not yet fully formed. It is this period that seems to be used for initiating translation of the *A* cistron under normal conditions (Robertson and Lodish, 1970; Kolakofsky and Weissmann, 1971). Since the mature virus particle contains only one molecule of the A protein per 180 molecules of the coat protein, the relatively brief period during which the initiation of the *A* cistron translation is possible appears to be sufficient for the required production of the A protein. Thereafter, the elongated MS2 RNA folds in such a way that the initiation region of the *A* cistron becomes involved in some three-dimensional structure which makes the RBS inaccessible to free ribosomes.

Thus, the translation of MS2 RNA provides examples of several different regulatory systems. First, the interaction of the RNA replicase molecule with the initiation site of the *C* cistron and the binding of the coat protein molecules to the initiation site of the *S* cistron represent a typical translational repression mechanism. In addition, translational coupling via opening of the RBS by translation of a preceding cistron (Section 16.3.1), translational coupling via reinitiation (Section 16.3.2), and coupled replication–translation are observed during expression of the polycistronic MS2 RNA.

16.4.2. Regulation of Translation of Ribosomal Protein mRNAs

The bacterial cell is known to avoid overproduction of the ribosomal proteins. Generally speaking, the ribosomal proteins are synthesized in amounts required just for ribosomal assembly, in accordance with the amount of ribosomal RNA formed; under normal conditions, the cell contains no significant excess of free ribosomal proteins. The coordinated levels of production of nearly all ribosomal proteins in equimolar amounts are achieved even though their genes are not organized as a single regulated block, but are represented by approximately 16 independent operons, which are scattered throughout the cell genome. The coordinated and virtually stoichiometric production of the ribosomal proteins and the prevention of their overproduction are maintained by a controlling mechanism which provides the repression of translation by protein excess (translational feedback control) (Dean and Nomura, 1980).

A large proportion of the genes coding for ribosomal proteins (31 out of 52) are present in two main clusters on the *E. coli* chromosome. One of these clusters is located in the *str-spc* region at the 72nd min, and the other in the *rif* region at the 89th min. The *str-spc* region contains four operons coding for 27 ribosomal proteins, EF-Tu, and EF-G, as well as for the α-subunit of RNA polymerase. The *rif* region contains two operons coding for four ribosomal proteins, as well as for the β- and β′-subunits of RNA polymerase. Each operon produces a polycistronic mRNA. The cistrons and their order in these polycistronic mRNAs are shown schematically in Fig. 16.7.

Studies conducted by several groups (for reviews, see Nomura, Jinks-Robertson, and Miura, 1982; Nomura *et al.*, 1984; Lindahl and Zengel, 1982, 1986; Draper, 1987) demonstrated that in the case of each polycistronic mRNA, one of the translation products, a ribosomal protein, serves as a repressor of the translation of a corresponding mRNA (these products are circled in Fig. 16.7). This effect has been demonstrated both in experiments *in vivo* and in cell-free systems. Experiments *in vivo* have shown that the synthesis of the ribosomal proteins coded by the corresponding mRNA is inhibited when the overproduction of one of the proteins circled in Fig. 16.7 is induced. Induction of protein S7, L4, S8, S4, L1, or L10 results in the inhibited synthesis of only those ribosomal proteins that are coded by the polycistronic mRNA possessing the cistron of the corresponding protein. Experiments *in vitro* have brought even more direct results: adding one of these proteins (e.g., S7, L4, S8, S4, L1, or L10) to the cell-free translation system leads to a selective inhibition of synthesis of only that set of proteins which is coded by the mRNA containing the cistron corresponding to the added protein.

Synthesis of some of the proteins coded by the listed polycistronic mRNAs, however, is not inhibited when the repressory ribosomal proteins are added. Protein S12, for example, continues to be synthesized after the protein S7 is added to the cell-free system or in the *in vivo* version of the experiment with the selective induction of protein S7. Similarly, the synthesis of proteins L14 and L24 does not stop in response to the addition or overproduction of protein S8. It is noteworthy that the cistrons of the proteins not controlled by proteins S7 or S8 are found to be proximal to the 5′-end of the polycistronic mRNA.

All of these observations may be best explained by assuming that the repressor protein binds specifically to the initiation region of one cistron and blocks the translation of all cistrons located in the direction of the 3′-end. Protein S8, for example, binds with the origin of the cistron of protein L5 and, as a result, the translation of all subsequent downstream-located (but not upstream-located) cistrons is repressed. The implication is that in these cases free ribosomes cannot initiate the translation of each cistron independently. The *sequential translation via reinitiation* appears to occur instead: ribosomes that have terminated the translation of the preceding cistron do not dissociate from the template but pass directly to reinitiation at the next cistron. Such sequential translation of cistrons provides for the equimolar production of ribosomal proteins coded by a given polycistronic mRNA.

There are some exceptions, however, in the same polycistronic mRNAs. For example, it has been shown that the translation of mRNA cistrons for EF-Tu and

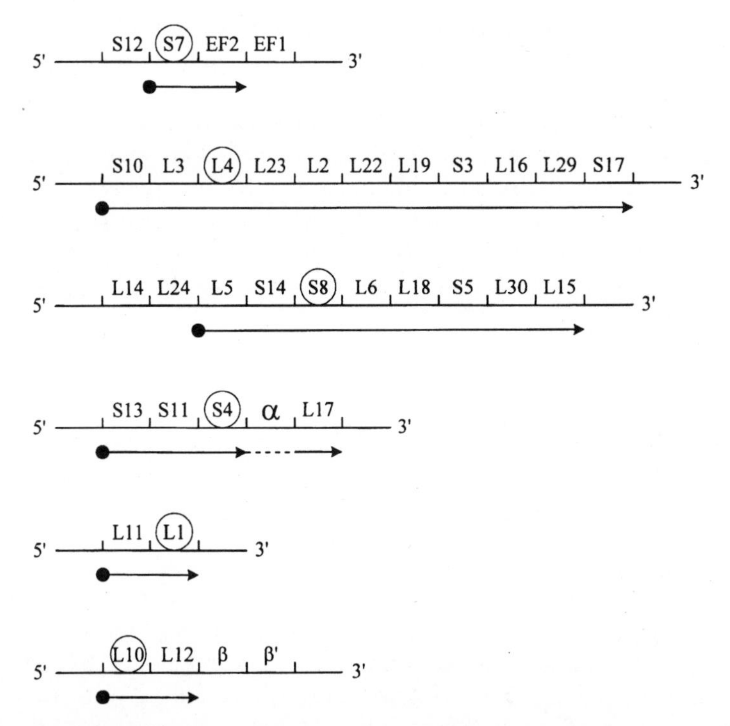

FIG. 16.7. Schematic representation of the sequential arrangement of ribosomal protein cistrons along polycistronic mRNA chains. See Dean D., and Nomura, M. (1980) *Proc. Natl. Acad. Sci. USA* **77**: 3590–33594; Nomura, M., Jinks–Robertson S., and Miura, A. (1982) in *Interaction of Translational and Transcriptional Controls in the Regulation of Gene Expression* (M. Grunberg–Manago and B. Safer, eds.), pp. 91–104, Elsevier, New York. α, β, and β' are the cistrons coding for the corresponding subunits of RNA polymerase. The origin of the arrow under the sequence designates the point of action of a repressor protein; the arrow extends to the cistron subject to the control. The circled ribosomal proteins serve as repressors; they bind to the regions of the polycistronic mRNAs corresponding to the origins of the arrows.

for RNA polymerase subunits β, β′, and α is not repressed by proteins S7, L10, and S4, respectively (Fig. 16.7). Therefore, it appears that the initiation regions of these mRNA cistrons are capable of binding the free ribosomes, providing for independent initiation of translation. On the other hand, it is known that in the case of the synthesis of proteins L10 and L7/L12, the production of protein L7/L12 is four times greater than the production of protein L10, in accordance with their stoichiometry in the ribosome; the independent initiation of translation of the L7/L12 cistron takes place here and the initiation rate for this cistron is far greater than that for the L10 cistron. At the same time, as has already been indicated, protein L10 represses the translation of both proteins L10 and L7/L12: the repression of its own cistron prevents the opening of the RBS of the next L7/L12 cistron (see Section 16.3.1).

Identifying the attachment sites (operators) of the repressory ribosomal proteins on polycistronic mRNAs is particularly interesting. It has been demonstrated that if the origin of the structural gene for protein S13 and the preceding nucleotide sequence is deleted, protein S4 is unable to repress translation of the corresponding polycistronic mRNA (fourth line in Fig. 16.7) (Nomura *et al.*, 1980). In contrast, protein S7 can repress its own synthesis if the leader of its polycistronic mRNA (first line in Fig. 16.7), including the S12 cistron, is absent (Dean, Yates, and Nomura, 1981a,b). Protein L1 has been shown to exert repressory action upon its bicistronic mRNA (fifth line in Fig. 16.7) only in the presence of the 5′-terminal sequence, preceding the L11 cistron (Yates and Nomura, 1981). It follows that the attachment sites of the repressor proteins should be located at the origins of the cistrons from which the repression of the sequential translation begins in polycistronic mRNA. Correspondingly, for protein S7 the operator site should lie somewhere between the cistrons of proteins S12 and S7 or at the origin of the S7 cistron; for protein S4 the operator should be located before or at the beginning of the S13 cistron; and for protein L1 this site should be before the L11 cistron. Continuing this line of argument, the site of the repressor action for protein L4 should be located prior to or at the beginning of the S10 cistron (Yates and Nomura, 1980); for protein L10, prior to or at the beginning of its own cistron (Yates *et al.*, 1981; Johnsen *et al.*, 1982); and for protein S8, between the cistrons of proteins L24 and L5 or at the beginning of the L5 cistron (Dean *et al.*, 1981).

It is known that proteins S4, S7, S8, L1, and L4 play an important part in ribosomal structure and self-assembly: they are core proteins that bind tightly to specific sites on the ribosomal RNA (see Section 7.5). The attachment sites of these proteins on ribosomal RNA have been identified (see Fig. 7.6). There are grounds to believe that these ribosomal proteins, playing the role of repressors, bind to mRNA through the same RNA-binding centers that participate in the binding to ribosomal RNA. Then, it may be expected that the structures of the RNA regions binding a given ribosomal protein should be similar in ribosomal RNA and in mRNA. Indeed, the comparisons of primary and predicted secondary structures of the protein-binding sites on the ribosomal RNA and at the operator regions of mRNAs have demonstrated, at least in some cases, their similarity: the operator regions of mRNAs may mimic the protein-binding sites of ribosomal proteins on ribosomal RNA (see Nomura *et al.*, 1980; Olins and Nomura, 1981).

As an example, Figure 16.8 presents a comparison of the primary and predicted secondary structures of the intercistronic L24–L5 region and the origin of the L5 cistron assumed to be the binding site of the repressory protein S8 (Fig. 16.7, third line), and the region of the ribosomal 16S RNA that binds protein S8. The ho-

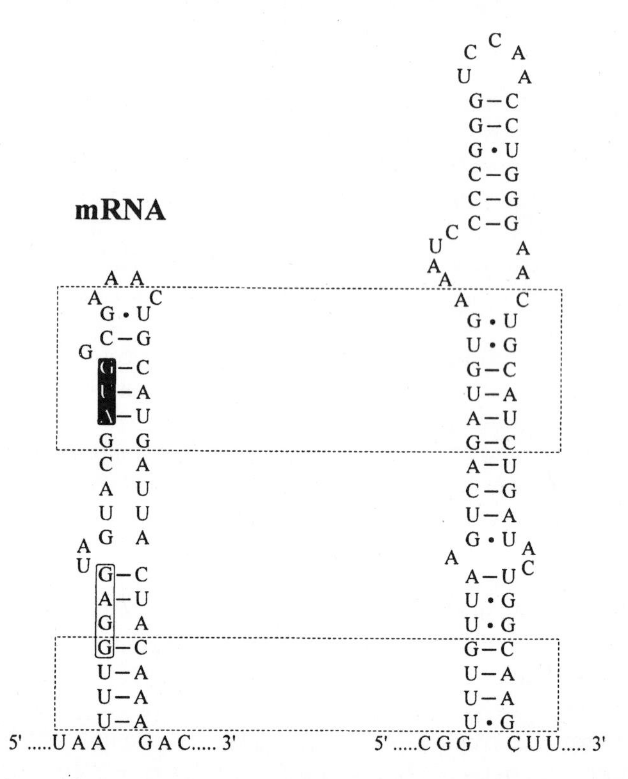

FIG. 16.8. Comparison of the predicted secondary structure of the polycistronic mRNA region (L24–L5 junction) recognized by ribosomal protein S8 acting as a translational repressor (*left;* initiation codon and SD sequence are boxed) and that of the 16S ribosomal RNA region recognized by protein S8 in ribosome assembly (*right*). Homologous helices are enclosed by the broken line. See Olins P.O., and Nomura, M. (1981) *Nucleic Acids Res.* **9:** 1757–1764.

mology is apparent. Another example, shown in Fig. 16.9, is the structural homology between the L11 mRNA leader (Fig. 16.7, fifth line) and the L1-protected region of the 23S ribosomal RNA (Yates and Nomura, 1981).

In all the above cases, the model of repressory action of the corresponding ribosomal protein seems to be analogous to the repression of the *S* cistron by the phage MS2 or R17 coat protein (see Section 16.4.1): the RBS is involved in a rather labile secondary and tertiary structure which does not, by itself, prevent initiation; however, when the specific repressor protein recognizes this structure and binds to it, the structure becomes stable and makes the RBS inaccessible for interactions with the ribosome and the initiator tRNA. Indeed, the regions of mRNA structurally homologous to the protein-binding regions of ribosomal RNA generally include the RBS (see Figs. 16.8 and 16.9).

In the case of the L10 mRNA (Fig. 16.7, sixth line) the operator region has been shown to be located rather distantly upstream of the RBS, at positions about −145 to −200 and in no way overlaps the SD sequence and the initiation codon. The secondary-structure element protected by protein L10 or the pentameric complex L10:$(L7/L12)_4$ on the mRNA, in comparison with the region of the 23S ribosomal RNA protected by the L10:$(L7/L12)_4$ complex, is shown in Fig. 16.10. Again their structural similarity can be observed. The mechanism of the repression, however, is somewhat different: instead of directly blocking the RBS, the binding of the repressory protein seems to induce a large-scale rearrangement of the secondary/tertiary structure of the leader sequence, resulting in the inaccessibility of the RBS to the initiating ribosomal particles.

The case of the control of initiation of the S13 mRNA by protein S4 (Fig. 16.7, fourth line) represents a more complicated mechanism. The operator comprises the beginning of the S13 protein cistron and the preceding 75-nucleotide-long sequence in mRNA; it folds into a complex tertiary structure consisting of two entangled pseudoknots, with the SD sequence and the initiation codon GUG not being involved in secondary and tertiary interactions (Fig. 16.11). The pseudoknot structure does not hinder the interaction of the initiating ribosomal particle with the RBS. No similarity of this structure with the binding site of protein S4 on the ribosomal 16S RNA has been noted. The most remarkable fact is that the binding of the repressory protein (S4) to the operator does not prevent the binding of the initiating ribosomal particle to the RBS. It seems that the binding of the repressor affects the structure of mRNA around the RBS, resulting in trapping the initiation complex at the initiation site, and no direct or indirect competitions between the repressor and the ribosomal particle exist in this case.

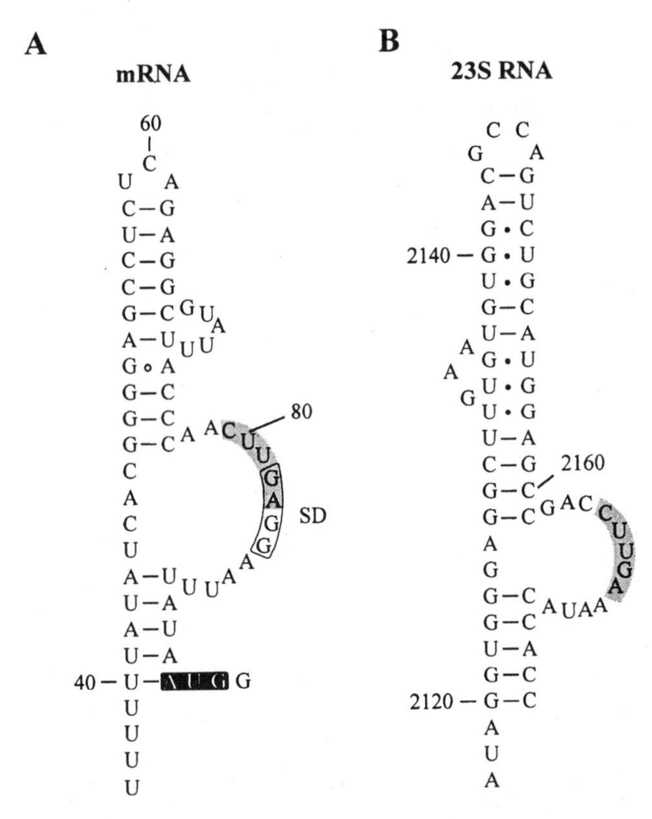

FIG. 16.9. Comparison of the predicted secondary structure of the bicistronic L11–L1 mRNA leader recognized by ribosomal protein L1 acting as a translational repressor *(left)* with that of the L1 protected region of 23S ribosomal RNA *(right)*. The identical sequence in the side loop of both RNAs are marked by the gray box. Reproduced, with minor modifications, from Draper, D.E. (1987) in *Translational Regulation of Gene Expression* (J. Ilan, ed.), pp. 1–26, Plenum Press, New York, with permission.

The ribosomal protein S15 is coded by the first cistron of a bicistronic message that also comprises the polynucleotide phosphorylase-encoding cistron. The translation of the S15 mRNA is regulated by protein S15, independently of the second cistron. Like some other ribosomal proteins, the protein S15 is a repressor of translation of its own mRNA (for review, see Portier and Grunberg-Manago, 1993). The operator region on the mRNA overlaps the RBS and extends upstream up to about position −60. Two hairpins can be formed within this region (Fig. 16.12A). The hairpins can interact with each other and be rearranged into a pseudoknot structure with the SD sequence, the initiation codon being thus freed (Fig. 16.12B). It seems that the two-hairpin structure and the pseudoknot structure are in equilibrium. Apparently, it is the pseudoknot conformation that binds the initiating ribosomal particle. At the same time, protein S15 also recognizes and thus stabilizes the pseudoknot structure. The 16S ribosomal RNA acts as an antirepressor of the S15 mRNA, suggesting the existence of a structural similarity between the operator on the mRNA and the protein S15 binding site on the 16S RNA. Indeed, such a similarity can be perceived from the comparison of the structures (Fig. 16.12C). From this it could be assumed that protein S15 and the initiating ribosomal particle would compete with each other for the common binding site. It has been found, however, that protein S15 does not prevent the binding of the 30S ribosomal particle to the RBS and the subsequent association of the initiator tRNA, but rather stabilizes the 30S:mRNA or 30S:mRNA:F-Met-tRNA complex. It has been suggested that the stabilization of the pseudoknot structure by protein S15 traps the initiating ribosomal particle at the RBS because it is difficult to overcome the helix adjacent to the initiation codon during transition from initiation to elongation,

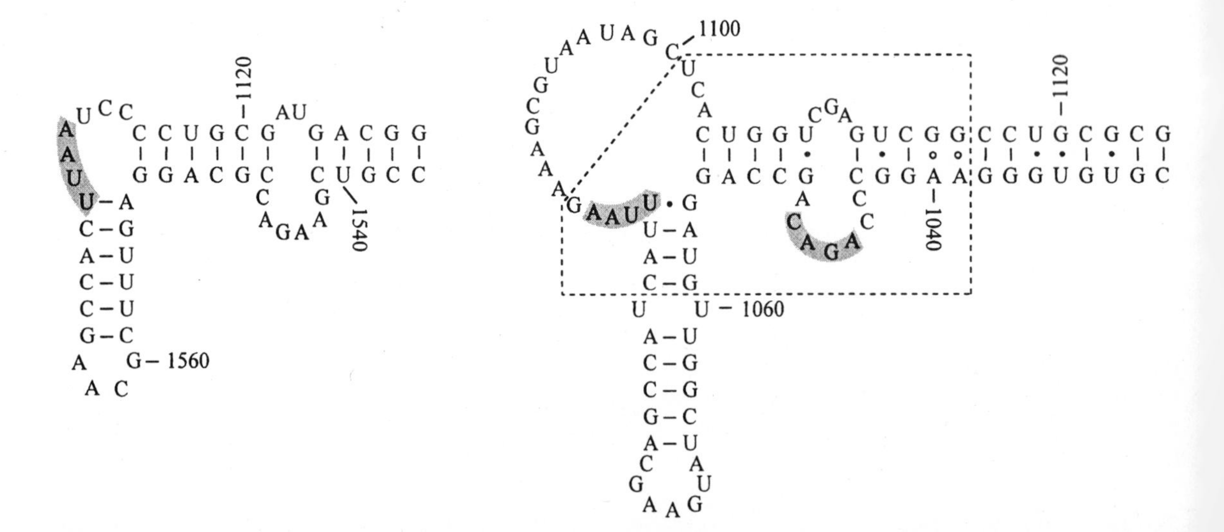

FIG. 16.10. Comparison of the predicted secondary structure of the leader region of the polycistronic L10-L7/L12 mRNA recognized by ribosomal protein L10 or pentameric complex L10:(L7/L12)$_4$ acting as translational repressors (*left*) with that of the 23S ribosomal RNA region protected by the pentameric complex L10:(L7/L12)$_4$ (*right*). The dashed box indicates the part of the ribosomal RNA structure most homologous to the mRNA operator. Identical sequences are marked by gray boxes. Reproduced, with minor modifications, from Draper, D. E. (1993) in *Translational Regulation of Gene Expression* (J. Ilan, ed.), pp. 1–26, Plenum Press, New York, with permission.

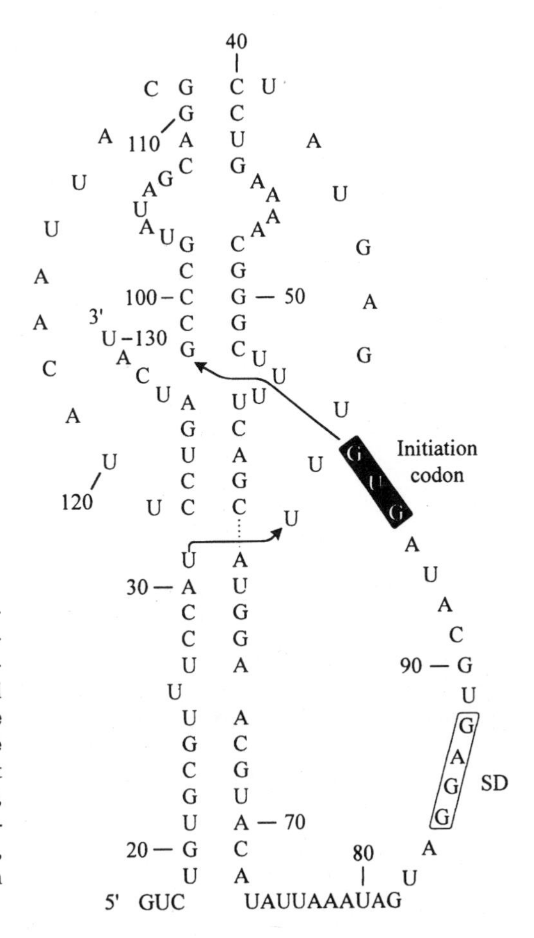

FIG. 16.11. Pseudoknot structure of the polycistronic S10 mRNA region recognized by ribosomal protein S4 acting as a translational repressor. The Shine-Dalgarno sequence and initiation codon (GUG) of S10 cistron are marked. Arrows indicate the polarity of the mRNA chain and the bond between two adjacent nucleotide residues. Reproduced from Portier, C., and Grunberg–Manago, M. (1993) in *Translational Regulation of Gene Expression 2*, (J. Ilan, ed.), pp. 23–47, Plenum Press, New York, with permission.

while in the absence of the repressor the pseudoknot structure is capable of rearranging into a less stable conformation.

The hypothesis that a repressory ribosomal protein, at least in most cases, employs a common active center for binding to ribosomal RNA in the course of ribosome self-assembly, and for binding to mRNA operators in the course of translational repression, has been confirmed by another series of facts. It has been demonstrated that ribosomal RNA added to the translation system removes the repression exerted by a corresponding ribosomal protein. Thus, in experiments *in vitro* the repressory action of protein L1 upon the synthesis of proteins L1 and L11, as well as the inhibition of the synthesis of proteins L10 and L7/L12 by the L10:(L7/L12)$_4$ complex, can be prevented specifically by adding ribosomal 23S RNA.

Proceeding from these observations, a model for the coordinated regulation of synthesis of ribosomal proteins can be proposed (Fig. 16.13) (see Nomura *et al.*, 1980, 1982). The model is based on the idea that there is competition between ribosomal RNA and mRNA for binding to the core ribosomal proteins. Such proteins as S4, S7, S8, L1, L4, as well as the protein complex L10:(L7/L12)$_4$, possess strong affinity for specific sites on ribosomal RNA. Therefore, after they have been synthesized, they become involved immediately in the assembly of ribosomal subunits through their direct binding to the 16S and 23S RNA. Intrinsic high affinity

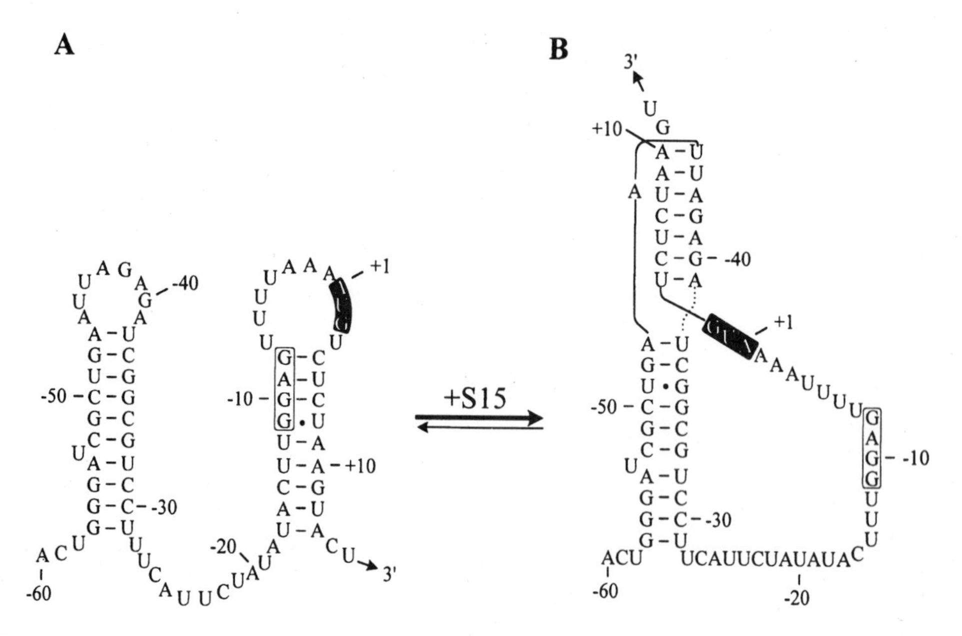

FIG. 16.12. Two conformations of the translational operator (leader sequence with the translation start sequence) of the mRNA coding for ribosomal protein S15. The two-hairpin conformation (A) seems to be inactive in ribosome binding and initiation because of the closure of the RBS. The spontaneous transition into the pseudoknot conformation (B), possibly stimulated by initiating ribosomal particles, opens the RBS. After initiation, the ribosomes exit from the pseudoknot due to the equilibrium between the conformations. Protein S15 has an affinity for the pseudoknot conformation (B), shifts the equilibrium toward it, and stabilizes the pseudoknot. As a result, the initiating ribosome is found to be trapped by the pseudoknot structure, and thus the translation is repressed. Reproduced, with modifications, from Ehresmann, C., Philippe, C., Westhof, E., Benard, L., Portier, C., and Ehresmann, B. (1995) in *Frontiers in Translation* (A. T. Matheson, J. E. Davies, P. P. Dennis, and W. E. Hill, eds.), *Biochem. Cell Biol.*, vol. 73, pp. 1131–1140, with permission.

for ribosomal RNA and the cooperativity of ribosomal assembly involving other ribosomal proteins result in the sequestration of the newly formed ribosomal proteins in the course of the particle assembly. Under these conditions, mRNA molecules are unable to compete for the binding of these proteins and, therefore, do not associate with them; so they can be translated normally. However, when the number of ribosomal proteins increases compared to the amount of available ribosomal RNA, a free pool of such proteins is formed. This leads to the binding of the corresponding key proteins to their mRNAs, and the result is inhibited initiation and thus repressed translation. The strict sequential translation of polycistronic mRNA coding for a set of ribosomal proteins enables just one repressor protein and one site of its attachment for each mRNA to be sufficient for the coordinated control of translation of the whole set of proteins coded by a given mRNA. This simple mechanism provides a direct regulatory relationship between the assembly of ribosomes and the synthesis of ribosomal proteins.

16.4.3. Translational Autoregulation of the Synthesis of Threonyl-tRNA Synthetase

Threonyl-tRNA synthetase (ThrRSase) is coded by the first cistron, *thrS*, of the polycistronic message comprising cistrons *infC* coding for IF3, *rplT* coding for the ribosomal protein L20, *pheS* and *pheT* coding for the two subunits of PheRSase, and *himA* coding for the protein called "host integration factor" (Springer and Grunberg-Manago, 1987). Translation of the *thrS* cistron has been shown to be repressed by the product of the translation, ThrRSase. The presence of an excess of the substrate, $tRNA^{Thr}$, abolishes the repressory action of the ThrRSase, and the enzyme is synthesized. Hence, only when the enzyme is in excess, it represses the translation of its own mRNA and thus stops its further production. Threonine starvation leading to the accumulation of deacylated $tRNA^{Thr}$ derepresses the *thrS* mRNA. Under normal growth conditions the sequestration of aminoacylated $tRNA^{Thr}$ in the complex with EF-Tu:GTP allows the free synthetase to repress the *thrS* mRNA translation.

ThrRSase has been demonstrated to bind directly to the region of *thrS* mRNA upstream of the initiation codon, adjacent to the RBS (Moine *et al.*, 1988, 1990). The enzyme-repressor covers about 130 nucleotides. The most remarkable feature of the mRNA structure where Thr-tRNA binds as a repressor is that it mimics some elements of the structure of the $tRNA^{Thr}$, and specifically the structure of its anticodon stem loop which is recognized by the enzyme. The predicted and experimentally tested secondary structure of the operator of the *thrS* mRNA is presented in Fig. 16.14. It is characterized by two compound hairpins. The hairpin adjacent to the SD sequence contains a seven-nucleotide end loop strongly resembling the anticodon loop of $tRNA^{Thr}$, including the anticodon-like triplet CGU. The five-base-pair portion of the helix underlying the loop is also similar to the anticodon helix of the $tRNA^{Thr}$ (see Fig. 16.14, upper insert). It is this stem-loop element of the *thrS* mRNA operator that is specifically recognized and strongly bound by the ThrRSase. In addition, the other compound hairpin has also been found to participate in the binding of the enzyme-repressor to the thrS mRNA op-

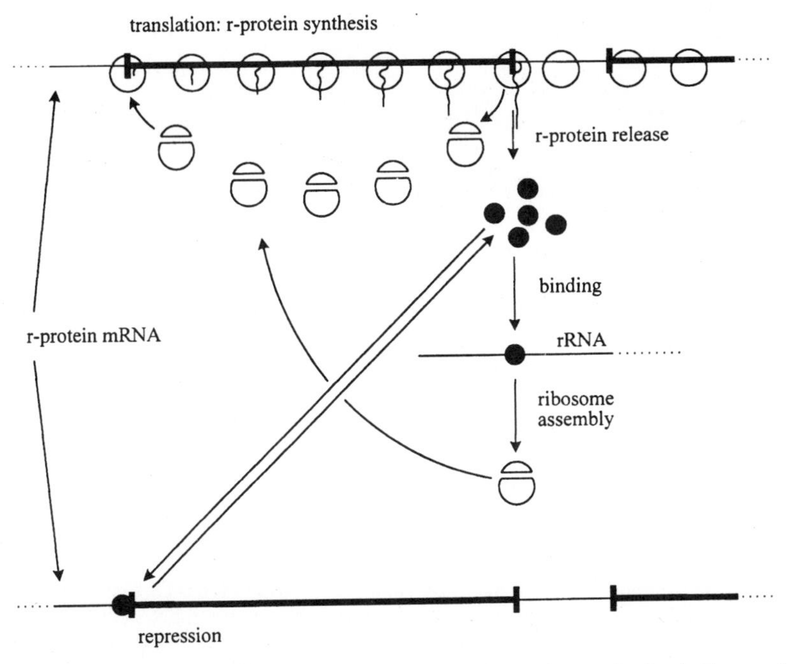

FIG. 16.13. Model for ribosomal protein autoregulation. The newly synthesized ribosomal proteins bind to their sites on ribosomal RNA and assemble into ribosomal particles (the right part of the scheme). When ribosomal proteins are in excess, some of them bind to a target site (translational operator) on their own polycistronic mRNA (the lower part of the scheme).

erator. Its basal part (the upper seven base pairs in Fig. 16.14) seems to resemble the acceptor stem of tRNA^{Thr}, with the ACCA continuation at the 3'-side (Fig. 16.14, lower insert). The binding of the ThrRSase to these recognition elements mimicking those of the tRNA^{Thr} results in blocking the initiation of translation of the *thrS* mRNA by preventing the binding of the ribosomal particles to the RBS. It is likely that the binding of the enzyme-repressor stabilizes the complementary, otherwise weak, interaction of the SD sequence UAAGGA with the sequence UUUUUA in positions −51 to −56, and hence makes the RBS inaccessible to ribosomal particles (Brunel *et al.*, 1995).

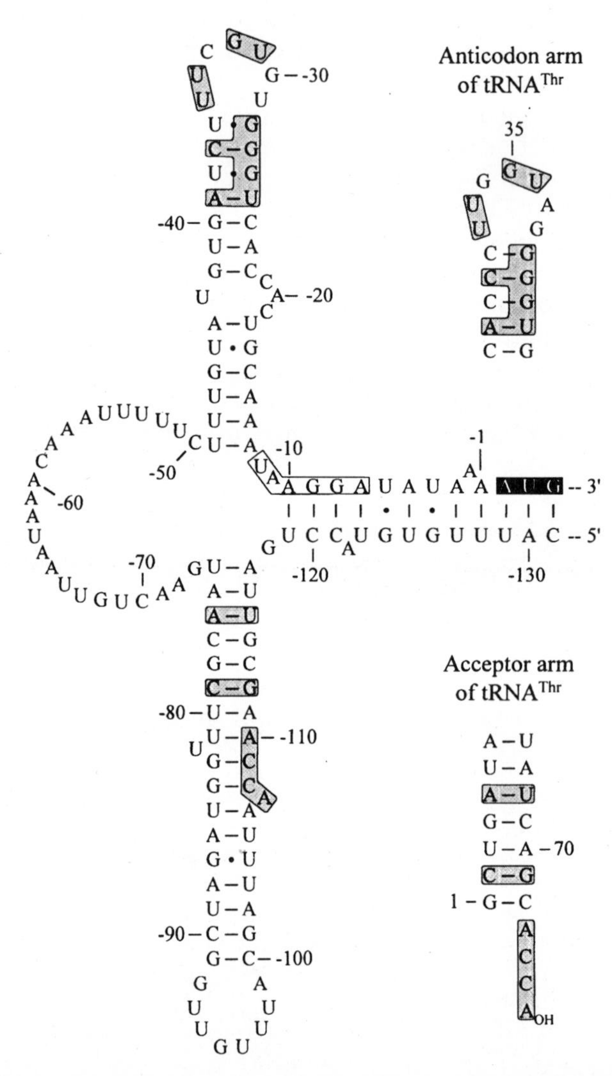

FIG. 16.14. Secondary structure of the operator region of the *thrS* mRNA. This structure has an affinity for the ThrRSase that serves as a translational repressor for its own mRNA (*thrS* mRNA). The elements identical to those in the anticodon and acceptor helices of tRNA^{Thr} are marked by gray boxes. The SD sequence and initiation codon of the *thrS* mRNA are in empty and black boxes, respectively. From Moine, H., Romby, P., Springer, M., Grunberg–Manago, M., Ebel, J.-P., Ehresmann, C., and Ehresmann, B. (1988) *Proc. Natl. Acad. Sci. USA* **85**:7892–7896; (1990) *J. Mol. Biol.* **216**:299–310.

Since $tRNA^{Thr}$ and the *thrS* operator are recognized by the ThrRSase in the analogous way, the excess $tRNA^{Thr}$ is able to displace the mRNA from the enzyme. Therefore, the $tRNA^{Thr}$ acts as an antirepressor, implying that the level of free tRNA can modulate the repressor activity of the ThrRSase. The repression/de-repression control of *thrS* mRNA translation ensures precise adjustment of the ThrRSase synthesis depending on the changes of $tRNA^{Thr}$ and ThrRSase levels in the cell.

16.4.4. Regulation of Translation of Bacteriophage T4 mRNAs

Several translational repressors govern the expression of different bacteriophage T4 mRNAs (reviewed by Gold, 1988). RegA protein represses the translation of a large number of early T4 messages. One such message is the T4 rIIB mRNA; its RBS includes the following sequence:

... U**AAGGA**AAAUU **AUG** UAC AAU AUU AAA ...

RegA protein specifically binds to the region just downstream of the SD sequence and covers the initiation codon and the adjacent four codons. As a result it prevents the formation of the ternary initiation complex.

Bacteriophage T4 DNA polymerase proved to be the repressor of its own synthesis: it binds to the initiation region and blocks the translation of the T4 DNA polymerase message. Thus the synthesis of the enzyme is regulated autogenously. This repressor covers the SD sequence and a section downstream, but not the initiation codon. It is likely that the repressor inhibits the initiation at the stage of the association of the ribosomal 30S particle with mRNA.

Another T4 protein that is autogenously regulated at the level of translation is the so-called single-stranded DNA-binding protein, or the gene 32 protein. Its binding site on the mRNA, however, is removed from the RBS far upstream and contains a pseudoknot (positions −40 to −67). The binding of the protein molecule induces a cooperative polymerization of many copies of the protein along an unstructured RNA stretch overlapping the RBS. As the concentration of the gene 32 protein increases, the polymer of multiple copies of the protein reaches the SD sequence and covers it, resulting in the repression of translation initiation.

16.5. Antisense Blockade

The RBS of mRNA may be blocked by interaction with a complementary RNA, called *antisense RNA* (for reviews, see Inouye, 1988; Wagner and Simons, 1994). This phenomenon is analogous to the translational repression. Several cases of such a blockade of an RBS by a natural antisense RNA are known in bacteria. It seems that transcripts of accessory genetic elements, such as plasmids and transposons, as well as bacteriophages, are more often controlled by antisense RNAs than genomic mRNAs.

The best known example of natural antisense RNA of *E.coli* controlling a genomic mRNA translation is the so-called *micF* RNA (Mizuno, Chou, and Inouye, 1984; Aiba *et al.*, 1987). There are two subspecies of this RNA, one 93 and the other 174 nucleotides long. The RNA regulates the synthesis of OmpF, a protein of the outer bacterial membrane that forms diffusion pores for small molecules. The an-

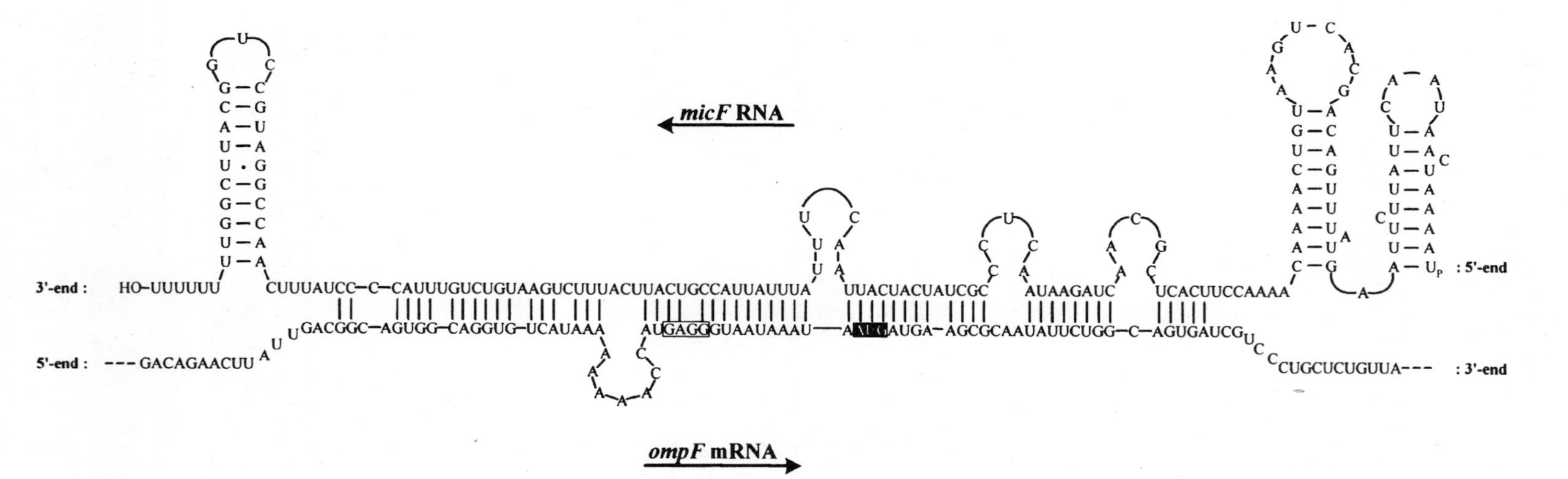

FIG. 16.15. Complementary interaction between the RBS region of the *ompF* mRNA coding for an outer membrane protein of *E.coli* (the lower sequence) and the antisense *micF* RNA (the upper sequence). The SD sequence and initiation codon of the *ompF* mRNA are in empty and black boxes, respectively. Reproduced from Mizuno, T., Chou, M.-Y., and Inouye, M. (1984) *Proc. Natl. Acad. Sci. USA* **81**:1966–1970, with permission.

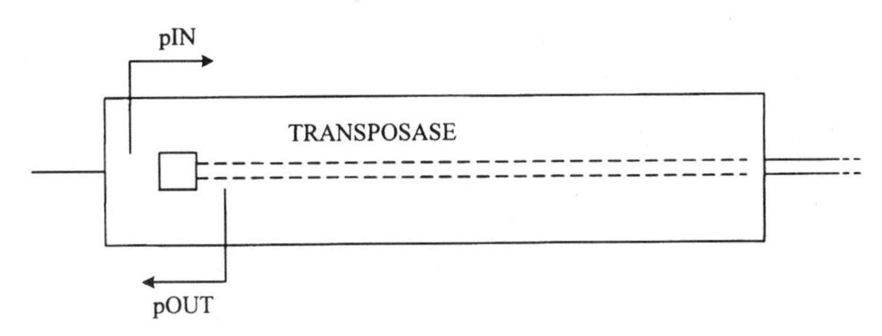

FIG. 16.16. Scheme of the transcription of the transposase gene of *IS10* transposon in two opposite directions using promoters *pIN* and *pOUT*. As a result, the transposase mRNA (*IN* mRNA) and the antisense mRNA (*OUT* mRNA) fully complementary to the RBS region of the transposase mRNA are produced. Reproduced from Simons, R., W., and Kleckner, N. (1983) *Cell* **34**:683–691, with permission.

tisense RNA (both subspecies) is transcribed from a special chromosomal site removed from the *ompF* gene. It is induced by an increased osmolarity of the medium. This RNA is partially complementary to the RBS of the *ompF* mRNA including the SD sequence and the initiator codon (Fig. 16.15). Thus, the production of the *micF* RNA induced by high osmolarity blocks the translation initiation of *ompF* mRNA due to their complementary interaction at the *ompF* RNA RBS. The result is the inhibition of OmpF production in bacteria under these conditions.

The synthesis of the plasmid replication initiator protein RepA encoded in the enterobacterial plasmid R1 is also controlled by an antisense RNA, but in a more complicated way. Translation of the *repA* mRNA is tightly coupled with the translation of the preceding open reading frame coding for a short leader polypeptide Tap: translation of the *tap* cistron is necessary for unwinding of a stable hairpin that prevents an independent *repA* initiation (see Section 16.3.1). The antisense RNA, called *CopA* (about 90 nucleotides long), is constitutively transcribed from the same plasmid and complementarily interacts with the region immediately upstream of the SD sequence of the *tap* cistron. The SD sequence itself and the downstream sequence, including the initiation codon of the *tap* message, are not involved in the complementary interaction. Nevertheless, the initiation of the *tap* message translation is blocked, resulting in the inability to initiate *repA* mRNA translation (Malmgren *et al.*, 1996).

The synthesis of the transposase encoded by the insertion sequence *IS10* of transposon *Tn10* is controlled by an antisense RNA, called *OUT* RNA (70 nucleotides long) (Simons and Kleckner, 1983; Ma and Simons, 1990). In this case, however, it is produced by the transcription of the RBS region of the transposase gene itself, but in the orientation opposite to the transposase mRNA transcription (Fig. 16.16, see promoters *pIN* and *pOUT*). Hence, the *OUT* RNA is fully complementary to the region of the transposase mRNA (*IN* RNA) containing its RBS and thus directly blocks the binding of ribosomes and initiation of mRNA translation. Similar regulation of mRNA translation by antisense RNAs transcribed from the target genes in the opposite orientation has been reported also for some bacteriophage and plasmid mRNAs.

References

Adhin, M. R., and van Duin, J. (1989) Translational regulation of the lysis gene in RNA bacteriophage fr requires a UUG initiation codon, *Mol. Gen. Genet.* **218**:137–142.

Adhin, M. R., and van Duin, J. (1990) Scanning model for translational reinitiation in Eubacteria, *J. Mol. Biol.* **213**:811–818.

Aiba, H., Matsuyama, S.-I., Mizumo, T., and Mizushima, S. (1987) Function of *micF* as an antisense RNA in osmoregulatory expression of the *ompF* gene in *Escherichia coli, J. Bacteriol.* **169**:3007–3012.

Bernardi, A. and Spahr, P. F. (1972) Nucleotide sequence at the binding site for coat protein on RNA of bacteriophage R17, *Proc. Natl. Acad. Sci. USA* **69**:3033–3037.

Brunel, C., Romby, P., Sacerdot, C., de Smit, M., Graffe, M., Dondon, J., van Duin, J., Ehresmann, B., Ehresmann, C., and Springer, M. (1995) Stabilised secondary structure at a ribosomal binding site enhances translational repression in *E. coli, J. Mol. Biol.* **253**:277–290.

Das, A., and Yanofsky, C. (1984) A ribosome binding site sequence is necessary for efficient expression of the distal gene of a translationally-coupled gene pair, *Nucl. Acids Res.* **12**:4757–4768.

de Smit, M. H., and van Duin, J. (1990) Control of prokaryotic translational initiation by mRNA secondary structure, *Progr. Nucleic Acid Res. Mol. Biol.* **38**:1–35.

Dean, D., and Nomura, M. (1980) Feedback regulation of ribosomal protein gene expression in *Escherichia coli, Proc. Natl. Acad. Sci. USA* **77**:3590–3594.

Dean, D., Yates, J. L., and Nomura, M. (1981a) *Escherichia coli* ribosomal protein S8 feedback regulates part of *spc* operon, *Nature* **289**:89–91.

Dean, D., Yates, J. L., and Nomura, M. (1981b) Identification of ribosomal protein S7 as a repressor of translation within the *str* operon of *E. coli, Cell* **24**:413–419.

Draper, D. E. (1987) Translational regulation of ribosomal proteins in *Escherichia coli:* Molecular mechanisms, in *Translational Regulation of Gene Expression* (J. Ilan, ed.), pp. 1–26, Plenum Press, New York and London.

Friesen, J. D., Fiil, N. P., Dennis, P. P., Downing, W. L., An, G., and Holowachuk, E. (1980) Biosynthetic regulation of *rplJ, rolL, rpoB* and *rpoC* in *Escherichia coli,* in *Ribosomes: Structure, Function, and Genetics* (G. Chamblis, G. R. Craven, J. Davies, K. Davis, L. Kahan, and M. Nomura, eds.), pp. 719–742, University Park Press, Baltimore.

Gold, L. (1988) Posttranscriptional regulatory mechanisms in *Escherichia coli, Annu. Rev. Biochem.* **57**:199–233.

Gualerzi, C., la Teana, A., Spurio, R., Canonaco, M. A., Severini, M., and Pon, C. L. (1990) Initiation of protein biosynthesis in procaryotes: Recognition of mRNA by ribosomes and molecular basis for the function of initiation factors, in *The Ribosome: Structure, Function, and Evolution* (W. E. Hill, A. Dahlberg, R. A. Garrett, P. B. Moore, D. Schlessinger, and J. R. Warner, eds.), pp. 281–291, ASM Press, Washington, DC.

Hellmuth, K., Rex, G., Surin, B., Zinck, R., and McCarthy, J. E. G. (1991) Translational coupling varying in efficiency between different pairs of genes in the central region of the *atp* operon of *Escherichia coli, Mol. Microbiol.* **5**:813–824.

Inouye, M. (1988) Antisense RNA: its functions and applications in gene regulation—a review, *Gene* **72**:25–34.

Ivey–Hoyle, M., and Steege, D. A. (1989) Translation of phage f1 gene VII occurs from an inherently defective initiation site made functional by coupling, *J. Mol. Biol.* **208**:233–244.

Johnsen, M., Christensen, T., Dennis, P. P., and Fiil, N. P. (1982) Autogenous control: Ribosomal protein L10-L12 complex binds to the leader sequence of its mRNA, *EMBO J.* **1**:999–1004.

Kolakofsky, D., and Weissmann, C. (1971) Possible mechanism for translation of viral RNA from polysome to replication complex, *Nature New Biol.* **231**:42–46.

Lindahl, L., and Zengel, J. (1982) Expression of ribosomal genes in bacteria, in *Advances in Genetics* (E. W. Caspari, ed.), vol. 21, pp. 53–111, Academic Press, New York.

Lindahl, L., and Zengel, J. M. (1986) Ribosomal genes in *Escherichia coli, Annu. Rev. Biochem.* **20**:297–326.

Lodish, H. F., and Robertson, H. D. (1969) Regulation of in vivo translation of bacteriophage f2 RNA, *Cold Spring Harbor Symp. Quant. Biol.* **34**:655–673.

Lodish, H. F., and Zinder, N. D. (1966) Mutants of the bacteriophage f2. VIII. Control mechanisms for phage-specific syntheses, *J. Mol. Biol.* **19**:333–348.

Ma, C., and Simons, R. W. (1990) The IS*10* antisense RNA blocks ribosome binding at the transposase translation initiation site, *EMBO J.* **9**:1267–1274.

Malmgren, C., Engdahl, H. M., Romby, P., and Wagner, E. G. H. (1996) An antisense/target RNA duplex or a strong intramolecular RNA structure 5' of a translation initiation signal blocks ribosome binding: The case of plasmid R1. *RNA* **2**:1022–1032.

McCarthy, J. E. G. (1988) Expression of the *unc* genes in *Escherichia coli, J. Bioenerg. Biomembr.* **20**:19–39.

McCarthy, J. E. G. (1990) Post-transcriptional control in the polycistronic operon environment: studies of the *atp* operon of *Escherichia coli, Mol. Microbiol.* **4**:1233–1240.

McCarthy, J. E. G., and Brimacombe, R. (1994) Prokaryotic translation: the interactive pathway leading to initiation, *Trends Genet.* **10**:402–407.

McCarthy, J. E. G., and Gualerzi, C. (1990) Translational control of prokaryotic gene expression, *Trends Genet.* **6**:78–85.

Meyer, F., Weber, H., and Weissmann, C. (1981) Interactions of Qβ replicase with Qβ RNA, *J. Mol. Biol.* **153**:631–660.

Mizuno, T., Chou, M.-Y., and Inouye, M. (1984) A unique mechanism regulating gene expression: translational inhibition by a complementary RNA transcript (micRNA), *Proc. Natl. Acad. Sci. USA* **81**:1966–1970.

Moine, H., Romby, P., Springer, M., Grunberg–Manago, M., Ebel, J. P., Ehresmann, C., and Ehresmann, B. (1988) Messenger RNA structure and gene regulation at the translational level in *Escherichia coli:* The case of threonine:tRNA^{The} ligase, *Proc. Natl. Acad. Sci. USA* **85**:7892–7896.

Moine, H., Romby, P., Springer, M., Grunberg–Manago, M., Ebel, J.-P., Ehresmann, B., and Ehresmann, C. (1990) *Escherichia coli* threonyl–tRNA synthetase and tRNA^{Thr} modulate the binding of the ribosome to the translational initiation site of the *ThrS* mRNA, *J. Mol. Biol.* **216**:299–310.

Nomura, M., Gourse, R., and Baughman, G. (1984) Regulation of the synthesis of ribosomes and ribosomal components, *Annu. Rev. Biochem.* **53**:75–117.

Nomura, M., Jinks–Robertson, S., and Miura, A. (1982) Regulation of ribosome biosynthesis in *Escherichia coli,* in *Interaction of Translational and Transcriptional Controls in the Regulation of Gene Expression* (M. Grunberg–Manago and B. Safer, eds.), pp. 91–104, Elsevier, New York.

Nomura, M., Yates, J. L., Dean, D., and Post, L. E. (1980) Feedback regulation of ribosomal protein gene expression in *Escherichia coli:* Structural homology of ribosomal RNA and ribosomal protein mRNA, *Proc. Natl. Acad. Sci. USA* **77**:7084–7088.

Olins, P. O., and Nomura, M. (1981) Translational regulation by ribosomal protein S8 in *Escherichia coli:* Structural homology between rRNA binding site and feedback target on mRNA, *Nucl. Acids Res.* **9**:1757–1764.

Oppenheim, D. S., and Yanofsky, C. (1980) Translational coupling during expression of the tryptophan operon of *Escherichia coli, Genetics* **95**:785–795.

Petersen, C. (1989) Long-range translational coupling in the *rplJL-rpoBC* operon of *Escherichia coli, J. Mol. Biol.* **206**:323–332.

Platt, T., and Yanofsky, C. (1975) An intercistronic region and ribosome-binding site in bacterial messenger RNA, *Proc. Natl. Acad. Sci. USA* **72**:2399–2403.

Portier, C., and Grunberg–Manago, M. (1993) Regulation of ribosomal protein mRNA translation in bacteria: The case of S15, in *Translational regulation of gene expression 2* (J. Ilan, ed.), pp. 23–47, Plenum Press, New York, London.

Robertson, H. D., and Lodish, H. F. (1970) Messenger characteristics of nascent bacteriophage RNA, *Proc. Natl. Acad. Sci. USA* **67**:710–716.

Sarabhai, A., and Brenner, S. (1967) A mutant which reinitiates the polypeptide chain after chain termination, *J. Mol. Biol.* **27**:145–162.

Schumperli, D., McKenney, K., Sobieski, D. A., and Rosenberg, M. (1982) Translational coupling at an intercistronic boundary of the Escherichia coli galactose operon, *Cell* **30**:865–871.

Simons, R. W., and Kleckner, N. (1983) Translational control of IS*10* transposon, *Cell* **34**:683–691.

Springer, M., and Grunberg–Manago, M. (1987) *Escherichia coli* threonyl-transfer RNA synthetase as a model system to study translational autoregulation in prokaryotes, in *Translational Regulation of Gene Expression* (J. Ilan, ed.), pp. 51–61, Plenum Press, New York, London.

Steitz, J. A. (1980) RNA·RNA interactions during polypeptide chain initiation, in *Ribosomes: Structure, Function, and Genetics* (G. Chambliss, G. R. Craven, J. Davies, K. Davis, L. Kahan, and M. Nomura, eds.), pp. 479–495, University Park Press, Baltimore.

Stormo, G. D. (1986) Translation initiation, in *Maximizing Gene Expression* (W. Reznikoff and L. Gold, eds.), pp. 195–224, Butterworths, Boston, London.

Stormo, G. D. (1987) Translational regulation of bacteriophages, in *Translational Regulation of Gene Expression* (J. Ilan, ed.), pp. 27–49, Plenum Press, New York, London.

Voorma, H. O. (1996) Control of translation initiation in prokaryotes, in *Translational Control* (J. W. B. Hershey, M. B. Mathews, and N. Sonenberg, eds.), pp. 759–777, Cold Spring Harbor Laboratory Press, New York.

Wagner, E. J., and Simons, R. W. (1994) Antisense RNA control in bacteria, phages, and plasmids, *Annu. Rev. Microbiol.* **48**:713–742.

Weber, H., Billeter, M. A., Kahane, S., Weissmann, C., Hindley, J., and Porter, A. (1972) Molecular basis for repressor activity of Qβ replicase, *Nature New Biol.* **237**:166–169.

Weissmann, C., Billeter, M. A., Goodman, H. M., Hindley, J., and Weber, H. (1973) Structure and function of phage RNA, *Annu. Rev. Biochem.* **42**:303–328.

Yates, J. L., Dean, D., Strycharz, W. A., and Nomura, M. (1981) *E. coli* ribosomal protein L10 inhibits translation of L10 and L7/L12 mRNAs by acting at a single site, *Nature* **294**:190–192.

Yates, J. L., and Nomura, M. (1980) *E.coli* ribosomal protein L4 is a feedback regulatory protein, *Cell* **21**:517–522.

Yates, J. L., and Nomura, M. (1981) Feedback regulation of ribosomal protein synthesis in *E. coli:* Localization of the mRNA target sites for repressor action of ribosomal protein L1, *Cell* **24**:243–249.

17

Translational Control in Eukaryotes

17.1. Importance of Translational Control in Eukaryotes

Generally, the protein production of the eukaryotic cell can be regulated at several levels: (1) issuing encoded genetic information in the form of RNA, that is, transcription; (2) processing of the RNA and its intracellular transport (mostly from the nucleus to the cytoplasm); (3) reading the messenger RNA formed, or translation; (4) degradation of the product of translation; and (5) degradation of messenger RNA. Ribosomes may be involved in some of these levels, such as control of mRNA degradation, but the translational regulation of protein production is the main level that directly concerns ribosomes. How important is this level of regulation of protein synthesis in eukaryotes?

The relative metabolic stability of most eukaryotic mRNAs makes translational control particularly important in the general pattern of protein synthesis regulation. Specifically, along with the signals for template activation, that is, for the initiation of translation, the signals for the arrest of translation become necessary. Hormonal regulation of translation provides examples of both the switching on and the shutting down of translation of certain mRNA species. Heat shock triggers the synthesis of a few special proteins, while translation of most of the preexisting cellular mRNAs ceases or decreases; cell recovery at a normal temperature is accompanied by reactivation of major mRNA translation and cessation of heat-shock protein synthesis. Oogenesis and spermatogenesis, as well as plant seed ripening, are accompanied by inactivation and storage of mRNA which further exists in oocytes, spermatocytes, or seeds in a nontranslatable, masked form. Fertilization, as well as seed germination, results in general and selective activation of translation of the stored mRNA. In processes of embryonic development and cell differentiation the synthesis of mRNA and the accumulation of mRNA in the cytoplasm may take place long before this RNA is used in translation; specific signals selectively activate the corresponding mRNA species at proper stages.

Several systems of translational regulation are known to exist. They can be subdivided, rather conditionally, into two groups: the systems for nonselective regulation of total level of translation, and the systems where the control is selective and mRNA-specific.

17.2. Total Translational Regulation

17.2.1. Regulation by Modifications of Initiation Factors

17.2.1.1. Phosphorylation of Met-tRNA$_i$/GTP-binding Factor (eIF2)

Phosphorylation of eIF2, and specifically of its α-subunit, is one of the most used mechanisms of global translational regulation in mammalian cells, as well as in yeast (for reviews, see Jackson, 1991; Chen, 1993; Kramer, Kudlicki, and Hardesty, 1993; Clemens, 1996). Two special protein kinases, both phosphorylating serine residue(s) at the N-terminal part (Ser-51, sometimes also Ser-48) of the α-subunit of eIF2, are known in mammalian cells. One, called "heme-controlled repressor" (HCR) or "heme-regulated inhibitor" (HRI), is a 90-kDa (625 aa) protein present in a soluble (ribosome-unbound) form in the cytoplasm of reticulocytes and, possibly, some other mammalian cells. The other, called "double-stranded RNA-activated inhibitor" (DAI or PKR), is a 68-kDa (550 aa) protein inducible in mammalian cells by interferon; this protein sticks to ribosomes. Both kinases have some sequence homology. An interesting feature of both kinases is their capability of multiple phosphorylation (by casein kinase II) and autophosphorylation at serine and threonine residues in response to some signals, which results in their activation. For example, the PKR is autophosphorylated and therefore activated in response to double-stranded RNAs or extended elements of RNA secondary structure. On the other hand, there exists a protein in mammalian cells that interacts with eIF2 and protects it from phosphorylation by the kinases; this is a p67 glycoprotein containing multiple O-linked GlcNAc residues. Phosphatases can also dephosphorylate eIF2. It is likely that all mammalian cells have some level of the eIF2-directed kinase activities, but at the same time they have antiphosphorylation protection mechanisms.

In reticulocytes, both kinase activities are quite noticeable and can be further enhanced by heme deficiency (HCR stimulation) or by double-stranded RNA (PKR stimulation). Heat shock, serum deprivation, amino acid starvation, and viral infections are known to promote the kinase activities in a variety of mammalian cells. Fungi (yeast) also possess an analogous eIF2-specific kinase called GCN2 which is activated in response to similar environmental factors. When eIF2 becomes phosphorylated due to the enhancement of the kinase activities, the protein synthesis is inhibited.

What is the mechanism of the inhibition? It is found that the eIF2 with the phosphorylated α-subunit is quite functional in the formation of the ternary Met-tRNA:GTP:eIF2 complex and in the interactions with ribosomal particles, including ribosome-induced GTP cleavage. However, when the eIF2:GDP complex of the phosphorylated eIF2 with GDP (eIF2αP:GDP) is released from the ribosome and interacts with eIF2B (see Fig. 15.17), a very stable, unexchangeable eIF2αP:eIF2B complex is formed. As a result, the eIF2B which is present in the cell in a limited amount becomes sequestered by the eIF2αP. Under conditions of eIF2B shortage the exchange of GDP for GTP on eIF2 is decelerated and the rate of initiation declines. In other words, the downregulation of the recycling of eIF2 takes place. Hence, mammalian cells react to heme deficiency, growth factors deprivation, amino acid starvation, heat shock, or virus infection by reducing the total protein synthesis via this mechanism of the initiation rate inhibition.

The inhibition of protein synthesis in reticulocytes in response to heme deficiency is the best studied example. Several models of inducing phosphorylation of eIF2 by HCR have been reported. In any case, the depletion of heme is known to lead to reduction of intramolecular disulfide bridges in latent HCR, its multiple phosphorylation and autophosphorylation, and therefore to its activation. There are indications that HCR can directly bind heme. According to one of the models, the inactive HCR contains heme and is assembled into inactive homodimer through the formation of disulfide bridges (Fig. 17.1A). The depletion of heme could be responsible for inducing reduction of the bridges, accompanying phosphorylation and autophosphorylation of the protein, and dissociation into active monomers. The phosphorylated active kinase further phosphorylates eIF2.

Another model is based on the observation that HCR can interact with the heat-shock protein HSP 90 and the interaction is directly involved in the process of activation of HCR under stress conditions. This model suggests that the heme-bound HCR is associated with HSP 90 into inactive heterodimer, with intramolecular disulfide bonds in each subunit (Fig. 17.1B). Upon removal of heme, the heterodimer reversibly dissociates, and the subsequent phosphorylation of the

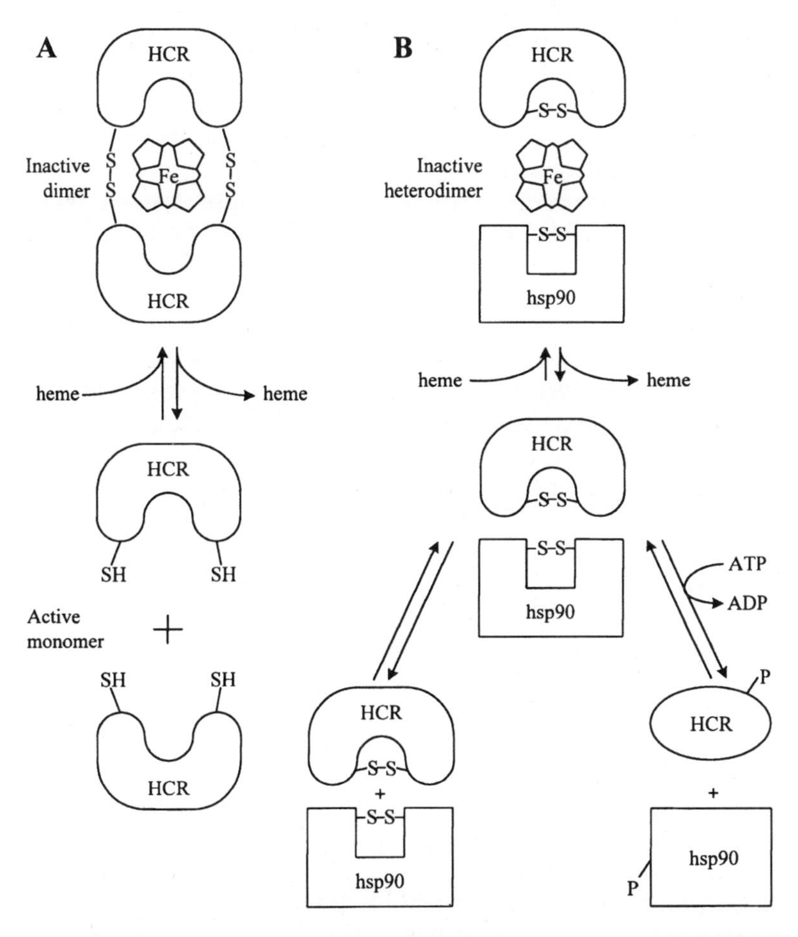

FIG. 17.1. Two models for the activation of the heme-regulated protein kinase (HCR). **(A)** Activation by homodimer dissociation. **(B)** Activation by heterodimer dissociation with subsequent phosphorylation. See the text for more details. Reproduced from Clemens, M. J. (1996) in *Translational Control* (J. W. B. Hershey, M. B. Mathews, and N. Sonenberg, eds.), pp. 139–172, CSHL Press, with permission.

monomeric HCR and HSP 90 results in irreversible dissociation and activation of HCR.

An alternative model suggests that HCR is already autophosphorylated and active in normal reticulocyte cytoplasm, but the glycoprotein p67 protects eIF2 from enzymatic attack by binding to it (Gupta *et al.*, 1993). In this case, the absence of heme induces deglycosylation of p67 and its subsequent degradation, thus permitting the attack of HCR on eIF2. All these models may be not mutually exclusive, and it may be that each mechanism works in reticulocytes to some extent.

It is interesting that p67 seems to be present in all mammalian cells and protects eIF2 from phosphorylation. It is also deglycosylated and subsequently degraded, for example, in response to growth factors deprivation; this leads to phosphorylation of eIF2 which can contribute to protein synthesis reduction under these conditions. On the contrary, mitogens induce an increase in p67, and this correlates with the increase in total protein synthesis. It is not clear yet which of the two kinases, HCR or PKR, is mainly responsible for the phosphorylation of eIF2 as a result of growth factors deprivation, amino acid starvation, or heat shock.

The inhibition of total protein synthesis by the PKR during viral infection is another well-studied example of global translational regulation (for reviews, see Mathews, 1996; Schneider, 1996; Katze, 1996). Many viruses induce the production of interferon in mammalian cells. In turn, interferon stimulates the synthesis of the p68-kinase (PKR) in the targeted cells. The cells acquire the so-called antiviral state. The penetration of a virus into such a cell and the appearance of long double-stranded RNA regions in the cell immediately activate the presynthesized PKR, seemingly as a result of direct interaction of a double-stranded region or fragment with the kinase. This leads to the phosphorylation of eIF2 and, thus, the reduction of the initiation rate. This event may be significant to the inhibition of synthesis of viral components, among other events induced by interferon.

As in the case of HCR, PKR is inactive until phosphorylated. In contrast to HRC, however, the inactive PKR is monomeric. Its N-terminal part contains two dsRNA-binding domains. When two PKR molecules are bound with dsRNA side by side they seem to interact with each other, resulting in their dimerization and mutual phosphorylation or autophosphorylation. This makes them active PKR molecules and induces their dissociation (Fig. 17.2). Once PKR is phosphorylated its kinase activity becomes independent of dsRNA. The active (phosphorylated) monomeric PKR attacks eIF2.

The phosphorylation of eIF2, seemingly a regulatory response of the cell to viral infection (downregulation of total protein synthesis), can in turn can be regulated by a virus (downregulation of phosphorylation). Indeed, many viruses encode or induce a factor that prevents the phosphorylation of eIF2 (for a review, see, for example, Hovanessian, 1993). For example, adenovirus produces small transcripts, VA-RNAs (I and II), that can bind to the latent PKR and block the activation of its kinase activity (Schneider, 1996). Epstein-Barr virus also generates small RNAs—EBER-1 and EBER-2—which bind to the kinase and prevent its activation (Clarke et al., 1991; Mathews, 1996). Poliovirus, at the early stage of the infection, induces proteolytic degradation of the kinase (Black et al., 1989). Influenza virus stimulates the production of a cellular, not a viral-encoded, protein, p58, possessing a strong inhibitory effect on the kinase (Katze, 1996). These mechanisms allow viruses to circumvent the protective downregulation exerted by the host and to restore the high rate of general protein synthesis required for high viral production.

17.2.1.2. Phosphorylation of Cap-binding Factor (eIF4E)

It is known that most of the eukaryotic initiation factors, especially in mammalian cells, are phosphorylated proteins. They include eIF2B, eIF3, eIF4B, eIF4E, eIF4F, and eIF5. The phosphorylations can be performed by different nonspecific (multipotential) protein kinases, such as protein kinase C, casein kinases, ribosomal protein S6 kinase, cAMP-dependent kinase, and others (reviewed by Humbelin and Thomas, 1991; Proud, 1992). The functional significance of these phosphorylations, however, is not clear. In any case, they are active in the phosphorylated state, and there are reports about the positive correlation between their enhanced phosphorylation and the enhanced protein synthetic activity. With two cap-binding factors, eIF4E and eIF4F, the phosphorylation has been directly shown to be either absolutely required for activity (phosphorylation at Ser-53 in eIF4E), or strongly stimulatory (multiple phosphorylation in the 220-kDa subunit of eIF4F) (Rhoads, Joshi-Barve, and Rinker-Schaffer, 1993; Frederickson and Sonenberg, 1993; Sonenberg, 1996). It cannot be excluded that these phosphorylations, and especially the critical one-site phosphorylation of eIF4E, may play a regulatory role in the rate of translational initiation in eukaryotes. The point is that eIF4E, as a cap-binding subunit of eIF4F, seems to be one of the limiting components in the initiation machinery, and thus the amount of the active subunit determines the amount of the complete functional eIF4F which is responsible for the cap-dependent initiation.

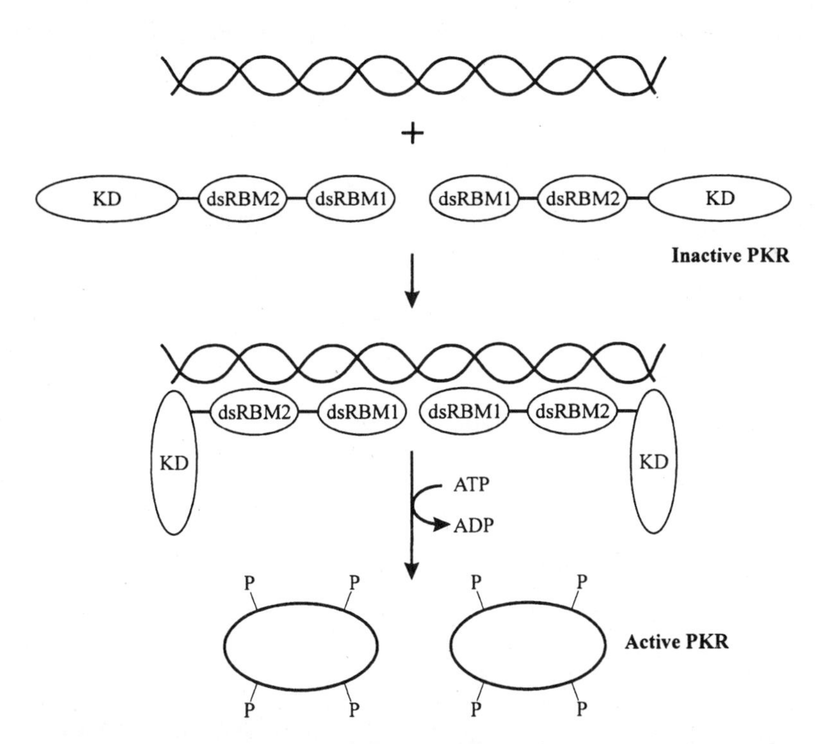

FIG. 17.2. A model for the activation of the dsRNA-regulated protein kinase (PKR): activation by dimerization on dsRNA with subsequent trans- and autophosphorylation and dissociation. KD is kinase domain, and dsRBM1 and dsRBM2 are dsRNA-binding motifs of the enzyme molecule. See the text for more details. Reproduced from Clemens, M. J. (1996) in *Translational Control* (J. W. B. Hershey, M. B. Mathews, and N. Sonenberg, eds.), pp. 139–172, CSHL Press, with permission.

At the same time, the main role in the regulation of the eIF4E level in the cell may be played by phosphorylation of a special protein called 4E-binding protein, or 4E-BP, rather than by phosphorylation of the eIF4E itself. The 4E-BP forms an inactive complex with eIF4E and thus depletes it from the medium. Phosphorylation of the 4E-BP by mitogen-activated kinase (MAP) induced by insulin or growth factors has been shown to result in the dissociation of the complex and the release of the active eIF4E (Sonenberg, 1996).

17.2.2. Regulation by mRNP Formation

In all considerations of translational aspects of protein synthesis in eukaryotes it should always be taken into account that eukaryotic mRNAs never exist as free polynucleotides but only in the form of messenger ribonucleoproteins (mRNPs). The problem of structural organization and functional significance of mRNPs, despite the research spanning 30 years, is still far from its solution. Functionally, the proteins complexed with mRNA can exert very different effects, from stimulation of translation to complete "masking" of mRNA. Some effects are total and will be discussed here, whereas other effects are mRNA-specific and will be considered in the corresponding sections below.

Among a variety of proteins in mRNPs of different origin and intracellular localization, there is the major protein component called p50 (for reviews, see Spirin, 1994, 1996; Wolffe, 1994; see also Sommerville, 1992; Murray, Schiller, and Franke, 1992; Minich, Maidebura, and Ovchinnikov, 1993; Evdokimova *et al.*, 1995). It is bound to heterogeneous mRNA sequences both in free cytoplasmic mRNP particles and in polyribosomes. It seems to have no preference for any special sequences in mRNAs, including coding and untranslated regions (though it probably has a higher affinity for G-rich regions), but poly(A) tails, double-stranded RNAs, tRNAs, and ribosomal RNAs are incapable of strongly binding to the p50. The p50 is present in many copies per mRNA chain, possesses the strongest affinity for heterogeneous mRNA sequences among other mRNA-binding proteins, and can be qualified as a core protein component of mRNPs.

The p50 component of free and polysomal mRNPs has several peculiar characteristics. First, this ubiquitous protein component of eukaryotic mRNPs is found to be not a single protein but a family of closely related proteins, with high sequence homology. Very often (but not always), a pair of the proteins of this family is present in mRNPs of a given origin. The family includes the DNA-binding transcription factors stimulating the synthesis of mRNAs from the so-called Y-box-containing promoters. In several cases the same proteins were revealed both as nuclear Y-box transcription factors and as major cytoplasmic mRNP-forming components. They are not present in nuclear hnRNPs and not accumulated in nuclei in a detectable amount, thus they are predominantly cytoplasmic proteins. The proteins are basic, have an isoelectric point (pI) of about 9.5, and generally contain a large proportion of charged amino acids. They are also characterized by an unusually high content of glycine. The unusual amino acid composition of the proteins may explain their anomalous electrophoretic behavior in the presence of sodium dodecylsulfate: the apparent molecular mass as estimated from SDS electrophoresis (50 kDa) is found to be significantly higher than the real molecular mass of about 35 kDa. In mRNPs they are multiply phosphorylated. In the free state they form large aggregates or particles sedimenting at 18S (molecular mass of about 10^6) and are also phosphorylatable. The interaction with mRNA results in rather extensive melting of the RNA secondary structure and increased exposure of the RNA to nucleolytic attack.

Thus, all mRNA in the cytoplasm is complexed with the p50, and the interaction seems to be rather strong. It means that the eukaryotic translational machinery deals with this matter, not with just polyribonucleotide sequences and their intrinsic secondary and tertiary structures. It can be expected that the presence of the bound protein molecules along the entire mRNA sequence and the partially melted state of mRNA secondary structure must exert serious effects on initiation and elongation mechanisms.

The presence of the p50 on the translatable mRNA in polyribosomes suggests the positive contribution of the protein to translational initiation and/or elongation. Indeed, stimulation of translation of exogenous mRNA in a cell-free system by relatively low concentrations of the p50 has been reported. On the other hand, the same *in vitro* experiments have demonstrated a negative contribution: increase in the ratio of the p50 to mRNA in cell-free systems inhibited translation. In agreement with this, the mRNPs isolated from polyribosomes that retain their p50 content are well translatable, whereas the free mRNPs with higher content of the p50 are either nontranslatable or translatable only with the help of a special activator system (probably including eIF4F). *In vivo* observations indicate that the increased accumulation of the p50 in the cell correlates with the inhibition of protein synthesis. Such accumulation is typical of germ cells, that is, oocytes and spermatocytes, known to contain inactive, or masked mRNA (see Section 17.5). Artificial overexpression of the p50 in somatic cells is also inhibitory for translation. From this a hypothesis has been proposed that the p50 can be responsible for global downregulation of translation by sequence-nonspecific interactions with mRNAs. Yet, the presence of the p50 on mRNA in translating polyribosomes does not fit well with the hypothesis in its straightforward form and requires elucidation. In any case, it is likely that the p50, the sequence-nonspecific mRNA-binding protein responsible for the formation of the cytoplasmic mRNPs, plays an important role in global regulation of translation in eukaryotic cells. The level of this protein in the cytoplasm may regulate the expressivity of mRNA. For example, the degree of saturation of mRNA with the protein may govern conformational transitions of mRNPs, thus modulating the accessibility of mRNA for translation factors and ribosomes. This is an open field for further investigations.

17.3. Discrimination of mRNAs

Like in prokaryotes, different mRNAs of eukaryotes possess different "strength" in initiation of their translation, that is, they initiate with different rates. This is basically determined by their structure, and first of all by the structure of their 5′-terminal regions and the regions around the initiation codon. Two main parameters may play a decisive role: (1) the availability and the affinity of these mRNA structures for ribosomes (for the initiating small ribosomal subunit), and (2) the affinity of these structures for a limiting RNA-binding initiation factor, such as eIF4F.

17.3.1. Discrimination by Initiating Ribosomal Particles

As discussed in Section 16.2., the discrimination of mRNAs by initiating ribosomal particles is typical of prokaryotes. As to eukaryotes, a similar phenomenon of mRNA discrimination by the initiating 40S ribosomal subunits may be possible in some cases. The classical example where this model of translational discrimination (Lodish, 1976) was applied is different translation rates of the mRNAs encoding for α- and β-globin chains in mammalian (e.g., rabbit) reticulocytes. The β-globin mRNA was shown to be a more effective message than the α-globin mRNA (Lodish, 1971). At the same time, the elongation rate was shown to be identical in both cases (Lodish and Jacobsen, 1972). It is the initiation rate that has been shown to be responsible for the difference in translatability: each molecule of β-globin mRNA initiates translation 1.7 times more frequently as does each α-globin mRNA. In other words, the β-globin mRNA is "stronger" in initiation. To compensate for this difference and to produce equimolar quantities of α- and β-globin chains, the cell contains a correspondingly higher amount of mRNA for the α-globin than for β-globin. It can be supposed that the 5′-terminal sequence of β-globin mRNA has some structural characteristic determining its higher affinity for the initiating 40S ribosomal subunits.

This model, however, cannot be accepted as ultimate: in the case of eukaryotic systems the competition between mRNAs for RNA-binding initiation factors, rather than for initiating ribosomal complexes, seems to be more realistic (see Section 17.3.2).

17.3.2. Discrimination by mRNA-Binding Initiation Factors

As already mentioned, eukaryotic initiation factors can be grouped into two main categories: ribosome-binding proteins and mRNA-binding proteins. In most eukaryotic cells the mRNA-binding initiation factors, eIF4F and eIF4B, seem to be present in molar amounts less than the total molar amount of mRNA, that is, there is a shortage of the mRNA-binding initiation factors. Hence, competition between mRNAs for the factors exists. This situation is different from that described for prokaryotes when the free competition between initiating ribosomal complexes for the mRNA initiation sites (RBS) is assumed. Here, in eukaryotes, the competition between mRNAs for an initiation component takes place *prior* to their binding to the initiating ribosomal complex. In such a case, if different mRNAs vary in their affinities for the component (an mRNA-binding initiation factor), this will determine different "strengths" of the messages in initiation of their translation (Walden, Godefroy-Colburn, and Thach, 1981; Brendler *et al.*, 1981a,b; Godefroy and Thach, 1981).

Two mRNA-binding initiation factors, eIF4F and eIF4B, can be considered as the candidates for the message-discriminatory components. Both factors are present in most eukaryotic cells in a limiting amount, and both were reported to vary in the affinities for different mRNAs or their 5′-terminal sequences. In many cases (e.g., in nonproliferating cells, such as rabbit reticulocytes) the situation with eIF4F is especially tight: its cap-binding 25-kDa subunit (eIF4E) is present in an amount of about 0.02 molecules per ribosome and thus seems to be the most limiting component of the translational initiation machinery. On the other hand, eIF4B possesses a fairly nonspecific RNA-binding activity and may be required in many copies per mRNA for effective initiation.

The mRNA encoding for ferritin, the iron-storing protein of eukaryotes, is an example of a very competitive cellular mRNA. When the ferritin mRNA is derepressed and translated *in vivo* or *in vitro,* it outcompetes all other mRNAs. It has been demonstrated that a special secondary/tertiary structural fold ("FR") at the

5′-untranslated region of the ferritin mRNA, near the cap, is responsible for a great part of this initiation rate "strength." This "positive control element" of the ferritin mRNA may be considered as a selectively strong binding site for a limiting RNA-binding initiation factor.

The problem of the competition between mRNAs for a limiting ("discriminatory") initiation factor was more thoroughly investigated, both theoretically and experimentally, for several cases of virus-infected cells and viral RNA-directed cell-free translation systems (e.g., Walden *et al.*, 1981). The conclusions were that (1) viral and host mRNAs compete for a message-discriminatory component prior to their binding to the initiation 40S ribosomal complex, (2) this component (initiation factor) is limiting in virus-infected cells, and (3) a hierarchy exists among mRNAs in terms of their affinity for this component. Encephalomyocarditis (EMC) virus RNA was shown to be "stronger" than host mRNA. (There is evidence that the affinity of the EMC virus RNA for eIF4B is one order of magnitude greater than that of the "average" host mRNA). The host mRNA, however, was found to be "stronger" than several species of reovirus mRNA. At the same time, the reovirus RNA competes successfully with the host mRNA due to unusually high production, and hence very high amount, of this RNA in infected cells.

Among cellular mRNAs there is a class of poorly competing messages which are found predominantly in small-size polyribosomes and free mRNP particles. The investigation of the problem has led to the conclusion that their low competitive activity is due to the presence of a tightly bound protein, probably the core mRNP-forming protein (p50) discussed previously (Section 17.2.2). The removal of the protein (e.g., by phenol treatment) makes them very competitive in cell-free translation systems. At the same time, the bound protein seems to be in equilibrium with free protein, and thus the situation simply reflects the fact that the mRNAs under consideration have lower affinity for initiation factors than for the mRNP protein which impedes the binding of the initiation factors. In any case, the competition between the mRNA-binding initiation factors and other mRNA-binding proteins must be taken into account when analyzing the problem of translational discrimination of eukaryotic mRNAs.

Another case is a strong mRNA-specific binding of a repressor protein which blocks the initiation of translation of a highly competitive mRNA, such as ferritin mRNA. It can do this due to the presence of a special secondary/tertiary structure at the 5′-untranslated region of a given mRNA with high affinity for a given protein. This case will be discussed in Section 17.4.1.

17.3.3. Modulation of Translational Discrimination by Inhibiting Elongation

The situation in eukaryotes where initiation of translation is determined by a limiting mRNA-binding initiation factor provides an interesting opportunity to regulate expression of "weak" mRNAs through alterations in elongation rate. Inhibition of elongation rate causes a limiting initiation factor to be in the free (not bound to mRNA) state for a longer time than in the case of fast elongation, thus reducing the competition (Fig. 17.3). Under these conditions "weak" mRNAs have better chances to initiate translation and therefore to be fairly expressed. Experiments with elongation inhibitors, such as cycloheximide, directly supported this expectation. For example, cycloheximide inhibited translation of "strong" host mRNA and stimulated translation of "weak" reovirus mRNAs in the virus-infected cells, as well as in cell-free systems. Likewise, the class of cellular mRNAs poorly translatable due to interference with mRNP-forming protein displays better translation upon the inhibition of total protein synthesis (elongation) by cycloheximide (Walden *et al.*, 1981).

The phenomenon of stimulation of translation of "weak" mRNAs when elongation rate is downregulated can play an important physiological role in the cell. mRNAs encoding for cytokines, protooncogenes, and other proteins responsible for cell activation or G_0–G_1 transition are usually "weak." Hence, Ca^{2+}/calmodulin-induced phosphorylation of eEF2 and resultant inhibition of elongation (Section 13.4.2) must result in stimulation of translational initiation of these mRNAs and thus give an impetus for cell activation and/or proliferation (see Spirin and Ryazanov, 1991; Ryazanov and Spirin, 1993). Indeed, it has been demonstrated that transient phosphorylation of eEF2, and probably immediate short-term reduction of protein synthesis, takes place at mitogenic stimulation of quiescent cells (Celis, Madsen, and Ryazanov, 1990).

17.3.4. Modulation of Translational Discrimination by Changing Initiation Rate

Stimulation of quiescent cells and even their malignant transformation can be obtained by overexpression of eIF4E (Lazaris–Karatzas, Montine, and Sonenberg, 1990). This may be explained exactly in the same terms as above: the most limiting initiation factor, complete eIF4F, becomes more abundant and thus available for "weak" mRNAs, including protooncogene (e.g., c-*fos*, c-*jun*, c-*myc*) mRNAs (see Rhoads *et al.*, 1993). In this way the upregulation of an initiation component may

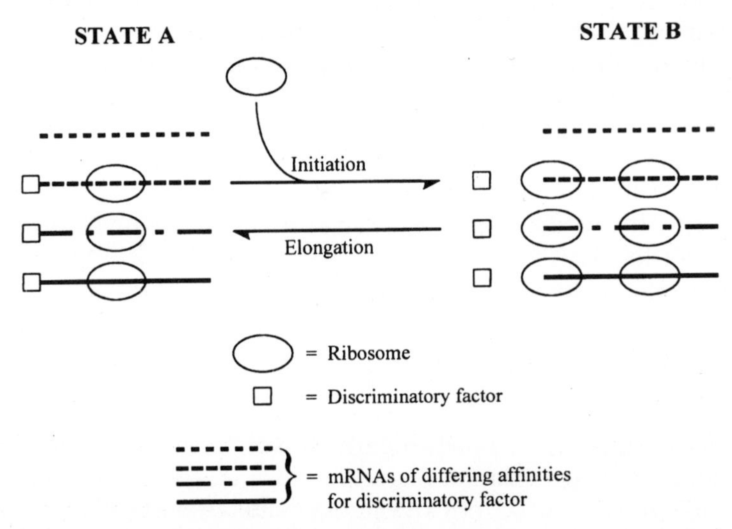

FIG. 17.3. Scheme illustrating the effect of the inhibition of elongation rate on the discrimination of mRNAs. The discriminatory initiation factor is shown to exist in one of the two states, either bound to mRNA **(A)** or free **(B)**. Slowing the ribosomal elongation rate increases the average amount of time that the system spends in state B. This increases the steady-state free factor concentration and hence the probability that the low-affinity mRNA (indicated by the *dotted line*) will bind the discriminatory factor and thus be translated. Reproduced from Walden, W. E., Godefroy–Colburn, T., and Thach, R. E. (1981) *J. Biol. Chem.* **256**:11739–11746, with permission.

result in the reduction of translational discrimination and lead to the effective expression of a new set of messages ("weak" mRNAs).

The phosphorylation of eIF2α, as a response to heme deficiency, growth factor deprivation, amino acid starvation, heat shock, and virus infection, can produce an opposite effect on translational discrimination. As a result of phosphorylation, eIF2 becomes less available for initiation. Since eIF2 is a ribosome-binding factor that converts passive 40S ribosomal subunits into active initiating 40S ribosomal complex, the less eIF2 is available, the less initiating ribosomes are present. Under these conditions "strong" mRNAs will trap all limiting initiation components such as eIF4F and eIF4B, and translation of "weak" mRNAs will be even more reduced.

Following the same logic, the modulation of translational discrimination of mRNAs at the level of initiation could be achieved also by changes in the amount or activity of the mRNP-forming protein (see Section 17.2.2). The increase in the amount (or the RNA-binding activity, e.g., by phosphorylation) of the protein should propel the displacement of the mRNA-binding initiation factors from mRNAs and thus lead to the reduction of the discrimination, that is, to the inhibition of translation of "strong" mRNAs and the stimulation of "weak" mRNAs. The addition of free "weak" mRNA or untranslatable RNA (e.g., antisense RNA) to a cell-free system, or overproduction of such an RNA in the cell should have an opposite effect: the mRNP protein would be bound and redistributed among more RNA molecules ("diluted"), and the limiting mRNA-binding initiation factors would serve almost exclusively "strong" mRNAs.

Some viruses use another, more radical way to change the discrimination profile and to switch translational initiation predominantly on viral mRNAs. For example, many picornaviruses (enteroviruses, rhinoviruses, and aphthoviruses) induce proteolytic degradation of the p220 subunit (eIF4G) of eIF4F and thus switch off cap-dependent initiation (see Ehrenfeld, 1996). As a result, the internal initiation on polioviral mRNA becomes advantageous. It is interesting that the picornavirus-encoded proteinases, such as the 2A proteinase of poliovirus and Lb proteinase of foot-and-mouth disease virus, split eIF4G into two parts, the N-terminal part which binds eIF4E and the C-terminal part which associates with eIF4A (and eIF3), thus disconnecting the cap-binding function and the helicase function of eIF4F. Correspondingly, the C-terminal product of the eIF4G cleavage, seemingly in association with eIF4A, has been shown to be inactive in the cap-dependent initiation, but very stimulatory in the internal (IRES-driven) initiation and generally in initiation of translation of uncapped mRNAs.

A similar mechanism may switch translation from "normal" mRNAs to heat-shock mRNAs during heat shock (for review, see Duncan, 1996). Indeed, many heat-shock mRNAs, especially in *Drosophila*, have long 5′-untranslated regions, suggesting that ribosomes may not scan all these sequences to reach the initiation codon but rather initiate internally. The mRNAs encoding for some heat-shock proteins (e.g., HSP 70 mRNA of HeLa cells) have been shown to be translated in poliovirus-infected cells, presumably by a cap-independent mechanism. The experiments with inhibition of expression of eIF4E by antisense RNA demonstrated that the decrease in the amount of eIF4E in the cell was accompanied, as expected, by the reduction in total protein synthesis, but the synthesis of heat-shock proteins was unaffected, again suggesting a cap-independent mechanism of translational initiation (Rhoads *et al.*, 1993). Extensive dephosphorylation of the eIF4E subunit during heat shock may be one of the possible mechanisms of inactivation of the whole eIF4F complex as a cap-binding initiation factor.

17.4. Regulation of Initiation by Upstream Open Reading Frames

There are eukaryotic mRNAs where the coding sequence is preceded by one or several short open reading frames (ORFs) with their own initiation and termination codons. Assuming the mechanism of scanning of a message from the 5′-end by initiating ribosomal particles in eukaryotes, the translation of the main coding sequence should proceed via termination of the upstream short ORF and reinitiation. Like in prokaryotes (see Section 16.3.2), the reinitiation can take place both downstream and upstream (when both coding sequences overlap) of the termination codon of the preceding ORF.

In contrast to prokaryotes, however, the presence of an upstream ORF in a eukaryotic mRNA usually weakens the expression of the following coding sequence (for review, see Geballe, 1996). The weakening can be a simple consequence of a low efficiency of reinitiation in eukaryotes: only a fraction of the terminated ribosomes remains bound to mRNA and can start a new scanning run. In such cases the sequence of a short ORF and the sequence of an ORF-encoded peptide may have no significance. In other cases the inefficiency of reinitiation can be enhanced by the peptide synthesized on ORF; it seems that some short ORFs encode peptides with special sequences inhibitory for ORF termination.

The possibilities of positive contributions of upstream short ORFs should be also mentioned. First of all, translation of upstream ORFs can unwind some stable secondary/tertiary structural elements of the 5′-UTR that inhibit nontranslating scanning ribosomes. Favorable translation-induced reorganizations of secondary/tertiary structure of the 5′-UTR should be attributed to the same category of possible positive effects. Second, some unique combinations of structural elements with short ORFs may exist that induce skipping (shunting) of long sequences and inhibitory stable structures within the 5′-UTR by scanning ribosomal initiation complex.

17.4.1. ORF Sequence-Dependent Inhibition of Initiation

As mentioned, in many cases the inhibitory effect strongly depends on the sequence of the upstream ORF and, more precisely, on the sequence of a short peptide synthesized on it. It seems that these peptides function only in *cis,* that is, as nascent ribosome-bound chains. A specific interaction with some ribosomal components and retardation of termination could be a possible mechanism of their action (*cf.* Section 13.2.4). The examples of mRNAs with upstream ORFs encoding for such inhibitory peptides are mammalian *S*-adenosylmethionine decarboxylase mRNA, yeast *CPA1* mRNA coding for an enzyme of arginine biosynthesis, and human cytomegalovirus gp48 mRNA (reviewed by Geballe, 1996). The length of the peptides are from 6 to 25 in the above cases; their amino acid sequences, rather than the nucleotide sequences of the ORFs, have been directly shown to be critical for the inhibitory effect.

The inhibition of the downstream translation by the upstream ORFs can be regulated by some *trans*-factors and environmental conditions. Thus, translation of *S*-adenosylmethionine decarboxylase mRNA, which is inhibited by the upstream ORF in resting T cells, becomes activated upon T-cell stimulation (Fig

17.4). Efficient translation of the *S*-adenosylmethionine decarboxylase mRNA in the stimulated T cells, as well as in non-T-cell lines, seems to result from an inefficient use (ignoring) of the ORF initiation codon by scanning ribosomal particles under these conditions.

Analogously, the *CPA1* ORF impedes the downstream initiation of translation only in the presence of arginine. The arginine depletion reduces the inhibitory effect of the ORF. The mechanism of this regulation is obscure. Either a reduced initiation at the ORF initiation codon or a decreased interaction of the peptide with the ribosome may be responsible for the effect of arginine depletion.

17.4.2. ORF-Mediated Regulation of Yeast Transcription Factor GCN4

Translational regulation of the synthesis of a yeast transcription factor, called GCN4, is a different and more complicated case (for reviews, see Hinnebusch and Klausner, 1991; Hinnebusch *et al.*, 1993; Hinnebusch, 1996). The *GCN4* mRNA

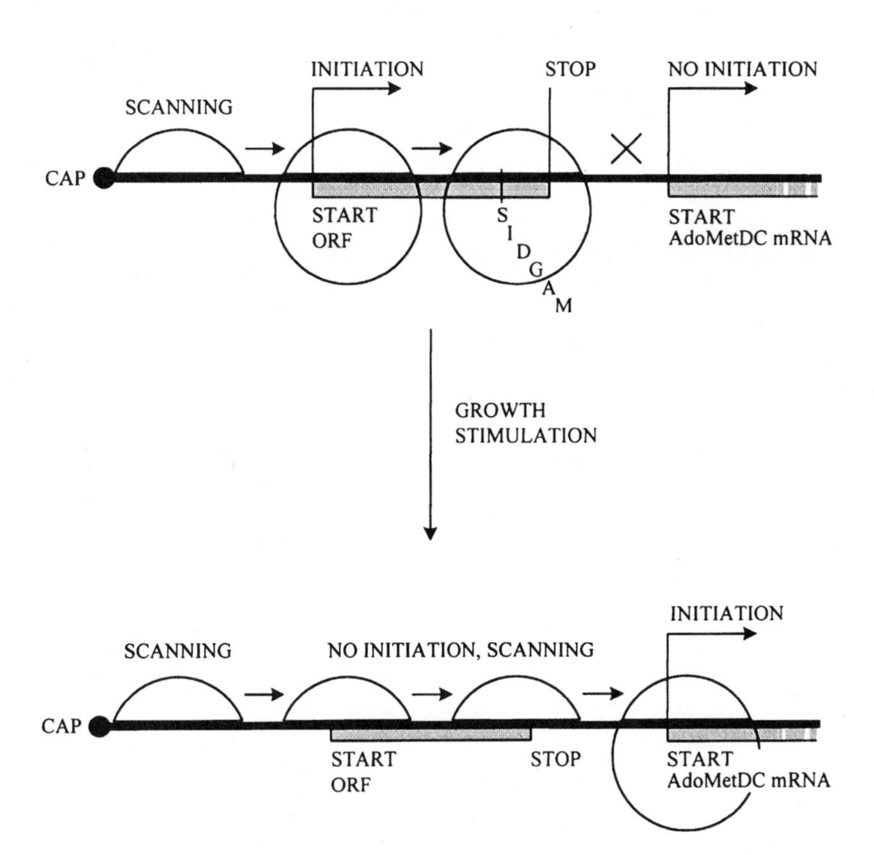

FIG. 17.4. Schematic representation of the mechanism of the ORF sequence-dependent inhibition of initiation, as exemplified by the case of the regulation of *S*-adenosylmethionine decarboxylase (AdoMetDC) mRNA translation. Translation of the upstream ORF results in the synthesis of the hexapeptide MetAlaGlyAspIleSer (MAGDIS) that seemingly remains bound to the ribosome and somehow blocks termination and further movement along mRNA (*upper*). Under some conditions, for example, upon growth stimulation, the scanning ribosomes ignore the ORF initiation signal, scan the mRNA further, and initiate at the start of the AdoMetDC coding sequence (*lower*). Adapted from Hill, J. R., and Morris, D. R. (1993) *J. Biol. Chem.* **268**:726–731; Ruan, H., Hill, J. R., Fatemie–Nainie, S., and Morris, D. R. (1994) *J. Biol. Chem.* **269**:17905–17910.

contains four very short upstream ORFs in its long leader sequence, located between 150 and 360 nucleotides upstream of the initiation codon (Fig. 17.5). The di- and tripeptide sequences synthesized on the ORFs are hardly of decisive importance for the inhibition of the initiation at the downstream *GCN4* coding sequence. The ORFs inhibit the initiation at the downstream *GCN4* start under normal conditions when nutrients are abundant. Under starvation conditions, however, the inhibition by the ORFs is reduced or abolished. The "derepression" of GCN4 correlates with the starvation-induced phosphorylation of eIF2. It has been found that the phosphorylation of eIF2 controls the effect of the ORFs on the efficiency of reinitiation at the *GCN4* start. The first and the fourth ORFs seem to take the most important part in the regulation. The deletion of the first ORF reduces the *GCN4* expression under derepressing (starvation) conditions and has no significant effect when the *GCN4* translation is inhibited; this suggests the role of the first ORF as a positive control element. On the contrary, the deletion of the fourth ORF, or both the third and the fourth ORFs, abolishes the inhibition of the *GCN4* initiation under repressing conditions, as expected if a negative control element is removed.

A model has been proposed (Fig. 17.6) in which the reinitiation at the *GCN4* start inversely depends on the level of the active (nonphosphorylated) eIF2. The ribosomes that bind to the 5′-end of the *GCN4* mRNA will initiate at the first ORF, translate it, and terminate. Under non-starvation conditions the active eIF2 is abundant, and so the terminated ribosomal particles will quickly reassociate with the eIF2:GTP:Met-$tRNA_i$ ternary complex. The part of them that remains to be bound with mRNA will reinitiate at the second ORF, and after termination again only a part will remain mRNA-bound and reinitiate, and so on. Thus, most of the ribosomes will be lost during termination and reinitiation at the four ORFs and, therefore, fail to reach the *GCN4* start. Under starvation conditions, the level of the eIF2:GTP:Met-$tRNA_i$ ternary complex is reduced due to phosphorylation of eIF2. From this it may follow that the ribosomal particles after termination at the first ORF will scan the downstream sequence but will not have time to catch the ternary complex and reinitiate at the second, the third, and the fourth ORFs. As a result, most of them may reach the initiation codon of the *GCN4* coding sequence. According to the model, most of the scanning particles will bind the ternary complex

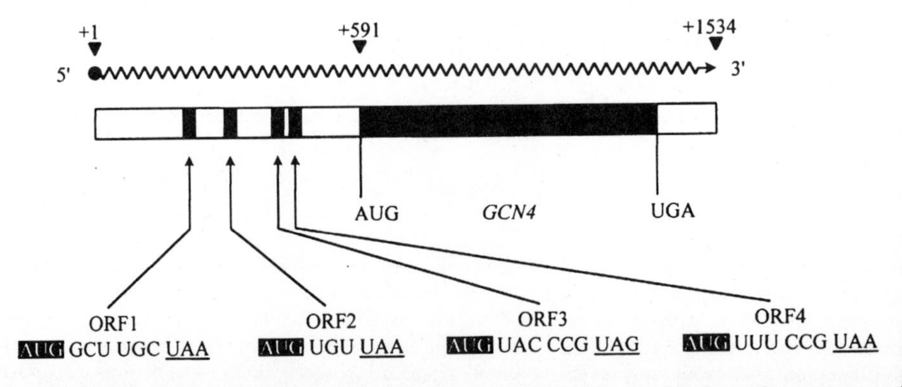

FIG. 17.5. Scheme of the yeast *GCN4* mRNA. Four very short ORFs precedes the coding sequence fo the GCN4. Reproduced from Hinnebusch, A. G., Wek, R. C., Dever, T. E., Cigan, A. M., Feng, L., an Donahue, T. E. (1993) in *Translational Regulation of Gene Expression 2* (J. Ilan, ed.), pp. 87–115 Plenum Press, New York, with permission.

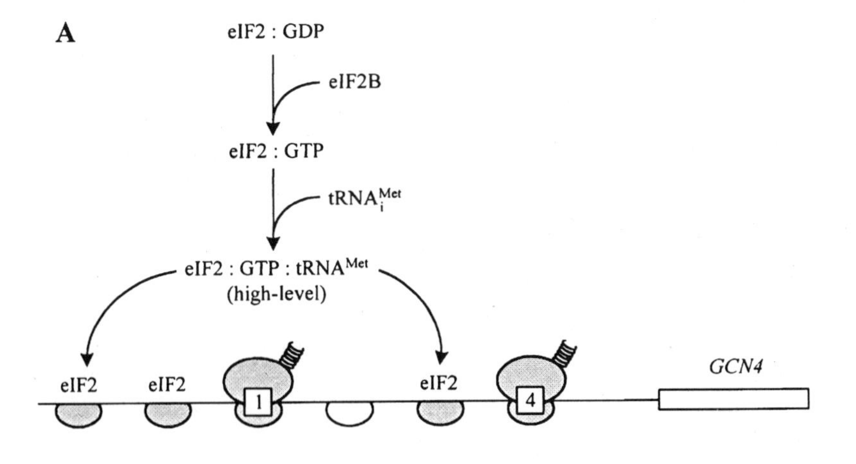

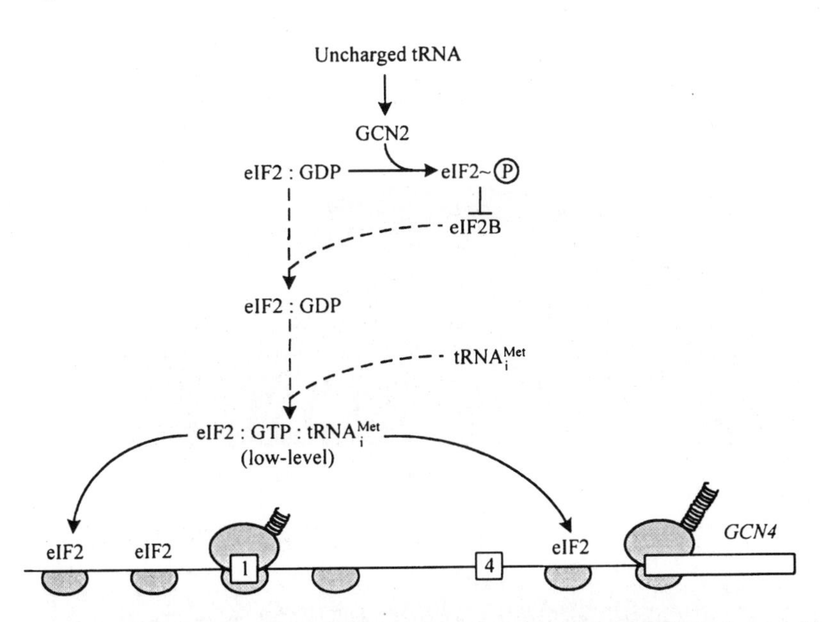

FIG. 17.6. Model for translational control of yeast GCN4 synthesis. *GCN4* mRNA is shown with ORFs 1 and 4 (boxed). 40S ribosomal subunits are shaded when they associated with the ternary Met-tRNA$_i$:eIF2:GTP complex and are thus competent to reinitiate translation; unshaded 40S subunits lack the ternary complex and therefore cannot reinitiate. **(A)** Under non-starvation conditions, eIF2:GDP is readily recycled to eIF2:GTP by eIF2B, leading to high levels of eIF2:GTP and ternary complex formation. The ternary complexes thus formed reassemble with 40S particles scanning downstream from ORF1, causing reinitiation to occur at ORF4. **(B)** Under starvation conditions, uncharged tRNA accumulates and activates the protein kinase GCN2 that phosphorylates eIF2α. The phosphorylated eIF2 traps a significant portion of eIF2B and thus reduces the rate of eIF2:GDP-to-eIF2:GTP recycling, resulting in a low level of ternary complex formation. In this situation the 40S ribosomal particles scanning from ORF1 have a low chance of reassembling with the ternary complex while scanning from ORF1 to ORF4, and most of them do not reinitiate at ORF4. While scanning further they have time to catch the ternary complex and so reinitiate at the start of the coding sequence of *GCN4* mRNA. Reproduced, with some modifications, from Hinnebusch, A. G. (1996) in *Translational Control* (J. W. B. Hershey, M. B. Mathews, and N. Sonenberg, eds.), pp. 199–244, CSHL Press, with permission.

while traversing the leader segment between the fourth ORF and the start of the *GCN4* coding sequence.

17.4.3. ORF-Involving Ribosome Shunting

Cauliflower mosaic virus (CaMV) 35S RNA has a long 600-nucleotide leader containing seven short ORFs (Fig. 17.7). Translation of the downstream viral polycistronic message requires the recognition of the capped 5′-end and the subsequent scanning of the leader by the ribosomal initiation complex. As expected, the ORFs within the leader are inhibitory for the initiation of translation of the main viral cistrons. In addition, according to the secondary structure prediction, almost all of the leader sequence is folded into a long compound hairpin (as shown in Fig. 17.7 where paired nucleotides are connected by arcs) that should be a high-energy barrier for the scanning complex. This makes the CaMV RNA hardly translatable both *in vitro* and *in vivo* in the plant systems that are not hosts for the virus. However, an alleviation of the translational inhibition is observed in host plant cells. It seems that the alleviation is provided by the presence (or higher concentration) of a special cellular factor in the cytoplasm, but in any case it has been found to depend on the first short ORFs.

It has been demonstrated that when the CaMV RNA is translationally active the initiating ribosomal particle does not scan all the leader sequence continuously

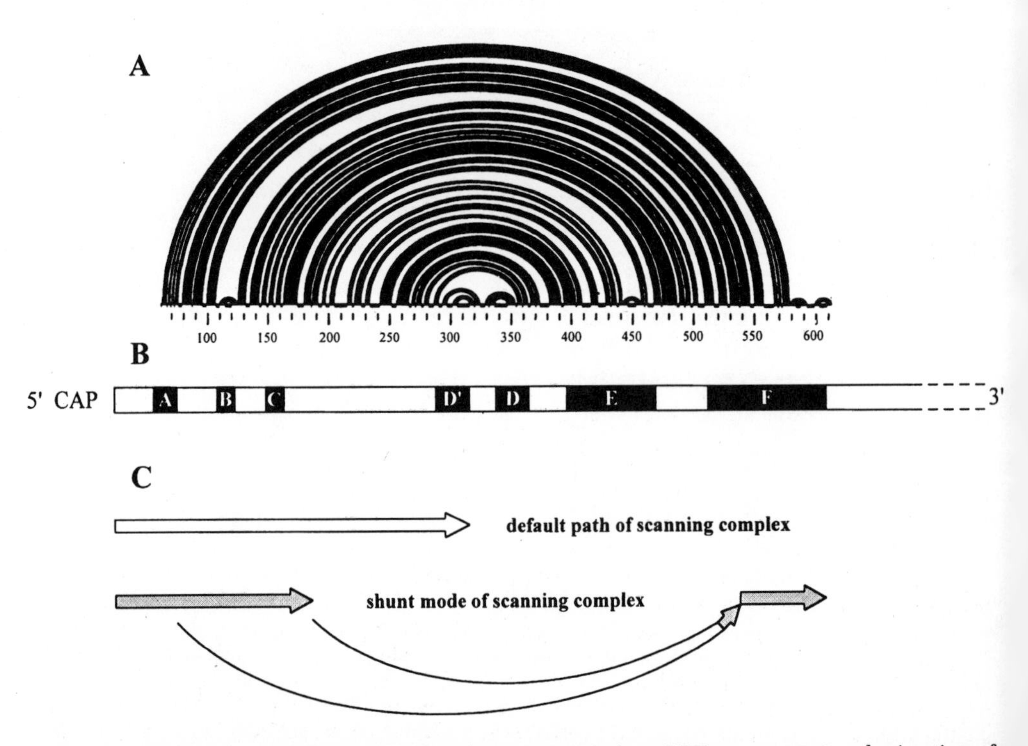

FIG. 17.7. Schematic representation of secondary structure, short ORF arrangement and migration of scanning ribosomal particles along the leader sequence of the cauliflower mosaic virus (CaMV) RNA. **(A)** Predicted secondary structure of the CaMV RNA leader: paired bases are connected by arcs. **(B)** Positions of seven short ORFs along the leader. **(C)** The routes of the scanning ribosomal complex: default path (*upper*) and shunting (*below*). Reproduced from Fütterer, J., Kiss–Laszlo, Z., and Hohn, T. (1993) *Cell* **73**:789–802, with permission.

but skips the central part of it by jumping it over, or shunting. For this the reading of the first three ORFs (A, B, and C in Fig. 17.7) seems to be necessary. It is likely that the translation of these ORFs causes unwinding of the compound hairpin base and thus entering of the ribosomal particle into the structured region. Something exists in the structure that induces the jump of the ribosomal initiating particle (seemingly 43S particle) from a region downstream of the third ORF to a region inside the seventh ORF (F in Fig. 17.7). Then the nontranslating (scanning) particle moves downstream and initiates translation on the first initiation codon of the main coding sequence of the RNA.

17.5. Translational Repression

As in the case of prokaryotes, translational repression is defined as the prevention of initiation of translation due to the interaction of a special protein repressor with a unique structural element in the mRNA upstream or at the beginning of the coding sequence. Indeed, stable binding of a protein to the 5′-untranslated region (5′-UTR) of mRNA would block either the association of the ribosomal initiation complex with the cap-adjacent mRNA sequence (or the internal initiation site), or the movement of the ribosomal initiation complex along mRNA toward the initiation site (Fig. 17.8). In fact, this classical initiation repression model has been supported by several observations concerning the role of 5′-UTRs in translational repression of eukaryotic mRNAs.

17.5.1. Iron-Responsive Regulation

The best studied example of translational control via interaction of a 5′-UTR element with a specific regulatable repressor protein is the case of ferritin mRNA (for reviews, see Hinnebusch and Klausner, 1991; Walden, 1993; Theil, 1987, 1993; Rouault, Klausner, and Harford, 1996). Here, a 98-kDa protein recognizes a 28-nucleotide structure ("iron responsive element," or IRE) in the 5′-UTR, in the spatial neighborhood of the cap (Fig. 17.9). In the absence of iron (Fe^{3+}), the protein (the so-called IRE-binding protein, or IRE-BP) has a strong affinity for IRE and acts as a repressor, blocking the initiation of translation of ferritin mRNA. It has been demonstrated that 5′- and 3′-flanking sequences of IRE which are mutually base-paired form a stem structure ("flanking region," or FL) changeable upon IRE-BP binding; this change seems to be important for the repressor power of IRE-BP. In the presence of iron ions, the affinity of the protein for IRE decreases, and this allows the initiation of translation (Fig. 17.10). The product of translation, ferritin, will bind and occlude the excess Fe^{3+}.

The repressor, IRE-BP, proved to be identical to aconitase, the enzyme converting citrate into isocitrate and possessing a 4Fe-4S (iron-sulfur) cluster necessary for its activity. It is likely that the affinity of the IRE-BP for the IRE is governed by the change in the iron-sulfur cluster, which is known to be reversibly convertible from the active 4Fe-4S form in aconitase into inactive iron-deficient form. Thus, the same protein with intact, iron-saturated 4Fe-4S cluster is aconitase and does not possess the IRE-binding activity, while its iron-deficient form with the disassembled iron-sulfur cluster is IRE-BP.

A similar negative regulation system with IRE and IRE-BP has been found in the case of erythroid-specific δ-aminolevulinic acid synthase (eALAS) mRNA. It is remarkable that among different ferritin and eALAS mRNAs the position of the

IRE relative to the 5′-terminus of mRNA is evolutionarily conserved: the IRE is always located within the first 40 nucleotides of the 5′-UTR. The introduction of a spacer between the cap structure and the IRE results in a reduction or abolishment of translational repression. In other words, the IRE must be located close enough to the capped 5′-terminus of mRNA in order to block initiation of translation upon

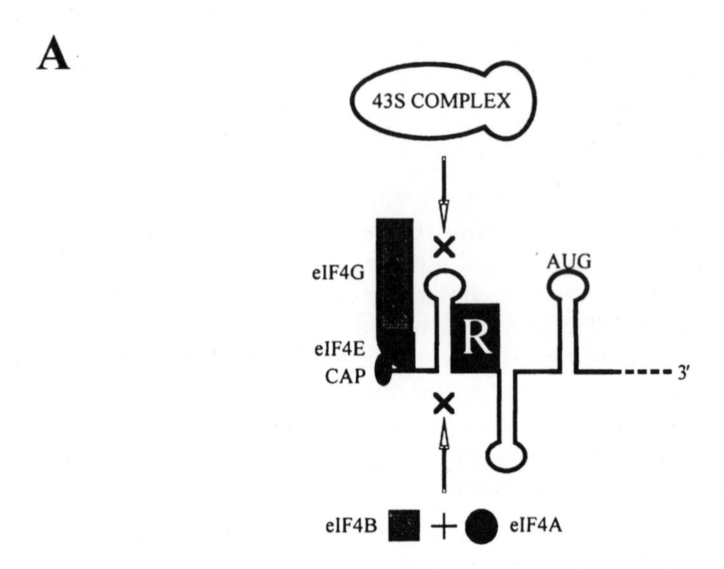

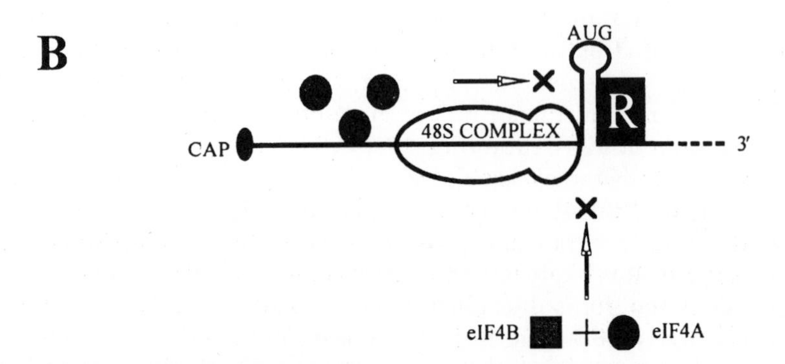

FIG. 17.8. Models for two mechanisms of blocking the initiation by translational repressor (R). **(A)** Repressor protein binds and stabilizes a hairpin near the cap structure, thus preventing its unwinding by eIF2A + eIF2B and the landing of the 43S initiation complex. **(B)** Repressor binds and stabilizes a hairpin remote from the cap and located before or at the beginning of the coding sequence; in such a case the protein prevents the unwinding of the helix by the scanning 48S initiation complex and its movement to initiation codon.

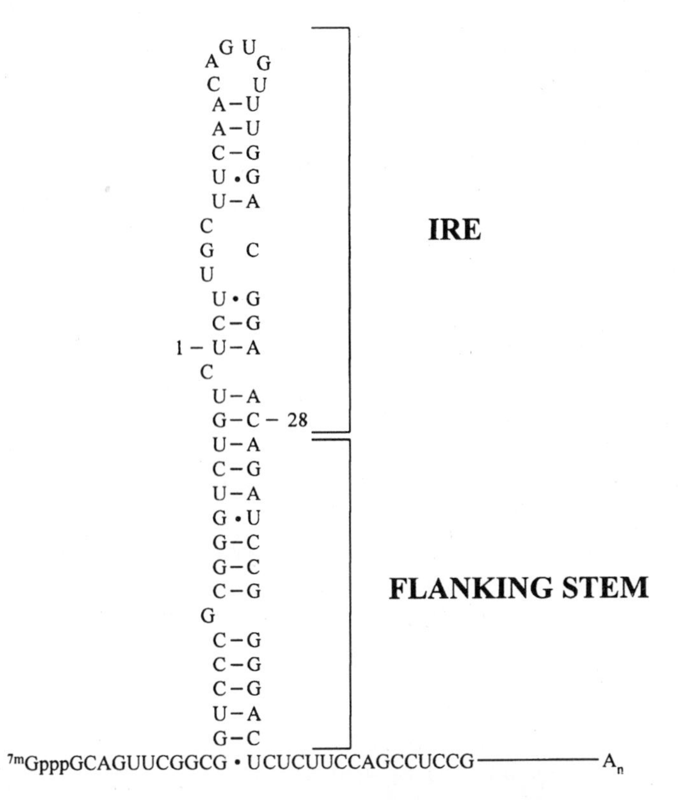

FIG. 17.9. Secondary structure of human L-chain ferritin IRE and flanking elements. The IRE is composed of 28 nucleotides, numbered 1 to 28, forming the upper stem loop. Reproduced from Walden, W. E. (1993) in *Translational Regulation of Gene Expression 2* (J. Ilan, ed.), pp. 321–334, Plenum Press, New York, with permission.

interaction with IRE-BP. This observation is in full agreement with the finding that the binding of the IRE-BP by the IRE prevents the association of the ribosomal 43S initiation complex with the cap-adjacent region of mRNA (Fig. 17.10; see also Fig. 17.8A).

17.5.2. Repression via Prevention of Helix Unwinding

The 5′-UTRs of natural eukaryotic mRNAs are often characterized by a considerable extent of folding, that is, secondary and, probably, tertiary interactions. RNA helicase activities during initiation of translation contribute to the unwinding (melting) of those structures. The role of repressor proteins can be the recognition and stabilization of those structures resulting in prevention of translation initiation, first of all by blocking the interaction of the ribosomal 43S initiation complex with the cap-adjacent sequence (Fig. 17.8A), as in the case of the ferritin mRNA repression (Fig. 17.10). Special stimuli, effectors, or environmental conditions should be able to displace a repressor and thus derepress such an mRNA. A mechanism of this kind can be envisioned in the cases of ornithine decarboxylase mRNA (Manzella and Blackshear, 1992) and c-*myc* mRNA (Parkin *et al.*, 1988; Lazarus, Parkin, and Sonenberg, 1988).

Ornithine decarboxylase (ODC) catalyzes the first step in polyamine biosynthesis. The *in vivo* induction of ODC synthesis at the translational level is observed in response to cell proliferation as well as to the level of polyamines. The 5′-UTR of the ODC mRNA contains a very stable secondary structure (the 5′-proximal long GC-rich hairpin) which is inhibitory for the initiation of translation by itself. Growth stimuli are able to relieve this constitutive translational inhibition, probably via stimulation of the activity of an RNA helicase or relevant initiation factors (eIF4A/eIF4B/eIF4F complex). At the same time a 58-kDa protein has been shown to specifically bind to the hairpin-adjacent region within the 5′-UTR. The protein is absent from tissues with high constitutive synthesis of ODC, and present in tissues where its translation is regulated. From this it can be thought that the 5′-UTR-binding protein is a repressor that prevents the melting of the 5′-proximal stable secondary structure by RNA helicases (initiation factors) in the process of initiation, but dissociates from mRNA under the action of some stimuli, thus permitting melting and initiation. The loss of the affinity of the protein for the 5′-UTR-binding site in response to oxidative conditions has been experimentally demonstrated.

The situation seems to be similar with c-*myc* mRNA, where the 5′-UTR structure is found to be inhibitory for translation. A stable hairpin is present at the beginning of the c-*myc* mRNA. The inhibition, however, can be observed in some but not all cell lines or translation systems. Moreover, the rate of translation of the c-*myc* mRNA undergoes dramatic changes depending on the developmental pro-

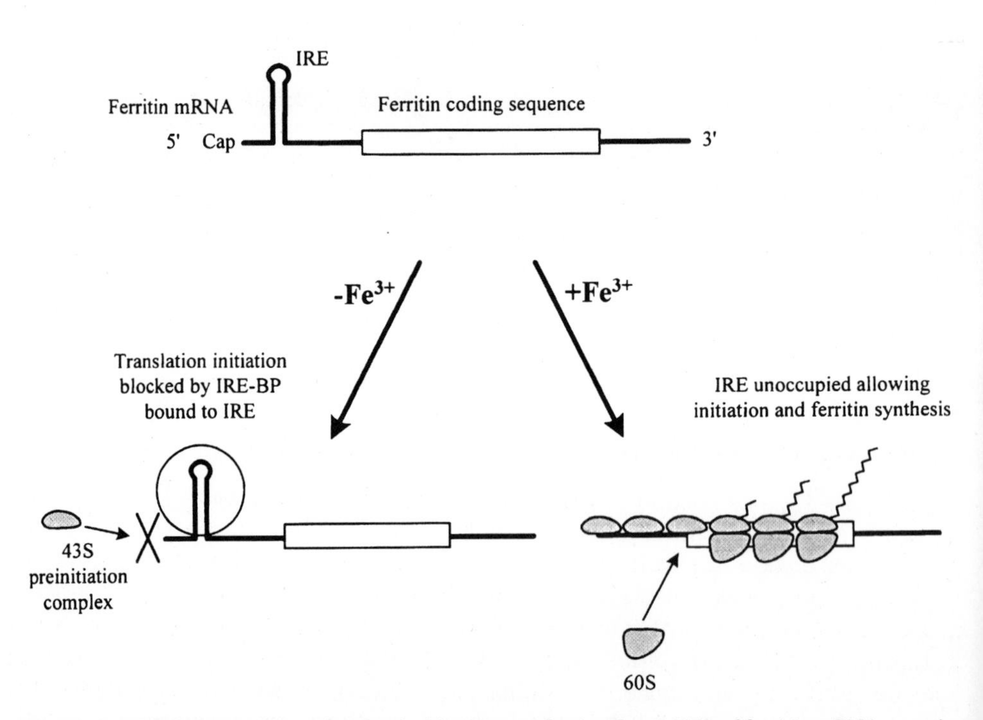

FIG. 17.10. Model for translational control of ferritin synthesis. The 5′-UTR of ferritin mRNA contains IRE (see Fig. 17.8). When Fe^{3+} is scarce, a cytoplasmic IRE-binding protein (IRE-BP) interacts with IRE and prevents the 43S initiation complex from associating with the cap structure of the ferritin mRNA (see Fig. 17.8A). When Fe^{3+} is abundant, the IRE-BP loses its affinity for IRE and dissociates, allowing the initiation and translation of the ferritin mRNA. Reproduced, with modifications, from Roualt, T. A., Klausner, R. D., and Harford, J. B. (1996) in *Translational Control* (J. W. B. Hershey, M. B. Mathews, and N. Sonenberg, eds.), pp. 335–362, CSHL Press, with permission.

cesses of oogenesis and embryogenesis in *Xenopus*. Two alternatives are possible: (1) either a putative repressor protein that stabilizes the 5′-UTR structure and prevents its unwinding is present in different amounts in different cells and regulated during development, or (2) some systems possess more active RNA helicase activity than others and these activities can be changed in developmental processes.

The repression of translation using this mechanism can be artificially reproduced by constructing chimeric mRNAs bearing a 5′-proximal hairpin structure in the 5′-UTR with a specific affinity for a given mRNA-binding protein. For example, when the hairpin structure with a high affinity for the bacteriophage MS2 coat protein (see Fig. 16.6) was introduced into the 5′-UTR of capped chloramphenicol acetyltransferase transcript, the coat protein was found to be an efficient repressor of translation of this mRNA both *in vitro*, in rabbit reticulocyte and wheat germ cell-free systems (Stripecke and Hentze, 1992), and *in vivo*, within human HeLa cells and yeast (Stripecke *et al.*, 1994). In the same way the spliceosomal protein U1A was converted into a translational repressor when the protein-binding site (weak hairpin) from the small U1 RNA was placed at the 5′-UTR of the chloramphenicol acetyltransferase mRNA (Stripecke and Hentze, 1992; Stripecke *et al.*, 1994). In both cases the translational repression seemed to be caused by the formation of a stable RNA/protein complex near the cap structure thus preventing the association of the initiating ribosomal particle (43S initiation complex) with mRNA. The experiments have demonstrated that this type of translational repression mechanism may indeed be generally used by eukaryotic cells.

17.5.3. Regulation of Ribosomal Protein mRNA Translation

The synthesis of ribosomal proteins during oogenesis and embryogenesis and in response to changes in cellular growth rate is also regulated at the translational level by a repression mechanism. The 5′-UTRs of ribosomal protein mRNAs have been shown to be involved in regulation of their translation in mammalian cells, *Xenopus* embryos, insects, and slime molds (for reviews, see Jacobs–Lorena and Fried, 1987; Kaspar, Morris, and White, 1993; Meyuhas, Avni, and Shama, 1996).

In vertebrates, the ribosomal protein mRNAs have a common sequence motif, an oligopyrimidine tract of 7 to 14 nucleotides starting with C at the capped 5′-terminus, that seems to be important for translational regulation. No hairpin structure, however, can be generated within the short pyrimidine-rich 5′-UTRs of the ribosomal protein mRNAs. In addition to the polypyrimidine 5′-terminus, a sequence immediately downstream seems to be required for the full manifestation of translational control. Conceivably, a repressor protein should be able to specifically bind to the 5′-terminal regulatory sequence ("translational regulatory element," or TRE) of ribosomal protein mRNA and prevent the initiating ribosomal particles from interacting with the mRNA. The affinity of such a repressor protein for TRE must change in response to cellular demands for ribosomal proteins. In any case, in contrast to prokaryotic systems, the eukaryotic ribosomal protein mRNAs are not autogenously regulated by ribosomal proteins.

No repressor protein has been unambiguously identified yet to bind to the polypyrimidine tract of the 5′-UTR in ribosomal protein mRNAs of vertebrates. Several polypyrimidine-binding proteins were suspected but without direct proof of their repressor action. If such a repressor does exist, the question arises whether the sequence-specific binding of the protein to the 5′-terminal sequence can block the ribosome/mRNA association by itself, without stabilization of a structured RNA element. In such a case the mechanism of translational repression should be considered somewhat different from that discussed in the preceding section. For example, direct prevention of interactions of the capped 5′-terminus with initiation factors or the initiating ribosomal complex is possible.

17.5.4. Repression by Prevention of Initiation Complex Movement along mRNA

A different repression mechanism is possible where the interaction of an RNA element of 5′-UTR with a repressor protein does not prevent the association of the ribosomal 43S complex with mRNA but forms a barrier which cannot be overcome (melted) by the 43S complex moving (scanning) from the cap structure to the initiation codon (Fig. 17.8B). Seemingly this is the case of the feedback translational repression of the human thymidylate synthase mRNA by the product of the translation, that is, by the thymidylate synthase (Chu *et al.*, 1993a). Here a 30-nucleotide stem-loop structure in the 5′-UTR specifically interacts with the thymidylate synthase probably resulting in (or contributing to) the repression of translation (Fig. 17.11). The stem-loop element is about 80 nucleotides apart from the cap and includes the initiation codon. There is also the second thymidylate synthase-binding site in the coding region of the same mRNA, the role of which is not clear. In any case the thymidylate synthase, the enzyme catalyzing the conversion of dUMP into dTMP, is found to be a specific mRNA-binding protein and translational repressor of its own mRNA.

There is an indication that human dihydrofolate reductase (DHFR) is also capable of specifically binding to its own mRNA and to repress translation (Chu *et al.*, 1993b), though no specific site for the binding of the protein within the 5′-UTR has been identified yet.

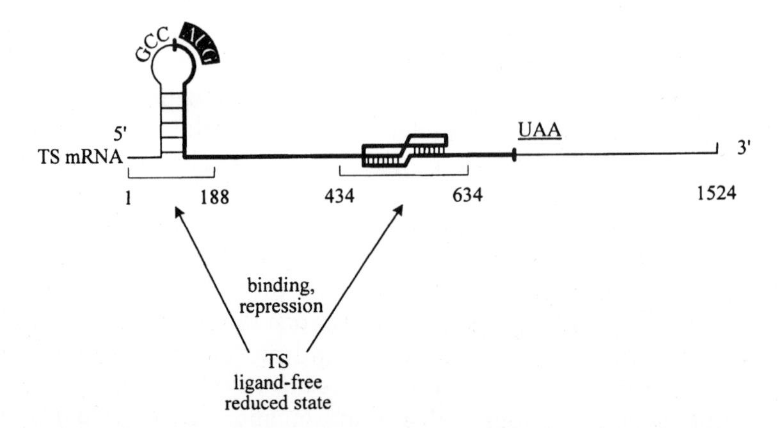

FIG. 17.11. Model for translational autoregulation of thymidylate synthase (TS) synthesis. Adapted from Chu, E., and Allegra, C. J. (1996) *BioEssays* **18**:191–198. The enzyme in a ligand-free (in the absence of thymidylate and folate) reduced state has an affinity for a stem-loop structure on the border between the 5′-UTR and the coding sequence of its own mRNA, as well as for a pseudoknot structure inside the coding sequence. The binding of TS to the upstream hairpin stabilizes it and thus blocks the movement of the scanning 48S initiation complex to the initiation AUG codon (see Fig. 17.8B).

The most typical examples of masked mRNAs are the messages stored in oocytes and spermatocytes. There are also striking examples of long-term storage of mRNAs in somatic cells. The stored mRNA is really masked against any processing events, including translation, degradation, and polyadenylation/deadenylation. This can be accepted as a definition of masking, in contrast to repression where just one function, translation, is blocked. The most remarkable discovery in recent years was the finding that the masking of mRNA involves primarily the 3′-UTRs of mRNAs and that the specific interactions of proteins ("masking proteins") with defined regions within the 3′-UTRs are responsible for switching off the functional activities of the respective mRNAs (Fig. 17.12) (for reviews, see Spirin, 1994, 1996).

17.6.1. Masked mRNA in Oocytes and Spermatocytes

The pioneer report in this field was made by Standart and co-workers (1990), who demonstrated that oocytes of a clamp *Spisula solidissima* contain large amounts of masked mRNA encoding for the small subunit of ribonucleotide reductase and for cyclin A. Fertilization triggers selective unmasking of these major stored mRNAs of the oocyte cytoplasm, while translation of a number of previously active mRNAs ceases. Under *in vitro* conditions, the unmasking of these two mRNAs was achieved by high salt treatment which presumably resulted in the dissociation of masking proteins. The most remarkable observation, however, was that a specific region in the middle of the 3′-UTR was responsible for binding a masking protein ("dissociable factor") (see Fig. 17.12). Those specific unmasking experiments were done with the use of antisense RNAs complementary to defined regions in the 3′-UTRs of both the mRNAs under study (the so-called competitive unmasking assay). The removal of the 3′-UTR sequence with the "masking box" from the mRNA also resulted in the promotion of translation. An oocyte protein of 82 kDa was found to bind specifically to the "masking element" within the 3′-UTR, and its presence in the masked mRNPs correlated with translational inactivity of the corresponding mRNAs. Evidence has been obtained that the unmasking of the maternal mRNAs is due to a maturation-dependent kinase which phosphorylates the protein (Walker, Dale, and Standart, 1996).

The decisive role of the 3′-UTR in masking mRNA during developmental processes was supported by genetic analysis of the switch from spermatogenesis to oogenesis in the hermaphrodite nematode *Caenorhabditis elegans* (Ahringer and Kimble, 1991). During development of the hermaphrodite, the germ cell precursors differentiate into sperms at the larval stage and then into oocytes in adult animals. The translation of the so-called *fem-3* mRNA is responsible for directing spermatogenesis, and the subsequent masking of this mRNA determines the switch to oogenesis. Point mutations in the middle of its 250 nt 3′-UTR, as well as the deletion of the central part of the 3′-UTR, abolish the capability of the *fem-3* mRNA to be masked and, hence, to allow oogenesis to occur. It seems that the mutations destroy a binding site for a protein whose function is to induce masking (see Fig. 17.12).

Another sex-determining mRNA of *C. elegans, tra-2,* which seems to be accumulated during oogenesis in a masked form and then activated to direct sexual differentiation, was also shown to be regulated by a specific RNA-binding protein interacting with a sequence (direct repeat) in the 3′-UTR.

Spermatogenesis is another developing system where masking/unmasking phenomena seem to play a decisive role (see Brown, 1990; Hecht, 1990; Schafer *et al.*, 1995; Kleene, 1996). Here transcription ceases during meiosis or at early postmeiotic stages, but translation of mRNA encoding for the most abundant spermatozoan proteins (e.g., protamines) is delayed for many days. Thus the mRNA is stored in a masked form as cytoplasmic mRNP particles and translated only during late spermiogenesis. In particular, protamine mRNA in mice is found to be accumulated as an untranslated mRNP during the stage of round spermatid, stored for up to one week, and translated at the stage of elongating spermatid. It is the 3′-UTR of protamine mRNA that proved to be responsible for this type of translational regulation. Though some proteins capable of specifically binding to the 3′-UTR sequences have been indicated, the function of the 3′-UTR-binding proteins in mRNA masking and what governs their binding or removal during spermatogenesis remains to be determined.

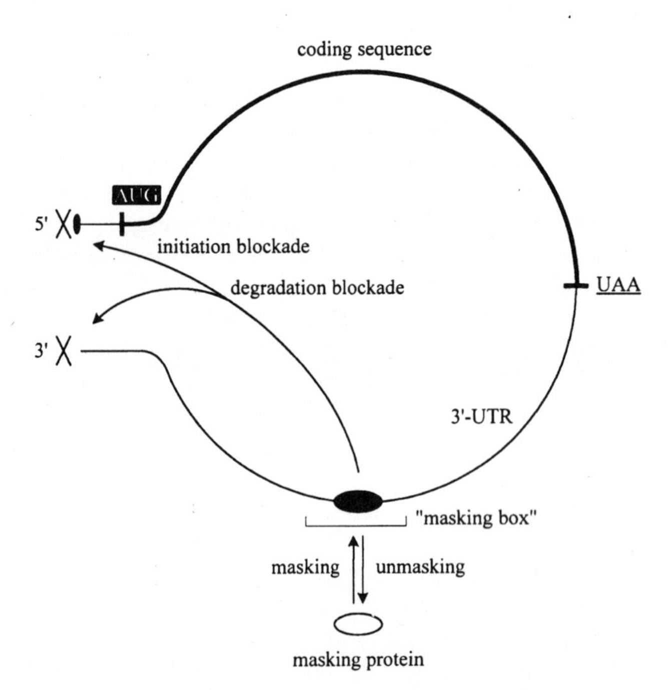

FIG. 17.12. Schematic representation of the masking induction effect on mRNA of a special protein ("masking protein") recognizing a middle section ("masking box") within 3′-UTR. The eventual result of the interaction of the 3′-UTR with the "masking protein" is the prevention of translation initiation (at 5′-end of the mRNA) and the stabilization of mRNA against enzymatic attacks (at the 3′-end especially). The classical examples are the masking of ribonucleotide reductase mRNA in oocytes of a clamp *Spisula solidissima* (Standart, N., Dale, M., Stewart, E., and Hunt, T. (1990) *Genes Develop.* **4**:2157–2168), *fem-3* mRNA during the switch from spermatogenesis to oogenesis in the hermaphrodite nematode *Caenorhabditis elegans* (Ahringer, J., and Kimble, J. (1991) *Nature* **349**:346–348), *nanos* mRNA of *Drosophila* eggs and early embryos (Gavis, E., and Lehmann, R. (1994) *Nature* **369**:315–318), and erythroid 15-lipoxygenase mRNA during erythropoiesis in mammals (Ostareck–Lederer, A., Ostareck, D., Standart, N., and Thiele, B. (1994) *EMBO J.* **13**:1476–1481). See also Wickens, M., Kimble, J., and Strickland, S. (1996) in *Translational Control* (J. W. B. Hershey, M. B. Mathews, and N. Sonenberg, eds.), pp. 411–450, CSHL Press, for further reading.

17.6.2. mRNA Masking and Unmasking during Embryonic Development

Early embryonic development and morphogenesis in *Drosophila* includes a number of interrelated events of mRNA masking and unmasking (for recent reviews, see Seydoux, 1995; Wickens, Kimble, and Strickland, 1996). One of the maternal masked mRNAs in *Drosophila* eggs is *nanos* mRNA encoding for Nos protein, a morphogen governing the abdominal segmentation in embryos by blocking expression of other mRNAs. After fertilization the *nanos* mRNA is localized at the posterior pole of the laid egg due to a special sequence element in its 3′-UTR which determines this anchoring. The sequences required for the posterior localization and for the masking overlap. While unlocalized *nanos* mRNA is really masked, that is, inactive and stable, the localization of the mRNA induces its unmasking. Thus, a novel trigger of unmasking has been revealed. Since the *nanos* mRNA is active in translation only locally, at the posterior pole of the egg, the Nos protein is synthesized just at the posterior pole and diffuses to the anterior pole thus forming a gradient along the egg and then along the anterior–posterior axis of the embryo.

The Nos protein is found to be an RNA-binding protein that recognizes similar sequences, the so-called NRE (for Nos response element), in the 3′-UTRs of two other maternal mRNAs, namely *hunchback* mRNA uniformly distributed in the egg, and *bicoid* mRNA localized at the anterior pole. The interaction of Nos with the NREs of these mRNAs blocks their translation. The consequence is that the expression of *hunchback* mRNA is gradually decreased from the anterior pole to the posterior, this being the main factor in the proper abdominal segmentation of the embryo. Normally the *bicoid* mRNA should not meet Nos because of the anterior localization of this mRNA, but in case of its delocalization it will be blocked by Nos. It is not clear, however, if the blocking effect of Nos is real masking of the mRNAs, or just a different type of translational inactivation without mRNA conservation. As to *hunchback* mRNA, its rapid degradation after translational inactivation has been reported.

17.6.3. mRNA Masking and Unmasking during Cell Differentiation

The best studied case of mRNA masking/unmasking during the final stages of cell differentiation, rather than at germ cell maturation and activation, or at early embryonic development, is the fate of the mRNA encoding for erythroid 15-lipoxygenase (LOX) (Ostareck-Lederer *et al.*, 1994). The LOX mRNA is synthesized at the early stages of erythropoiesis and becomes masked, or stored in the form of untranslatable and stable cytoplasmic mRNPs, for all the subsequent stages, until the late stage of peripheral reticulocytes. The unmasking of the LOX mRNA and the synthesis of the enzyme takes place during maturation of reticulocytes into erythrocytes. The enzyme attacks phospholipids, inducing the degradation of mitochondria during final erythrocyte maturation. The long 3′-UTR of the reticulocyte LOX mRNA contains a characteristic sequence where a pyrimidine-rich 19-nucleotide motif is repeated ten times. It is the 3′-UTR repeat region that was found to be responsible for the masking of the LOX mRNA through specific binding of a 48-kDa protein. The 48-kDa protein (LOX-BP) is a part of the translationally inactive (masked) LOX-mRNP in bone marrow cells and reticulocytes. The protein is also capable of selectively inhibiting translation of hybrid foreign mRNAs, such as chloramphenicol acetyltransferase or luciferase mRNAs, that contain the same 3′-UTR regulatory element. The minimal binding site for LOX-BP required for the effect is two to four repeats. It is noteworthy that the LOX-BP seems to act independently of the 5′-cap and 5′-UTR, so the involvement of a "cross-talk" between the two ends of mRNA (see below) is unlikely in this case.

17.6.4. Masking and Unmasking of mRNA in Differentiated Cells

There are also other examples of the involvement of the 3′-UTR in translational control (reviewed in Spirin, 1994, 1996). Creatine kinase B mRNA was shown to be regulated due to its 3′-UTR, and a specific RNAase-resistant, gel-retarded complex was demonstrated to be formed from this mRNA or its 3′-UTR and some components (presumably proteins) of the cell extract. This suggests that some portion of the 3′-UTR of creatine kinase B mRNA binds to a protein resulting in block of translation, and that the block can be relieved in response to stage-specific, tissue-specific, or hormonal regulatory signals.

Translation of human interferon β mRNA was also reported to be dependent on its 3′-UTR. The translation was inhibited in animal cell extracts but not in the wheat germ extract, suggesting the existence of a specific 3′-UTR-binding "inhibitor" in animal cells. The sequence responsible for the translational inhibition was found to be rich in uridines and adenosines and to contain several AUUUA repeats. It is remarkable that the sequence is effective in mRNA repression independent of its position within the 3′-UTR, but is no longer effective when inserted upstream from the AUG initiation codon, that is, in the 5′-UTR.

Similar AU-rich elements in the 3′-UTRs were reported to be recognition signals for selective rapid degradation of several other mRNAs, such as lymphokine, cytokine, and protooncogene mRNAs. Correspondingly, all these mRNAs also seem to be subject to the translational control under consideration. The translational control imparted by the presence of the AU-rich element in the 3′-UTR seems to be regulated (released) by inducing stimuli. For example, the tumor necrosis factor (TNF) mRNA or some artificial chimeric mRNAs can exist within macrophages in a translationally inactive or stored form, seemingly due to the presence of the AU-rich element in the 3′-UTR. Endotoxin (LPS) specifically induces their translation. Just as it is important in suppressing translation, the AU-rich element has been found to be critical for response to endotoxin. All this suggests the existence of cytoplasmic protein factors (proteins) that can recognize the AU-rich elements in the 3′-UTR, thus inducing mRNA masking, and also respond to a signal by releasing its masking activity.

Indeed, several groups have reported on a number of cytoplasmic RNA-binding proteins that specifically recognize the AU-rich and U-rich sequences in the 3′-UTRs of cytokine, lymphokine, and oncogene mRNAs. The proteins identified by different groups vary in their size from 15 to 70 kDa. It is not clear which of them are mRNA-destabilizing factors and which can participate in mRNA masking. A remarkable observation is that some of the proteins earlier known as constituents of heterogeneous nuclear ribonucleoproteins (hnRNPs), namely proteins A1 and C, are found in the cytoplasm to be associated with the AUUUA sequences in the 3′-UTRs of mRNAs. In agreement with this, the proteins were shown to shuttle between the nucleus and the cytoplasm. Both proteins, A1 and C, possess typical RNA-binding domains (RBD) of the "RNP consequence sequence" (RNP-CS) type, with two conserved RNP-1 and RNP-2 sequence motifs. About ninety amino acid residues of the protein C RBD are arranged into a two-layer βαββαβ structure where a four-stranded antiparallel β-sheet with the conserved RNP-1 and RNP-2 sequences is laid on two antiparallel packed α-helices.

There are experimental indications that some special sequences in 3′-UTR may be required for unmasking of mRNA. Thus, it was reported that the removal of the 3′-proximal 100-nucleotide sequence of the tissue plasminogen activator (t-PA) mRNA prevented both translational activation and destabilization of the masked mRNA, as well as its polyadenylation. The existence of some protein factors recognizing specific structures of the 3′-UTRs and thus inducing mRNA (mRNP) unmasking has been suspected. Such a *trans*-acting unmasking factor or activator relieving the masking effect of the TNF mRNA 3′-UTR has been detected in one of the human cell lines. The factor seems to abolish the effect of the 3′-UTR-binding masking factor which was discussed above and found to be more universal for different human and mammalian cell lines. It is not known if the unmasking factor directly or indirectly competes with the masking factor and displaces it from mRNA, or the unmasking effect overpowers the masking.

17.6.5. Models of mRNA Masking

In virtually all cases of mRNA masking, the effect of the 3′-UTR and the 3′-UTR-binding protein(s) is displayed as the block of the initiation step of translation. At the same time the initiation takes place at the 5′-part of mRNA. The question arises: How can the 3′-located events affect the 5′-located processes? One hypothesis could be that the 3′-part and the 5′-part of mRNA are in a protein-mediated contact with each other, providing some kind of a "cross-talk" or even a noncovalent "circularization" (see Section 15.3.7 and Fig. 15.14).

An alternative model can be based on the fact that during its lifetime mRNA is complexed with a large amount of protein and thus organized in mRNP structures. Global structural reorganizations of mRNP (e.g., similarly to condensation/decondensation of chromatin) could result in mRNA masking/unmasking. Interactions of the 3′-UTR with signal-regulatable proteins can be supposed to induce structural reorganizations of mRNP particles. Masked mRNA may be considered a condensed form of mRNP (Fig. 17.13) where RNA is not available for the functional interactions with other macromolecules, including ribosomes and/or translational initiation factors, poly(A) polymerase, and ribonucleases. This structural masking theory does not necessarily exclude the idea of the 3′-part being in proximity to the 5′-part of masked mRNP.

In addition to the key role of 3′-UTR-binding proteins in mRNA masking, the core mRNP protein p50 (see Section 17.2.2) has been mentioned as a principal participant of masking processes (Sommerville, 1992; Sommerville and Ladomery, 1996; Ranjan, Tafuri, and Wolffe, 1993; Tafuri, Familari, and Wolffe, 1993; Bouvet and Wolffe, 1994; Wolffe, 1994; Matsumoto, Meric, and Wolffe, 1996). Indeed, the extensive masking of mRNA during oogenesis and spermatogenesis is always accompanied by massive accumulation of p50 in the cell cytoplasm. Overproduction of p50 in the cell leads to enhanced masking of cytoplasmic mRNAs. Masked mRNAs of all types are found in association with large amounts of p50. It may be that, while a specific "masking protein" bound at the 3′-UTR in a single or few copies triggers the masking process, the sequence-nonspecific protein p50 loading the full sequence of mRNA completes the masking and forms a proper quaternary structure of the masked mRNP particle (Spirin, 1994). The proposed structural re-

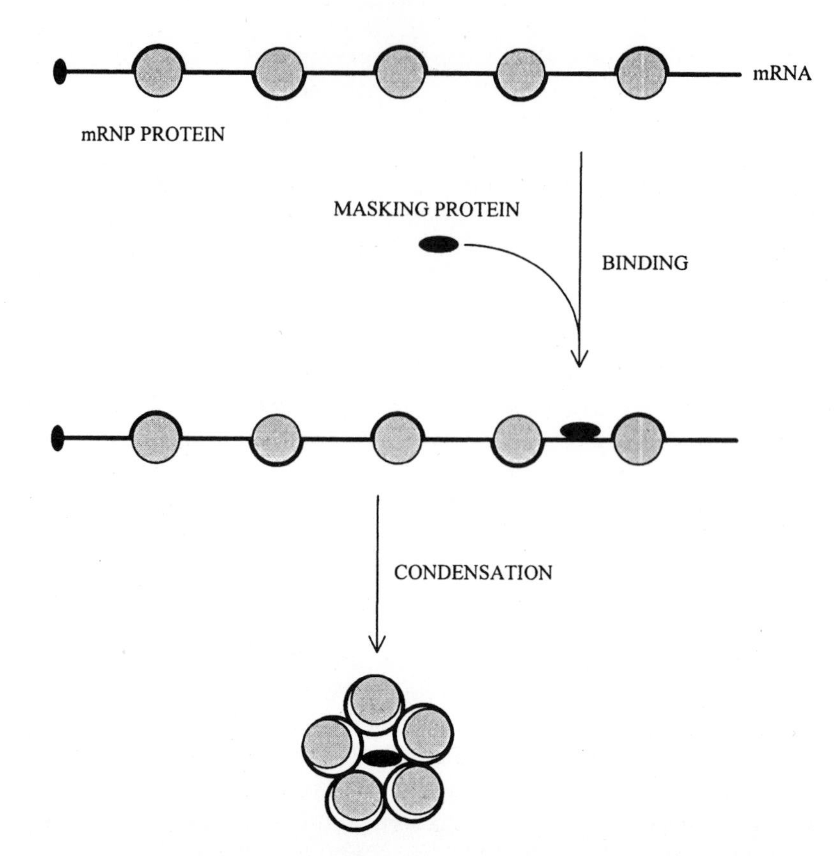

FIG. 17.13. Model for a possible mechanism of mRNA masking ("structural masking theory"). See Spirin, A. S. (1994) *Mol. Reprod. Develop.* **38**:107–117. The eukaryotic mRNA is complexed with a large amount of protein, thus forming mRNPs (or "informosomes"). See Spirin, A. S. (1996) *Curr. Topics Develop. Biol.* **1**:1–38; (1969) *Eur. J. Biochem.* **10**:20–35. Binding of a masking protein to a defined section ("masking box"; see Fig. 17.12) within the 3′-UTR may induce a structural rearrangement of the mRNP particle into a condensed form that may be inaccessible to initiation factors, ribosomes, ribonucleases, and poly(A) polymerase (masked mRNP). The sequence of events may vary in different situations; for example, first a masking protein is bound to a nascent mRNA chain and then the loading of a large number of core proteins result in a compact packing of the mRNP. Reproduced from Spirin, A. S. (1996) in *Translational Control* (J. W. B. Hershey, M. B. Mathews, and N. Sonenberg, eds.), pp. 319–334, CSHL Press.

organization of mRNP of the condensation type (Fig. 17.13) may be a function of p50. A cooperative interaction of multiple copies of this protein within mRNP may be induced by the specific 3′-UTR-bound "masking" protein.

References

Ahringer, J., and Kimble, J. (1991) Control of the sperm-oocyte switch in *Caenorhabditis elegans* hermaphrodites by the *fem-3* 3' untranslated regions, *Nature* **349**:346–348.

Black, T. L., Safer, B., Hovanessian, A., and Katze, M. G. (1989) The cellular 68,000-M_r protein kinase is highly autophosphorylated and activated yet significantly degraded during poliovirus infection: Implications for translational regulation, *J. Virol.* **63**:2244–2251.

Bouvet, P., and Wolffe, A. P. (1994) A role for transcription and FRGY2 in masking maternal mRNA within Xenopus oocytes, *Cell* **77**:931–941.

Braun, R. E. (1990) Temporal translational regulation of the protamine 1 gene during mouse spermatogenesis, *Enzyme* **44:**120–128.

Brendler, T., Godefroy-Colburn, T., Garhill, R. D., and Dhach, R. E. (1981a) The role of mRNA competition in regulating translation. II. Development of a quantitative *in vitro* assay, *J. Biol. Chem.* **256:**11747–11754.

Brendler, T., Godefroy-Colburn, T., Garhill, R. D., and Dhach, R. E. (1981b) The role of mRNA competition in regulating translation. II. Comparison of *in vitro* and *in vivo* results, *J. Biol. Chem.* **256:**11755–11761.

Celis, J. E., Madsen, P., and Ryazanov, A. G. (1990) Increased phosphorylation variants of elongation factor 2 during mitosis in transformed human amnion cells correlates with a decreased rate of protein synthesis, *Proc. Natl. Acad. Sci. USA* **87:**4231–4235.

Chen, J.-J. (1993) Translational regulation in reticulocytes: The role of heme-regulated eIF-2α kinase, in *Translational Regulation of Gene Expression 2* (J. Ilan, ed.), pp. 349–372, Plenum Press, New York, London.

Chu, E., Takimoto, C. H., Voeller, D., Grem, J. L., and Allegra, C. J. (1993a) Specific binding of human dihydrofolate reductase protein to dihydrofolate reductase messenger RNA in vitro, *Biochemistry* **32:**4756–4760.

Chu, E., Voeller, D., Koeller, D. M., Drake, J. C., Takimoto, C. H., Maley, G. F., Maley, F., and Allegra, C. J. (1993b) Identification of an RNA binding site for human thymidylate synthase, *Proc. Natl. Acad. Sci. USA* **90:**517–521.

Clarke, P. A., Schwemmle, M., Schickinger, J., Hilse, K., and Clemens, M. J. (1991) Binding of Epstein–Barr virus small RNA EBER-1 to the double-stranded RNA-activated protein kinase DAI, *Nucleic Acids Res.* **19:**243–248.

Clemens, M. J. (1996) Protein kinases that phosphorylate eIF2 and eIF2B, and their role in eukaryotic cell translational control, in *Translational Control* (J. W. B. Hershey, M. B. Mathews, and N. Sonenberg, eds.), pp. 139–172, Cold Spring Harbor Laboratory Press, New York.

Duncan, R. F. (1996) Translational control during heat shock, in *Translational Control* (J. W. B. Hershey, M. B. Mathews, and N. Sonenberg, eds.), pp. 271–293, Cold Spring Harbor Laboratory Press, New York.

Ehrenfeld, E. (1996) Initiation of translation by picornavirus RNS, in *Translational Control* (J. V. B. Hershey, M. B. Mathews, and N. Sonenberg, eds.), pp. 549–574, CSHL Press, New York.

Evdokimova, V. M., Wei, C.-L., Sitikov, A. S., Simonenko, P. N., Lazarev, O. A., Vasilenko, K. S., Ustinov, V. A., Hershey, J. W. B., and Ovchinnikov, L. P. (1995) The major protein of messenger ribonucleoprotein particles in somatic cells is a member of the Y-box binding transcription factor family, *J. Biol. Chem.* **270:**3186–3192.

Frederickson, R. M., and Sonenberg, N. (1993) eIF-4E phosphorylation and the regulation of protein synthesis, in *Translational Regulation of Gene Expression 2* (J. Ilan, ed.), pp. 143–162, Plenum Press, New York, London.

Geballe, A. P. (1996) Translational control mediated by upstream AUG codons, in *Translational Control* (J. W. B. Hershey, M. B. Mathews, and N. Sonenberg, eds.), pp. 173–197, Cold Spring Harbor Laboratory Press, New York.

Godefroy-Colburn, T., and Dhach, R. E. (1981) The role of mRNA competition in regulating translation. IV. Kinetic model, *J. Biol. Chem.* **256:**11762–11773.

Gupta, N. K., Datta, B., Ray, M. K., and Roy, A. L. (1993) Protein synthesis initiation in animal cells: Mechanism of ternary and Met-tRNA$_f$·40S·mRNA complex formation and the regulatory role of an eIF-2-associated 67-kDa polypeptide, in *Translational Regulation of Gene Expression 2* (J. Ilan, ed.), pp. 405–431, Plenum Press, New York, London.

Hecht, N. B. (1990) Regulation of "haploid expressed genes" in male germ cells, *J. Reprod. Fert.* **88:**679–693.

Hinnebusch, A. G. (1996) Translational control of *GCN4:* Gene-specific regulation by phosphorylation of eIF2, in *Translational Control* (J. W. B. Hershey, M. B. Mathews, and N. Sonenberg, eds.), pp. 199–244, Cold Spring Harbor Laboratory Press, New York.

Hinnebusch, A. G., and Klausner, R. D. (1991) Examples of eukaryotic translational control: GCN4 and ferritin, in *Translation in Eukaryotes* (H. Trachsel, ed.), pp. 243–272, CRC Press, Boca Raton, Ann Arbor, Boston, London.

Hinnebusch, A. G., Wek, R. C., Dever, T. E., Cigan, A. M., Feng, L., and Donahue, T. F. (1993) Regulation of *GCN4* expression in yeast: Gene-specific translational control by phosphorylation of eIF-2α, in *Translational Regulation of Gene Expression 2* (J. Ilan, ed.), pp. 87–115, Plenum Press, New York, London.

Hovanessian, A. G. (1993) Interferon-induced and double-stranded RNA-activated proteins as key enzymes regulating protein synthesis, in *Translational Regulation of Gene Expression 2* (J. Ilan, ed.), pp. 163–185, Plenum Press, New York, London.

Humbelin, M., and Thomas, G. (1991) Covalent modification of translational components, in *Translation in Eukaryotes* (H. Trachsel, ed.), pp. 231–242, CRC Press, Boca Raton, Ann Arbor, Boston, London.

Jackson, R. J. (1991) Binding of Met-tRNA, in *Translation in Eukaryotes* (H. Trachsel, ed.), pp. 193–230, CRC Press, Boca Raton, Ann Arbor, Boston, London.

Jacobs-Lorena, M., and Fried, H. M. (1987) Translational regulation of ribosomal protein gene expression in eukaryotes, in *Translational Regulation of Gene Expression* (J. Ilan, ed.), pp. 63–85, Plenum Press, New York, London.

Kaspar, R. L., Morris, D. R., and White, M. W. (1993) Control of ribosomal protein synthesis in eukaryotic cells, in *Translational Regulation of Gene Expression 2* (J. Ilan, ed.), pp. 335–348, Plenum Press, New York, London.

Katze, M. G. (1996) Translational control in cells infected with influenza virus and reovirus, in *Translational Control* (J. W. B. Hershey, M. B. Mathews, and N. Sonenberg, eds.), pp. 607–630, Cold Spring Harbor Laboratory Press, New York.

Kleene, K. C. (1996) Patterns of translational regulation in the mammalian testis, *Mol. Rep. Develop.* **43**:268–281.

Kramer, G., Kudlicki, W., and Hardesty, B. (1993) Regulation of reticulocyte eIF-2α kinases by phosphorylation, in *Translational Regulation of Gene Expression 2* (J. Ilan, ed.), pp. 373–390, Plenum Press, New York, London.

Lazaris–Karatzas, A., Montine, K. S., and Sonenberg, N. (1990) Malignant transformation by a eukaryotic initiation factor subunit that binds to mRNA 5′ cap, *Nature* **345**:544–547.

Lazarus, P., Parkin, N., and Sonenberg, N. (1988) Developmental regulation of translation by the 5′ noncoding region of murine c-*myc* mRNA in *Xenopus laevis, Oncogene* **3**:517–521.

Lodish, H. F. (1971) Alpha and beta globin messenger ribonucleic acid: Different amounts and rates of initiation of translation, *J. Biol. Chem.* **246**:7131–7138.

Lodish, H. F. (1976) Translational control of protein synthesis, *Annu. Rev. Biochem.* **45**:39–72.

Lodish, H. F., and Jacobsen, M. (1972) Regulation of hemoglobin synthesis: Equal rates of translation and termination of α- and β-globin chains, *J. Biol. Chem.* **247**:3622–3629.

Manzella, J. M., and Blackshear, P. J. (1992) Specific protein binding to a conserved region of the ornithine decarboxylase mRNA 5′-untranslated region, *J. Biol. Chem.* **267**:7077–7082.

Mathews, M. B. (1996) Interactions between viruses and the cellular machinery for protein synthesis, in *Translational Control* (J. W. B. Hershey, M. B. Mathews, and N. Sonenberg, eds.), pp. 505–548, Cold Spring Harbor Laboratory Press, New York.

Matsumoto, K., Meric, F., and Wolffe, A. P. (1996) Translational repression dependent on the interaction of the *Xenopus* Y-box protein FRGY2 with mRNA, *J. Biol. Chem.* **271**:22706–22712.

Meyuhas, O., Avni, D., and Shama, S. (1996) Translational control of ribosomal protein mRNAs in eukaryotes, in *Translational Control* (J. W. B. Hershey, M. B. Mathews, and N. Sonenberg, eds.), pp. 363–388, Cold Spring Harbor Laboratory Press, New York.

Minich, W. B., Maidebura, I. P., and Ovchinnikov, L. P. (1993) Purification and characterization of the major 50-kDa repressor protein from cytoplasmic mRNP of rabbit reticulocytes, *Eur. J. Biochem.* **212**:633–638.

Murray, M. T., Schiller, D. L., and Franke, W. W. (1992) Sequence analysis of cytoplasmic mRNA-binding proteins of *Xenopus* oocytes identifies a family of RNA-binding proteins, *Proc. Natl. Acad. Sci. USA* **89**:11–15.

Ostareck–Lederer, A., Ostareck, D. H., Standart, N., and Thiele, B. J. (1994) Translation of 15-lipoxygenase mRNA is inhibited by a protein that binds to a repeated sequence in the 3' untranslated region, *EMBO J.* **13**:1476–1481.

Parkin, N., Darveau, A., Nicholson, R., and Sonenberg, N. (1988) *cos*-Acting translational effects of the 5' noncoding region of c-*myc* mRNA, *Mol. Cell. Biol.* **8**:2875–2883.

Proud, C. G. (1992) Protein phosphorylation in translational control, *Curr. Top. Cell. Reg.* **32**:243–369.

Ranjan, M., Tafuri, S. R., and Wolffe, A. P. (1993) Masking mRNA from translation in somatic cells, *Genes Develop.* **7**:1725–1736.

Rhoads, R. E., Joshi–Barve, S., and Rinker–Schaffer, C. (1993) Mechanism of action and regulation of protein synthesis initiation factor 4E: Effects on mRNA discrimination, cellular growth rate, and oncogenesis, *Progr. Nucleic Acid Res. Mol. Biol.* **46**:183–219.

Rouault, T. A., Klausner, R. D., and Harford, J. B. (1996) Translational control of ferritin, in *Translational Control* (J. W. B. Hershey, M. B. Mathews, and N. Sonenberg, eds.), pp. 335–362, Cold Spring Harbor Laboratory Press, New York.

Ryazanov, A. G., and Spirin, A. S. (1993) Phosphorylation of elongation factor 2: A mechanism of shut off protein synthesis for reprogramming gene expression, in *Translational Regulation of Gene Expression 2* (J. Ilan, ed.), pp. 433–455, Plenum Press, New York, London.

Schafer, M., Nayernia, K., Engel, W., and Schafer, U. (1995) Translational control in spermatogenesis, *Develop. Biol.* **172**:344–352.

Schneider, R. J. (1996) Adenovirus and vaccinia virus translational control, in *Translational Control* (J. W. B. Hershey, M. B. Mathews, and N. Sonenberg, eds.), pp. 575–605, Cold Spring Harbor Laboratory Press, New York.

Seydoux, G. (1996) Mechanisms of translational control in early development, *Curr. Opinion Genet. Develop.* **6**:555–561.

Sommerville, J. (1992) RNA-binding proteins: masking proteins revealed, *BioEssays* **14**:337–339.

Sommerville, J., and Ladomery, M. (1996) Masking of mRNA by Y-box proteins, *FASEB J.* **10**:435–443.

Sonenberg, N. (1996) mRNA 5′ cap-binding protein eIF4E and control of cell growth, in *Translational Control* (J. W. B. Hershey, M. B. Mathews, and N. Sonenberg, eds.), pp. 245–269, Cold Spring Harbor Laboratory Press, New York.

Spirin, A. S. (1994) Storage of messenger RNA in eukaryotes: envelopment with protein, translational barrier at 5′ side, or conformational masking by 3′ side?, *Mol. Reprod. Develop.* **38**:107–117.

Spirin, A. S. (1996) Masked and translatable messenger ribonucleoproteins in higher eukaryotes, in *Translational Control* (J. W. B. Hershey, M. B. Mathews, and N. Sonenberg, eds.), pp. 319–334, Cold Spring Harbor Laboratory Press, New York.

Spirin, A. S., and Ryazanov, A. G. (1991) Regulation of elongation rate, in *Translation in Eukaryotes* (H. Trachsel, ed.), pp. 325–350, CRC Press, Boca Raton, Ann Arbor, Boston, London.

Standart, N., Dale, M., Stewart, E., and Hunt, T. (1990) Maternal mRNA from clam oocytes can be specifically unmasked in vitro by antisense RNA complementary to the 3′-untranslated region, *Genes Develop.* **4**:2157–2168.

Stripecke, R., and Hentze, M. W. (1992) Bacteriophage and spliceosomal proteins function as position-dependent *cis/trans* repressors of mRNA translation *in vitro, Nucleic Acids Res.* **20**:5555–5564.

Stripecke, R., Oliveira, C. C., McCarthy, J. E. G., and Hentze, M. W. (1994) Proteins binding to 5′ untranslated region sites: a general mechanism for translational regulation of mRNAs in human and yeast cells, *Mol. Cell. Biol.* **14**:5898–5909.

Tafuri, S. R., Familari, M., and Wolffe, A. P. (1993) A mouse Y box protein, MSY1, is associated with paternal mRNA in spermatocytes, *J. Biol. Chem.* **268**:12213–12220.

Theil, E. C. (1987) Storage and translation of ferritin messenger RNA, in *Translational Regulation of Gene Expression* (J. Ilan, ed.), pp. 141–163, Plenum Press, New York, London.

Theil, E. C. (1993) The IRE (iron regulatory element) family: structures which regulate mRNA translation or stability, *BioFactors* **4**:87–93.

Walden, W. E. (1993) Repressor-mediated translational control: The regulation of ferritin synthesis by iron, in *Translational Regulation of Gene Expression 2* (J. Ilan, ed.), pp. 321–334, Plenum Press, New York, London.

Walden, W. E., Godefroy–Colburn, T., and Thach, R. E. (1981) The role of mRNA competition in regulating translation. I. Demonstration of competition *in vivo, J. Biol. Chem.* **256**:11739–11746.

Walker, J., Dale, M., and Standart, N. (1996) Unmasking mRNA in clam oocytes: Role of phosphorylation of a 3′ UTR masking element-binding protein at fertilization, *Develop. Biol.* **173**:292–305.

Wickens, M., Kimble, J., and Strickland, S. (1996) Translational control of developmental decisions, in *Translational Control* (J. W. B. Hershey, M. B. Mathews, and N. Sonenberg, eds.), pp. 411–450, Cold Spring Harbor Laboratory Press, New York.

Wolffe, A. P. (1994) Structural and functional properties of the evolutionarily ancient Y-box family of nucleic acid binding proteins, *BioEssays* **16**:245–251.

18

Cotranslational Folding and Transmembrane Transport of Proteins

18.1. Contribution of Ribosomes to Protein Folding

The polypeptide chain on the ribosome elongates by consecutive growth from the N-terminus to the C-terminus. During growth, the C-terminus is covalently fixed in the ribosomal PTC, whereas the N-terminus is free. Obviously, the free N-terminal region of the nascent polypeptide chain must acquire some conformation. This implies that while the protein is being synthesized on the ribosome, it undergoes some folding that begins from its N-terminal part.

Factors governing the folding of the N-terminal part of the growing polypeptide chain, however, are different depending on the distance from the fixed carboxyl end. Three zones can be considered: PTC region; a space in the ribosome body ("intraribosomal tunnel") or on the ribosome surface (channel or groove for nascent peptide); and the space outside the ribosome, but in the immediate vicinity.

18.1.1. Starting Conformation Set by the Peptidyl Transferase Center

In the ribosomal PTC two amino acid residues, the donor and the acceptor, should be positioned in a certain standard orientation with respect to each other (see Section 11.3). The likely mutual orientation of the residues in the PTC corresponds to the α-helical conformation of the polypeptide backbone (Lim and Spirin, 1986). Hence, after transpeptidation, the conformation of the C-terminal dipeptide section is universal as determined by the stereospecificity of the PTC. This implies that, as further amino acid residues are added to the C-terminus, further folding occurs from a definite *starting conformation,* rather than from a random chain. Thus, the folding on the ribosome should begin as a rearrangement of the starting conformation of the C-terminal section of the nascent polypeptide (Spirin and Lim, 1986).

18.1.2. Intraribosomal Tunnel for Nascent Peptide: Does It Exist?

Translocation moves the aminoacyl residue preceding the C-terminal residue out of the ribosomal PTC. Subsequent additions of new aminoacyl residues at the

C-terminus move the N-terminal part of the nascent polypeptide chain further and further away from the PTC. According to the popular "ribosomal tunnel model," the polypeptide chain section, which has a length of about 30 to 40 residues beginning from the PTC (i.e., from the growing carboxyl end), is still screened by the ribosome in a putative intraribosomal tunnel and is not exposed to the medium. If this is the case, the most preferable conformation of the nascent polypeptide inside the ribosome seems to be the α-helix (Lim and Spirin, 1986): (1) it is the conformation set up in the PTC that will be stabilized along some distance by the fixation of the C-end; (2) it will be further stabilized within a limited space having a dimension comparable to the size of the statistical segment of the polypeptide chain; (3) it is rigid, which is required for pushing the nascent peptide through the ribosome; (4) the α-helix is the most saturated with hydrogen bonds and hence the least "sticky" structure of a polypeptide, which is also important for passage through the putative tunnel or channel; (5) the α-helical conformation can be universally adopted by any amino acid sequence.

The concept of an intraribosomal tunnel for the nascent peptide is based mainly on the fact that the C-proximal sequence of about 30 to 40 amino acid residues long is protected against proteinases in crude ribosome preparations (Malkin and Rich, 1967; Blobel and Sabatini, 1970; Smith, Tai, and Davis, 1978a,b). Visualization of structural canals and areas of low density inside the large ribosomal subunit has tempted some researchers to speculate about accommodation of the nascent peptide in an intraribosomal tunnel (see, e.g., Yonath, Leonard, and Wittman, 1987; Frank *et al.*, 1995). The length of the tunnel for the nascent peptide should be around 50 Å if the nascent peptide of the above-mentioned length were in α-helical conformation (or more than 100 Å for an extended conformation of the peptide).

It should be mentioned, however, that two other enzymes—bacterial deformylase and mammalian aminopeptidase—have been shown to attack the nascent polypeptide chain at the point removed by just 15 to 20 amino acid residues from the PTC (Yoshida, Watanabe, and Morris, 1970; Jackson and Hunter, 1970; Palmiter, Gagnon, and Walsh, 1978). Moreover, factor Xa protease cleaves the ribosome-associated nascent polypeptide at the 12th amino acid residue from the PTC (Wang, Sakai, and Wiedmann, 1995). These observations indicate that the putative intraribosomal tunnel for nascent peptide, if it exists, should be much shorter than that assumed from the results mentioned above. Another interpretation of the experimental data on the protection of the nascent peptide by ribosomes against enzymatic attack can be proposed: the nascent polypeptide goes from the PTC through a channel or groove on the ribosome surface, rather than through an intraribosomal tunnel. In such a case, a different point on the polypeptide pathway on the ribosome surface may be accessible for different enzymes.

In accordance with the above, immune-electron microscopy has revealed the exit of the hapten-labeled N-terminus of the nascent peptide on the ribosome surface immediately in the region of the PTC and nearby, that is, in the cleft between the central protuberance and the L1 side protuberance of the large ribosomal subunit (Ryabova *et al.*, 1988). These results argue against an intraribosomal tunnel for nascent peptide.

The discovery of a cytosolic "nascent peptide-associated complex," or NAC that is capable of binding to the C-proximal section of the ribosome-associated nascent peptide has provided an alternative interpretation of the results cited above: no intraribosomal tunnel for nascent peptide exists, and it is NAC that, by covering a nascent chain, gives the appearance of the existence of the putative ribosomal tunnel (Wang *et al.*, 1995). According to a recent model, the nascent polypeptide chain protrudes from the ribosome immediately at the PTC, but is seized and screened by special soluble proteins. This protein complex, called NAC in the eukaryotic cell, consists of two polypeptides, α (33 kDa), and β (21 kDa) (Wiedmann *et al.*, 1994). NAC associates with the peptide-carrying ribosome and makes the section of the growing polypeptide (up to 30 to 40 amino acid residues) inaccessible for interactions with surrounding macromolecules. No experimental information is available about the conformation of the nascent polypeptide in the complex with NAC. It may be that the α-helix is also a preferable conformation due to its generation by the PTC, its stabilization by the C-end fixation, and its universality. Thus, the discovery of the NAC which is capable of binding to the C-proximal section of the ribosome-associated nascent peptide has provided a principally different interpretation of the results mentioned above: it is NAC, but not the ribosome, that may cover a nascent chain section adjacent to the PTC. At the same time, due to the dynamic character of NAC association with ribosomes, the protection of the C-terminal section of the nascent peptide may depend on interactions with other macromolecules and on cotranslational folding of the peptide.

The recent studies of cotranslational protein folding has shown that the folding into tertiary structure can involve the C-proximal part of the nascent peptide, including the amino acid residues immediately adjacent to the PTC (see, e.g., Komar *et al.*, 1997). These findings also contradict the concept of the intraribosomal tunnel for nascent peptide and are consistent with the idea that the formation of a stable tertiary structure can outcompete the NAC binding.

18.1.3. Cotranslational Folding of Nascent Polypeptide

Beyond the section of 30 to 40 residues, or less, adjacent to the C-terminus (PTC), the N-terminal part of the nascent peptide becomes accessible to the environment. From this point, all the external factors responsible for polypeptide folding come into effect. However, several circumstances make this situation different from that observed in the case of the spontaneous renaturation of unfolded protein *in vitro*. First, if the ribosome creates and maintains a certain universal conformation of the nascent peptide, that is, the α-helix, then protein folding may begin not from an extended or a random coiled state of the chain, but from the given initial (starting) conformation. Second, the search for a folding pathway does not begin randomly from any region of the polypeptide chain but proceeds sequentially from the N-terminal part. Third, in the course of folding, the C-terminus is fixed at a bulky particle and, therefore, its mobility is limited; this should result in the intermediate conformations having a higher stability than the analogous structures of the free polypeptide chain. Fourth, the environmental conditions such as ionic composition, charges, and polarity in the vicinity of the ribosome surface may be very different from those in solution. Fifth, special protein factors, including some molecular chaperones (see below), may interact with the ribosome and thus modify the process of cotranslational folding.

According to several pieces of evidence, the correct folding of the polypeptide chain into the protein can occur during its synthesis on the ribosomes, that is, *cotranslationally* (reviewed by Kolb *et al.*, 1995). Thus, nascent peptides attached to ribosomes have been repeatedly reported to acquire activities characteristic of completed proteins with a formed tertiary structure. The synthesis of β-galactosi-

FIG. 18.1. Stereoscopic representation of a wire frame model of the incomplete globin chain (86 amino acid residues) with its C-terminus at the peptidyl transferase center (PTC) of the ribosome. The growing chain is assumed to acquire the same folding pattern as the corresponding N-terminal section of the complete globin; E and F designate the helices of the globin mainly responsible for heme binding. See Komar, A. A., Kommer, A., Krasheninnikov, I. A., and Spirin, A. S. (1997) *J. Biol. Chem.* **272:**10646–10651.

dase is a well-known example (Kiho and Rich, 1964; Hamlin and Zabin, 1972). In addition to the folding of the polypeptide chain into a corresponding tertiary structure, the assembly into a tetrameric quaternary structure is required for the enzymatic activity of this protein. It has been proved that the nascent chain, prior to its completion and when it is still attached to the ribosome, is capable of associating with the free subunits of this protein, and the ribosome-attached complex exhibits β-galactosidase activity (Zipser and Perrin, 1963). More recently a single-polypeptide enzyme, firefly luciferase, has been directly shown to fold correctly during its synthesis (Kolb, Makeyer, and Spirin, 1994; Frydman *et al.*, 1994) and acquire its native structure with enzymatic activity, while still covalently bound with its C-terminus to the ribosome (Makeyev, Kolb, and Spirin, 1996).

The cotranslational folding has also been demonstrated and studied in more detail in the case of the ribosomal synthesis of α-globin (Komar *et al.*, 1997). Here the process of folding was followed by the capacity of the nascent globin chain to bind specifically its ligand, hemin. It is known that hemin binding depends mainly on two properly arranged α-helices, E and F. The capacity of the ribosome-bound nascent polypeptide to bind hemin appears immediately upon the addition of the 12-amino acid sequence of the F-helix section (Fig. 18.1), suggesting both the formation of the F helix adjacent to the PTC and its correct arrangement relative to the preceding E-helix. In this case no indications have been obtained on the shielding of the C-terminal section of the nascent peptide by the ribosomal tunnel or by NAC. Thus the folding into tertiary structure can involve the C-proximal part of the nascent peptide, including the amino acid residues immediately adjacent to the PTC.

It is likely that, in contrast to renaturation of the free unfolded polypeptide *in vitro,* attaining the correct final conformation of protein on the ribosome proceeds more directionally and is therefore quicker and more reliable. In other words, the ribosome may contribute to a certain *folding pathway.* This contribution from the ribosome could include at least such factors as the determination of the folding sequence from the N-terminus to the C-terminus; the setting of a certain initial (starting) conformation, specifically φ and ψ angles, for each amino acid residue; and the stabilization of intermediate local conformations due to the fixation of the C-terminus.

18.2. Ribosome-Associated Molecular Chaperones

An important role in the cotranslational polypeptide folding may be played by cytosolic molecular chaperones (for reviews, see Ellis and van der Vies, 1991; Gething and Sambrook, 1992; Jaenicke, 1993; Hartl, 1996). There is evidence that in eukaryotes ribosome-bound nascent chains can be associated with two heat-shock proteins, HSP 70 and HSP 40, known to be chaperones that prevent polypeptide aggregation and maintain them in a folding-competent form. The ribosome-bound nascent chains are also associated with the large hetero-oligomeric ring complex (TRiC) ensuring ultimate folding of released polypeptide chains (Beckmann, Mizzen, and Welch, 1990; Nelson *et al.*, 1992; Hendrick *et al.*, 1993; Langer *et al.*, 1992; Frydman *et al.*, 1994; Hartl, 1996). Corresponding analogs in *E. coli* cells are the DnaK/DnaJ pair and the GroEL oligomer, respectively. At least in some cases the removal of any constituent of this entire chaperone/chaperonin machinery results in misfolded polypeptide chains lacking natural physiological activity. It has been shown that a growing polypeptide chain binds initially with HSP 70 in cooperation with HSP 40. The binding occurs as soon as a section of about 50 amino acid residues emerges beyond the NAC binding site of the nascent chain. After an additional 100–150 steps of elongation the polypeptide chain can be further captured by the TRiC oligomer. As a result, a transient complex including all the above proteins (i.e., the ribosome-bound nascent chain, HSP 70, HSP 40, and TRiC) can be formed (Fig. 18.2). After termination of translation, the polypeptide chain released from the TRiC complex is assumed to be ultimately folded. This model admits that nascent chains can undergo folding domain-wise (Netzer and Hartl, 1997). A minimal polypeptide domain folded cotranslationally may include

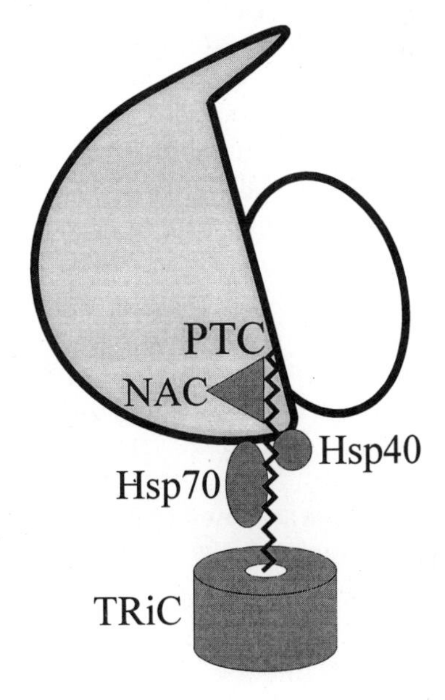

FIG. 18.2. Model for complex formation of the eukaryotic translating ribosome with NAC (See Wiedmann, B., Sakai, H., Davis, T. A., and Wiedmann, M. (1994) *Nature* **370:**434–440), Hsp40/Hsp70, and TRiC during the cotranslational folding of cytosolic proteins (See Hartl, F. U. (1996) *Nature* **381:**571–580). Polypeptides consisting of more than one domain may be folded by stages (See Netzer, W. J., and Hartl, F. U. (1997) *Nature* **388:**343–349).

about 100 amino acid residues or somewhat less. The interaction of nascent polypeptide chains with HSP 70/HSP 40 (DnaK/DnaJ) and TRiC (GroEL) is an ATP-dependent process involving some additional regulatory proteins. This matter, however, is beyond the scope of this book.

Thus, a nascent polypeptide chain emerges from the ribosome and enters into a putative "tunnel" formed by NAC very soon after initiation of its growth (after 12 initial steps of elongation or less) (Wiedmann *et al.*, 1994; Wang *et al.*, 1995). The polypeptide chain can be further captured by the chaperone/chaperonin machinery (Hartl, 1996). However, this model holds only for the proteins synthesized in the cytosol on so-called *free polyribosomes,* that is, polyribosomes that are not associated with any intracellular membrane.

18.3. Synthesis of Proteins by Free and Membrane-Bound Polyribosomes

In both prokaryotic and eukaryotic cells, a significant portion of the ribosomes organized into polyribosomes can be attached to membranes. In prokaryotes, the polyribosomes may reside on the inner surface of the cell plasma membrane, whereas in eukaryotes the membrane-bound ribosomes are located on the rough endoplasmic reticulum (RER) of the cytoplasm (see Figs. 4.2 and 4.5). It was demonstrated long ago that eukaryotic ribosomes attach to the membrane through their large (60S) subunit (Sabatini, Tashiro, and Palade, 1966). The 60S subunit appears to have a special site with affinity for the membrane of the endoplasmic reticulum. Therefore, all ribosomes attach to the membrane through a strictly fixed point in the same orientation. In this orientation the axis connecting the large and the small subunits is roughly parallel to the membrane surface (Fig. 18.3). Electron microscopy of eukaryotic ribosomes demonstrates that the long axis of the small subunit is roughly parallel to the membrane surface (Unwin, 1979; Christensen, 1994). This leads to the assumption that the attachment of the ribosome to the membrane takes place at the side of the lateral protuberances of the subunits (equivalent to the L1 ridge of the 50S subunit and the platform of the 30S subunit of *E. coli* ribosomes; see Section 5.3); the region of the presumed "pocket" for tRNA and the stalk of the large subunit should be at the side turned away from the membrane. In such a case, according to the presumed trajectory of mRNA movement through the translating ribosome (Fig. 9.5), the mRNA chain should go more or less parallel to the membrane surface. Schematic representation of a membrane-bound polyribosome is given in Fig. 18.4.

Apparently, the nascent peptide emerges from the ribosome somewhere at the side contacting the membrane, and the attached ribosomes donate the nascent polypeptides directly to the membrane. Correspondingly, depending on the localization, the primary cotranslational folding of the nascent polypeptide proceeds either in the aqueous medium of the cytoplasm in the case of free polyribosomes, as described above (Section 18.2), or in the hydrophobic environment of the membrane lipid bilayer for membrane-bound polyribosomes.

As was noted years ago, free polyribosomes synthesize primarily water-soluble proteins for housekeeping use in the cytoplasm, whereas the membrane-bound particles synthesize either the proteins for incorporation into membranes or the secretory proteins that are transported out of the cell through the membranes (Siekevitz and Palade, 1960; Redman, Siekevitz, and Palade, 1966; Redman, 1969; Ganoza

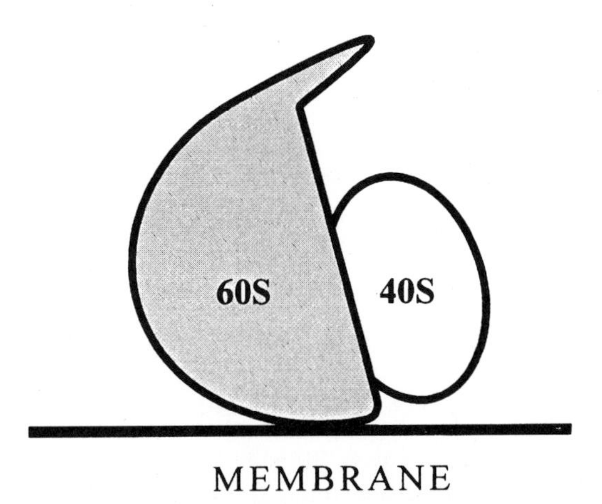

FIG. 18.3. Scheme illustrating the orientation of the subunits of the membrane-bound 80S ribosome relative to the surface of the endoplasmic reticulum membrane. See Unwin, P. N. T. (1979) *J. Mol. Biol.* **132**:69–84; Christensen, A. K. (1994) *Cell Tissue Res.* **276**:439–444.

and Williams, 1969; Morrison and Lodish, 1975). The soluble cytoplasmic proteins synthesized on free polyribosomes are folded in the aqueous medium as they emerge from the ribosomes. As a consequence, they form a typically globular structure, with a more or less polar surface and a hydrophobic core. In contrast, protein synthesis on membrane-bound ribosomes causes the growing polypeptide to come into contact with the hydrophobic milieu of the membrane lipid bilayer. In the case of proteins destined to become components of a given membrane (the endoplasmic reticulum membrane of eukaryotes or the plasma membrane of bacteria), the hydrophobic environment dictates the mode of their folding, with numerous hydrophobic residues being exposed outside the molecule. The transmembrane hydrophobic sequences of such proteins often exist in the α-helical conformation (consider, for example, the case of bacteriorhodopsin, Henderson and Unwin, 1975; see also Fig. 13.3 for the membrane-bound chloroplast reaction center protein D1).

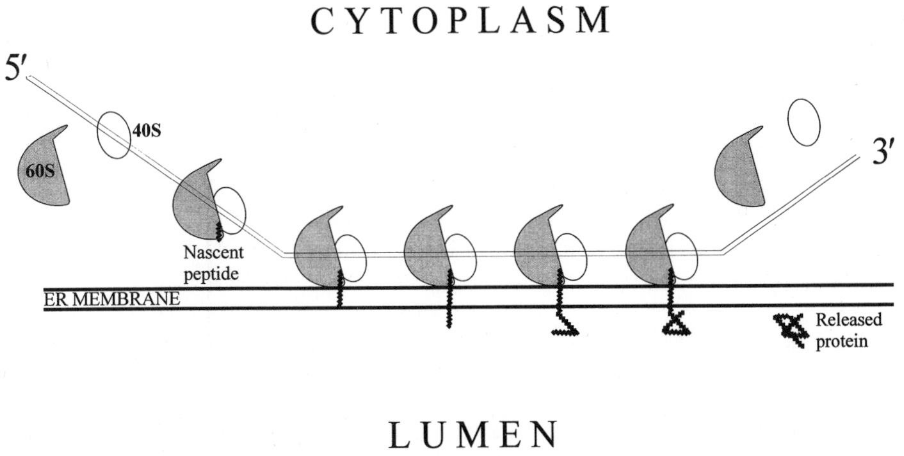

FIG. 18.4. Schematic representation of a membrane-bound polyribosome. In fact, the eukaryotic membrane-bound polyribosomes often have a circular shape, as seen in Fig. 4.5, with the small subunits oriented toward the inside of the polyribosomal curve; see Christensen, A. K. (1994) *Cell Tissue Res.* **276**:439–444.

With proteins transported through the membrane, however, the picture appears to be more complex. The nascent chain passing through the membrane is finally folded in the aqueous milieu of the endoplasmic reticulum lumen in eukaryotes, or in the periplasmic space of Gram-negative bacteria, or in the external medium for other bacteria. The transmembrane translocation of such polypeptides is accompanied by their multistage folding coupled with cotranslational processing and covalent modifications.

18.4. Interaction of Translating Ribosomes with Membranes

18.4.1. Early Observations

The idea that protein synthesis on membrane-bound ribosomes is coupled with transmembrane protein translocation emerged from observations on the intimate association of nascent polypeptide chains with membranes of the rough endoplasmic reticulum in eukaryotic cells (Sabatini and Blobel, 1970) and with the plasma membrane in bacteria (Smith *et al.*, 1978a). The translating ribosomes were found to be anchored firmly on the membrane by the growing peptide. Only puromycin treatment, which resulted in the abortion of the peptide from the ribosomes, allowed the complex to dissociate into free ribosomes and membranes, leaving the peptide in the membrane. Thus, it became clear that the growing peptide significantly contributes to the association between the translating ribosome and the membrane. A rupture of this anchor by puromycin in the bacteria results in the immediate release of the ribosomes from the membranes.

In contrast, with eukaryotic cells, after the peptide anchor is broken, the ribosomes still show a marked affinity for endoplasmic reticulum membranes. The complete dissociation of the ribosomes from endoplasmic reticulum membranes *in vitro* may be achieved only by combining the treatment of microsomes with puromycin and high ionic strength solutions (Adelman, Sabatini, and Blobel, 1973; Harrison, Brownlee, and Milstein, 1974). It also can be demonstrated that nontranslating ribosomes, or ribosomes that have just started translation and contain only a short peptide, have a certain affinity for endoplasmic reticulum membranes. From all this, it was assumed that the membranes of the rough endoplasmic reticulum contain specialized *receptors* that are responsible for the reversible association of ribosomes with a membrane; these receptors were also thought to help the membrane accept the ribosome-bound nascent peptides and thus form membrane "pores" (or intramembrane tunnels) for the growing polypeptide chains.

By now, numerous experimental facts concerning interactions of translating ribosomes and ribosome-bound nascent polypeptides with membranes and their structural elements have been accumulated, and well-substantiated models for ribosome-membrane and nascent-peptide-membrane recognition, as well as cotranslational transmembrane translocation of nascent polypeptide have been proposed (reviewed in Harwood, 1980; Inouye and Halegoua, 1980; von Heijne, 1988; High and Stirling, 1993; Walter and Johnson, 1994; Rapoport, Jungnickel, and Kitay, 1996; Martoglio and Dobberstein, 1996; Corsi and Schekman, 1996; Johnson, 1997).

18.4.2. Initial Nascent Peptide Interactions

As mentioned above, in eukaryotic cells the NAC seems to be the first among cytoplasmic protein factors that interacts with the ribosome-bound nascent chain emerging from the ribosomal PTC (Fig. 18.5A). It has been demonstrated that the

NAC covers both the C-terminal section of the nascent polypeptide and a ribosomal site characterized by high affinity for a special receptor on the endoplasmic reticulum membrane (Lauring *et al.,* 1995; Lauring, Kreibich, and Wiedmann, 1995). As a consequence the ribosome bearing the nascent chain fails to bind to the membrane until the NAC leaves it. *In vitro* the NAC removal can be achieved by treatment of the ribosome/nascent chain complex with concentrated salt solutions, that is under nonphysiological conditions. Such treatment disturbs the separation of the nascent polypeptide chains into those destined to be ultimately located in the cytosol and those normally directed into the exocytic pathway. In fact, in the absence of NAC the ribosomes synthesizing polypeptide chains of both types bind to the endoplasmic reticulum membrane thus allowing the resulting polypeptides to be translocated across the membrane albeit with efficiency radically depending on the presence of specific amino acid sequences.

In a living eukaryotic cell the NAC is expelled from the ribosome by an 11S ribonucleoprotein complex termed *signal recognition particle,* or SRP (see Section 18.4.4) (Fig. 18.5B) that binds selectively with the nascent chains encompassing the so-called *signal sequence* (see Section 18.4.3). As a result the ribosome becomes capable of binding with the membrane and putting the nascent polypeptide into the translocation channel (Fig. 18.5C). On the other hand, the polyribosomes synthesizing polypeptide chains lacking the signal sequence have no chance to be attached to the membrane, and resulting polypeptides become resident cytosolic proteins. Thus, the binding of the nascent chain-bearing ribosome to the endoplasmic reticulum surface predetermines translocation of polypeptides across the membrane, this process being initially controlled by the NAC and SRP relationships (Lauring, Kreibich, and Wiedmann, 1995; Powers and Walter, 1996).

18.4.3. Signal Amino Acid Sequence

In 1971, Blobel and Sabatini proposed that mRNAs, which are to be translated by membrane-bound ribosomes, have a special sequence immediately follow-

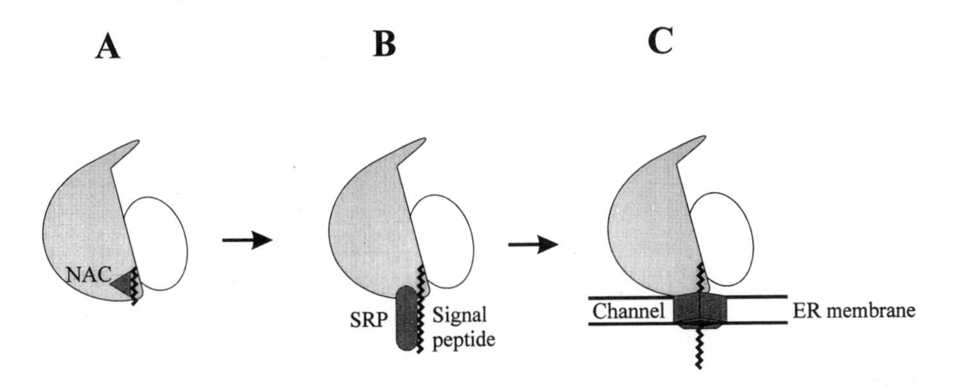

FIG. 18.5. Model for the sequence of events and the role of NAC and SRP during cotranslational targeting of the ribosome-bound nascent polypeptide chain. **(A)** NAC binds the nascent polypeptide chain immediately upon its exit from the PTC. In the absence of a signal sequence, NAC prevents targeting of the nascent chain by blocking the membrane-binding site on the ribosome. **(B)** In the presence of the signal sequence, SRP binds tightly to the ribosome and effectively competes with NAC, thus displacing it from the nascent polypeptide chain. **(C)** When SRP is present, the translating ribosome gains the capacity for binding with endoplasmic reticulum membrane and putting the nascent chain in the transmembrane translocation channel. See Powers, T., and Walter, P. (1996) *Curr. Biol.* **6**:331–338.

ing the initiation codon. This sequence codes for the characteristic N-terminal sequence of the nascent polypeptide, which provides a signal for recognition either by the membrane or by a factor mediating the association between the ribosome and the membrane. Thus, if a ribosome translates mRNA coding for the secretory or membrane protein, some time after initiation the N-terminal part of the growing peptide will protrude from the ribosome and signal the attachment to the membrane. In this case the ribosome becomes membrane bound. If the N-terminus of the growing peptide does not carry such a signal sequence, as is the case with water-soluble cytoplasmic proteins, the corresponding ribosomes (polyribosomes) will remain free throughout elongation (see also Blobel, 1980).

Independent experimental evidence for the existence of special N-terminal sequences in newly synthesized polypeptide chains of certain secretory proteins of eukaryotic cells appeared a short time after. The translation of mRNA coding for the light chain of immunoglobulin in the cell-free system on free ribosomes yielded a polypeptide which had an additional sequence at the N-terminus compared to the authentic light chain of the immunoglobulin; translation of the same mRNA in the presence of microsomes yielded a normal product (Milstein *et al.,* 1972). It was concluded that translation of the first portion of mRNA resulted in the synthesis of the N-terminal signal sequence which determined the attachment of the ribosome to the membrane; in the course of further elongation this sequence was cleaved off, and the final product did not contain it. More refined work using purified components of the same *in vitro* system (Blobel and Dobberstein, 1975) laid the groundwork for the elucidation of the matter. It was also demonstrated that such an N-terminal sequence was rich in hydrophobic aminoacyl residues (Schechter *et al.,* 1975); therefore, it was natural to conclude that the hydrophobic nature of the signal sequence of the nascent polypeptide chain contributed to the interaction of the polypeptide with the membrane.

Similar additional N-terminal sequences are also characteristic of the nascent chains of a number of bacterial proteins exported from the cytoplasm (Inouye and Beckwith, 1977). In Gram-negative bacteria the export of proteins takes place either into the periplasmic space (as for alkaline phosphatase, maltose-binding protein, arabinose-binding protein, and penicillinase) or into the outer membrane (as for outer membrane lipoprotein and λ-receptor). The beginning of the synthesis of proteins destined for exportation from the cell seems to be followed by their hydrophobic N-terminal sequences interacting with the inner cytoplasmic membrane of the bacterial cell, so that the subsequent steps of their synthesis proceed on the membrane-bound ribosomes. Cleavage of the N-terminal sequences may take place during or, in some cases, following elongation. After synthesis has been completed and after the termination of translation, the protein passes into the periplasmic space and then, depending on the hydrophobicity or hydrophilicity of its surface either remains in the periplasmic space as a water-soluble protein or is integrated into the outer membrane. All this is very similar to the situation with secreted proteins in eukaryotic cells.

18.4.3.1. Cleavable Signal Sequence

Thus, numerous secretory proteins of eukaryotes have been reported to be synthesized with an additional N-terminal sequence that is later cleaved off the membrane during protein synthesis. If the corresponding mRNAs are translated in the cell-free system in the absence of membranes, then elongated products referred to s *pre-proteins* are formed. Cleavage of the N-terminal signal sequence usually loes not take place when the membranes (the microsomal fraction) are added to he system after synthesis of the pre-protein is completed or even at the late stages f elongation. Only if membranes are present in the system from the beginning, or re added soon after the emergence of the N-terminal section of the polypeptide rom the ribosome, does the ribosome attach to the membrane, the nascent peptide nter the membrane, and cleaving of the signal sequence take place. Elongation of he peptide above a certain size in an aqueous medium seems to result in peptide olding, which screens the hydrophobic N-terminal region and thus prevents its nteraction with the membrane.

The signal sequences of this type usually include 15 to 30 amino acid residues t the amino terminus of a nascent polypeptide chain (von Heijne, 1988). Three re- ions with different properties can be distinguished in the following order: a short *n* (amino-terminal) region, positively charged in most cases; a hydrophobic *h* re- ion in the middle of the signal sequence; and a conservative *c* region preceding he cleavage site in the nascent chain. The *h* region, composed mainly of Leu, Ala, le, Phe, and Trp, is 10±3 residues long. In this respect it differs from membrane- panning sequences (24±2 residues) and from hydrophobic segments of globular roteins (6–8) residues. The major feature of the *h* region is its overall hydropho- icity; there is also evidence for its helicity. The composition of the *c* region (5 to residues) obeys the "−1, −3 rule," according to which the −1 position at the leavage site has to be occupied by Ala, Ser, Gly, Cys, Thr, or Gln only, and the −3 osition must be free of aromatic (Phe, His, Tyr, Trp), charged (Asp, Glu, Arg, Lys), r large polar (Asn, Gln) residues. Moreover, no Pro exists in the −3 to +1 region. Obviously, these severe restrictions in the amino acid composition are compatible vith the fact that signal sequences of a variety of polypeptide chains involved in he cotranslational import into the endoplasmic reticulum are cleaved by a com- non signal peptidase.

Though signal sequences are not homologous to each other and variable with espect to their size, they have a common basic feature, an ability to guide even a oreign passenger.

18.4.3.2. Uncleavable Signal Sequence

Some integral membrane proteins (type II proteins, see also Fig. 18.11A) en- ompass an internal signal sequence located at some distance (30–150 residues) rom the amino terminus. Such a sequence usually ranges from 20 to 30 hy- lrophobic and nonpolar residues (23±3 residues most often) preceded by a clus- er of positively charged residues at its amino-terminal end. The proteins with uch an internal signal are cotranslationally imported into the endoplasmic retic- ılum using SRP, SRP receptor, and the same translocation apparatus as the pro- eins with the cleavable signal sequence. In this particular case, proteolytic pro- essing of a nascent chain does not occur because of the absence of the cleavage ite matching the "−1, −3 rule," and the completed polypeptide chain does not eave the membrane. Another possible cause for long hydrophobic signal se- quences to be uncleavable is the remoteness of their C-termini from the signal pep- idase located in the membrane.

A general scheme of the process of the attachment of the translating ribosome/ nascent polypeptide to the ER membrane is depicted in Fig. 18.5. If the nascent polypeptide contains the signal sequence, the ribosome with the nascent chain interacts in the cytosol with SRP (Fig. 18.5B). The resulting complex, in which elongation of the nascent chain is arrested by SRP (see Section 13.2.4), migrates in the cytoplasm until it finds a specific receptor (SRP receptor) on the endoplasmic reticulum surface. This interaction results in the release of the ribosome with nascent chain from SRP and in the insertion of the nascent chain into a translocation channel in the membrane (Fig. 18.5C). From this point on, the nascent chain continues to grow into the lumen of the endoplasmic reticulum. Then the signal sequence either is cleaved by a specific endopeptidase or becomes inserted into the endoplasmic reticulum membrane. In the former case the completed chain is folded, in parallel with its covalent modifications (see Section 18.6), and continues to travel to the cell surface along the exocytic pathway as a soluble protein. The latter case is specific to membrane proteins located in the endoplasmic reticulum, the Golgi apparatus, the plasma membrane, and so on.

18.4.4. Interaction of the Ribosome/Nascent Chain Complex with the Signal Recognition Particle

Shortly after emerging from the ribosome (i.e., after 35–40 initial steps of elongation), all nascent polypeptide chains, whether they bear the signal sequence or not, interact with the NAC involved in their folding. However, in the nascent chains with a signal sequence, NAC is replaced by the SRP after 60–70 steps of elongation (Powers and Walter, 1996). Both the signal sequence and the ribosome are involved in SRP binding, which results in the arrest of elongation (see Section 13.2.4).

Both structural and functional aspects of SRPs have been under intensive investigation during recent years (see reviews by Siegel and Walter, 1988; Walter and Johnson, 1994; Luirink and Dobberstein, 1994; Luetcke, 1995). In higher eukaryotes SRP is a rod-shaped 11S ribonucleoprotein particle, 5–6 nm wide and 23–24 nm long. It is composed of 7S RNA and six protein subunits of 9, 14, 19, 54, 68, and 72 kDa. Four of them are grouped in heterodimers (SRP9/14 and SRP68/72), while two others exist as monomers located together in the same region of the SRP (Fig. 18.6).

The most extensively studied are the structure and the functions of the 54-kDa subunit (SRP54). First, the SRP54 serves to bind the nascent chain through its signal sequence. This subunit has two functional domains. One of them (SRP54M), interacting directly with the signal sequence, includes three amphiphatic helices enriched in methionine residues. It is suggested that the helices form a binding groove on the surface of the SRP54M, while the methionine residues, which are unique among other hydrophobic amino acids in the higher flexibility of their side chains, ensure the interaction with a multitude of the hydrophobic signal sequences of quite different composition and structure, conforming nevertheless to the steric restrictions imposed by the groove. This model is compatible with the fact that the hydrophobic *h* region of the signal sequence is crucial for its binding with SRP. The other domain of the same SRP subunit (SRP54G) has a GDP/GTP binding site seemingly involved in the regulation of SRP binding with a special receptor in the endoplasmic reticulum membrane. The SRP54 subunit is attached to the 7S RNA molecule through its M domain which has a specific RNA-binding amino acid motif. Figure 18.7 shows the schematic two-domain model for SRP54 describing the arrangement of its M-domain and G-domain as well as the interaction of the whole subunit with the RNA moiety and the signal sequence of the nascent chain.

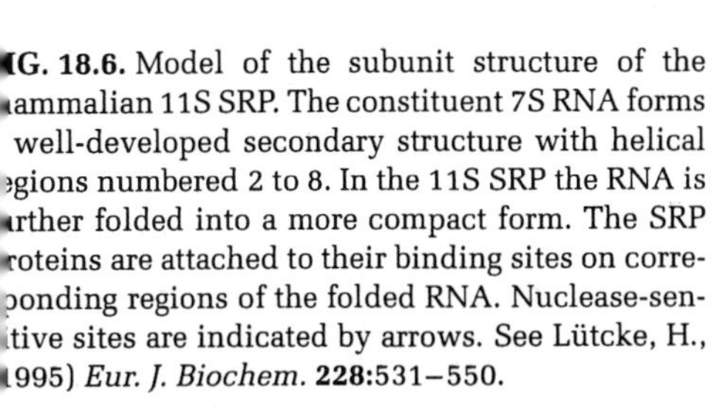

IG. 18.6. Model of the subunit structure of the ammalian 11S SRP. The constituent 7S RNA forms well-developed secondary structure with helical egions numbered 2 to 8. In the 11S SRP the RNA is rther folded into a more compact form. The SRP roteins are attached to their binding sites on corre- ponding regions of the folded RNA. Nuclease-sen- itive sites are indicated by arrows. See Lütcke, H., 1995) *Eur. J. Biochem.* **228**:531–550.

As to the functions of other SRP subunits, the 9/14 dimer located in the SRP t the other end of the elongated particle, opposite to the SRP54 subunit (Fig. 18.6), eems to be responsible for the arrest of elongation. This may be realized through lirect interaction of the 9/14 dimer with the ribosome concurrent with the bind- ng of the signal sequence by the SRP54 subunit. Such two-point binding with the ibosome/nascent chain complex could hinder structural rearrangements normal- y occurring in the ribosome during translation. A putative physiological signifi- ance of this particular function of the SRP is that the nascent chains are prevent- d from completion and ultimate folding prior to their translocation across the ndoplasmic reticulum membrane. Although posttranslational translocation is al- owed in the lower eukaryotic cells that have a specific translocation apparatus for his purpose, it seems to be rather unusual in higher eukaryotic cells.

The 68/72 dimer, occupying the central part of the SRP, somehow mediates nsertion of the nascent chain into the translocation apparatus. The SRPs contain- ng damaged SRP68 and SRP72 subunits still bind the ribosome/nascent chain omplex and arrest elongation, but they fail to provide transmembrane transloca- ion of the nascent chains. This is considered to be a result of the inability of such modified SRP to interact with its receptor on the endoplasmic reticulum mem- rane. In other words, the 68/72 dimer is obligatory for translocation of the nascent hain across the membrane at the stage of the docking of the ribosome/nascent hain/SRP complex to the membrane. Moreover, as SRP devoid of SRP68 and

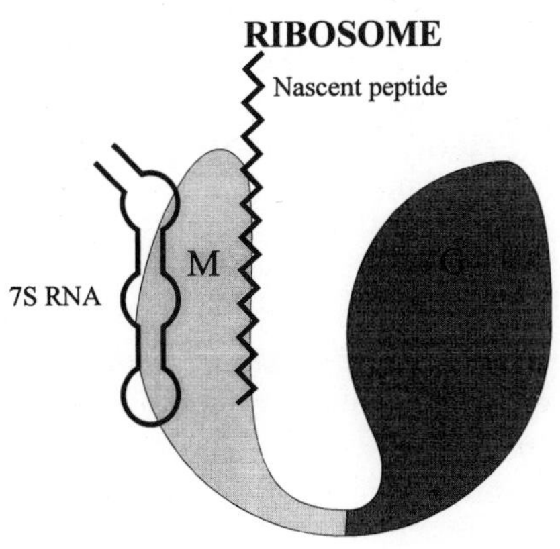

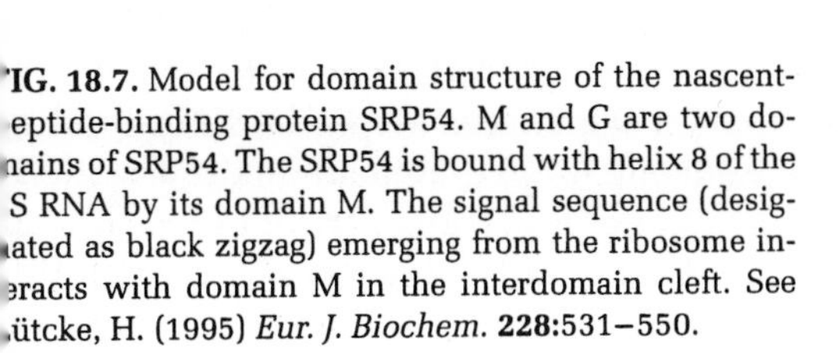

IG. 18.7. Model for domain structure of the nascent- eptide-binding protein SRP54. M and G are two do- nains of SRP54. The SRP54 is bound with helix 8 of the S RNA by its domain M. The signal sequence (desig- ated as black zigzag) emerging from the ribosome in- eracts with domain M in the interdomain cleft. See ütcke, H. (1995) *Eur. J. Biochem.* **228**:531–550.

SRP72 cannot arrest elongation, this bulky dimer seems to be crucial for main taining the correct conformation of the SRP as a whole.

The existence of the three regions in the SRP responsible for its three func tions (signal sequence reception, translation arrest, and transmembrane transloca tion) corresponds nicely to its shape observed in the electron microphotograph

Yeast cells contain very similar SRPs in terms of their size and constituent pro tein subunits. In contrast, the bacterial analog of the SRP is much smaller and in cludes only 4.5S RNA and a SRP54 homolog, the so-called Ffh. Compared with th eukaryotic 7S RNA, the Alu-like region is absent from the bacterial 4.5S RNA. Con sequently, the Ffh/4.5S RNA complex is able to bind nascent chains with signa sequences, but it cannot arrest elongation, seemingly because of the absence of cor responding protein components bound to the Alu-like domain (see Fig. 18.6). Th Ffh/4.5S RNA complex serves to maintain a nascent chain in a translocation-com petent form and to target it to the plasma membrane. Membrane protein FtsY is putative receptor of the Ffh/4.5S RNA complex that is involved in cotranslation al transmembrane translocation of some bacterial proteins.

18.4.5. Interaction of the Ribosome/Nascent Chain/SRP Complex with SRP Receptor on the Membrane

The mammalian ribosome/nascent chain/SRP complex is targeted to a uniqu receptor located exclusively in the endoplasmic reticulum membrane. The SRP re ceptor (SR) is composed of two subunits. The α-subunit is a 69-kDa protein, th bulk of which (a 52-kDa fragment) faces the cytoplasm, while the amino-termina part spans the membrane. The cytoplasmic region of SRα, interacting with SRP has a site for GTP binding. The β-subunit (SRβ) is a 30-kDa integral membrane pro tein and shows some affinity for SRP and an ability for GTP binding, but its func tion in the reception of SRP is still vague. Both subunits are counterparts provid ing a reliable SR anchoring in the membrane.

Two parts of SRP—SRP54 and SRP68/72—are involved in the interactio with the receptor. There is evidence that the SRP54 binds directly with SR throug its G-domain (SRP54G). The role of SRP68/72 consists rather in imparting a cor rect conformation to SRP: defects in this region result in the impaired insertion o a nascent chain into the endoplasmic reticulum membrane.

The relationships between the ribosome/nascent chain complex, SRP, an SR are regulated by GTP (Bacher *et al.*, 1996). As the scheme in Fig. 18.8 shows free SRP contains GDP. Initially, the attachment of the ribosome/nascent chai complex to SRP promotes binding of GTP to the 54-kDa subunit (SRP54G), wit a ribosomal component involved in the process. In the resulting triple complex GTP is not hydrolyzed until the complex binds with SRα, also loaded with GTP The interaction of SRP (bearing the ribosome/nascent chain complex) with SRα causes hydrolysis of GTP. Since this process occurs in the SRP54 subunit asso ciated both with the signal sequence through the M-domain and with SRα through the G-domain, it results in dissociation of the ribosome/nascent chai complex from SRP and SRP from SRα. Ultimately, SRP leaves the membrane whereas the ribosome/nascent chain complex binds to the structure calle *translocation channel* (TC) that has to be located on the membrane in the vicin ity of the SR. The details of this scheme, particularly the order of the events oc curring after the binding of the ribosome/nascent chain/SRP complex with SRα are still unclear. As soon as the ribosome/nascent chain complex becomes fre of SRP and associated with the endoplasmic reticulum membrane, the elongation arrest is recalled and the growing nascent chain penetrates into a translocation channel.

18.5. Cotranslational Transmembrane Translocation of Nascent Polypeptide Chains

18.5.1. Translocation Channel of the Endoplasmic Reticulum Membrane

The idea of a transient translocation channel, or tunnel, in the endoplasmic reticulum membrane through which the nascent polypeptide chain is transferred cotranslationally was a direct inference from the signal peptide hypothesis. Since then, many experimental facts and observations have provided a strong support for this model. Moreover, components of the protein complex serving as the intramembrane "pore" or tunnel (the so-called *translocon*) have been analyzed in detail (see reviews by Rapoport *et al.*, 1996; Corsi and Schekman, 1996; Martoglio and Dobberstein, 1996), and the complex has been visualized by electron microscopy (Hanein *et al.*, 1996; Beckmann *et al.*, 1997).

In mammals, the dominant constituent of the translocating channel is the Sec61p complex composed of subunits α (40 kDa), β (14 kDa), and γ (8 kDa). In fact, proteoliposomes that include this complex and SR as sole protein constituents are able to perform SRP-dependent translocation of nascent polypeptide chains. The imported chains become resistant to exogenous proteases and can even be processed, if the proteoliposomes contain a signal peptidase in addition to the Sec61p complex and SR. However, some protein precursors need one more protein constituent to be translocated across the lipid bilayer—the so-called

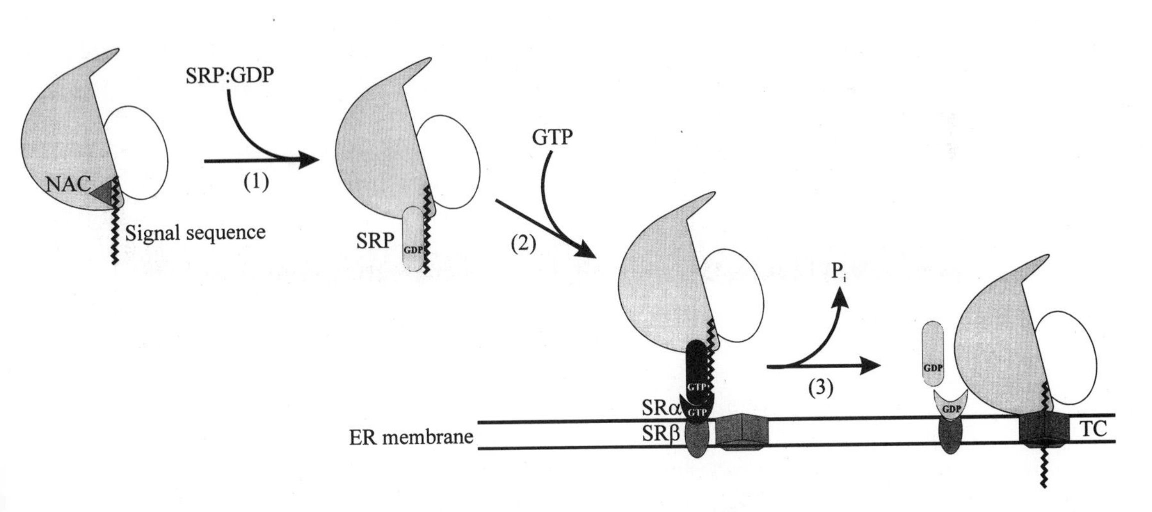

FIG. 18.8. Sequence of events initiating the cotranslational transmembrane translocation. **(1)** The ribosome/nascent polypeptide/NAC complex interacts with the GDP-bound form of SRP, resulting in binding of SRP and removal of NAC. **(2)** The substitution of GTP for GDP in SRP is induced by the binding. This leads to the appearance of the affinity of the SRP to the SRP receptor (SR, also in the GTP-bound form) on the membrane. **(3)** The interaction of SRP with SR induces GTP hydrolysis which results in dissociation of the ribosome/nascent polypeptide complex from SRP and its attachment to the translocation complex (TC) of the membrane. Concurrently, GDP forms of SRP and SR dissociate from each other. See Corsi, A. K., and Schekman, R. (1996) *J. Biol. Chem.* **271**:30299–30202.

translocating chain associated membrane protein, or TRAM, which probably assists in the proper arrangement of the signal sequence in the channel.

The mammalian Sec61α is a transmembrane protein; as can be predicted from its amino acid sequence with alternating hydrophobic and hydrophilic stretches, its polypeptide chain spans the membrane ten times. It is noteworthy that the hydrophobic, presumably transmembrane stretches of this protein contain up to 30% of hydrophilic, polar, and charged residues, which makes them capable of assuming the amphiphilic conformation under certain conditions. The mammalian Sec61α is homologous to the yeast Sec61p and the bacterial SecYp (Goerlich *et al.*, 1992; High and Stirling, 1993). All three proteins are very similar in their membrane topology. As to the other constituents of the mammalian Sec61p complex, Sec61β and Sec61γ are homologous to the Sbh1p and Sss1p yeast proteins, respectively. Generally, the mammalian Sec61(αβγ) complex resembles the Sec61p/Sbh1p/Sss1p complex in yeast. Yeast contain another trimeric complex, termed Ssh1p/Sbh2p/Sss1p, where Ssh1p is a distant relative of Sec61p, Sbh2p is a homolog of the Sbh1p subunit, and Sss1p is the common subunit in both complexes.

The proper insertion of the nascent polypeptide chains with the short *n* (polar) and/or *h* (hydrophobic) regions in their signal sequences requires involvement of the TRAM protein, which is as abundant in the endoplasmic reticulum membrane as the Sec61p complex. The TRAM protein is a 34-kDa glycoprotein presumably spanning the membrane eight times (as predicted from its amino acid sequence) with hydrophilic and charged residues located in the membrane interior, similarly to Sec61α.

The morphology of the translocation channel was studied by both traditional electron microscopy with negative staining (Hanein *et al.*, 1996) and cryo-electron microscopy (Beckmann *et al.*, 1997). It has been found that the Sec61p heterotrimers in one projection appear as rings with characteristic pentagonal contours of about 85 Å in diameter and with a central pore of about 20 Å. The rings (or, more exactly, pentagons) are shown to represent a top view of cylinders (or, more correctly, toroidal structures) 50 to 60 Å in height. By applying the cylinder model and taking the dimensions into account, the "molecular mass" of the particle was estimated to be about 250 kDa. Hence, several (probably, five) Sec61p heterotrimers appear to participate in the formation of the particle.

Interestingly, the rings are not seen in the proteoliposomes containing the Sec61p heterotrimers. However, they arise after addition of ribosomes that play the role of ligands stimulating specific association of the heterotrimers. Similar pentagonal rings can be identified on the surface of microsomes (freeze-fracture electron microscopy). There are good reasons to believe that Sec61p is an obligatory constituent of such structures. The microsomal rings are larger than those in the detergent-solubilized Sec61p preparations (100 Å versus 85 Å in diameter), which may be due either to different conformations of the Sec61p oligomers or the involvement of some additional subunits (e.g., the TRAM protein).

Thus, the pentagon-shaped toroidal particles formed by the Sec61p heterotrimers are likely to be the protein-translocating channels in the endoplasmic reticulum membrane. The involvement of several heterotrimers in the functioning of such channels, along with potentially amphiphilic properties of Sec61p(α), allows the coexistence of hydrophobic and hydrophilic sections of a growing polypeptide chain in the channel.

Figure 18.9 represents a model for the assembly of a translocation channel composed of Sec61p heterotrimers. According to this model, in the endoplasmic reticulum membrane, monomeric and oligomeric forms of the heterotrimer are in an equilibrium shifted toward the former ones. The ribosomes favor the formation of the oligomeric cylindrical particles that are still closed at the end (assembly stage). While the ribosome with a nascent polypeptide chain is sitting on the cytoplasmic end of the particle, its interior opens itself, allowing the polypeptide chain to penetrate through the membrane (gating stage).

18.5.2. Insertion of the Nascent Polypeptide Chain into the Endoplasmic Reticulum Membrane

As indicated above, after dissociation from SRP, the ribosome/nascent chain complex remains attached to the endoplasmic reticulum membrane, the nascent chain being inserted into the translocation channel. Such insertion, however, is not tightly coupled to the SRP/SR interaction. In fact, proteoliposomes containing Sec61p as the only protein constituent bind the ribosome/short nascent chain (80–90 residues) complexes in a signal-sequence-dependent manner, whereupon the polypeptide chains become protease-resistant. Mutations in the hydrophobic *h* region of the signal sequence result in disturbance of such binding. Thus, the signal sequence is recognized at least twice: first, when polypeptide chains to be exo-

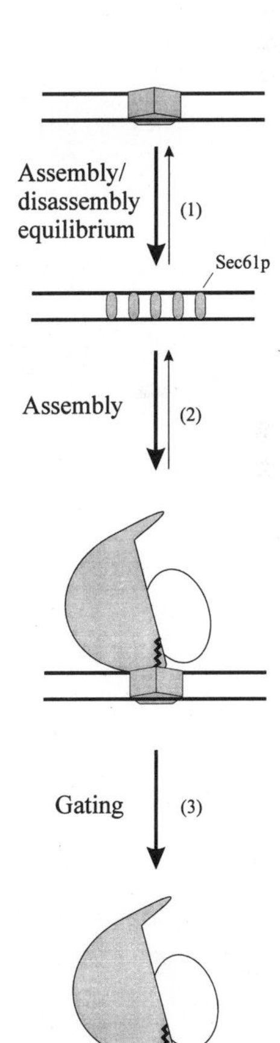

FIG. 18.9. Tentative sequence of events during formation of the translocating channel through the membrane. **(1)** In the absence of ligands there is an equilibrium between monomeric and oligomeric forms of the membrane protein Sec61p (assembly–disassembly equilibrium), shifted toward the monomeric form. **(2)** The binding of the ribosome/nascent polypeptide complex (see Fig. 18.8) stimulates the assembly of the Sec61p oligomer in the membrane. **(3)** The interaction of the Sec61p oligomer with the signal sequence of the nascent polypeptide chain opens the channel for transmembrane translocation (gating stage). See Hanein, D., Matlack, K. E. S., Jungnickel, B., Plath, K., Kalies, K.-U., Miller, K. R., Rapoport, T. A., and Akey, C. W. (1996) *Cell* **87**:721–732.

cyted are selected by means of their binding with SRPs; second, when they enter into the translocation channel. In other words, a cell exercises double control over penetration of polypeptide chains into the endoplasmic reticulum and further into the exocytic pathway. Thus, although the nascent polypeptide chain can interact directly with the translocation channel, as it follows from the model experiments with proteoliposomes, the interaction is most commonly mediated by SRP and SR in the case of natural microsomes.

As to the ribosomal moiety of the ribosome/nascent chain complex, it binds tightly to the translocation channel, most probably through direct interaction with the Sec61p complex. It has been found that the Sec61α protein can be detached from ribosomes, that is, freed of nascent polypeptide chains by treatment with puromycin only in the presence of detergents (e.g., digitonin) and concentrated (*ca.* 1.2 M) salt solutions. Thus, Sec61α falls into the category of the endoplasmic reticulum membrane proteins most firmly bound to the ribosomes. Currently, the involvement of other membrane proteins in ribosome binding (e.g., putative 34-kDa and 180-kDa ribosome receptors) is debatable.

It is believed that the nascent chain penetrates into the channel in the form of a loop exposing the polar (usually positively charged) amino terminus to the cytoplasm, with both the signal sequence and the mature part contacting the Sec61p protein until cleavage of the signal sequence by the signal peptidase (Fig. 18.10A). The nascent chains with negatively charged amino-terminal residues probably do not form the loop structure in the translocation channel. Their amino termini face the lumen of the endoplasmic reticulum while the signal sequences span the membrane once, thereupon playing the role of an anchor (Fig. 18.10B). Such an arrangement of the nascent chain in the channel can be changed to the loop-shaped one just by substituting positively charged amino-terminal residues for negatively charged ones.

As a result of the interaction of the nascent polypeptide chain with the endoplasmic reticulum membrane, the ribosome stays attached to the channel structure in such a way that it seals up the cytoplasmic end of the channel thus making it inaccessible for even relatively small ions (e.g., iodide ion), to say nothing of proteases. The channel still remains closed also to the lumen if the polypeptide contains less than 70 amino acid residues (Crowley *et al.*, 1994). The open state of the channel is induced upon interaction of its constituents with the nascent polypeptide chain of a sufficient length (more than 70 residues). Such construction of the channel and features of its interaction with the ribosome/nascent chain complex predetermine the growing polypeptide chain to move unidirectionally through the channel toward the lumen of the endoplasmic reticulum.

In yeast, an additional mechanism of polypeptide chain translocation across the endoplasmic reticulum membrane has been revealed. In the case of translocation of some protein precursors (pre-proteins), their movement through the channel composed of the Sec61p heterotrimers is stimulated by special ATP-binding proteins (Kar2p or Lhs1p) on the luminal side of the membrane, in combination with a transmembrane protein (Sec63p). In these cases a "driving force" for translocation is ensured by interaction of the trapped polypeptide chain with the Kar2p protein on the luminal end of the channel. This ATP-dependent step is controlled by the luminal DnaJ-like domain of protein Sec63p stimulating ATPase activity of Kar2p. Kar2p in its ADP-bound form associates with the incoming polypeptide chain, thus making translocation unidirectional. In other words, Kar2p in combination with Sec63p acts as a "molecular ratchet." Mutations in this protein disturbing its capability of ATP binding or interaction with Sec63p result in reduced cotranslational import of pre-proteins into the microsomal fraction of yeast cells. A similar role is probably played by Lhs1p, another yeast protein that also functions in combination with Sec63p. An Lhs1p homolog (Grp170) has recently been found in mammalian microsomes. This protein has been shown to stimulate insertion of nascent polypeptide chains into the translocation apparatus of the endoplasmic reticulum membrane in a nucleoside triphosphate-dependent manner.

18.5.3. Arrangement of the Nascent Polypeptide Chain in the Endoplasmic Reticulum Membrane

The nascent polypeptide chains penetrating into the endoplasmic reticulum lumen undergo various cotranslational modifications (see Section 18.6). The cleavage of the signal sequence occurs when the nascent chain becomes approximately 150 residues long. The chain continues to grow through the channel re-

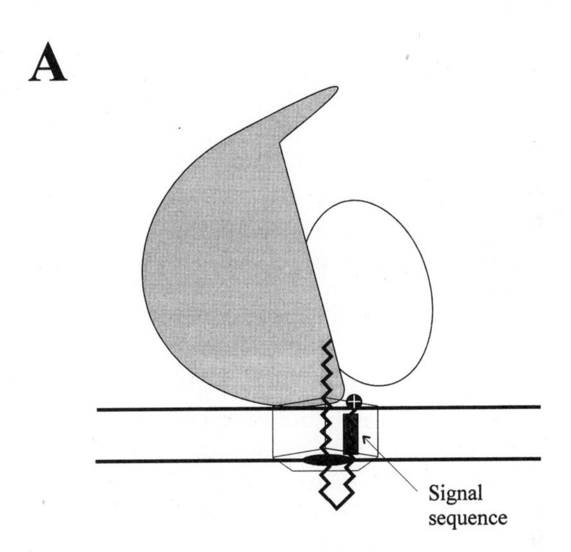

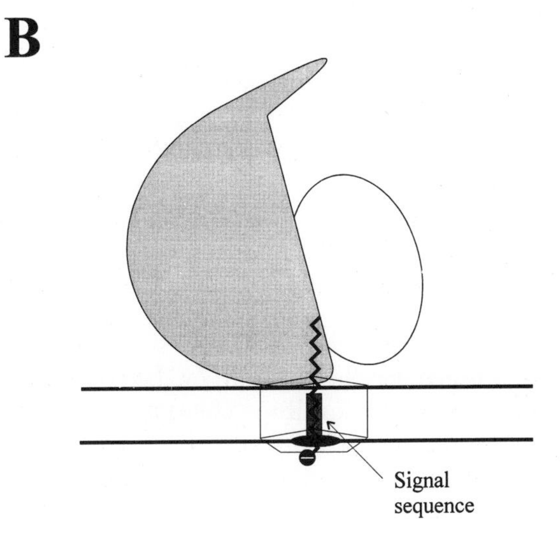

FIG. 18.10. Schematic representation of two ways of the insertion of the nascent polypeptide chain into the translocation channel. **(A)** When the N-terminus of the nascent polypeptide chains is positively charged, it remains on the cytoplasmic side of the endoplasmic reticulum membrane, the signal sequence is anchored in the membrane (until it is cut off by signal peptidase), and the growing polypeptide chain forms a loop protruding through the channel into the intramembrane lumen. **(B)** In the case of the negatively charged N-terminus, it goes through the channel and faces the lumen, the signal sequence is anchored in the membrane, and thus the polypeptide chain spans the membrane once.

maining in contact with the Sec61p protein until termination of elongation. The proteins destined for secretion are folded in the lumen of the endoplasmic reticulum, this process being governed by special molecular chaperones (GRP78 and GRP94), and then enters the exocytic flow.

Polypeptide chains bearing an uncleavable signal peptide remain ultimately bound to the endoplasmic reticulum membrane with the amino and carboxyl termini facing the cytoplasm and the lumen, respectively (Fig. 18.11A). Such polypeptide chains span the lipid bilayer (type II membrane proteins). In this case both the hydrophobic signal peptide and the subsequently formed hydrophilic parts of the nascent chain occupy the same channel (presumably being two ends of a loop, like in Fig. 18.10A) until termination of elongation occurs, that is, for the relatively long interval. Such a situation suggests an inducible amphiphilic character of the channel. After termination the hydrophilic part of the newly synthesized protein is released into the lumen, as depicted in Fig. 11A.

Another type (type I) of membrane-bound proteins exposing the carboxyl and amino termini to the cytoplasm and the lumen, respectively, and spanning the lipid bilayer once, is formed when the polypeptide chain contains, in addition to the cleavable signal sequence, a specific region (uncleavable stop-transport signal) of about 20 nonpolar amino acid residues located close to the carboxyl terminus and usually flanked by a cluster of basic residues at its cytoplasmic end (Fig. 18.11B). The change from a hydrophilic content of the channel to a hydrophobic one, taking place during translocation of the growing chain, implies again potential amphiphilic properties of the channel. There is evidence that the nascent chain leaves the channel to be embedded into the lipid bilayer only after termination of elongation, and from this moment its stop-transport region no longer contacts the Sec61p and TRAM proteins.

One more type of targeting signal is the so-called start–anchor or start–stop sequence that bears negative charge at its N-terminal part (Fig. 18.11C). Polypeptide chains with such a signal are membrane-bound and characterized by a bulky C-terminal part facing the cytoplasm.

The arrangement of polytopic polypeptide chains spanning the membrane several times is achieved owing to the presence of a repeating insertion sequence in addition to the signal sequence (cleavable, uncleavable, or start–anchor). As experiments with recombinant proteins have shown, the function of the insertion signal can be performed by an amino acid sequence of about 20 nonpolar amino acids. The efficiency of such a signal is augmented by positively charged residues flanking it at the carboxy-terminal side. Thus, one can see that the insertion signal is similar, if not identical, to the stop-transport signal (see above). It has been shown that as each transmembrane segment in a polytopic membrane protein emerges from the ribosome, it sequentially translocates across the membrane. Reinsertion of a growing polypeptide chain into the membrane depends on the distance between the preceding transmembrane region and the newly formed insertion signal; hydrophilic amino acid sequences longer than 100 residues hinder this process.

There are grounds to assume that a growing polypeptide chain does not leave the translocation channel, or the translocation channel preserves its integrity in the membrane, until termination of elongation occurs. It has been found that the hydrophilic environment of the chain in the channel is not changed to a hydrophobic one until the chain remains bound to the ribosome. Termination of translation and release of the nascent polypeptide from the ribosome seems to allow a rearrangement of the membrane environment of the transmembrane polypeptide such that its hydrophobic segments are trapped by the lipid bilayer of the membrane, and the folding of the transmembrane protein concludes (see Borel and Simon, 1996).

Although the nascent-chain-bearing ribosome is generally docked at the en-

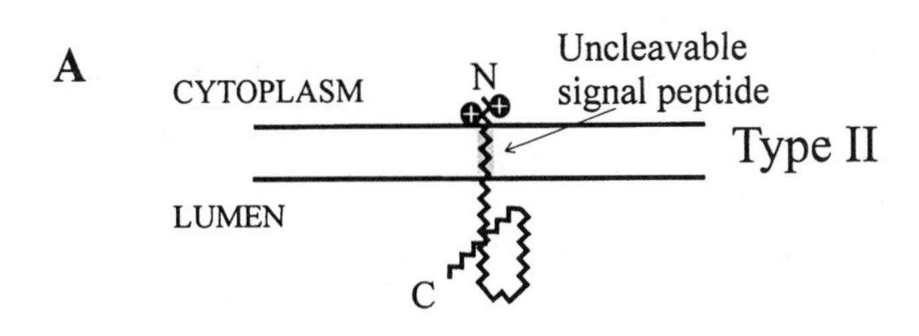

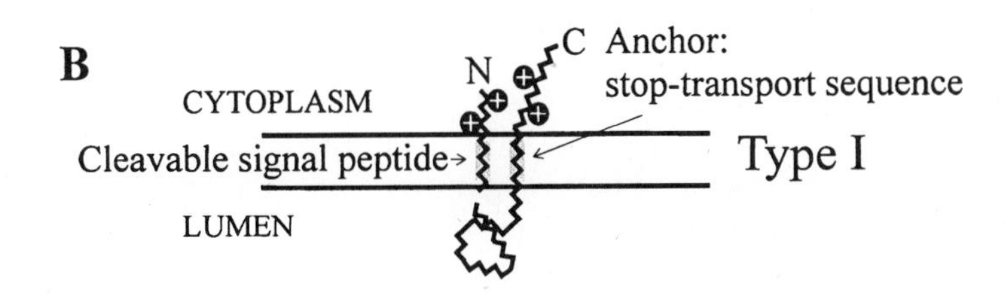

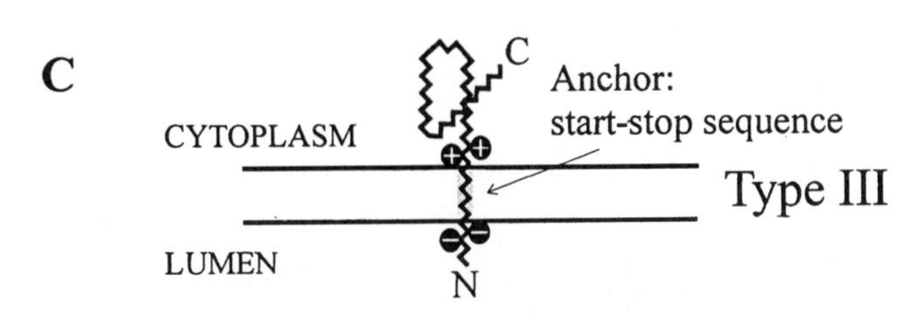

FIG. 18.11. Different types of transmembrane proteins classified according to von Heijne, G. (1988) *Biochim. Biophys. Acta* **947**:307–333. **(A)** Proteins with uncleavable signal peptide (USP, type II). The positively charged N-terminus remains on the cytoplasmic side of the membrane (see Fig. 18.10A), the signal peptide is anchored in the membrane, and the rest of the protein falls through the membrane into the lumen. **(B)** Proteins with cleavable signal peptide, but containing uncleavable nonpolar "stop-transport" (anchor) sequence closer to the C-terminus (ST, type I). After formation of the loop (Fig. 18.10A), the signal peptide is split off, and the polypeptide chain grows into the lumen until the "stop-transport" sequence is anchored in the membrane; the C-terminus remains on the cytoplasmic side and usually is positively charged. **(C)** Proteins with nonpolar "start-stop" sequence near the N-terminus (SST, type III). The negatively charged N-terminus is translocated to the lumen side of the membrane (Fig. 18.10B) and followed by a nonpolar sequence that becomes anchored in the membrane. As a result, further translocation of the growing polypeptide through the channel is stopped, and thus the rest of the protein (a bulky C-terminal part) remains on the cytoplasmic side.

doplasmic reticulum membrane in such a way that the translocation channel is isolated from the cytoplasmic environment as long as the translocation proceeds, there are some exceptions to this rule. For example, cotranslational translocation of apolipoprotein B (apoB) was shown to be discontinuous, that is, the apoB polypeptide chain is exposed transiently to the cytoplasm during its elongation (Hegde and Lingappa, 1996). Translocation pauses occur when a specific amino acid sequence (pause transfer sequence) in apoB meets a membrane component and binds transiently with it. At this moment the tight ribosome-membrane junction opens, and the growing chain becomes sensitive to exogenous proteases and antibodies. The exposure of the nascent chain is followed by its reentrance into the translocation channel and by the restoration of the ribosome-membrane junction (Fig. 18.12). A part of the nascent chain temporarily protruding into the cytoplasm may be 50–60 residues long.

18.6. Cotranslational Covalent Modifications and Folding of the Nascent Polypeptide Chain in the Endoplasmic Reticulum

As the growing ribosome-bound polypeptide chain penetrates into the lumen of the endoplasmic reticulum it undergoes covalent modifications, including cleavage of the signal sequence (reviewed by Harwood, 1980), primary or "core" N-glycosylation [linking the Glc_3 $Man_9(GlcNAc)_2$ oligosaccharide chains with Asn residues in Asn-X-Thr/Ser motifs] (for reviews, see Phelps, 1980; Hubbard and Ivatt, 1981), and formation of intramolecular disulfide bridges (reviewed by Freedman and Hillson, 1980). These types of modifications are indicated because of their obligatory character for virtually all proteins involved in the exocytic pathway. It is obvious that proteins destined for secretion would never leave the endoplasmic reticulum membrane without proteolytic processing. Formation of disulfide bridges is considered to be necessary for protein stabilization in order to

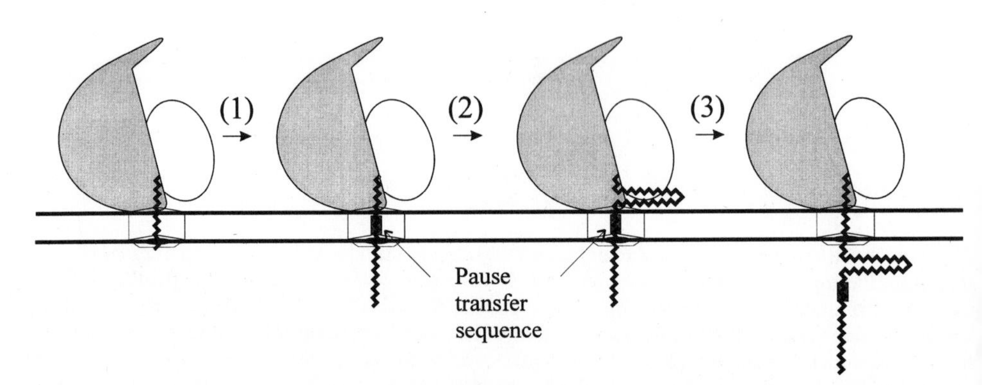

FIG. 18.12. Sequence-specific pausing in the cotranslational transmembrane translocation of the growing polypeptide chain. See Hegde, R. S., and Lingappa, V. R. (1996) *Cell* **85**:217–228. **(1)** During elongation of the nascent polypeptide a specific sequence, called "pause transfer sequence" (PT, boxed), enters into the translocation channel and becomes temporarily anchored there. **(2)** The continuing elongation results in exposing the following section of the growing polypeptide to the cytosol. **(3)** As the polypeptide chain continues to grow further, the PT is forced through the channel and the transmembrane translocation becomes restored.

prevent protein unfolding during exocytosis and thus escaping the exocytic pathway. N-glycosylation, at least in some cases, may be crucial for proper protein folding, as well as for protein stabilization.

Regarding the proteolytic processing and the N-glycosylation, cotranslational character of these processes is a consequence of the location of the enzymes responsible for them (i.e., signal peptidase and glycosyl transferase complex) in the vicinity of the translocation channel. In contrast to proteolytic processing and glycosylation, formation of intramolecular disulfide bonds occurs due to the action of protein disulfide isomerase, a soluble resident protein of the endoplasmic reticulum.

Different types of nascent chain modification are interdependent and intimately connected with polypeptide folding, which in turn is mediated by both membrane-bound and soluble chaperones of the endoplasmic reticulum. This point can be exemplified by the data for influenza virus hemagglutinin (Chen *et al.*, 1995). The protein has the N-terminal cleavable signal sequence and the transmembrane region (residues 514 to 540) close to its C-terminus. The body of the protein faces the lumen with seven N-linked oligosaccharides and six interchain disulfide bridges contributing to its structure. It has been documented that glycosylation of Asn165 and Asn286 as well as formation of disulfide bond 52–277 proceed before the completion of polypeptide chain elongation. Interestingly, inhibition of glycosylation with tunicamycin prevents proper folding of hemagglutinin and causes its aggregation due to formation of aberrant intermolecular disulfide bridges.

Glycosylation of the hemagglutinin polypeptide chain and trimming of N-linked oligosaccharides (cleavage of two glucose residues by glucosidase I) precedes the interaction of the protein with calnexin, a lectin-like chaperone located in the endoplasmic reticulum membrane. Association of influenza hemagglutinin with calnexin begins as soon as four of its seven N-terminal glycans are formed. It has been shown that not only calnexin but also soluble chaperones of the lumen (e.g., BiP, GRP94, and calreticulin) are involved in hemagglutinin folding (Tatu and Helenius, 1997). On the whole, upon transmembrane translocation of the nascent hemagglutinin polypeptide chain its different parts are concurrently engaged in several processes—glycosylation, glycan trimming, disulfide bridge formation, and interaction with both membrane and soluble chaperones in the endoplasmic reticulum.

References

Adelman, M. R., Sabatini, D. D., and Blobel, G. (1973) Ribosome-membrane interaction: Nondestructive disassembly of rat liver rough microsomes into ribosomal and membranous components, *J. Cell Biol.* **56**:206–229.

Bacher, G., Luetcke, H., Jungnickel, B., Rapoport, T. A., and Dobberstein, B. (1996) Regulation by the ribosome of the GTPase of the signal-recognition particle during protein targeting, *Nature* **381**:248–251.

Beckmann, R., Bubeck, D., Grassucci, R., Penczek, P., Verschoor, A., Blobel, G., and Frank, J. (1997) Alignment of conduits for the nascent polypeptide chain in the ribosome-Sec61 complex, *Science* **278**:2123–2126.

Beckmann, R. P., Mizzen, L. A., and Welch, W. J. (1990) Interaction of Hsp 70 with newly synthesized proteins: Implications for protein folding and assembly, *Science* **248**:850–854.

Blobel, G. (1980) Intracellular protein topogenesis, *Proc. Natl. Acad. Sci. USA* **77**:1496–1500.

Blobel, G., and Dobberstein, B. (1975) Transfer of protein across membranes. I. Presence of proteolytically processed and unprocessed nascent immunoglobulin light chains on membrane-bound ribosomes by murine myeloma, *J. Cell Biol.* **67**:835–851.

Blobel, G., and Sabatini, D. D. (1970) Controlled proteolysis of nascent polypeptides in rat liver cell fractions. I. Location of the polypeptides within ribosomes, *J. Cell Biol.* **45**:130–145.

Blobel, G., and Sabatini, D. D. (1971) Ribosome-membrane interaction in eukaryotic cells, in *Biomembranes* (L. A. Manson, ed.), vol. 2, pp. 193–195, Plenum Press, New York, London.

Borel, A. C., and Simon, S. M. (1996) Biogenesis of polytopic membrane proteins: membrane segments assemble within translocation channels prior to membrane integration, *Cell* **85**:379–389.

Chen, W., Helenius, J., Braakman, I., and Helenius, A. (1995) Co-translational folding and calnexin binding during glycoprotein synthesis, *Proc. Natl. Acad. Sci. USA* **92**:6229–6233.

Christensen, A. K. (1994) Negatively-stained polysomes on rough microsome vesicles viewed by electron microscopy: further evidence regarding the orientation of attached ribosomes, *Cell Tissue Res.* **276**:439–444.

Corsi, A. K., and Schekman, R. (1996) Mechanism of polypeptide translocation into the endoplasmic reticulum, *J. Biol. Chem.* **271**:30299–30302.

Crowley, K. S., Liao, S., Worrell, V. E., Reinhart, G. D., and Johnson, A. E. (1994) Secretory proteins move through the endoplasmic reticulum membrane via an aqueous, gated pore, *Cell* **78**:461–471.

Ellis, R. J., and van der Vies, S. M. (1991) Molecular chaperones, *Annu. Rev. Biochem.* **60**:321–347.

Frank, J., Verschoor, A., Li, Y., Zhu, J., Lata, R. K., Radermacher, M., Penczek, P., Grassucci, R., Agrawal, R. K., and Srivastava, S. (1995) A model of the translational apparatus based on a three-dimensional reconstruction of the *Escherichia coli* ribosome, in *Frontiers in Translation* (A. T. Matheson, J. E. Davies, P. P. Dennis, and W. E. Hill, eds.), *Biochem. Cell Biol.* **73**:757–765.

Freedman, R. B., and Hillson, D. A. (1980) Formation of disulfide bonds, in *The Enzymology of Post-translational Modifications of Proteins* (R. B. Freedman and H. C. Hawkins, eds.), vol. 1, pp. 157–212, Academic Press, London, New York.

Frydman, J., Nimmesgern, E., Ohtsuka, K., and Hartl, F. U. (1994) Folding of nascent polypeptide chains in a high molecular mass assembly with molecular chaperones, *Nature* **370**:111–117.

Ganoza, M. C., and Williams, C. A. (1969) *In vitro* synthesis of different categories of specific protein by membrane-bound and free ribosomes, *Proc. Natl. Acad. Sci. USA* **63**:1370–1376.

Gething, M.-J., and Sambrook, J. (1992) Protein folding in the cell, *Nature* **355**:33–45.

Goerlich, D., Prehn, S., Hartmann, E., Kalies, K.-U., and Rapoport, T. A. (1992) A mammalian homolog of SEC61p and SECYp is associated with ribosomes and nascent polypeptides during translocation, *Cell* **71**:489–503.

Hamlin, J., and Zabin, I. (1972) β-Galactosidase: Immunological activity of ribosome-bound, growing polypeptide chains, *Proc. Natl. Acad. Sci. USA* **69**:412–416.

Hanein, D., Matlack, K. E. S., Jungnickel, B., Plath, K., Kalies, K.-U., Miller, K. R., Rapoport, T. A., and Akey, C. W. (1996) Oligomeric rings of the Sec61p complex induced by ligands required for protein translocation, *Cell* **87**:721–732.

Harrison, T. M., Brownlee, G. G., and Milstein, C. (1974) Studies on polysome-membrane interactions in mouse myeloma cells, *Eur. J. Biochem.* **47**:613–620.

Hartl, F. U. (1996) Molecular chaperones in cellular protein folding, *Nature* **381**:571–580.

Harwood, R. (1980) Protein transfer across membranes: The role of signal sequences and signal peptidase activity, in *The Enzymology of Post-translational Modifications of Proteins* (R. B. Freedman and H. C. Hawkins, eds.), vol. 1, pp. 3–52, Academic Press, London, New York.

Hegde, R. S., and Lingappa, V. R. (1996) Sequence-specific alteration of the ribosome-membrane junction exposes nascent secretory proteins to the cytosol, *Cell* **85**:217–228.

Henderson, R., and Unwin, P. N. T. (1975) Three-dimensional model of purple membrane obtained by electron microscopy, *Nature* **257**:28–32.

Hendrick, J. P., Langer, T., Davis, T. A., Hartl, F. U., and Wiedman, M. (1993) Control of folding and membrane translocation by binding of the chaperone DnaJ to nascent polypeptides, *Proc. Natl. Acad. Sci. USA* **90**:10216–10220.

High, S., and Stirling, C. J. (1993) Protein translocation across membranes: common themes in divergent organisms, *Trends Cell Biol.* **3**:335–339.

Hubbard, S. C., and Ivatt, R. J. (1981) Synthesis and processing of asparagine-linked oligosaccharides, *Annu. Rev. Biochem.* **50**:555–583.

Inouye, H., and Beckwith, J. (1977) Synthesis and processing of an *Escherichia coli* alkaline phosphatase precursor *in vitro, Proc. Natl. Acad. Sci. USA* **74**:1440–1444.

Inouye, M., and Halegoua, S. (1980) Secretion and membrane localization of proteins in *Escherichia coli, CRC Crit. Rev. Biochem.* **7**:339–371.

Jackson, R., and Hunter, T. (1970) Role of methionine in the initiation of protein synthesis, *Nature* **227**:672–676.

Jaenicke, R. (1993) Role of accessory proteins in protein folding, *Curr. Opinion Struct. Biol.* **3**:104–112.

Johnson, A. E. (1997) Protein translocation at the ER membrane: a complex process becomes more so, *Trends Cell Biol.* **7**:90–95.

Kiho, Y., and Rich, A. (1964) Induced enzyme formed on bacterial polyribosomes, *Proc. Natl. Acad. Sci. USA* **51**:111–118.

Kolb, V. A., Makeyev, E. V., Kommer, A., and Spirin, A. S. (1995) Cotranslational folding of proteins, *Biochem. Cell Biol.* **73**:1217–1220.

Kolb, V. A., Makeyev, E. V., and Spirin, A. S. (1994) Folding of firefly luciferase during translation in a cell-free system, *EMBO J.* **13**:3631–3637.

Komar, A. A., Kommer, A., Krasheninnikov, I. A., and Spirin, A. S. (1997) Cotranslational folding of globin, *J. Biol. Chem.* **272**:10646–10651.

Langer, T., Lu, C., Echols, H., Flanagan, J., Hayer, M. K., and Hartl, F. U. (1992) Successive action of DnaK, DnaJ and GroEL along the pathway of chaperone-mediated protein folding, *Nature* **356**:683–688.

Lauring, B., Kreibich, G., and Wiedmann, M. (1995) The intrinsic ability of ribosomes to bind to endoplasmic reticulum membranes is regulated by signal recognition particle and nascent-polypeptide-associated complex, *Proc. Natl. Acad. Sci. USA* **92**:9435–9439.

Lauring B., Sakai, H., Kreibich, G., and Wiedmann, M. (1995) Nascent polypeptide-associated complex prevents mistargeting of nascent chains to the endoplasmic reticulum, *Proc. Natl. Acad. Sci. USA* **92**:5411–5415.

Lim, V. I., and Spirin, A. S. (1986) Stereochemical analysis of ribosomal transpeptidation conformation of nascent peptide. APPENDIX. Lim, V. I. Disallowed conformations of the tetrahedral intermediate, *J. Mol. Biol.* **188**:565–577.

Luetcke, H. (1995) Signal recognition particle (SRP), a ubiquitous initiator of protein translocation, *Eur. J. Biochem.* **228**:531–550.

Luirink, J., and Dobberstein, B. (1994) Mammalian and *Escherichia coli* signal recognition particles, *Mol. Microbiol.* **11**:9–13.

Makeyev, E. V., Kolb, V. A., and Spirin, A. S. (1996) Enzymatic activity of the ribosome-bound nascent polypeptide, *FEBS Letters* **378**:166–170.

Malkin, L. I., and Rich, A. (1967) Partial resistance of nascent polypeptide chains to proteolytic digestion due to ribosomal shielding, *J. Mol. Biol.* **26**:329–346.

Martoglio, B., and Dobberstein, B. (1996) Snapshots of membrane-translocating proteins, *Trends Cell Biol.* **6**:142–147.

Milstein, C., Brownlee, G. G., Harrison, T. H., and Mathews, M. B. (1972) A possible precursor of immunoglobulin light chains, *Nature New Biol.* **239**:117–120.

Morrison, T. G., and Lodish, H. F. (1975) Site of synthesis of membrane and nonmembrane proteins of vesicular stomatitis virus, *J. Biol. Chem.* **250**:6955–6962.

Nelson, R. J., Ziegelhoffer, T., Nicolet, C., Werner–Washburne, M., and Craig, E. A. (1992) The translational machinery and 70 kd heat shock protein cooperate in protein synthesis, *Cell* **71**:97–105.

Netzer, W. J., and Hartl, F. U. (1997) Recombination of protein domains facilitated by co-translational folding in eukaryotes, *Nature* **388**:343–349.

Palmiter, R. D., Gagnon, J., and Walsh, K. A. (1978) Ovalbumin: A secreted protein without a transient hydrophobic leader sequence, *Proc. Natl. Acad. Sci. USA* **75**:94–98.

Phelps, C. F. (1980) Glycosylation, in *The Enzymology of Post-translational Modifications of Proteins* (R. B. Freedman and H. C. Hawkins, eds.), vol. 1, pp. 105–155, Academic Press, London, New York.

Powers, T., and Walter, P. (1996) The nascent polypeptide-associated complex modulates interactions between the signal recognition particle and the ribosome, *Curr. Biol.* **6**:331–338.

Rapoport, T.A., Jungnickel, B., and Kutay, U. (1996) Protein transport across the endoplasmic reticulum and bacterial inner membranes, *Annu. Rev. Biochem.* **65**:271–303.

Redman, C. M. (1969) Biosynthesis of serum proteins and ferritin by free and attached ribosomes of rat liver, *J. Biol. Chem.* **244**:4308–4315.

Redman, C. M., Siekevitz, P., and Palade, G. E. (1966) Synthesis and transfer of amylase in pigeon pancreas microsomes, *J. Biol. Chem.* **241**:1150–1160.

Ryabova, L. A., Selivanova, O. M., Baranov, V. I., Vasiliev, V. D., and Spirin, A. S. (1988) Does the channel for nascent peptide exist inside the ribosome? Immune electron microscopy study, *FEBS Letters* **226**:255–260.

Sabatini, D. D., and Blobel, G. (1970) Controlled proteolysis of nascent polypeptides in rat liver cell fractions. II. Location of the polypeptides in rough microsomes, *J. Cell Biol.* **45**:146–157.

Sabatini, D. D., Tashiro, Y., and Palade, G. E. (1966) On the attachment of ribosomes to microsomal membranes, *J. Mol. Biol.* **19**:503–524.

Schechter, I., McKean, D. J., Guyer, R., and Terry, W. (1975) Partial amino acid sequence of the precursor of immunoglobulin light chain programmed by messenger RNA *in vitro, Science* **188**:160–162.

Siegel, V., and Walter, P. (1988) Functional dissection of the signal recognition particle, *Trends Biochem. Sci.* **13**:314–316.

Siekevitz, P., and Palade, G. E. (1960) A cytochemical study on the pancreas of the guinea pig. V. *In vivo* incorporation of leucine-1-C^{14} into the chymotrypsinogen of various cell fractions, *J. Biophys. Biochem. Cytol.* **7**:619–630.

Smith, W. P., Tai, P.-C., and Davis, B. D. (1978a) Nascent peptide as sole attachment of polysomes to membranes in bacteria, *Proc. Natl. Acad. Sci. USA* **75**:814–817.

Smith, W. P., Tai, P. C., and Davis, B. D. (1978b) Interaction of secreted nascent chains with surrounding membrane in *Bacillus subtilis, Proc. Natl. Acad. Sci. USA* **75**:5922–5925.

Spirin, A. S., and Lim, V. I. (1986) Stereochemical analysis of ribosomal transpeptidation. Translocation, and nascent peptide folding, in *Structure, Function, and Genetics of Ribosomes* (B. Hardesty and G. Kramer, eds.), pp. 556–572, Springer-Verlag, New York, Berlin.

Tatu, U., and Helenius, A. (1997) Interactions between newly synthesized glycoproteins, calnexin and a network of resident chaperones in the endoplasmic reticulum, *J. Cell Biol.* **136**:555–565.

Unwin, P. N. T. (1979) Attachment of ribosome crystals to intracellular membranes, *J. Mol. Biol.* **132:** 69–84.

von Heijne, G. (1988) Transcending the impenetrable: how proteins come to terms with membranes, *Biochim. Biophys. Acta* **947**:307–333.

Walter, P., and Johnson, A. E. (1994) Signal sequence recognition and protein targeting to the endoplasmic reticulum membrane, *Annu. Rev. Cell Biol.* **10**:87–119.

Wang, S., Sakai, H., and Wiedmann, M. (1995) NAC covers ribosome-associated nascent chains thereby forming a protective environment for regions of nascent chains just emerging from the peptidyl transferase center, *J. Cell Biol.* **130**:519–528.

Wiedmann, B., Sakai, H., Davis, T. A., and Wiedmann, M. (1994) A protein complex required for signal-sequence-specific sorting and translocation, *Nature* **370**:434–440.

Yonath, A., Leonard, K. R., and Wittmann, H. G. (1987) A tunnel in the large ribosomal subunit revealed by three-dimensional image reconstruction, *Science* **236**:813–817.

Yoshida, A., Watanabe, S., and Morris, J. (1970) Initiation of rabbit hemoglobin synthesis: methionine and formylmethionine at the N-terminal, *Proc. Natl. Acad. Sci. USA* **67**:1600–1607.

Zipser, D., and Perrin, D. (1963) Complementation on ribosomes, *Cold Spring Harbor Symp. Quant. Biol.* **28**:533–537.

19

Conclusion:
General Principles of Ribosome Structure and Function

19.1. Introduction

In this concluding chapter an attempt to formulate several general principles of the structure and function of the ribosome is undertaken. Our understanding of the ribosome is far from complete, and the formulations reflect only the current level of knowledge in this area. Concerning the ribosome structure, the principles formulated can be considered as a summary of factual information and its generalization, whereas the principles of the function are rather hypothetical and represent just plausible models. Nevertheless, this seems to be the first attempt to give a generalized conceptual vision of the ribosome and to coordinate the structure and the function. From both a scientific and educational point of view, such tentative formulations may be a useful appendix to the main course of experimental ribosomology.

19.2. Basic Features of Ribosome Structure

19.2.1. Two Disparate Subparticles (Ribosomal Subunits)

The ribosome is a compact particle that can be roughly approximated by a sphere with a diameter of about 30 to 35 nm. Its structure lacks any internal or external symmetry. The most prominent physical feature of the particle is its *subdivision into two unequal asymmetric subparticles, or ribosomal subunits.*

The subdivision can be easily demonstrated by electron microscopy: a deep groove along the ribosome separates the large and the small subunits (Fig. 5.2). Upon decrease of Mg^{2+} concentration in the medium, the ribosome dissociates into the two subunits. The dissociation can be recorded by ultracentrifugation (Fig. 5.3). During dissociation the homogeneous ribosomal particles with sedimentation coefficient of 70S, in the case of prokaryotic ribosomes, or 80S in the case of eukaryotic ribosomes, convert into a two-component mixture of 50S and 30S, or 60S and 40S particles, respectively:

$$70S \rightarrow 50S + 30S$$

$$80S \rightarrow 60S + 40S$$

The dissociation can be also induced by high monovalent ion concentration, urea, elevated temperature, and other factors. The dissociation is reversible, and the restoration of the conditions optimal for ribosome stability results in the reassociation of the ribosomal subunits (Fig. 5.4).

Thus, the first unique structural principle of the ribosome is that it is always constructed from two unequal blocks, called the large and small ribosomal subunits. The blocks (subunits) of the ribosome are rather loosely associated with each other. The two ribosomal subunits perform different functions in translation, and their labile association may be required for mutual mobility of the ribosomal blocks in the working process of the ribosome (see Section 19.3.4).

19.2.2. Self-folding of Ribosomal RNA into a Compact Core

Each ribosomal subunit contains one molecule of a high-polymer ribosomal RNA that comprises one half to two thirds of the subunit mass. The large subunit contains RNA that is approximately two times longer rRNA than that contained in the small subunit rRNA. The lengths of the two eubacterial high-polymer rRNAs are about 3,000 nucleotides (23S rRNA) and about 1,500 nucleotides (16S rRNA), whereas higher animals have rRNA up to 4,800 nucleotides in length (28S rRNA) and 1,900 nucleotides in length (18S rRNA).

Under conditions suppressing the electrostatic repulsion of phosphate groups (at high concentrations of salts, especially magnesium ions), the chains of isolated high-polymer rRNAs are capable of folding into compact particles of a specific shape. The shape of the compactly folded 23S rRNA is similar to that of the hemispheric large ribosomal subunit, and the 16S rRNA in the compact state resembles the elongated small ribosomal subunit (Fig. 6.8). Analogous folding and compacting is observed in the presence of ribosomal proteins. This suggests that (1) each of the chains of the respective high-polymer rRNAs (large and small) *specifically self-folds* during the formation of ribosomal particles in the cell, similar to the specific self-folding of polypeptide chains into globular proteins, and (2) it is the specific compact structure of the rRNA that determines the main features of the final morphology of the corresponding ribosomal subunits.

Physical measurements demonstrate that most of the rRNA mass tends to be closer to the center of a ribosomal particle, whereas the mass of ribosomal proteins occupies on average a more peripheral position. The conclusion can be made that the compactly folded molecule of a high-polymer rRNA forms the *structural core* of a ribosomal particle (ribosomal subunit). Thus, the rRNA core of each ribosomal subunit determines its compactness, its specific shape, and the organization of ribosomal proteins on it. The scaffold role of the high-polymer rRNA for the specific arrangement of ribosomal proteins is discussed in the next section.

In addition to the high-polymer rRNAs the large ribosomal subunit contains one or two molecules of relatively low-molecular-mass rRNAs; these are 5S rRNA in the case of bacteria and other prokaryotes, and 5S rRNA and 5.8S rRNA in eukaryotes. The small rRNAs are comparable in size with ribosomal proteins and arranged with them on the high-polymer rRNA core as a scaffold.

19.2.3. Assembly of Various Ribosomal Proteins on RNA Core

Each ribosomal subunit contains many molecules of ribosomal proteins, and they all are different. In this respect the ribosomal ribonucleoprotein particle principally differs from a viral ribonucleoprotein, where the protein coat is built from uniform proteins by means of their symmetric arrangement (self-assembly) on the RNA surface. The symmetric packing of identical units cannot be realized in the case of diverse ribosomal proteins. Thus, another mechanism is used in the assembly of ribosomal particles: each ribosomal protein has its own landing site on rRNA. A protein specifically recognizes only that local structure of rRNA and attaches to it. In this way various ribosomal proteins take seats on rRNA. In bacterial (*E. coli*) ribosomes, 32 different ribosomal proteins are arranged on 23S rRNA (Fig. 7.2, Table 7.1), and 21 proteins sit on 16S rRNA (Fig. 7.1).

As already mentioned, each of the two high-polymer rRNAs forms the core of the respective ribosomal subunit, and proteins in general are arranged more on the periphery of the particles. At the same time, many rRNA sections are also found on the particle periphery. Unlike viral ribonucleoproteins, the protein of the ribosome does not form a coat around RNA. First, the protein proportion in ribosomal particles is much less than in viral ribonucleoprotein, and the amount of ribosomal protein in the particles is not sufficient to cover all of the rRNA; rRNA is rather just "decorated" with proteins. Second, ribosomal proteins do not form surface layers but are organized rather in groups or three-dimensional clusters where some proteins are covered by others and thus not exposed on the surface. Third, peripheral structures of rRNA may be involved in the formation of the protein clusters and cover some proteins from the surface.

Ribosomal proteins can play a dual role in the ribosome. On one hand, some of them may directly participate in binding and catalytic functions of the ribosome; in any case, all functional centers of the ribosome seem to contain proteins. On the other hand, ribosomal proteins serve as stabilizers or modifiers of some local structures of rRNA and thus maintain them in a functionally active state; proteins may be also responsible for conformational switches of rRNA. In particular, the main catalytic function of the ribosome, its peptidyl transferase activity, is principally ensured by a local structure of the rRNA of the large ribosomal subunit (Fig. 9.7), but some ribosomal proteins may be required for maintenance (stabilization) of the active conformation of the structure.

19.3. Structural and Biochemical Grounds of Ribosome Function

19.3.1. Structural Pockets for Functional Centers

Substrate binding and enzymatic catalysis on macromolecules, including proteins and supramolecular complexes, proceed, as a rule, in grooves, hollows, cavities, and crevices between subunits or domains, that is, in *structural pockets,* rather than on smooth molecular surfaces. Taking this rule into consideration and analyzing the morphological features of the ribosome, together with the results of some direct experiments, a number of conclusions can be made concerning some structural details of the ribosome in their relation to the localization of ribosomal functional centers.

The most prominent morphological feature of electron-microscopic images of ribosomes is the groove that separates the two ribosomal subunits (Figs. 5.2 and 5.10B). The groove is widened at a certain point: the so-called "eye" of the ribosome is revealed by electron microscopy. This peculiarity reflects a real fact: a significant cavity exists between the ribosomal subunits. As demonstrated recently by high-resolution electron microscopy, it is this cavity that accommodates main substrates of the ribosome, namely peptidyl-tRNA and aminoacyl-tRNA, participating in the transpeptidation reaction. Thus, the *tRNA-binding center* (or centers) is located in the cavity (the "eye" region) between the two ribosomal subunits (Fig. 9.15).

The small ribosomal subunit is subdivided by a deep groove into the "head" and the "body" (Figs. 5.5–5.7). The groove of the small subunit (the "neck") is the site where the *mRNA-binding center* is located (Figs. 9.4 and 9.5). An mRNA chain is drawn through this site from the 5′-end to the 3′-end and read out by triplets during translation.

The large ribosomal subunit also has its "head": this is the central protuberance among three visible protrusions in the crown-like projection of the subunit (Figs. 5.8 and 5.9). The "neck," or the groove separating the "head" from the "body" of the large subunit, harbors the main catalytic center of the ribosome, the *peptidyl transferase center* (Fig. 9.8), responsible for the formation of peptide bonds during elongation.

The two subunits are associated in such a way that the two "necks" are opposite each other, forming the large space or cavity between the subunits; this cavity is visible as the "eye" in the corresponding (nonoverlap) projection of the ribosome (Figs. 5.10–5.13). Two substrate tRNA molecules bound in this intersubunit pocket must interact simultaneously with adjacent mRNA codons on the small subunit by their anticodons, and with the peptidyl transferase center on the large subunit by their acceptor ends. Thus, the orientation of the tRNAs relative to the subunits in the ribosome is explicitly determined: the anticodons are in the "neck" of the small subunit, whereas the acceptor ends contact the "neck" of the large subunit (Figs. 9.13 and 9.15).

The important characteristic features of the ribosome are the moveable rod-like stalk of the large subunit visible as the lateral protuberance on the right from the "head" in the overlap projection of the ribosome, and the uncovered area of the large subunit at the base of the stalk (Fig. 5.10A). A number of observations suggest that the uncovered area accepts the complex of EF1 with newly coming aminoacyl-tRNA, and the rod-like stalk interacts with EF1 and becomes fixed by this interaction, possibly in the perpendicular orientation relative to the subunit interface (Fig. 9.17). The same large pocket formed by the uncovered area of the large subunit, the side surface of the small subunit, and the rod-like stalk may accept the other elongation factor, EF2, which binds to the ribosome in order to catalyze translocation (Fig. 9.17).

The other lateral protuberance that is to the left of the "head" of the large subunit in the overlap projection (Fig. 5.10A; see also the left protuberance in Figs. 9.8 and 9.13), together with the adjacent side ridge of the large subunit, seems to be directly involved in the association of the two ribosomal subunits. The side bulge of the small subunit (Fig. 5.5) appears to be the area of contact with the large subunit in the association.

19.3.2. Division of Labor between Ribosomal Subunits

In the process of protein biosynthesis the ribosome accepts encoded genetic information in the form of mRNA and decodes it, catalyzes the formation of peptide bonds in the transpeptidation reaction, and moves mRNA and tRNAs. Correspondingly, there are three functions served by the ribosome: (1) a decoding apparatus (genetic function), (2) an enzyme peptidyl transferase (biochemical function), and (3) a molecular machine (mechanical function). It is remarkable that the decoding function and the enzymatic function are distinctly separated between the two ribosomal subunits.

19.3.2.1. Genetic Functions of the Small Subunit

Translation of mRNA is initiated by the small ribosomal subunit, rather than by the complete ribosome. This implies that in order to initiate translation the ribosome must first dissociate into the subunits (Fig. 15.2). It is the small subunit that binds mRNA at the start of translation (Figs. 15.6 and 15.8) and thus serves as the *primary acceptor of genetic information* for the protein-synthesizing machinery. Only later, by the end of the initiation stage, the large ribosomal subunit joins to begin elongation of the polypeptide.

In the process of elongation the ribosome retains mRNA and moves along its chain in the direction from the 5′- to the 3′-end. The small ribosomal subunit of the complete translating ribosome is fully responsible for the retention of mRNA on the ribosome during elongation, whereas the large subunit seems to be devoid of any contact with mRNA. Hence, the scanning of a coding sequence of mRNA, that is, the readout of genetic information during elongation, is performed on the small subunit of the translating ribosome.

The mechanism of the triplet-by-triplet scanning of mRNA involves tRNA molecules that interact first with the small ribosomal subunit. It seems that both the A and the P sites of the ribosome are organized predominantly by surfaces of the small subunit, except the sites where the acceptor ends of tRNAs are bound on the large subunit. The anticodons of tRNAs specifically recognize cognate codons of mRNA in the "neck" region of the small subunit. The codon–anticodon interaction on the small subunit is maintained up to the translocation step of the elongation cycle (Fig. 9.1), and the movement of tRNA residues during translocation results in the shift of the mRNA chain along the "neck" of the small subunit by one triplet.

Thus, the small ribosomal subunit in the individual (dissociated) state accepts genetic information in the form of mRNA and initiates the translation process. In the course of translation the small subunit of the complete ribosome retains mRNA, decodes it with the use of tRNAs, and successively replaces mRNA codons and tRNAs by the translocation mechanism. Since all these operations are performed with genetic material, the small ribosomal subunit proves to be specialized in performing genetic functions of the ribosome.

19.3.2.2. Enzymatic Functions of the Large Subunit

When aminoacyl-tRNA occupies the P site and aminoacyl-tRNA is found in the A site of the ribosome (Fig. 9.1, state II), the acceptor ends of the tRNAs with their aminoacyl residues interact with the large ribosomal subunit. The site of this interaction is the peptidyl transferase center (Fig. 11.6). This interaction induces the transpeptidation reaction between the two substrates, peptidyl-tRNA and aminoacyl-tRNA: the carboxyl group of the peptidyl-tRNA is transferred to the amino group of the aminoacyl-tRNA (Fig. 11.1). As a result, a new peptide bond is formed, and the peptide residue becomes elongated by one amino acid residue. Thus, the large ribosomal subunit of the translating ribosome serves as an enzyme responsible for the formation of peptide bonds and generally for the synthesis (elongation) of a polypeptide chain. This is the main enzymatic function of the ribosome. No special protein with such an activity is found among ribosomal proteins, and the peptidyl transferase center is considered to be an integral part of the large subunit organized mainly by a special domain of the large rRNA (Fig. 9.7).

In addition to the catalysis of the transpeptidation reaction, the large ribosomal subunit participates in the enzymatic hydrolysis of GTP during translation. The participation is not direct: the enzymatic centers are localized rather on the translation factors (including elongation factors EF1 and EF2, initiation factor IF2, and termination factor RF3), and it is the interaction of the factor with the factor-binding site on the large subunit (Fig. 9.10) that activates the GTPase center of the factor. Such a cooperation of the large ribosomal subunit and the translation factors in performing the enzymatic hydrolysis of GTP is required for the promotion (catalysis) of noncovalent transitions in the translating ribosome involving aminoacyl-tRNA binding and peptidyl-tRNA translocation. In any case, the temporary association of the large ribosomal subunit with the translation factor is found to be essential for the formation of an enzymatically active GTPase center.

Thus, a "division of labor" between the two ribosomal subunits is observed: the small subunit is responsible for receiving and decoding genetic information, whereas the large subunit participates in enzymatic reactions in the process of translation.

19.3.3. Large-Block Mobility of the Ribosome

The capability of the ribosome to translocate the relatively large molecular masses of tRNAs in each elongation cycle (Fig. 9.1) and to draw over the mRNA chain during elongation (Fig. 4.4) suggests its own mechanical mobility. Periodical alterations of mutual positions of the ribosomal subunits relative to each other may be a principal type of large-block mobility of the ribosome. There are some experimental indications in favor of such an intersubunit mobility. In addition, the mobility of the rod-like stalk of the large ribosomal subunit relative to the rest of the subunit body, and the flexibility of the head-to-body junction in the small subunit have been mentioned.

Theoretically, a conformational mobility is needed to solve the problem of transition from one stable functional state to another through an intermediate state. In the case of the translating ribosome (Fig. 9.1), the transitions from state I to state II (codon-dependent binding of aminoacyl-tRNA), and especially from state III to state I (translocation) seem to be difficult without intermediate states. The point is that such large ligands as tRNAs are bound to the ribosome by several contact areas of their surfaces (multicenter binding), and simultaneous formation or breakage of many contacts would be inevitably accompanied by very high kinetic barriers resulting in very slow rates. Moreover, the tRNA-binding sites and the tRNA molecules in all the states indicated (Fig. 9.1) appear to be clamped or closed between the subunits. Intermediate states could provide some freedom of intraribosomal movements or partial delocalization of the ribosomal ligands (tRNAs) and thus reduce the kinetic barriers.

Fig. 19.1 presents the hypothetical model according to which the ribosome in the elongation cycle oscillates between two conformational states, the closed (locked) and the open (unlocked) ones. In the locked state the ribosomal ligands (tRNAs) are clamped between the subunits, bound by the maximum number of contacts with the ribosome and devoid of intraribosomal mobility. In the unlocked state the ligands are more mobile, have less contacts with the ribosome, and can enter or quit the ribosome. Thus, at the first stage of the aminoacyl-tRNA-binding process the ribosome must be open (unlocked) for receiving the ligand. This open state is fixed by EF1 (with GTP). Then, after GTP hydrolysis, EF1 is released, the ribosomal subunits are closed, and the aminoacyl end of the newly bound aminoacyl-tRNA comes into contact with the peptidyl transferase center of the large sub-

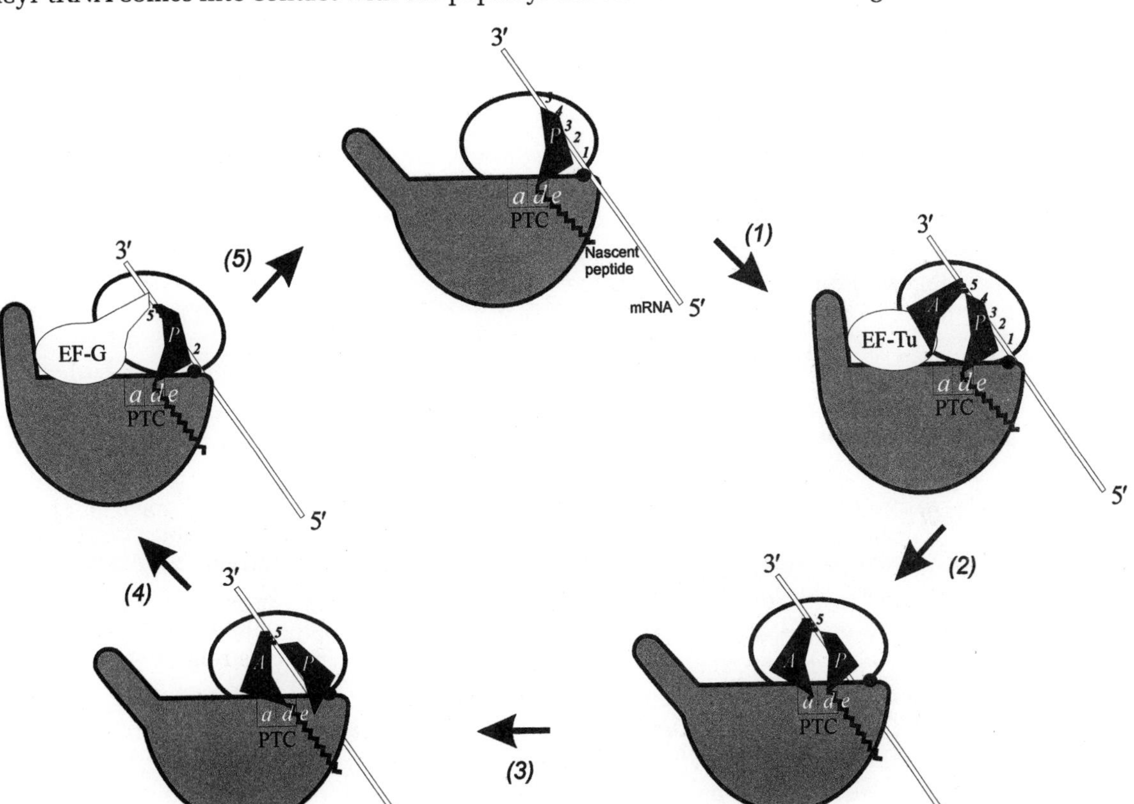

FIG. 19.1. Model of the dynamic ribosome in the elongation cycle. (The "heads" of both ribosomal subunits are turned to the viewer). The model postulates that the two ribosomal subunits are joined in a mobile way and capable of drawing apart (unlocking) and coming together (locking). The unlocking opens the functional centers on the subunit interface, such as the A site for receiving an aminoacyl-tRNA, and facilitates the ligand displacements, such as those during translocation. The locking of the subunits fixes the ligands inside the ribosome and brings the substrates of the transpeptidation reaction into a tight contact. It is proposed that the unlocking is induced by the association of elongation factors with GTP, and the hydrolysis of GTP and the resultant release of the factors allow the ribosome to lock again.

The figure presents the elongation cycle (see Fig. 9.1) in terms of the locking–unlocking model. (1) Binding of aminoacyl-tRNA with the ribosome requires the unlocked state, and EF1 complexed with GTP is destined to "open" the ribosome. (2) After GTP cleavage, EF1 leaves the ribosome, and the aminoacyl residue of the newly bound aminoacyl-tRNA interacts with the peptidyl transferase center of the large subunit and thus contributes to the locking of the ribosome. (3) The reaction of transpeptidation between the closely approaching groups of the two substrates (peptidyl-tRNA and aminoacyl-tRNA) proceeds in the closed (locked) ribosome. (4) The unlocking of the pre-translocation ribosome is promoted by the association of EF2 with GTP, resulting in the exit of the deacylated tRNA and the displacement of the tRNA residue of the peptidyl-tRNA molecule together with mRNA. (5) The hydrolysis of GTP and, as a consequence, the release of EF2, allows the ribosome to lock again. Thus, according to the model, the translating ribosome oscillates between the locked and unlocked states.

unit. In the locked state the peptidyl-tRNA and the aminoacyl-tRNA are brought together, and they react with each other (transpeptidation reaction). Now, in order to discard the deacylated tRNA and give the freedom for the displacement of the tRNA residue of the peptidyl-tRNA molecule from the A site to the P site, the ribosome must be open. This is promoted by EF2 (with GTP). After GTP hydrolysis and EF2 release the ribosome can be locked again, until the next aminoacyl-tRNA with EF1 comes.

According to the model, it is the process of periodic locking and unlocking of the ribosome that is energy-dependent: the elongation factors EF1 and EF2 in their GTP form fix the open conformation, and the factor–ribosome interaction activates the GTPase, resulting in the release of the elongation factor and the ribosome closing. Thus, one molecule of GTP is expended for each unlocking–locking event. As the ribosome unlocks and locks twice in each elongation cycle (Fig. 19.1), two GTP molecules are split per cycle. This is the energy payment for the effective (fast and reliable) functioning of the translating ribosome as a molecular machine.

19.3.4. GTP-Dependent Catalysis of Conformational Transitions

At the same time, thermodynamically all the three principal steps of the elongation cycle (Fig. 9.1)—aminoacyl-tRNA-binding, transpeptidation, and translocation—are spontaneous, downhill processes. Indeed, aminoacyl-tRNA can specifically (in a codon-dependent way) bind to the ribosome and form a normal functional complex (state II in Fig. 9.1) in the absence of EF1 and GTP (Fig. 12.5, nonenzymic binding), though in this case the process is significantly slower than in the presence of EF1 and GTP. The transpeptidation catalyzed by the ribosome itself is known as a typical "exergonic" reaction proceeding with the release of free energy. The pre-translocation-state ribosome (state III in Fig. 9.1) is thermodynamically unstable and can slowly slip to the post-translocation state without EF2 and GTP (Fig. 12.5, nonenzymic translocation). On the whole, slow "nonenzymic," or factor-free translation is possible in a cell-free system where the ribosomes are provided with just aminoacyl-tRNAs and a message polynucleotide, without any additional energy sources such as GTP or ATP.

The point is that the main source of free energy for performing the *useful work* (the synthesis of polypeptide with a strictly determined amino acid sequence) in the process of elongation seems to be the reaction of transpeptidation (Fig. 11.1). The net gain of free energy of the reaction is about -7 kcal/mol, which is comparable to the free energy of hydrolysis of a high-energy bond in ATP or GTP. It is clear that the nonenzymic translation is performed at the expense of the transpeptidation reaction. This implies that the transpeptidation reaction can supply the entire elongation cycle with energy. Hence, the translation process, using aminoacyl-tRNAs as substrates and an energy source, is spontaneous, in the thermodynamic sense of the word.

Why are two additional high-energy compounds (GTP molecules) per cycle expended, if not for useful work? The GTP hydrolysis in the elongation cycle with the participation of the ribosome and the elongation factors has been demonstrated to be direct, rather than coupled with any other covalent reaction and formation of a phosphorylated intermediate. This is the attack of the water molecule on the pyrophosphate bond of GTP, with the dissipation of all the free energy released into heat. At the same time, the expenditure of two additional high-energy compounds (GTP molecules) per cycle makes elongation much faster. This suggests that the role of GTP hydrolysis in translation is kinetic, rather than thermodynamic. Indeed, since the elongation factors EF1 and EF2 are considered as catalysts of two noncovalent steps of the elongation cycle, aminoacyl-tRNA binding and translocation, respectively (Fig. 9.1), the GTP cleavage with their participation is obviously involved in the catalysis of the processes.

The requirement of the expenditure of a high-energy compound for catalysis looks unusual from the enzymological point of view. In typical cases of enzymatic catalysis a spontaneous reaction is accelerated by a protein without consumption of any additional energy. Free energy is released in the course of the catalyzed reaction itself. The catalysis, that is, overcoming (reducing) an activation energy barrier of the reaction, results from the affinity of an enzyme for the so-called transition state of a substrate. The complex of the enzyme with the transition intermediate, however, would be a deadlock if the intermediate did not spontaneously split into reaction products with the release of free energy that compensated for the energy of the enzyme–intermediate interaction. Thus, the release of free energy in a catalyzed covalent reaction is necessary for the desorption of products from an enzyme.

The situation can be more complicated in the case of the catalysis of conformational (noncovalent) changes with large macromolecules. This is the case of the EF1-promoted binding of aminoacyl-tRNA and the EF2-promoted translocation in the elongation cycle (Figs. 9.1, 12.5, and 19.1). Although both steps (aminoacyl-tRNA binding and translocation) can proceed spontaneously, they are slow due to high activation barriers. A catalytic protein, such as an elongation factor, having an affinity for a *conformational transition state* of the ribosomal complex may decrease the barrier for intraribosomal movements of large ligands (tRNAs) while it fixes this intermediate. However, the subsequent desorption of the catalyst itself from the ribosome may be a problem because of a high activation barrier of the rupture of multicenter interactions between the protein and the ribosome. Thus, the completion of the catalysis would be retarded by the delay in the release of the catalyst from the ribosome.

This problem can be solved by involving a covalent downhill reaction in the catalysis of conformational, noncovalent transitions. In the case of the elongation factors as catalysts, GTP hydrolysis is used for the desorption (release) of the factor from the ribosome. The model proposed is as follows. It is well known that the elongation factor, EF1 or EF2, can interact with the ribosome only after its association with GTP: GTP, at the expense of its affinity for the factor, alters its conformation in such a way that the protein acquires an affinity for the ribosome. According to the model, it is the elongation factor with GTP that fixes the conformational transition state of the ribosome. But the attachment of the elongation factor to the ribosome induces the GTPase activity of the protein, GTP is hydrolyzed, and, as a consequence, the factor loses its affinity for the ribosome and goes out. The ribosome without the factor falls down from the barrier to a thermodynamically stable state. In other words, the energy of the affinity of the factor for the conformational transition state of the ribosome is compensated by an accompanying covalent, downhill reaction, namely GTP hydrolysis.

Thus, the direct hydrolysis of GTP by water seems to be necessary for the enzymic (factor-promoted) catalysis of conformational, noncovalent transitions in the elongation cycle. The main role of such a hydrolysis is the destruction of a ligand that induces the affinity of a catalyst for the conformational transition state, in order to induce the exit of the ribosome from the intermediate complex and the transition to the next, productive state.

Index